16th International Symposium on Capture Gamma-Ray Spectroscopy and Related Topics (CGS16)

EPJ Web of Conferences Volume 178 (2018)

Shanghai, China
18 – 22 September 2017

Editor:

Yang Sun

ISBN: 978-1-5108-6309-5

Printed from e-media with permission by:

Curran Associates, Inc.
57 Morehouse Lane
Red Hook, NY 12571

Some format issues inherent in the e-media version may also appear in this print version.

Printed by Curran Associates, Inc. (2018)

For additional information, please contact EDP Sciences – Web of Conferences at the address below.

EDP Sciences – Web of Conferences
17, Avenue du Hoggar
Parc d'Activité de Courtabœuf
BP 112
F-91944 Les Ulis Cedex A
France

Phone: +33 (0) 1 69 18 75 75
Fax: +33 (0) 1 69 28 84 91

contact-edps@webofconferences.org

Additional copies of this publication are available from:

Curran Associates, Inc.
57 Morehouse Lane
Red Hook, NY 12571 USA
Phone: 845-758-0400
Fax: 845-758-2633
Email: curran@proceedings.com
Web: www.proceedings.com

TABLE OF CONTENTS

The 16th International Symposium on Capture Gamma-Ray Spectroscopy and Related Topics

Shanghai, September 18-22, 2017

Preface

The Sixteenth International Symposium on *Capture Gamma-Ray Spectroscopy and Related Topics* (CGS16) was held on the quiet historical campus of Shanghai Jiao Tong University, located in the heart of the metropolis of Shanghai, from September 18 to 22, 2017.

CGS16 was the sixteenth convention of the series that started in 1969 in Studsvik, Sweden, and continued once every three years in Europe and North America. It was previously held in Petten, The Netherlands (1974); Brookhaven, USA (1978); Grenoble, France (1981); Knoxville, USA (1984); Leuven, Belgium (1987); Asilomar, USA (1990); Fribourg, Switzerland (1993); Budapest, Hungary (1996); Santa Fe, USA (1999); Prague, Czech Republic (2002); South Bend, USA (2005); Cologne, Germany (2008); Guelph, Canada (2011); and Dresden, Germany (2014). For the first time in its history of nearly half a century, the CGS conference moved to Asia.

The CGS conference, being one of the largest series of meetings in nuclear physics, initially focused on topics related to gamma-ray spectroscopy as used broadly in nuclear structure, nuclear reactions, nuclear probes for fundamental physics, nuclear data, and novel techniques and applications. Later it extended the scope to include nuclear astrophysics and interdisciplinary studies. At CGS16, nuclear astrophysics became one of the major subjects.

The International Advisory Committee (IAC) recommended many outstanding speakers, and the Organizing Committee (OC) received many contributed abstracts. The members of the Program Committee (PC) reviewed the abstracts to produce the conference program. Contributed talks were selected from both well-established researchers and the younger generation. In the five-day scientific program, a total of 128 talks were presented in 32 sessions. Female nuclear physicists played notable roles in CGS16; about one quarter of the invited speakers and one quarter of the session chairs were women.

CGS16 attracted in total 164 participants from 22 countries. Some participants were regular attendees of the CGS series, including several previous CGS chairs. Unfortunately, Professor Till von Egidy of TU Munich, who had attended all 15 previous conferences, could not travel this time. Nevertheless, he sent his best wishes for CGS16.

The social program included a welcoming reception on Monday evening in an engaging atmosphere. A half-day excursion on Wednesday afternoon was arranged to visit the traditional Shanghai-style Yu Garden. It followed with a Huangpu River cruise at dusk. Unfortunately, the weather did not cooperate and it rained heavily. Despite that, the gorgeous lights from tall buildings along the banks gave an exciting view. The conference banquet was held on Thursday evening in the SJTU campus restaurant. The distinguished Professor Steve Yates and others gave lovely speeches. Another well-known senior researcher in attendance, Professor Yongshou Chen, was among the first group of Chinese scientists sent to the Niels-Bohr Institute after the Cultural Revolution in China (marking the opening of China to the western world). To show his happiness with the reunion in China with his foreign friends, Professor Chen brought a con affetto song for the banquet.

In a special lunch meeting with the IAC and PC members on Thursday, Dr. Caterina Michelagnoli represented the Institut Laue-Langevin (ILL) to propose hosting CGS17. After considerable discussion, the decision was made that the next CGS symposium, CGS17, will take place in 2020 in Grenoble, France.

The coordination of CGS16 was a large effort for the Organization Committee. I want to thank all my colleagues and my group members who offered tremendous help to make the symposium successful. In particular, I want to mention our secretary Ms. Gang He, who worked hard during the hot summer prior to the conference to write more than one hundred invitation letters and handled every correspondence. Dr. Surja Ghorui was responsible for creation and maintenance of the CGS16 website, and Dr. Wendell Misch provided the main guide on the website regarding transportation, social activities, and local restaurants and bars. Dr. Yuan Qiu, and Ms. Cuijuan Lv put forth a great effort preparing the booklet for the conference. I am also very thankful to Dr. Jianmin Dong and Dr. Xinle Shang of the Institute of Modern Physics from Lanzhou, who brought four other graduate students to help. All these people offered their assistance during the conference sessions. I should particularly acknowledge our financial sponsors: the *Institute of Modern Physics*; the *Institute of Theoretical Physics*; *Huzhou University*, and the *School of Physics and Astrophysics, SJTU.* The provided funding was mainly used as financial aid for students and some researchers.

Organizing CGS16 was a very special experience for me. For nearly two years leading up to and after the conference, I enjoyed writing hundreds emails to my colleagues and friends, old and new. I deeply appreciate the IAC and PC members, who gave me very valuable advice. I would particularly like to mention the past CGS chairs Ronald Schwengner (CGS15), Paul Garrett (CGS14), and Ani Aprahamian (CGS12). Finally, I want to express my best wishes to our colleagues at ILL who have taken up the torch. I look forward to CGS17 in Grenoble and hope to see all of you there.

Yang Sun, Chair of CGS16
Professor of Physics
School of Physics and Astronomy
Shanghai Jiao Tong University

Email: sunyang@sjtu.edu.cn

Local Organization Committee

Yang Sun (**Chair**)
Surja Kiran Ghorui (**Coordinator**)
Gang He (**Secretary**)
Yan-Xin Liu
Cui-Juan Lv
Wendell Misch
Yuan Qiu
Ying-Chun Yang

International Advisory Committee

Ani Aprahamian (Notre Dame, USA)
Lee Bernstein (Livermore, USA)
Klaus Blaum (Heidelberg, Germany)
Mike Carpenter (Argonne, USA)
Jolie Cizewski (Rutgers, USA)
Giacomo de Angelis (Legnaro, Italy)
J. Luis Egido (Madrid, Spain)
Paul Garrett (Guelph, Canada)
Jan Jolie (Koeln, Germany)
Rostislav Jolos (Dubna, Russia)
Rauno Julin (Jyvaskyla, Finland)
Guinyun Kim (Daegu, Korea)
Wei-Ping Liu (Beijing, China)
Augusto Macchiavelli (Berkeley, USA)
Jie Meng (Beijing, China)
Takaharu Otsuka (Tokyo, Japan)
Norbert Pietralla (Darmstadt, Germany)
Peter Ring (Munich, Germany)
Hiroyoshi Sakurai (Tokyo, Japan)
Hendrik Schatz (Michigan State, USA)
Ronald Schwengner (Dresden, Germany)
Javid Sheikh (Kashmir, India)
Pieter Van Isacker (GANIL, France)
Philip Walker (Surrey, UK)
Michael Wiescher (Notre Dame, USA)
Steven Yates (Kentucky, USA)
Victor Zamfir (Bucharest, Romania)
Yu-Hu Zhang (Lanzhou, China)

Program Committee

Alejandro Algora (Valencia, Spain)
Alison Bruce (Brighton, UK)
Aaron Couture (Los Alamos, USA)
Jens Dilling (TRIUMF, Canada)
Jerzy Dudek (Strasbourg, France)
Alfredo Galindo-Uribarri (Oak Ridge, USA)
Alejandro Garcia (Seattle, USA)
Rolf-Dietmar Herzberg (Liverpool, UK)
Michael Jentschel (ILL Grenoble, France)
Jenny Lee (Hong Kong, China)
Filip Kondev (Argonne, USA)
Yuri Litvinov (GSI, Germany)
Gabriel Martinez-Pinedo (Darmstadt, Germany)
Tohru Motobayashi (RIKEN, Japan)
Takashi Nakatsukasa (Tsukuba, Japan)
Peter von Neumann-Cosel (Darmstadt, Germany)
Rudrajyoti Palit (Mumbai, India)
Patrick Regan (Surrey, UK)
Grigory Rogachev (Texas A&M, USA)
Andrew Stuchbery (Canberra, Australia)
Xiao-Dong Tang (Lanzhou, China)
Anton Tonchev (Livermore, USA)
Robert Wadsworth (York, UK)
Fu-Rong Xu (Beijing, China)
Shan-Gui Zhou (Beijing, China)
Andreas Zilges (Koeln, Germany)

List of Participants

Allmond, Mitch
Oak Ridge National Laboratory, USA
jmallmond@gmail.com

Andreyev, Andrei
University of York, United Kingdom
andrei.andreyev@york.ac.uk

Aprahamian, Ani
University of Notre Dame, USA
aapraham@nd.edu

Banu, Adriana
James Madison University, USA
banula@jmu.edu

Bao, Man
Shanghai Jiao Tong University, China
bymmy@sjtu.edu.cn

Bark, Robert
iThemba Laboratory for Accelerator Based Sciences, South Africa
bark@tlabs.ac.za

Bassauer, Sergej
Technische Universität Darmstadt, Germany
sbassauer@ikp.tu-darmstadt.de

Beck, Tobias
Technische Universität Darmstadt, Germany
tbeck@ikp.tu-darmstadt.de

Bemmer, Daniel
Helmholtz-Zentrum Dresden-Rossendorf (HZDR), Germany
d.bemmerer@hzdr.de

Bentley, Michael

University of York, United Kingdom
michael.bentley@york.ac.uk

Bruce, Alison

University of Brighton, United Kingdom
Alison.Bruce@brighton.ac.uk

Burbadge, Christina

University of Guelph, Canada
cburbadg@uoguelph.ca

Cai, Fangjing

Shanghai Jiao Tong University, China
caifj@sjtu.edu.cn

Cakirli, R. Burcu

Istanbul University, Turkey
rburcu@istanbul.edu.tr

Casanovas, Adrià

Technical University of Catalonia (UPC), Spain
ksas896@gmail.com

Casten, Richard

Yale University and Michigan State University, USA
richard.casten@yale.edu

Cavanna, Francesca

Istituto Nazionale di Fisica Nucleare, Italy
francesca.cavanna@ge.infn.it

Chen, Fangqi

Peking University, China
fangqi0591@126.com

Chen, Penghui

Institute of Modern Physics, Chinese Academy of Sciences, China
chenpenghui@impcas.ac.cn

Chen, Ruijiu

Institute of Modern Physics, Chinese Academy of Sciences, China
chenrj13@impcas.ac.cn

Chen, Yongshou

China Institute of Atomic Energy, China
yongshouchen@yahoo.com

Cheng, Yiyuan

Shanghai Jiao Tong University, China
yiyuancheng@sjtu.edu.cn

Cizewski, Jolie A.

Rutgers University, USA
cizewski@rutgers.edu

Crawford, Heather

Lawrence Berkeley National Laboratory, USA
hlcrawford@lbl.gov

de Boer, Richard

University of Notre Dame, USA
rdeboer1@nd.edu

de France, Gilles

Grand Accélérateur National d'Ions Lourds, France
defrance@ganil.fr

Devlin, Matthew

Los Alamos National Laboratory, USA
devlin@lanl.gov

Dimitriou, Paraskevi

International Atomic Energy Agency, Austria
p.dimitriou@iaea.org

Dong, Guoxiang

Huzhou University, China
gxdong@zjhu.edu.cn

Dong, Jianmin

Institute of Modern Physics, Chinese Academy of Sciences, China
dongjm07@impcas.ac.cn

Doornenbal, Pieter

RIKEN Nishina Center, Japan
pieter@ribf.riken.jp

Egido, J. Luis

Universidad Autónoma de Madrid, Spain
j.luis.egido@uam.es

Escher, Jutta E.

Lawrence Livermore National Laboratory, USA
escher1@llnl.gov

Fang, Xiao

Sun Yat-Sen University, China
xfang.9@gmail.com

Fan, Xiaohua

Institute of Modern Physics, Chinese Academy of Sciences, China
fanxiaohua@impcas.ac.cn

Fortunato, Lorenzo

University of Padova, Italy
fortunat@pd.infn.it

Fu, Guanjian

Tongji University, China
gjfu@tongji.edu.cn

Fujita, Yoshitaka

Osaka University, Japan
fujita@rcnp.osaka-u.ac.jp

Gao, Zaochun

China Institute of Atomic Energy, China
15311738118@163.com

Garcia-Ramos, Jose-enrique

University of Huelva, Spain
enrique.ramos@dfaie.uhu.es

Garg, Swati

Indian Institute of Technology Roorkee, India
gargidph@iitr.ac.in

Garrett, Paul

University of Guelph, Canada
pgarrett@physics.uoguelph.ca

Ghorui, Surja K.

Shanghai Jiao Tong University, China
surja@sjtu.edu.cn

Goriely, Stephane

Université Libre de Bruxelles, Belgium
sgoriely@ulb.ac.be

Grzywacz, Robert

University of Tennessee, USA
rgrzywac@utk.edu

Guttormsen, Magne

University of Oslo, Norway

magne.guttormsen@fys.uio.no

Hales, Brian

Japan Atomic Energy Agency, Japan

hales.brian@jaea.go.jp

Han, Sophia

University of Tennessee, USA

jhan18@utk.edu

Hayes, Anna

Los Alamos National Laboratory, USA

anna_hayes@lanl.gov

He, Gang

Shanghai Jiao Tong University, China

hegang@sjtu.edu.cn

He, Guozhu

China Institute of Atomic Energy, China

heguozhu818@126.com

He, Jianjun

National Astronomical Observatories, Chinese Academy of Sciences, China

hejianjun@nao.cas.cn

He, Xiaotao

Nanjing University of Aeronautics and Astronautics, China

hext@nuaa.edu.cn

Holt, Jason

TRI University Meson Facility (TRIUMF), Canada

jholt@triumf.ca

Huang, Meirong

Inner Mongolia University for Nationalities, China
huangmeirong05@163.com

Hu, Qingli

Shanghai Jiao Tong University, China
461538600@qq.com

Isaak, Johann

Osaka University, Toyonaka, Japan
jisaak@rcnp.osaka-u.ac.jp

Iwamoto, Nobuyuki

Japan Atomic Energy Agency, Japan
iwamoto.nobuyuki@jaea.go.jp

Iwasaki, Hiro

Michigan State University, USA
iwasaki@nscl.msu.edu

Jain, Ashok Kumar

Indian Institute of Technology Roorkee, India
ashkumarjain@yahoo.com

Jigmeddorj, Badamsambuu

University of Guelph, Canada
bjigmedd@uoguelph.ca

Kajino, Taka

Beihang University, China / The University of Tokyo, Japan
kajino@nao.ac.jp

Katabuchi, Tatsuya

Tokyo Institute of Technology, Japan
buchi@nr.titech.ac.jp

Kim, Guinyun

Kyungpook National University, South Korea
gnkim@knu.ac.kr

Kobayashi, Nobuyuki

Osaka University, Japan
kobayash@rcnp.osaka-u.ac.jp

Kondev, Filip G.

Argonne National Laboratory, USA
kondev@anl.gov

Kumar, Pawan

Indian Institute of Technology Ropar, India
pawan.kumar@iitrpr.ac.in

Lam, Yihua

Institute of Modern Physics, Chinese Academy of Sciences, China
lamyihua@yahoo.com

Lane, Greg

Australian National University, Australia
Gregory.Lane@anu.edu.au

Lee, Riedinger

University of Tennessee, USA
lrieding@utk.edu

Lenzi, Silvia

University of Padova and INFN, Italy
lenzi@pd.infn.it

Leviatan, Amiram

The Hebrew University, Israel
ami@phys.huji.ac.il

Li, Zhihong
China Institute of Atomic Energy, China
zhhli@foxmail.com

Li, Zhiquan
Shandong University, China
zhiquan921@yahoo.com

Litvinov, Yuri
GSI Helmholtzzentrum für Schwerionenforschung, Germany
Y.Litvinov@gsi.de

Liu, Chen
Shandong University, China
cliu@sdu.edu.cn

Liu, Lei
Shandong Agriculture and Engineering University, China
liulei20062007@126.com

Liu, Weiping
China Institute of Atomic Energy, China
wpliu@ciae.ac.cn

Liu, Yanxin
Huzhou University, China
hzlyx@zjhu.edu.cn

Liu, Zhong
Institute of Modern Physics, Chinese Academy of Sciences, China
liuzhong@impcas.ac.cn

Lozeva, Radomira
Centre de Sciences Nucléaires et de Sciences de la Matière (CSNSM/CNRS), France
radomira.lozeva@csnsm.in2p3.fr

Lu, Yi

Qufu Normal University, China
luyi19881007@126.com

Lu, Zhengtian

University of Science and Technology of China, China
ztlu@ustc.edu.cn

Luo, Yixiao

Vanderbilt University / Lawrence Berkeley National Laboratory, USA
yxluo@lbl.gov

Lv, Cuijuan

Shanghai Jiao Tong University, China
cuijuanlv0403@163.com

Macchiavelli, Augusto

Lawrence Berkeley National Laboratory, USA
aom@lbl.gov

Mei, Bo

Extreme Light Infrastructure-Nuclear Physics(ELI-NP), Romania
bo.mei@eli-np.ro

Meisel, Zach

Ohio University, USA
meisel@ohio.edu

Michelagnoli, Caterina

Institut Laue-Langevin, France
michelagnolic@ill.fr

Misch, G. Wendell

Shanghai Jiao Tong University, China
misch@sjtu.edu.cn

Motobayashi, Tohru

RIKEN Nishina Center, Japan
motobaya@riken.jp

Nunes, Filomena

Michigan State University, USA
nunes@nscl.msu.edu

Otsuka, Takaharu

RIKEN Nishina Center/ University of Tokyo/ KU Leuven/ MSU
takaharuotsuka@gmail.com

Pakarinen, Janne

University of Jyväskylä, Finland
janne.pakarinen@jyu.fi

Pei, Junchen

Peking University, China
peij@pku.edu.cn

Peters, Erin E.

University of Kentucky, USA
fe.peters@uky.edu

Pietralla, Norbert

Technische Universität Darmstadt, Germany
pietralla@ikp.tu-darmstadt.de

Piot, Julien

Grand Accélérateur National d'Ions Lourds (GANIL), France
julien.piot@ganil.fr

Qi, Bin

Shandong University, China
bqi@sdu.edu.cn

Qi, Chong

KTH Royal Institute of Technology, Sweden
chongq@kth.se

Qian, Yibin

Nanjing University of Science and Technology, China
qyibin@njust.edu.cn

Qiu, Yuan

Shanghai Jiao Tong University, China
ymsfd9816@sjtu.edu.cn

Ranerjee, Projjwal

Shanghai Jiao Tong University, China
projjwal@sjtu.edu.cn

Reifarth, Rene

Goethe University Frankfurt, Germany
reifarth@physik.uni-frankfurt.de

Ren, Zhongzhou

Nanjing University, China
zren@nju.edu.cn

Ring, Peter

Technical University Munich, Germany
ring@ph.tum.de

Sakurai, Hiroyoshi

RIKEN Nishina Center / Univ. of Tokyo, Japan
sakurai@ribf.riken.jp

Savran, Deniz

GSI Helmholtzzentrum fur Schwerionenforschung, Germany
d.savran@gsi.de

Shang, Xinle

Institute of Modern Physics, Chinese Academy of Sciences, China
shangxinle@impcas.ac.cn

Sherrill, Bradley

Michigan State University, USA
sherrill@frib.msu.edu

Shimizu, Noritaka

Center for Nuclear study, the University of Tokyo, Japan
shimizu@cns.s.u-tokyo.ac.jp

Shi, Yue

Harbin Institute of Technology, China
yueshi@hit.edu.cn

Shneidman, Timur

Joint Institute for Nuclear Research, Russia
shneyd@theor.jinr.ru

Shuai, Peng

Institute of Modern Physics, Chinese Academy of Sciences, China
shuai620@impcas.ac.cn

Spyrou, Artemis

Michigan State University, USA
spyrou@nscl.msu.edu

Stuchbery, Andrew

Australian National University, Australia
andrew.stuchbery@anu.edu.au

Sun, Yang

Shanghai Jiao Tong University, China
sunyang@sjtu.edu.cn

Surman, Rebecca

University of Notre Dame, USA

rsurman@nd.edu

Tan, Wanpeng

University of Notre Dame, USA

wtan@nd.edu

Tang, Xiaodong

Institute of Modern Physics, Chinese Academy of Sciences, China

xtang@impcas.ac.cn

Tonchev, Anton P.

Lawrence Livermore National Laboratory, USA

tonchev2@llnl.gov

Utsunomiya, Hiroaki

Konan University, Japan

hiro@konan-u.ac.jp

Van Isacker, Piet

Grand Accélérateur National d'Ions Lourds, France

isacker@ganil.fr

Von Neumann-Cosel, Peter

Technische Universität Darmstadt, Germany

ikp.tu-darmstadt.de

Wang, Hankui

Zhou Kou Normal Universiy, China

whk2007@163.com

Wang, He

RIKEN Nishina Center, Japan

wanghe@ribf.riken.jp

Wang, Jiaqi

China Institute of Atomic Energy, China
wangjq_m@163.com

Wang, Kai

Peking University, China
wangkai2u@163.com

Wang, Meng

Institute of Modern Physics, Chinese Academy of Sciences, China
wangm@impcas.ac.cn

Wang, Ning

Guangxi Normal University, China
wangning@gxnu.edu.cn

Wang, Xiaobao

Huzhou University, China
xbwang@zjhu.edu.cn

Watanabe, Hiroshi

Beihang University, China
hiroshi@ribf.riken.jp

Werner, Volker

Technische Universität Darmstadt, Germany
vw@ikp.tu-darmstadt.de

Wiedeking, Mathis

iThemba Laboratory for Accelerator Based Sciences, South Africa
wiedeking@tlabs.ac.za

Wienholtz, Frank

CERN, Switzerland
Frank.Wienholtz@cern.ch

Wiescher, Michael

University of Notre Dame, USA
mwiesche@nd.edu

Wilhelmy, Julius

University of Cologne, Germany
wilhelmy@ikp.uni-koeln.de

Wimmer, Kathrin

The University of Tokyo, Japan
wimmer@phys.s.u-tokyo.ac.jp

Wu, Xinyi

Shanghai Jiao Tong University, China
wxy0120729016@sjtu.edu.cn

Xiao, Guoqing

Institute of Modern Physics, Chinese Academy of Sciences, China
xiaogq@impcas.ac.cn

Xing, Yuanming

Institute of Modern Physics, Chinese Academy of Sciences, China
xym@impcas.ac.cn

Xu, Chang

Nanjing University, China
cxu@nju.edu.cn

Xu, Chuan

Peking University, China
chuan@pku.edu.cn

Xu, Yi

Extreme Light Infrastructure - Nuclear Physics (ELI-NP), Romania
yi.xu@eli-np.ro

Yang, Zuxing
Institute of Modern Physics, Chinese Academy of Sciences, China
yangzuxing16@impcas.ac.cn

Yates, Steven
University of Kentucky, USA
yates@uky.edu

Ye, Yanlin
Peking University, China
yeyl@pku.edu.cn

Yin, Peng
Institute of Modern Physics, Chinese Academy of Sciences, China
yinpeng@impcas.ac.cn

Yu, Changhong
Oak Ridge National Laboratory, USA
yuc@ornl.gov

Yuan, Cenxi
Sun Yat-Sen University, China
yuancx@mail.sysu.edu.cn

Zamfir, Victor
National Institute for Physics and Nuclear Engineering (IFIN-HH), Romania
victor.zamfir@eli-np.ro

Zhang, Gaolong
Beihang University, China
zgl@buaa.edu.cn

Zhang, Guangxin
Beihang University, China
skyzhanggx@buaa.edu.cn

Zhang, Ningtao

Institute of Modern Physics, Chinese Academy of Sciences, China
zhangningtao@impcas.ac.cn

Zhang, Shisheng

Beihang University, China
zss76@buaa.edu.cn

Zhang, Yu

Liaoning Normal University, China
dlzhangyu_physics@163.com

Zhang, Yuhu

Institute of Modern Physics, Chinese Academy of Sciences, China
yhzhang@impcas.ac.cn

Zhao, Juan

Shanghai Jiao Tong University, China
zhaojuan@sjtu.edu.cn

Zhou, Xianrong

East China Normal University, China
xrzhou@phy.ecnu.edu.cn

Zhu, Shaofei

Argonne National Laboratory, USA
zhu@anl.gov

Zhu, Yi

Peking University, China
zhuy13@pku.edu.cn

Zhu, Zhiyuan

Shanghai Tech University, China
zhuzhiyuan@sinap.ac.cn

Zielinska, Magda
CEA Saclay, France
magda@slcj.uw.edu.pl

Zilges, Andreas
University of Cologne, Germany
zilges@ikp.uni-koeln.de

edp sciences

Web of Conferences

Statement of Peer review

In submitting conference proceedings to *Web of Conferences*, the editors of the proceedings certify to the Publisher that

1. They adhere to its **Policy on Publishing Integrity** in order to safeguard good scientific practice in publishing.
2. All articles have been subjected to peer review administered by the proceedings editors.
3. Reviews have been conducted by expert referees, who have been requested to provide unbiased and constructive comments aimed, whenever possible, at improving the work.
4. Proceedings editors have taken all reasonable steps to ensure the quality of the materials they publish and their decision to accept or reject a paper for publication has been based only on the merits of the work and the relevance to the journal.

Title, date and place of the conference

16th International Symposium on Capture Gamma-Ray Spectroscopy and Related Topics
September 18-22, 2017, Shanghai, China

Proceedings editor(s):

Yang Sun

Date and editor's signature April 27, 2018

EPJ Web of Conferences **178**, 01001 (2018) https://doi.org/10.1051/epjconf/201817801001
CGS16

Future Opportunities at the Facility for Rare Isotope Beams

Bradley M. Sherrill[1,*]

[1]Facility for Rare Isotope Beams, Michigan State University, East Lansing, MI 48824, USA

Abstract. This paper overviews the Facility for Rare Isotope Beams, FRIB, its construction status at the time of the conference, and its scientific program. FRIB is based on a high-power, heavy-ion, superconducting linear accelerator that is designed to deliver at least 400kW at 200 MeV/u for all stable-ion beams and produce a large fraction of all possible isotopes of the elements. A three-stage fragment separator will separate rare isotope beams for use in experiments at high energy or stopped and reaccelerated to up to 10MeV/u. The facility is expected to have first beams in 2021. An overview of the planned scientific program, experimental capabilities, and equipment initiatives are presented.

1 Introduction and Overview of the Facility for Rare Isotope Beams

Creation and study of rare isotopes, isotopes of the elements that are not normally found on Earth, is a key element of the goal to accurately model the atomic nucleus. Rare isotopes provide insights into nuclear structure, benchmark models, and can be used to determine model parameters. Accurate models of nuclear properties and the reactions of atomic nuclei, with sufficient precision, would be revolutionary in our ability to understand the universe and in our ability to apply nuclei to solve societal problems. It would remove nuclear physics uncertainties in the application of nuclei to probe for the fundamental interactions of nature.

One of the major new facilities to make beams of rare isotopes is the Facility for Rare Isotope Beams, FRIB [1]. Establishment of FRIB as a user facility open to all researchers is supported by the Department of Energy Office of Science. FRIB is based on a high-power, heavy-ion, superconducting linear accelerator that is designed to deliver at least 400 kW at 200 MeV/u for all stable-ion beams. For reference, 400 kW of a ^{238}U beam corresponds to 8pμA or $5\text{x}10^{13}$ ions/s. A three-stage fragment separator [2] will separate rare isotope beams in-flight. Beams can be used in experiments at high energy or stopped and reaccelerated to up to 10MeV/u. The availability of rare isotope beams at energies from tens of kV to 200 MeV/u is one of the key features of FRIB. Particularly unique, is the plan to reaccelerate beams after in-flight separation and stopping.

An overview of the facility is shown in Figure 1. The callout circles illustrate various parts of the facility. Ions are created in a pair of ECR ion sources and are accelerated in a

* Corresponding author: sherrill@frib.msu.edu

https://doi.org/10.1051/epjconf/201817801001

superconducting linear accelerator that is folded two times. After the first section the ions are stripped in a liquid lithium charge stripper. Rare isotopes are produced in a rotating graphite target and the unused beam is stopped in a rotating drum of water. An interesting possibility is to harvest otherwise unused isotopes from the water drum [3]. Isotopes are separated, in-flight with a fragment separator [2] and delivered to experimental areas for fast, stopped, and reaccelerated beams. Beams are stopped in pure helium and reaccelerated in the ReA3/6 reaccelerators [4]. The reaccelerators were developed at the National Superconducting Cyclotron Laboratory supported by the USA National Science Foundation [5].

FRIB is expected to have first beams in 2021. Once operational the facility is projected to take five years to increase the beam power from 10 kW in the first year to 400 kW in year five. FRIB is planned to have ^{238}U, ^{48}Ca, ^{78}Kr, ^{82}Se, ^{124}Xe, ^{18}O, ^{86}Kr, ^{16}O, and ^{36}Ar primary beams in the first year. Four to five new primary beams will be added each year.

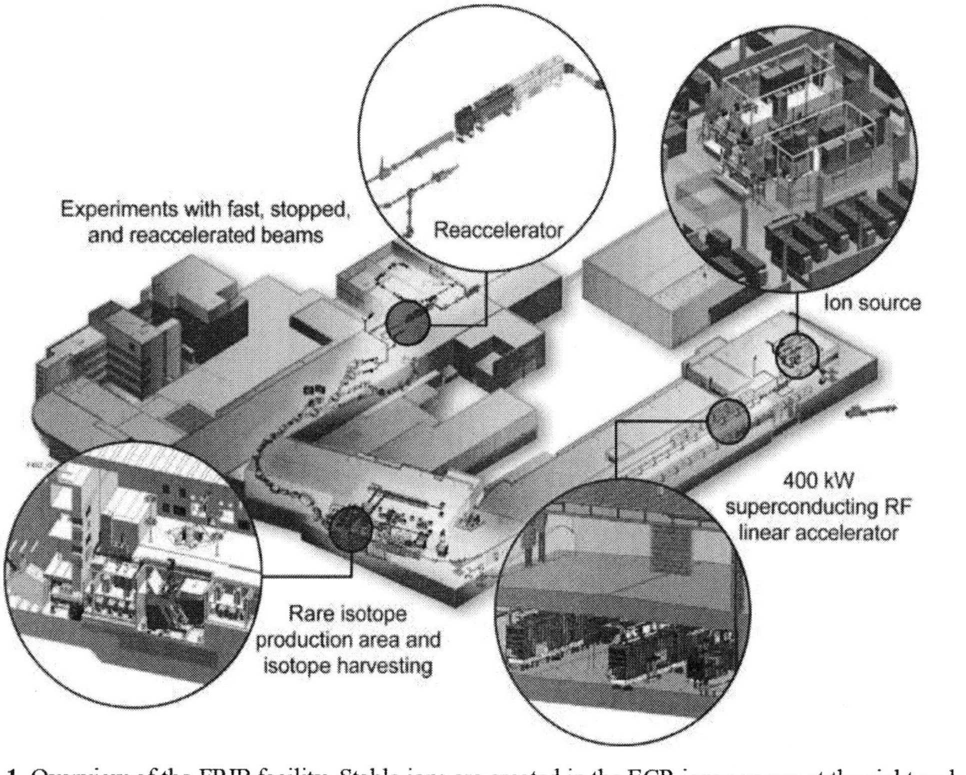

Fig. 1. Overview of the FRIB facility. Stable ions are created in the ECR ions sources at the right and accelerated in the FRIB superconducting LINAC up to 200 MeV/u for uranium ions with beam power up to 400 kW. The beam strikes a production target and the rare isotopes produced are separated in-flight and delivered to an experimental area with capabilities for study of fast, stopped, and reaccelerated beams. Reaccelerated beams will be available at FRIB at up to 10 MeV/u based on the current ReA6 facility.

2 Scientific Program

The scientific program at FRIB will have four main areas of study; nuclear structure, exploration of the role of nuclei in the cosmos, use of atomic nuclei to provide tests of nature's fundamental symmetries, and the applications of atomic nuclei to address societal issues [6]. FRIB will offer researchers the possibility to achieve the overarching goal to accurately model atomic nuclei, mentioned above. Furthermore, FRIB will make possible the measurement of a majority of key nuclear reactions needed for a quantitative understanding of the nuclear processes leading to the chemical makeup of the universe [7]. FRIB will enable tests of Nature's fundamental symmetries by providing the largest quantifies of key isotopes to measure nuclear EDMs, search for interactions beyond V-A, or measure annapole moments [8].

The scientific potential of FRIB is made possible by the vast range of isotope beams that the facility will produce. The intensity, displayed as the LOG of the rate per second, of the projected beams is illustrated in Figure 2. The useful range of intensities varies from 10^{12}/s near the stable beam isotopes to production rates of atoms/week. The rates are calculated with the program LISE++ using the EPAX2.15 cross section systematics for fragmentation and the LISE-FIS model for fission [9]. For reference the dark line shows the approximate region of known nuclei. FRIB should be able to reach the dripline for elements up to approximately 30 and many of the isotopes relevant for the r-process. The high reaccelerated rates of proton rich nuclei up to mass 100 will allow many of the key reactions important for modelling X-ray bursts to be measured directly using SECAR [7].

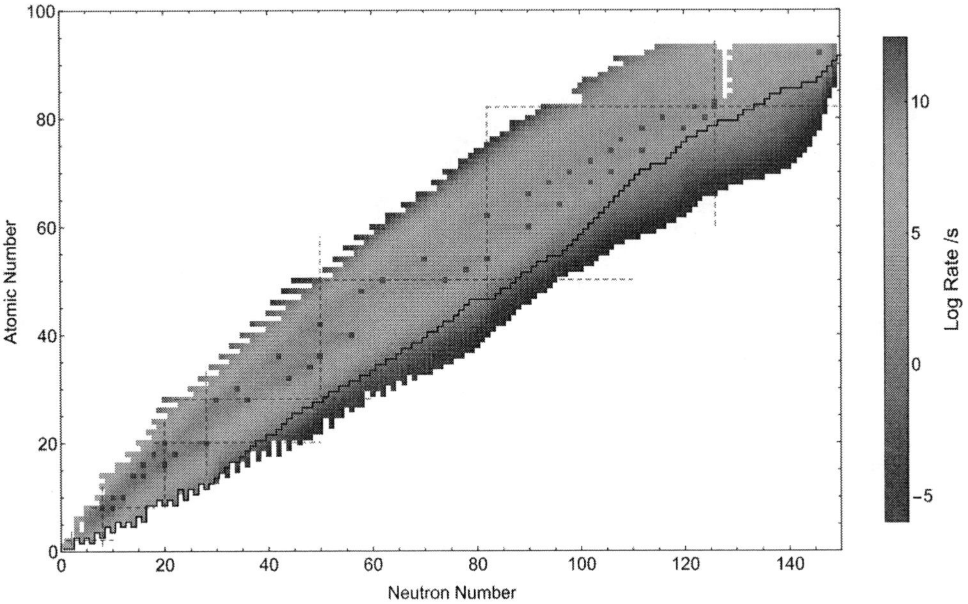

Fig. 2. Projected rare isotope beam intensities for fast beams that will be available at FRIB. Rates for stopped and reaccelerated beams will be reduced by the stopping and reacceleration efficiencies. The rates have been calculated by the program LISE++ with the assumptions described in the text. The rates for very neutron-rich nuclei are uncertain.

During the production of an isotope for a particular experiment, there will be the opportunity to collect unused isotopes by harvesting them from the beam dump or at various locations in the fragment separator. For special applications, such as collection of a sample for neutron irradiation where isotopic purity is important it is possible to have dedicated runs for collection. These options would allow micro-mole samples of certain isotopes to be collected and used at other facilities, such as neutron-irradiation facilities to study neutron-capture reactions. The harvested isotopes will also likely be of interest for a wide variety of applications [3].

3 Experimental Equipment

A wide variety of experimental equipment has been developed or is being proposed by the user community. These initiatives are described at the FRIB User Organization, FRIBUO web site [10]. The FRIBUO has approximately 1,400 members, with 250 institutions and 51 countries represented as of September 2017. A summary of the major experimental equipment items is listed in Table 1. A schematic layout of the FRIB equipment areas is shown in Figure 3. FRIB builds on the experiment that has been developed at the National Superconducting Cyclotron Laboratory, which is supported by the USA National Science Foundation.

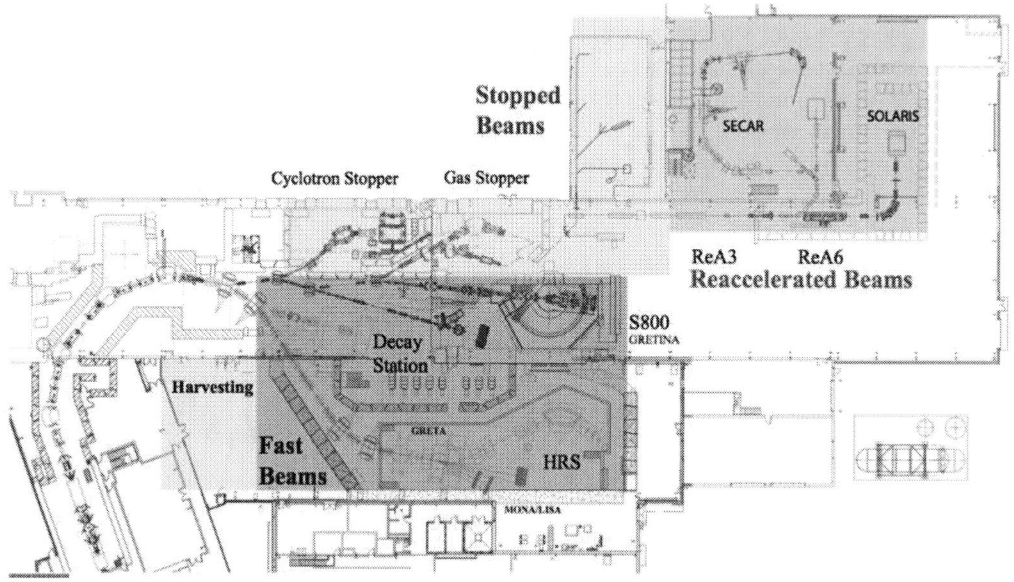

Fig. 3. Schematic layout of the FRIB facility showing the various experimental equipment planned.

Table 1. List of key experimental equipment planned for FRIB.

Equipment Item	Description and Program	Status	Ref
HRS	Bend reaction products with rigidities up to 8 Tm, which allows efficient use of FRIB beams and luminosity gains of 20x	Conceptual Design Report in 2018	[11]
Beta Decay Station	Study of radioactive decay of rare isotopes provide a sensitive test of nuclear structure and key for r-process modelling	White Paper in 2018	[12]
GRETA	Germanium-based gamma-ray array with 4pi coverage and position sensitivity to allow Doppler reconstruction	Funded, completion in 2025	[13]
SECAR	Recoil mass separator to allow direct measurement of key nuclear reaction rates	Funded, completion in 2020	[14]
SOLARIS	Solenoid spectrometer for measurement of nuclear reactions with reaccelerated beams	White Paper in 2018	[15]
ISLA	Time-of-flight mass separator for tagging of nuclear reactions with reaccelerated beams	White Paper in 2015	[16]
Existing Equipment	Many pieces of experimental equipment already exist and will be used at FRIB for day-one experiments. Examples are the S800, GRETINA, LEBIT, and BECOLA	Existing	[17]

4 Summary

FRIB will provide a vast array of rare isotopes that will allow scientists to create a comprehensive model of nuclei, determine nuclear reactions and properties key to modelling the universe, provide rare isotopes to test fundamental symmetries, and provide a fast path to many isotopes that can be important for developing new techniques in other fields. The reach of FRIB will be enabled by the 400 kW heavy ion, superconducting driver LINAC. It is anticipated that first experiments will begin by 2022. A wide array of experimental equipment is under development by the FRIB user community. Existing equipment will be ready for day-one experiments.

Acknowledgements: Establishment of the Facility for Rare Isotope Beams is supported by the Department of Energy Office of Science. The project is led by the FRIB Laboratory Director Prof. Thomas Glasmacher and the Project Manager Prof. Paul Mantica. The work described here is carried out by the FRIB project team and not the author. Plans for the scientific program described here has been developed by the FRIB user community represented by the FRIB User Organization, FRIBUO.

References

1. J. Wei et al., LINAC12: Proceedings of the 26th International Linear Accelerator Conference (2012); http://accelconf.web.cern.ch/accelconf/linac2012/index.htm
2. M. Hausmann, A. M. Aaron, A. M. Amthor, M. Avilov, L. Bandura, R. Bennett, G. Bollen, T. Borden, T. W. Burgess, S. S. Chouhan, V. B. Graves, W. Mittig, D. J. Morrissey, F. Pellemoine, M. Portillo, R. M. Ronningen, M. Schein, B. M. Sherrill, A. Zeller, Nucl. Instr. and Methods Section B **317(PART B)** 349 (2013)
3. T. Mastren, A. Pen, G.F. Peaslee, N. Wozniak, S. Loveless, S. Essenmacher, L.G. Sobotka, D. J. Morrissey, S.E. Lapi, Nature Scientific Reports **4** 6706 (2014)
4. D. Leitner, Nucl. Instr. and Methods Section B 317 (PART B) 235 (2013)
5. A. Gade and B. M. Sherrill, *Physica Scripta* **91** 033006 (2016)
6. A.B. Balantekin, J. Carlson, D.J. Dean, G.M. Fuller, R.J. Furnstahl, M. Hjorth-Jensen, R.V.F. Janssens, B.-A. Li, W. Nazarewicz, F.M. Nunes, W.E. Ormand, S. Reddy, B.M. Sherrill, Mod. Phys. Lett. A **29** 11 (2014)
7. J. Carlson et al., Prog. Part. Nucl. Physics **94** 68 (2017)
8. Fundamental Symmetry Tests with Nucleons, Nuclei, and Atoms: A Snowmass Report, K. Kumar et al., arXiv:1312.5416v1 [hep-ph] 19 Dec 2013
9. https://groups.nscl.msu.edu/frib/rates/fribrates.html
10. http://fribusers.org/
11. http://hrs.lbl.gov/science
12. https://groups.nscl.msu.edu/fribdecaystation/
13. http://greta.lbl.gov/documents/GRETA-WhitePaper.pdf
14. http://fribastro.org/SECAR/REFERENCE/SECAR-design.pdf
15. https://www.anl.gov/phy/group/solaris
16. http://www.lecmeeting.org/whitepapers/White paper Isla recoil complete.pdf
17. http://fribusers.org/documents/2014/FRIB_EQUIPMENT_whitepaper.pdf

Nuclear Physics Research at ELI-NP

N.V. Zamfir*

ELI-NP, "Horia Hulubei" National Institute for Physics and Nuclear Engineering (IFIN-HH), Bucharest-Magurele, Romania

Abstract. The new research facility Extreme Light Infrastructure – Nuclear Physics (ELI-NP) is under construction in Romania, on the Magurele Physics campus. Valued more than 300 Meuros the center will be operational in 2019. The research center will use a high brilliance Gamma Beam and a High-power Laser beam, with unprecedented characteristics worldwide, to investigate the interaction of very intense radiation with matter with specific focus on nuclear phenomena and their applications. The energetic particle beams and radiation produced by the 2x10 PW laser beam interacting with matter will be studied. The precisely tunable energy and excellent bandwidth of the gamma-ray beam will allow for new experimental approaches regarding nuclear astrophysics, nuclear resonance fluorescence, and applications. The experimental equipment is presented, together with the main directions of the research envisioned with special emphasizes on nuclear physics studies.

1 Overview

Extreme Light Infrastructure (ELI) Project was initiated in 2005, aiming to build the most powerful laser in history and to use it for generating applications unprecedented in science, as well as in industry and medicine. ELI became part of the 2006 Roadmap of ESFRI (36 European mega-projects of scientific research) [1], as a project aiming at the construction of a European and international research center at the highest level, focused on ultra high power lasers, laser-matter interaction and secondary sources of radiation. In 2009 the European Commission decided to implement this project in Czech Republic, Hungary and Romania, three complementary pillars, using Structural Funds.

In 2012 European Commission approved the Romanian application to host the nuclear pillar of the ELI project and the implementation started in January 2013. The new Center, Extreme Light Infrastructure – Nuclear Physics (ELI-NP) will be operational in 2019.

Selected by the most important science committees in Nuclear Physics in Europe - NuPECC - in the Nuclear Physics Long Range Plan in Europe [2] as a major facility, the ELI-NP infrastructure comprises two main components:

- A very High Power Laser System (HPLS), with two 10 PW laser arms based on Optical Parametric Chirped Pulse Amplification (OPCPA) able to reach intensities of 10^{23} W/cm^2 and electrical fields of 10^{15} V/m, constructed by an association between Thales Optronique France and Thales Romania;

- A very intense (10^{13} γ/s), brilliant Gamma Beam System (GBS), ~ 0.1 % bandwidth, with E_γ up to 19.5 MeV, which is obtained by Compton back scattering of a laser light off a very brilliant, intense, classical electron beam (E_e up to 720 MeV) produced by a warm LINAC. GBS is constructed by EuroGammas Consortium composed of leading research institutions and private companies in accelerator and laser technologies: INFN - Italy, University "La Sapienza" - Italy, CNRS - France, ALSYOM - France, ACP Systems S.A.S.U. - France, COMEB Srl -Italy, ScandiNova Systems – Sweden.

The combination of these equipment will create a new European laboratory with a broad range of science covering frontier fundamental physics, new nuclear physics and astrophysics, as well as applications in various fields [3].

2 Current Status

The ELI-NP project, valued at 311 M Euros, is financed by Structural Funds and Romanian national budget.

The first phase of the project (2013-2016) was finalized with the construction of the facility, shown in Fig. 1, which covers approximately 33,000 m^2 of high quality, energy efficient buildings, mostly dedicated to the high-power laser system, gamma-beam system and instrumentation for experiments, distributed in eight experimental areas:

- E1: laser induced nuclear reactions;
- E2: nuclear resonance fluorescence and applications;
- E3: positrons beams;

* Corresponding author: victor.zamfir@eli-np.ro

- E4 and E5: accelerated particle beams induced by HPLS (0.1 and 1 PW, respectively) at high repetition rates;
- E6: intense electron and gamma beams induced by high power laser beams;
- E7: experiments with combined laser and gamma beams;
- E8: nuclear reactions induced by high energy gamma beams.

Fig. 1. ELI-NP buildings.

The civil construction comprises an office building, a guest house and a canteen.

Considering the unique features of the high–power laser and gamma beams, the scientific program was ellaborated : *ELI–NP White Book [4]* is a result of a large international collaboration of more than 100 scientists from 30 countries. Following the general program, the Technical Design Reports (TDRs) were finalized and approved by ELI-NP International Scientific Advisory Board, in June 2015, and published in 2016 [5].

The second phase of the implementation of the project has received the approval of the European Comission on February 5th, 2016.

ELI-NP entered the second phase (2016-2019) with the installation and testing, in a progressive manner, of various components of the big equipment and experimental set-ups. The first laser beam, of a lower power, will be obtained by the end of 2017. The tests are going to last for two years, the infrastructure becoming fully operational in the summer of 2019, when all the parameters and full capacity of the large equipment will be reached. Figure 2 shows the laser hall with 2 arms of 10 PW each installed.

Fig. 2. The ELI-NP Laser Room with the two arms of 10PW each.

From the point of view of Human Resources, ELI-NP is also halfway along the road: in 2019, the research teams will consist in 324 employees, out of which 275 researchers. At this time (September 2017), our team consists of 149 employees, out of which 100 research and technical personnel from Romania and 20 other countries. Various positions are advertised all over the world in order to attract the best competencies, while the social impact of the project in Romania is also important to counteract the country's brain drain, by providing motivation for top young researchers to remain in the country.

By the end of the implementation phase (in 2019 and even earlier), there will be established a European Research Infrastructure Consortium - ELI-ERIC, a legal entity functioning in accordance with EU legal framework and benefiting from certain flexibility in regard with the rules of functioning. ELI-ERIC is meant to contribute significantly to strengthening and promoting leading European scientific research worldwide. Currently, the foundations of ELI-ERIC are being set-up through the preparation and submission of the documentation that will be the basis for issuing the European Commission Decision on establishing the Consortium (the Consortium Statutes and the Technical-Scientific Description of ELI infrastructure). The Romanian Government has approved, on 2 June 2016, the Memorandum on Romania's participation as a host Member State in the European Consortium for Research Infrastructure "Extreme Light Infrastructure" - ELI-ERIC.

3 Scientific Program

ELI-NP has the potential to be, for many years, in the forefront of worldwide science from theoretical physics to biology. The project has a great flexibility to cover various interdisciplinary areas, as a consequence of the possibility to employ, simultaneously in experiments, multiple radiation types, produced by equipment that will be unique at the moment of entering operation.

The most advanced research infrastructure in the world focusing on photonuclear physics studies and applications, ELI-NP will provide new opportunities for the study of fundamental processes that occur in light-matter interaction [6] or in photonuclear reactions [7] and will actively promote its applications for the benefit of society, among which we can mention:

- Investigation of the high-power laser-matter interactions using nuclear physics methods in order to study the possibilities of obtaining electron, proton and heavy ion accelerated beams using lasers.
- The extremely high intensity of the laser beam will allow the study of fundamental physics phenomena anticipated by theory, such as vacuum birefringence and pair creation in intense electric fields.
- Investigation of nuclear structure via Nuclear Fluorecence Resonance (NRF) experiments.

- Measurements of reactions cross sections of interest for astrophysics using photonuclear reactions (γ,n), (γ,p) or (γ,α).
- New methods of identification and remote characterisation of nuclear materials will be investigated with application for homeland security (remote automatic scanning of transport conteiners) and nuclear material management.
- New ways of producing more efficiently radioisotopes currently used in medicine and the producing of newly proposed ones.
- Simultaneous use of the high intensity gamma and laser beams will enable fundamental physics studies as pair production in vacuum.
- Material studies using intense beam of moderated positrons produced by the $(\gamma, e^{+}e^{-})$ reaction.
- Other possible applications of the scientific results of the ELI-NP research activity could be: industrial tomography, active interrogation of materials and nuclear forensic with high-energy gamma beam and so on.

For the planned experimental set-ups a large variety of detector systems for γ, n, e, and p are planned. For example, for the experiments with gamma beam the following detectors are under construction:

- multi-detector array ELIADE (ELI-NP Array of DEtectors) will comprise 8 CLOVER Ge detectors plus additional 4 CLOVER type or LaBr3(Ce) scintillator detectors mainly for the NRF studies [8];
- for experiments with gamma beam energy above neutron threshold – Gamma Above Neutron Threshold (GANT) [9], two different detector setups are proposed. The first is based on 3He tubes embedded in polyethylene for measurements of thermal neutrons (ELI-GANT-TN detector array), the second is an array of scintillators for measurements of gamma rays and fast neutrons (ELI-GANT-GN detector array) consisting of LaBr3:Ce and CeBr3 scintillation detectors for detection of gamma rays and BC501A liquid and GS20 6Li glass scintillators for detection of neutrons;
- High Efficiency Bragg Ionization Chamber (ELI-BIC) with Si DSSD Detector System and Thick Gas Electron Multiplier (ELITHGEM) Detector System [10] for photofission studies;
- Large area Silicon Strip Detector (SSD) and a gas Time Projection Chamber detector read by an electronic readout system (e-TPC) [11].

Since the most advanced research topics are nowadays usually pursued in very large facilities, developed at international scale in order to serve a broader community of scientists, and given the unparalleled possibilities opened by this project, ELI-NP is establishing an Academic Forum open to universities and research institutions from Romania and abroad, having signed so far 65 Memorandums of Understanding for collaboration in implementing the project with prestigious institutions from all over the world.

ELI-NP is also aiming to become a catalyst for innovation in industry and the private sector, able to enhance transfer technology both to small and medium-sized enterprises and to large firms, and to train scientists and engineers in numerous disciplines associated with Extreme Light. In order to develop relations with the economic environment and high-tech industries, it was created Magurele High Tech Cluster (MHTC) which currently comprises more than 80 members, small and medium size companies, involved in the development of high technology products. ELI-NP maintains an ongoing dialogue with the business community, in order to identify the best levels of collaboration between researchers and producers.

This work is supported by the Project Extreme Light Infrastructure - Nuclear Physics (ELI-NP) - Phase I, a project co-financed by the Romanian Government and European Union through the European Regional Development Fund.

References

1. *European Roadmap for Reserach Infrastructures Report2006* https://ec.europa.eu/research/infrastructures/pdf/esfri/esfri_roadmap/roadmap_2006/esfri_roadmap_2006_en.pdf
2. NuPECC Long Range Plan 2010, *Nuclear Physics: Perspectives for research in Europe, Science and Applications* http://www.nupecc.org/lrp2010/Documents/lrp2010_booklet_final.pdf
3. N.V. Zamfir, EPJ Web of Conferences **66**, 11043 (2014)
4. http://www.eli-np.ro/documents/ELI-NP-WhiteBook.pdf
5. Romanian Reports in Physics vol. **68** Supplement I,II (2016) (online: www.rrp.infim.ro)
6. N.V. Zamfir, Eur. Phys. J. Special Topics **223**, 1221–1227 (2014)
7. D. Filipescu et al., Eur. Phys. J. A **51**, 85(2015)
8. C.A. Ur et al., Romanian Reports in Physics, Vol. **68**, Supplement, P. S483–S538 (2016)
9. F. Camera et al., Romanian Reports in Physics, Vol. **68**, Supplement, P. S539–S619 (2016)
10. D.L. Balabanski et al., Romanian Reports in Physics, Vol. **68**, Supplement, P. S621–S698 (2016)
11. O. Tesileanu et al., Romanian Reports in Physics, Vol. **68**, Supplement, P. S699–S734 (2016)

Electroweak Decays of Highly Charged Ions

Ragandeep Singh Sidhu[1,2,*], *J.* Glorius[1], *Yu. A.* Litvinov[1], *M. K.* Pavićević[3], *M. S.* Sanjari[1], *T.* Stöhlker[1,4,5], and *the ILIMA, SPARC, LOREX, and NucAR* Collaborations

[1]GSI Helmholtzzentrum für Schwerionneforschung, 64291 Darmstadt, Germany
[2]Physikalisches Institut, Ruprecht-Karls-Universität, 69120 Heidelberg, Germany
[3]University of Salzburg, Department of Chemistry and Physics of Materials, 5020 Salzburg, Austria
[4]Helmholtz-Institut Jena, 07743 Jena, Germany
[5]Friedrich-Schiller-Universität Jena, 07743 Jena, Germany

Abstract. In this contribution we review the present status of experimental studies of electroweak decays of highly charged ions. A particular focus will be given on the bound state beta decay measurement of ^{205}Tl.

1 Introduction

Investigations of decays of highly charged radionuclides (HCR) have been among the physics motivations for the construction of the Experimental Storage Ring (ESR) at GSI Helmholtz Center in Darmstadt, Germany [1]. The very reason for the interest in such studies is that HCRs with none or a few bound electrons, like, for instance, hydrogen-like (H-like), helium-like (He-like) or lithium-like (Li-like) ions, represent well-defined nucleus plus lepton(s) systems with well-defined quantum numbers. Owing to the latter, HCRs offer unparalleled opportunities for studying the interplay of atomic structure and nuclear decay properties [2–9]. Apart from basic understanding of decay phenomena, such studies are essential for nuclear astrophysics where the nuclei involved into nucleosynthesis are highly ionized in hot and dense stellar environments [10–12].

The question on whether the nuclear decay probability is a fundamental constant or can be modified is addressed since the beginning of nuclear physics [13]. Only minor modifications of below about 1% have been observed by varying temperature, pressure, chemical environments, etc. [14–16]. However, it has been realized that nuclear half-lives ($T_{1/2}$) can change dramatically in HCRs [17–19]. On the one hand, this is straightforward for fully-ionized atoms where the decay branches involving atomic electrons, such as orbital electron capture (EC) or internal conversion (IC), are disabled. On the other hand, new decay channels disabled in neutral atoms can open up. One example of such exotic decay modes is the bound-state internal conversion (BIC) [20–22]. In the BIC process, an excited nuclear state resonantly transfers its excitation energy to a bound electron which is excited to a higher energy, but still bound, atomic level. This process has been discovered in single-pass experiments which are suitable for fast decays ($T_{1/2}$ <~ 500 ns).

Since typical weak lifetimes are longer than about a ms, for their investigations one requires facilities which are capable to store the produced radionuclides in a preserved high charged state for a sufficiently long time. Apart from the recent studies in the Electron Beam

*e-mail: R.S.Sidhu@gsi.de

Ion Trap (EBIT) at TRIUMF [23, 24], very recent measurements of weak decays of fully-ionized $^{49}Cr^{24+}$, $^{53}Fe^{26+}$ atoms [25] and the de-excitation of the $^{94m}Ru^{44+}$ isomeric state [26] at the experimental cooler storage ring CSRe at Institute for Modern Physics (IMP), Lanzhou, all other investigations of decays of HCRs were performed in the ESR of GSI [2–4, 27].

2 Heavy-Ions Storage Rings

There are three storage ring facilities in the world which allow for storing highly charged exotic nuclei [28]. These are the ESR at GSI [29], the CSRe at IMP [30] and the R3-storage ring at RIKEN Nishina Center in Wako [31–33].

Magnetic rigidity is defined $B\rho = mv\gamma/q$, where m/q is mass to charge ratio of a particle, v its velocity, and γ relativistic Lorentz factor. The high energy part of the GSI facility consists of an 18-Tm heavy-ion synchrotron SIS, the in-flight projectile fragment separator FRS [34] and the 10-Tm cooler-storage ring ESR [35]. Exotic nuclei are produced at relativistic energies of a few hundreds MeV/u through projectile fragmentation or in-flight fission nuclear reactions [36]. The FRS is employed to select and transport the ions of interest to the ESR [37]. The flight time through the FRS is merely 300 ns. In general the FRS can transport all produced exotic nuclei, the "cocktail" beam, lying within its $B\rho$ acceptance. By adding specially shaped degraders, one employs the atomic energy loss, ΔE, in the so-called $B\rho - \Delta E - B\rho$ separation method to achieve the separation of mono-isotopic beams [38].

Due to the high kinetic energies, orbital electrons are efficiently stripped off the fragments emerging from the target material [39–42]. By tuning the projectile energy, target material and thickness, ion beams in a specific charge state of interest can selectively be produced and stored in the ESR. Fully-ionised and up to 4-electron ions are routinely produced at energies of about 100 – 400 MeV/u, see, e.g., Refs. [43–54].

The storage ring mass spectrometry makes use of cocktail beams and is applied successfully at the ESR [56–67] and CSRe [68–77]. The R3 is being commissioned now and is expected to begin with mass measurements already in 2018.

3 Decay studies of highly charged radionuclides

The decay studies are typically conducted with mono-isotopic beams to avoid contaminations or secondary feedings. Furthermore, mono-isotopic beams are often needed in reaction studies with internal targets or electron cooler (see, e.g., [43, 78–80]).

The ESR has a circumference of 108 m. This is an ultra-high vacuum machine with an average residual gas pressure of about 10^{-10} – 10^{-11} mbar. The latter is an essential prerequisite for experiments with highly charged ions. The beam cooling is needed to reduce the inevitable momentum spread of the ions, $\Delta p/p$, which is mainly due to nuclear reaction process. This is of utmost importance to achieve sufficient resolving power for an unambiguous identification of parent and daughter ions. Stochastic [81] and electron [82] cooling methods are used to reduce $\Delta p/p$ to $\approx 10^{-5} - 10^{-7}$ within a few seconds. The achievable minimum $\Delta p/p$ depends critically on the number of stored ions and is determined by the balance between the cooling force and the intra-beam scattering effects. For electron-cooled ions, a transition to ordered beams is observed at low particle intensities of below about 2000 ions [83, 84]. In this mode, the mass resolving power of about 750000 is reached, which is sufficient to separate isobars and even low-lying isomers by their revolution frequencies in the ring. The intensities of stored ions are continuously monitored with non-destructive time-resolved Schottky spectrometry [85–89]. In addition, the decay/reaction products can be intercepted by dedicated particle detectors [90–93].

Numerous experiments have been performed at the ESR and since recently also at the CSRe for investigations of continuum beta decay and de-excitation of nuclear isomers [25, 26, 94–103]. In the last years a special attention is given to two-body beta decays. These are the orbital electron capture (EC) and bound-state β^--decay (β_b^-), which can be described in a general form $n + \nu_e \leftrightarrow p + e_b^-$, where p, n, e_b^-, ν_e are proton, neutron, bound electron and electron neutrino, respectively. Concerning the experiments on EC, the reader is referred to Refs. [104–121]. Here we focus on the β_b^- decay studies.

In the β_b^-, one of the neutrons in the nucleus is transmuted into a proton with an emission of an electron and an electron antineutrino. However, different from an ordinary β_c^- decay, the electron is not emitted to the continuum but occupies one of the bound orbitals [17]. This decay was first predicted by Daudel *et al.*. [122] in 1947 and then discussed in detail by Bahcall [123]. Similar to EC, in which the electron from any shell can be captured, the created electron in the β_b^--decay can occupy different shells in a daughter ion. The scaling of the probabilities to capture (generate) s-electron from (in) electron shells with different principal quantum number n scales as $1/n^3$ [124]. Since the inner orbitals in neutral atoms are Pauli-blocked, β_b^- is only a marginal decay branch in neutral atoms.

The first direct observation of the bound-state beta decay (β_b decay) was done in the ESR in 1992 [125] with the use of bare $^{163}Dy^{66+}$ ions. Neutral ^{163}Dy is stable. However, without electrons its half-life is merely $\sim$ 50 days. This result yielded the temperature T for the branching point of the s-process at $A = 163$ [2]. Another example is ^{187}Re atom, which has a very long half-life of 42 Gy. However, the increased Q-value in $^{187}Re^{75+}$ ions enables the decay to the first exited state in ^{187}Os. The measured $T_{1/2}$ is merely 33 years [126], which constrains the applicability of the $^{187}Re/^{187}Os$ pair in cosmo-chronometry [127].

Fully-ionised $^{205}Hg^{80+}$ and $^{206,207}Tl^{81+}$ nuclei have sufficiently large decay Q-values. The parent and daughter ions could thus be directly resolved by their revolution frequencies in the ESR. Both β_b^-- and β_c^--decay branches were measured [128, 129]. In contrast to numerous measurements of EC/β_c^+ branching ratios, these are the only measured β_b^-/β_c^- ratios, which are in fair agreement with theoretical estimations [18, 129].

4 Bound-State Beta Decay of $^{205}Tl^{81+}$

The measurement of the bound-state beta-decay rate of fully-ionized ^{205}Tl has been proposed more than 20 years ago [130, 131]. Its decay rate is needed to determine the matrix element for the EC of the 2.3 keV excited state in ^{205}Pb to the ground state of ^{205}Tl. This matrix element is important for the determination of the pp-solar neutrino capture probability into the 2.3 keV state of ^{205}Pb [132, 133]. It can also be used to constrain the very end of the s-process nucleosynthesis [134–137]. The proposal for this measurement has been approved at GSI.

A projectile beam of enriched ^{206}Pb will be accelerated by SIS to $\sim$ 700 MeV/u. The intensity of $2 \cdot 10^9$ particles/spill on target is expected. The production target will be ^9Be. The masses of parent $^{205}Tl^{81+}$ and daughter $^{205}Pb^{81+}$ ions differ by $\sim$ 31 keV. This tiny difference does not allow to separate them by revolution frequencies in the ESR. Therefore, the essential condition for this experiment is that the amount of $^{205}Pb^{81+}$ ions directly injected into the ESR is as small as possible. For this purpose we will employ the $B\rho - \Delta E - B\rho$ separation. The thicknesses of all matter in the FRS will be optimized such that the final energy of $^{205}Tl^{81+}$ ions equals 400 MeV/u. The ESR stochastic cooling system is matched to this energy. Beam stacking will be applied to accumulate about 10^6 parent ions.

In order to count the number of produced β_b^--daughters, the bound electron will be stripped from ^{205}Pb by employing a strong Ar gas-jet. The gas jet has a density of about 10^{13} argon atoms/cm^2. It will be turned-on for a fixed period of time. The change of the atomic charge state causes a significant alteration in the trajectory and hence the revolution frequency. The number of daughter ions can be counted in two ways either by Schottky diagnosis or by a particle detector on the corresponding orbit.

For the design number of 10^6 stored bare ^{205}Tl ions and a very cautiously estimated bound-state beta decay half-life of 1 year, we expect about 40 bound-state beta decays within a storage time of 1 hour. A waiting time is therefore necessary to accumulate daughter ions before the start of the gas-jet. We plan to have several waiting times of up to several hours. During the waiting time, the electron current in the cooler will be set to the minimum to reduce the recombination rate. The intensity of the parent ions will constantly be monitored by the Schottky analysis. This is essential to account for various beam losses.

The amount of the ^{205}Pb^{81+} ions directly injected from the FRS will be measured by switching the gas-jet immediately after injection. The primary beam of ^{206}Pb in charge states 82+ and 81+ will be used to calibrate the efficiency of the gas-jet stripping and beam losses.

5 Outlook

Heavy-ion storage-cooler rings have proven to be excellent tools for performing high-precision decay experiments on HCRs. There are many ideas for unique investigations of decays of HCRs which guarantee the rich perspectives of the field, see [112, 116, 138–142].

New storage ring projects are launched worldwide. The CRYRING has been installed behind the ESR [143], where HCRs decelerated to energies down to a few hundreds of A keV will be available. This will enable unique experiments at the interface between nuclear structure, atomic and astrophysics. Since the TSR@ISOLDE project at CERN [144] has been postponed, the TSR will probably be installed at IMP. Decays of HCRs belong to major physics cases for the TSR [144]. New storage rings are in the focus of the next-generation radioactive-ion beam facilities FAIR in Germany and HIAF in China. At FAIR, the stored HCRs will be available at energies covering 10 orders of magnitude from nearly at rest to about 5 A GeV [145–147]. Some discussions on the research with HCRs at FAIR and HIAF can be found in [148, 149].

This short paper is dedicated to our friends Paul Kienle and Fritz Bosch who made indispensable contributions to this field of research. This review is based on previous publications of our colleagues from several storage ring collaborations. To all of them we are deeply obliged. This work is in part supported by the Helmholtz/CAS Joint Research Group (HCJRG-108), and by the European Research Council (ERC) under the EU Horizon 2020 research and innovation programme (ERC CG 682841 "ASTRUm").

References

[1] P. Kienle *Forschung im Focus* (Interfrom, 1993).

[2] Yu. A. Litvinov and F. Bosch, *Rep. Prog. Phys.* **74**, 016301 (2011).

[3] Yu. A. Litvinov *et al.*, *Phys. Scripta* T **144**, 014001 (2011).

[4] F. Bosch *et al.*, *Prog. Part. Nucl. Phys.* **73**, 8 (2013).

[5] F. Bosch *et al.*, *Phys. Scripta* T **156**, 014025 (2013).

[6] D. R. Atanasov *et al.*, *Phys. Scripta* T **156**, 014026 (2013).

[7] F. Bosch and Yu. A. Litvinov, *Int. J. Mass Spectrom.* **349-350**, 151 (2013).

[8] D. R. Atanasov *et al.*, *J. Phys.* B **48**, 144024 (2015).

[9] D. R. Atanasov *et al.*, *J. Phys. Conf. Series* **875**, 012008 (2017).

[10] E. M. Burbidge *et al.*, *Rev. Mod. Phys.* **29**, 547 (1957).

[11] H. Schatz, *Int. J. Mass Spectrom.* **349-350**, 181 (2013).

[12] K. Langanke and G. Pinedo, *Phys. Scripta* T **166**, 014001 (2015).
[13] E. Rutherford and F. Soddy, *J. Chem. Soc. Trans.* **81**, 837 (1902).
[14] G. T. Emery, *Ann. Rev. Nucl. Sci.* **22**, 165 (1972).
[15] B. Wang *et al.*, *Eur. Phys. J.* A **28**, 375 (2006).
[16] T. Ohtsuki *et al.*, *Phys. Rev. Lett.* **98**, 252501 (2004).
[17] J. N. Bahcall, *Phys. Rev.* **126**, 1143 (1962).
[18] K. Takahashi and K. Yokoi, *Nucl. Phys.* A **404**, 578 (1983).
[19] K. Takahashi and K. Yokoi, *At. Data Nucl. Data Tables* **36**, 375 (1987).
[20] W. R. Phillips *et al.*, *Phys. Rev. Lett.* **62**, 1025 (1989).
[21] W. R. Phillips *et al.*, *Phys. Rev.* A **47** 3682 (1993).
[22] F. Attallah *et al.*, *Phys. Rev.* C **55**, 1665 (1997).
[23] K. G. Leach *et al.*, *Nucl. Instr. Meth.* A **780**, 91 (2015).
[24] K. G. Leach *et al.*, *Atoms* **5**, 14 (2017).
[25] X. L. Tu *et al.*, *Phys. Rev.* C **97**, 014321 (2018).
[26] Q. Zeng *et al.*, *Phys. Rev.* C **96**, 031303(R) (2017).
[27] D. Atanasov *et al.*, *J. Phys. Conf. Series* **875**, 012008 (2017).
[28] Y. H. Zhang *et al.*, *Phys. Scripta* **91**, 073002 (2016).
[29] B. Franzke, H. Geissel and G. Münzenberg, *Mass Spectrometry Reviews* **27**, 428 (2008).
[30] J. W. Xia *et al.*, *Nucl. Instr. Meth.* A **488**, 11 (2002).
[31] Y. Yamaguchi *et al.*, *Nucl. Instr. Meth.* **B 266**, 4575 (2008).
[32] A. Ozawa *et al.*, *Prog. Theor. Exper. Phys.* **2012**, 03C009 (2012).
[33] T. Yamaguchi *et al.*, *Int. J. Mass Spectrom.* **349-350**, 240 (2013).
[34] H. Geissel *et al.*, *Nucl. Instr. Meth.* B **70**, 286 (1992).
[35] B. Franzke, *Nucl. Instr. Meth.* B **24/25**, 18 (1987).
[36] H. Geissel, *Progress Part. Nucl. Phys.* **42**, 3 (1999).
[37] H. Geissel *et al.*, *Phys. Rev. Lett.* **68**, 3412 (1992).
[38] G. Münzenberg, *Nucl. Instr. Meth.* B **70**, 265 (1992).
[39] C. Scheidenberger *et al.*, *Nucl. Instr. Meth.* B **142**, 441 (1998).
[40] H. Ogawa *et al.*, *Phys. Rev.* A **75**, 020703(R) (2007).
[41] I. Tolstikhina *et al.*, *Phys. Usp.* **56**, 213 (2013). [2013 *Usp. Fiz. Nauk* **183** 225]
[42] V. P. Shevelko *et al.*, *Nucl. Instr. Meth.* B **421**, 45 (2018).
[43] C. Brandau *et al.*, *J. Phys. Conf. Series* **194**, 012023 (2009).
[44] C. Brandau *et al.*, *J. Phys. Conf. Series* **388**, 062042 (2012).
[45] W. Nörtershäuser *et al.*, *Phys. Scripta* T **156**, 014016 (2013).
[46] C. Brandau *et al.*, *Phys. Scripta* T **156**, 014050 (2013).
[47] S. Tashenov *et al.*, *Phys. Rev. Lett.* **113**, 113001 (2014).
[48] M. Lochmann *et al.*, *Phys. Rev.* A **90** 030501(R) (2014).
[49] P-M Hillenbrand *et al.*, *Phys. Rev.* A **91**, 022705 (2015).
[50] P-M Hillenbrand *et al.*, *Phys. Rev.* A **93**, 042709 (2016).
[51] H. Beyer *et al.*, J. Phys. B **48**, 144010 (2015).
[52] J. Ullmann *et al.*, J. Phys. B **48**, 144022 (2015).
[53] R. Sanchez *et al.*, J. Phys. B **50**, 085004 (2017).
[54] J. Ullmann *et al.*, *Nature Comm.* **8**, 15484 (2017).
[55] A. Gumberidze *et al.*, *Nucl. Instr. Meth.* B **408**, 27 (2017).
[56] T. Radon *et al.*, *Nucl. Phys.* **A 677**, 75 (2000).
[57] Yu. N. Novikov *et al.*, *Nucl. Phys.* A **697**, 92 (2002).

[58] J. Stadlmann *et al.*, *Phys. Lett.* B **586**, 27 (2004).
[59] Yu. A. Litvinov *et al.*, *Phys. Rev. Lett.* **95**, 042501 (2005).
[60] H. Geissel and Yu. A. Litvinov, *J. Phys.* G **31**, S1779 (2005).
[61] Yu. A. Litvinov *et al.*, *Nucl. Phys.* A **756**, 3 (2005).
[62] H. Geissel *et al.*, *Eur. Phys. J Special Topics* **150**, 109 (2007).
[63] Yu. A. Litvinov *et al.*, *Nucl. Phys.* A **787**, 315c (2007).
[64] B. H. Sun *et al.*, *Nucl. Phys.* A **812**, 1 (2008).
[65] B. H. Sun *et al.*, *Int. J. Mod. Phys* E **18** 346 (2009).
[66] L. X. Chen *et al.*, *Phys. Rev. Lett.* **102**, 122503 (2009).
[67] D. Shubina *et al.*, *Phys. Rev.* C **88**, 024310 (2013).
[68] M. Wang *et al.*, *Int. J. Mod. Phys.* E **18**, 352 (2009).
[69] X. L. Tu *et al.*, *Nucl. Instr. Meth.* A **654**, 213 (2011).
[70] X. L. Tu *et al.*, *Phys. Rev. Lett.* **106**, 112501 (2011).
[71] Y. H. Zhang, *Phys. Rev. Lett.* **109**, 102501 (2012).
[72] H. S. Xu *et al.*, *Int. J. Mass. Spectr.* **349-350**, 162 (2013).
[73] X. L. Yan *et al.*, *Astrophys. J. Lett.* **766**, L8 (2013).
[74] X. L. Tu *et al.*, *J. Phys.* G **41**, 025104 (2014).
[75] P. Shuai *et al.*, *Phys. Lett.* B **735**, 327 (2014).
[76] X. Xu *et al.*, *Phys. Rev. Lett.* **117**, 182503 (2016).
[77] P. Zhang *et al.*, *Phys. Lett.* B **767**, 20 (2017).
[78] N. Kalantar-Nayestanaki *et al.*, *Int. J. Mod. Phys.* E **18**, 524 (2009).
[79] B. Mei *et al.*, *Phys. Rev.* C **92**, 035803 (2015).
[80] H. Moeini *et al.*, *Nucl. Instr. Meth.* A **634**, 77 (2011).
[81] F. Nolden *et al.*, *Nucl. Instr. Meth.* A **532**, 329 (2004).
[82] M. Steck *et al.*, *Nucl. Instr. Meth.* A **532**, 357 (2004).
[83] M. Steck *et al.*, *Phys. Rev. Lett.* **77**, 3803 (1996).
[84] J. P. Schiffer and P. Kienle, *Z. Phys.* A **321**, 181 (1985).
[85] Yu. A. Litvinov *et al.*, *Nucl. Phys.* A **734**, 473 (2004).
[86] H. Geissel *et al.*, *Nucl. Phys.* A **746**, 150c (2004).
[87] H. Geissel *et al.*, *Eur. Phys. J. Special Topics* **150**, 109 (2007).
[88] F. Nolden *et al.*, *Nucl. Instr. Meth.* A **659**, 69 (2011).
[89] X. C. Chen *et al.*, *Nucl. Instr. Meth.* A **826**, 39 (2016).
[90] M. A. Najafi *et al.*, *Nucl. Instr. Meth.* A **836**, 1 (2016).
[91] B. Mei *et al.*, *Phys. Rev.* C **94**, 044615 (2016).
[92] O. Klepper *et al.*, *Nucl. Instr. Meth.* B **70**, 427 (1992).
[93] O. Klepper and C. Kozhuharov, *Nucl. Instr. Meth.* B **204**, 553 (2003).
[94] H. Irnich *et al.*, *Phys. Rev. Lett.* **75**, 4182 (1995).
[95] F. Attallah *et al.*, *Nucl. Phys.* A **701**, 561c (2002).
[96] Yu. A. Litvinov *et al.*, *Phys. Lett.* B **573**, 80 (2003).
[97] B. H. Sun *et al.*, *Eur. Phys. J.* A **31**, 393 (2007).
[98] B. H. Sun *et al.*, *Phys. Lett.* B **688**, 294 (2010).
[99] M. W. Reed *et al.*, *Phys. Rev. Lett.* **105**, 172501 (2010).
[100] L. X. Chen *et al.*, *Phys. Lett.* B **691**, 234 (2010).
[101] M. W. Reed *et al.*, *Phys. Rev.* C **86**, 054321 (2012).
[102] L. X. Chen *et al.*, *Phys. Rev. Lett.* **110**, 122502 (2013).
[103] A. Akber *et al.*, *Phys. Rev.* C **91**, 031301(R) (2015).

[104] L. M. Folan and V. I. Tsifrinovich, *Phys. Rev. Lett.* **74**, 499 (1995).
[105] Yu. A. Litvinov *et al.*, *Phys. Rev. Lett.* **99**, 262501 (2007).
[106] Z. Patyk *et al.*, *Phys. Rev.* C **77**, 014306 (2008).
[107] Yu. A. Litvinov, *Nucl. Phys.* A **805**, 260c (2008).
[108] Yu. A. Litvinov *et al.*, *Phys. Lett.* B **664**, 162 (2008).
[109] P. A. Vetter *et al.*, *Phys. Lett.* B **670**, 196 (2008).
[110] A. N. Ivanov *et al.*, *Phys. Rev.* C **78**, 025503 (2008).
[111] T. Faestermann *et al.*, *Phys. Lett.* B **672**, 227 (2009).
[112] Yu. A. Litvinov, *Int. J. Mod. Phys.* E **18**, 323 (2009).
[113] N. Winckler *et al.*, *Phys. Lett.* B **679**, 36 (2009).
[114] C. Giunti, *Nucl. Phys.* B (Proc. Suppl.) **188**, 43 (2009).
[115] A. Merle, *Phys. Rev.* C **80**, 054616 (2009).
[116] N. Winckler *et al.*, *Nucl. Phys.* A **834**, 432c (2010).
[117] F. Bosch and Yu. A. Litvinov, *Prog. Part. Nucl. Phys.* **64**, 435 (2010).
[118] K. Siegien-Iwaniuk *et al.*, *Phys. Rev.* C **84**, 014301 (2011).
[119] D. R. Atanasov *et al.*, *Eur. Phys. J.* A **48**, 22 (2012).
[120] P. Kienle *et al.*, *Phys. Lett.* B **726**, 638 (2013).
[121] A. Gal, *Symmetry* **8**, 49 (2016).
[122] R. Daudel *et al.*, *J. Phys. Radium* **8**, 238 (1948).
[123] J. N. Bahcall, *Phys. Rev.* **124**, 495 (1961).
[124] W. Bambynek *et al.*, *Rev. Mod. Phys.* **49**, 77 (1977).
[125] M. Jung *et al.*, *Phys. Rev. Lett.* **69**, 2164 (1992).
[126] F. Bosch *et al.*, *Phys. Rev. Lett.* **77**, 5190 (1996).
[127] F. Bosch *et al.*, *AIP Conf. Proc.* **477**, 344 (1999).
[128] T. Ohtsubo *et al.*, *Phys. Rev. Lett.* **95**, 052501 (2005).
[129] J. Kurcewicz *et al.*, *Acta Phys. Polonica* B **41**, 525 (2010).
[130] W. F. Henning *et al.*, *AIP Conf. Proc.* **126**, 203 91985).
[131] M. K. Pavićević, *Nucl. Instr. Meth.* A **271**, 287 (1988).
[132] M. K. Pavićević *et al.*, *Nucl. Instr. Meth.* A **621**, 282 (2010).
[133] M. K. Pavićević *et al.*, *J. Phys. Conf. Series* **888**, 012192 (2017).
[134] J. B. Blake, T. Lee and D. N. Schramm, *Nature Phys. Sci.* **242**, 98 (1973).
[135] J. B. Blake and D. N. Schramm, *Astroph. J.* **197**, 615 (1975).
[136] G. Huss *et al.*, *Geoch. Cosmoch. Acta* **73**, 4922 (2009).
[137] R. Baker *et al.*, *Earth and Planet. Sci. Lett.* **291**, 39 (2010).
[138] C. Scheidenberger *et al.*, *Hyperf. Interact.* **173**, 61 (2006).
[139] F. Bosch *et al.*, *EPJ Web Conf.* **123**, 04003 (2016).
[140] C. Nociforo *et al.*, *Phys. Scripta* T **150**, 014028 (2012).
[141] Ch. J. Cerjan *et al.*, *J. Phys.* G **45**, 033003 (2018).
[142] I. Dillmann and Yu. A. Litvinov, *Prog. Part. Nucl. Phys.* **66**, 358 (2011).
[143] M. Lestinsky *et al.*, *Eur. Phys. J. Special Topics* **225**, 797 (2016).
[144] M. Grieser *et al.*, *Eur. Phys. J. Special Topics* **207**, 1 (2012).
[145] P. M. Walker *et al.*, *Int. J. Mass Spectrom.* **349-350**, 247 (2013).
[146] T. Stöhlker *et al.*, *Hyperf. Interact.* **227**, 45 (2014).
[147] T. Stöhlker *et al.*, *Nucl. Instr. Meth.* B **365**, 680 (2015).
[148] Yu. A. Litvinov *et al.*, *Nucl. Instr. Meth.* B **317**, 603 (2013).
[149] P. J. Woods *et al.*, *Phys. Scripta* T **166**, 014002 (2015)

EXOGAM at the ILL: the EXILL campaign

Gilles de France[1,*], *Aurélien* Blanc[2], *Michael* Jentschel[2], *Ulli* Köster[2], *Silvia* Leoni[3], *Paolo* Mutti[4], *Gary* Simpson[2], *Torsten* Soldner[2], and *Waldemar* Urban[5]

[1]GANIL, CEA/DRF-CNRS/IN2P3, BP 55027, 14076 Caen cedex 5, France
[2]ILL, Institut Laue-Langevin, BP 156, 6, rue Jules Horowitz, 38042 Grenoble Cedex 9, France
[3]University of Milano and INFN sez. Milano,Via celoria 16, 20133 Milano, Italy
[4]LPSC, Université Grenoble Alpes, 38026 Grenoble Cedex, France
[5]Faculty of Physics, University of Warsaw, ul. Hoża 69, 00-681 Warsaw, Poland

Abstract. Radiative capture reactions on rare targets and prompt spectroscopy of neutron-rich nuclei produced in the neutron-capture induced-fission of ^{235}U and ^{241}Pu have been performed at the Institut Laue-Langevin (ILL). The high-efficiency germanium (Ge) EXOGAM array has been used, complemented with Ge detectors from GASP and the ILL. $LaBr_3$ detectors from the FATIMA collaboration have also been coupled to EXOGAM to measure lifetimes of low-lying excited states. The measured characteristics indicate very good performances of the overall setup. Some recent results will be discussed.

1 Introduction

Fission produces neutron-rich nuclei in a very effective way. In particular, one of the best mechanism is to induce fission of actinide targets using cold neutrons from a neutron guide. In this regime, the incident neutrons bring just enough energy into the system to trig fission, whilst preserving the neutron-richness of the fragments. The average spin generated in this reaction is around 6-7$\hbar$ [1] and states with spins as high as 20$\hbar$ can be observed. Of particular interest is the use of ^{235}U and ^{241}Pu fissile targets which give access to nuclei that are too weakly produced using spontaneous fission sources or another production mean to perform their prompt spectroscopy. In addition to the spectroscopy of nuclei produced in such a cold fission process, neutron capture reactions using rare targets have also been performed. In this contribution we will describe the installation of the EXOGAM array, usually standing at GANIL, at the Institut Laue Langevin (ILL), Grenoble, and the experimental campaign which took place in 2012 and 2013.

2 Physics case

The basic idea of EXILL was to install a large number of efficient germanium detectors around a fissile target irradiated by a well collimated cold neutron beam. Large arrays like Euroball or Gammasphere were already used to study fission fragments produced in the spontaneous fission of ^{252}Cf and ^{248}Cm sources but it was the first time that such a large array was installed in a reactor facility, around a fissile target irradiated by cold neutrons. Traditional

*e-mail: defrance@ganil.fr

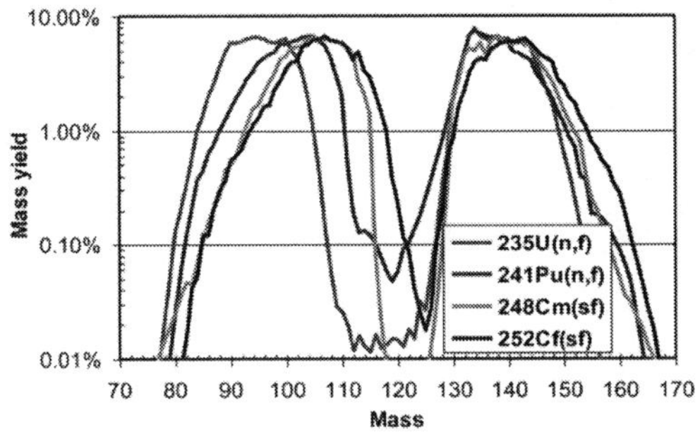

Figure 1. Mass yield distributions measured for the spontaneous fission of ^{252}Cf and ^{248}Cm compared to the thermal neutron induced fission of ^{235}U and ^{241}Pu (data from [2]).

nuclear physics at the ILL consists essentially in decay spectroscopy studies at the focal plane of the Lohengrin spectrometer. In the EXILL campaign instead we performed prompt spectroscopy. We used ^{235}U and ^{241}Pu targets giving access to many nuclei where little was known, especially in the regions north-east of ^{78}Ni and beyond ^{132}Sn. This is what is shown in Figure 1 which compares the mass yield distribution from the thermal neutron induced fission of ^{235}U and ^{241}Pu and the yield from the spontaneous fission of ^{252}Cf and ^{248}Cm. The production of nuclei in the mass A$\sim$90 region is an order of magnitude larger using a ^{235}U target compared to the two spontaneous fission sources. The same is true for nuclei in the mass A$\sim$125 region using a ^{241}Pu target. These two mass regions were the focus of the EXILL physics case.

In addition the high photopeak efficiency of the array allowed the study of samples that have low neutron capture cross sections of tens of millibarn or where only small milligram quantities of enriched material are available. An important property of capture reaction is its non selectivity which ensures the feeding of all available states in a certain energy and spin range for complete spectroscopy. Therefore, studies on collective models, symmetries and statistical properties of nuclei close to stability become possible.

The complete physics case of the EXILL campaign will not be detailed here but, to summarize it, we concentrated on a few topics which are:

- Shell Model (SM) nuclei with a few particles or holes outside a double shell closure
- Shapes and shape transitions
- Symmetry studies
- Statistical properties of nuclei
- Level densities
- Application to reactor physics
- Radioisotope for nuclear medicine

In this paper we will discuss a few examples illustrating the opportunities offered with the EXILL campaign.

3 Setups

Two setups have been operated (see reference [3] for a complete description). The first one consisted in 8 EXOGAM large clovers [4], 6 large coaxial detectors from GASP [5] and the 2 clovers from the ILL (see picture 2). In the second setup, the GASP and ILL detectors were replaced by 16 $LaBr_3$ crystals from the FATIMA collaboration [6].

The two setups were installed around the target point of the PF1B line of the ILL. The ILL reactor is the world's brightest continuous neutron source with an in-pile flux up to 1.5×10^{15}

Figure 2. Picture of the spectroscopy setup of the EXILL campaign in the PF1B area. The neutron beam is coming from the right hand side.

neutrons per s.cm^2. In order to produce around 10^4 fissions per second in a 400 μg/cm^2 thick, ^{235}U target a flux of 10^8 neutrons per s.cm^2 is required. An efficient collimation system was developped which details can be found in [3]. The campaign covered 93 days of effective beam time for experiments.

In the absence of any fragment identification, it is crucial to acquire high-fold events (3 or larger) to possibly associate the observed γ-rays to a given fragment. This technique relies on the knowledge of the transitions deexciting the first excited states in the two fragments and the measurements of the evolution of the unknown γ-rays intensity gated by known transitions in a series of isotopes. ^{235}U and ^{241}Pu targets have been used leading for a given nucleus to two distinct complementary fragments, which greatly helps to confirm the assignment. The method relies therefore primarily on the efficient detection of high-fold data, which is achieved using a large number of efficient HPGe detectors. Such a technique is now well established and has been proven to work very well to analyze fission data [8, 9]. The total photopeak efficiency for the 16 Ge detectors setup was about 6% at E_γ=1.3 MeV. The second setup using the $LaBr_3$ detectors was dedicated to lifetime measurements. In this configuration, the EXOGAM detectors are used to select the fragment of interest and the $LaBr_3$ crystals give the lifetimes via the electronics fast timing technique using $LaBr_3$-$LaBr_3$ coincidences [10–13]. This method requires therefore Ge-$LaBr_3$-$LaBr_3$ or Ge-Ge-$LaBr_3$-$LaBr_3$ coincidences. Recently, a novel method to analyze such data has been developed [14] and lead to the determination of lifetimes with an accuracy of 5 ps between 40 keV and 6.7 MeV.

4 Example of results

4.1 ^{133}Sb: Coupling collective and single-particle states in the vicinity of a doubly closed shell nucleus

Shell Model nuclei like those located one or few particles away from ^{132}Sn are cornerstones of nuclear structure theories. In this context we have studied particle-core excitations in the one-valence proton nucleus ^{133}Sb. For this we exploited the full complementarity of the various runs within the EXILL campaign i.e. spectroscopy and lifetime measurements using both ^{235}U and ^{241}Pu targets. In the spectroscopy part we looked for excited states above the $21/2^+$ isomer. This was possible thanks to the capability of EXILL to build prompt-delayed and prompt-prompt coincidences (see Figure 3). Several prompt transitions were found in coincidence with known delayed transitions in ^{133}Sb. Using the prompt-prompt coincidence relationships, we then could place these new transitions in the level scheme, above the isomer. We also used fission events from ^{235}U and ^{241}Pu targets to ensure that the newly observed transitions were observed in both cases hence not associated to a complementary fragment of ^{133}Sb.

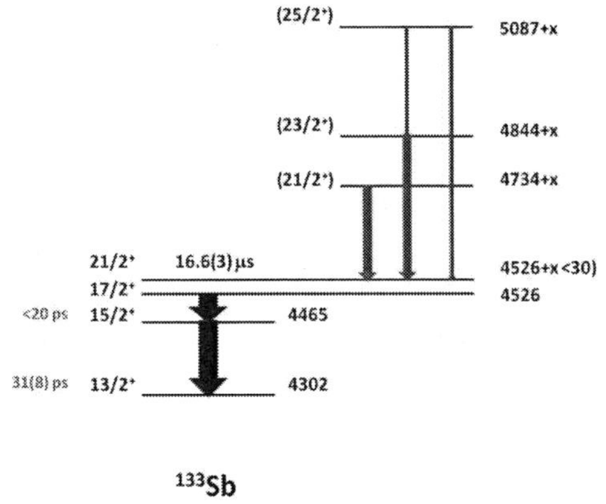

Figure 3. Partial level scheme of ^{133}Sb showing only states discussed in this paper with the newly observed γ-rays above the $21/2^+$ isomer in red and the new lifetimes in blue.

Lifetime measurements were then used to investigate collective properties of the $15/2^+$ and $13/2^+$ states located just below the isomer. These lifetimes were extracted using the centroid shift technique [12] from which the reduced transition probabilities of $B(M1 : 15/2^+ \rightarrow 13/2^+) > 0.24$ W.u and $B(M1 : 13/2^+ \rightarrow 11/2^+) = 0.0042(15)$ W.u were deduced. This difference of two orders of magnitude clearly reveals a profound modification in the structure of the two levels. In order to get insight into this puzzle a new model called the Hybrid Configuration Mixing Model was developed [15, 16] with the aim to couple collective and single-particle excitations in a fully self-consistent way. This model suggests that the low spin states are built from a $g_{7/2}$ protons coupled to the 2^+ ^{132}Sn core excitation while the higher lying levels arise from the valence proton coupled to single-neutron $h^{-1}_{11/2}f_{7/2}$ excitations. As a result the model gives 0.021 W.u and 0.001 W.u for the $15/2^+ \rightarrow 13/2^+$ and $13/2^+ \rightarrow 11/2^+$ transitions respectively, in fair agreement with data, which is not the case for shell model calculations [17].

4.2 Symmetry studies

Testing the validity of the SO(6) limit of the sd Interacting Boson Model ([18, 19]) is a long quest in symmetry studies. Using the fast timing part of the EXILL campaign, the Cologne group led by J Jolie proposed to study ^{196}Pt known to be a good candidate for such studies. In the SO(6) limit, the E2 transition operator is a generator of SO(6) and strictly speaking only transitions between states with $\Delta\sigma$=0 are allowed (σ being one of the quantum number associated to the irreducible representations of SO(6)). Therefore, E2 transitions between different SO(6) representations ($\Delta\sigma \neq 0$) are forbidden. The states of interest have low spin and high excitation energy and are well populated in (n,γ) reactions. This was realized during the EXILL campaign using a ^{195}Pt target and using Ge-LaBr$_3$-LaBr$_3$ coincidences to deduce the lifetimes of the third 0^+ state at 1402 keV and the first 2^+ state at 356 keV (which lifetime was already known at 49.2(2) ps). The measurements yielded the values of 50(5) ps for the 2^+_1 state and an upper limit of 12 ps for the 0^+_3 state. Using also the lower limit from Börner et al [19], it was established that 0.56 W.u $< B(E2; 0^+_3 \rightarrow 2^+_1) < 5$ W.u and 0.05 W.u $< B(E2; 0^+_3 \rightarrow 2^+_2) < 0.41$ W.u. The absence of collective B(E2) values demonstrates for the first time the validity of the SO(6) symmetry [20].

4.3 ^{195}Pt: the link between nuclear structure and medical physics

The structure of ^{195}Pt was also studied and more specifically the population of the $21/2^+$ isomer using the radiative capture ^{194}Pt(n,γ) reaction. This nucleus has been recognized since long as a good candidate to probe multi-j supersymmetry ([21, 22]). In addition to this fundamental interest, ^{195}Pt is also a promising candidate for nuclear medicine since it combines therapy and diagnosis capabilities. Indeed its $T_{1/2}$=4.01 days isomer is compatible with such a usage (long enough for transportation and short enough for biology). This isomer decays via a highly converted M4 transition which provides an interesting possibility for treatment, with a very high linear energy transfer deposited within a few nm^3, capable to induce double strand DNA breaks. In addition, the 99 keV gamma-ray and 65-78 keV X-rays emitted in the cascade from the isomer allow imaging using Single Photon Emission Computed Tomography and gamma cameras. During the EXILL campaign we searched for new doorway states that would significantly feed the isomer and decay to the ground state. In this study several new states have been identified in the decay from the $1/2^+$ neutron capturing state at 6.1 MeV. To envisage an abundant production of the isomer for nuclear medicine, a method to populate specifically the doorway state of interest has to be applied. For this, another experiment was realized at the HIGS facility (TUNL) where we tried to use a gamma beam at energies tuned around the newly identified states in order to measure the population of the isomer of interest. This work is still under analysis and is part of the thesis work of D. Wilmsen.

5 Conclusion

The EXILL campaign ran smoothly for two reactor cycles (~100 days) at the ILL, harvesting results from 73 proposals and gathering together 18 different laboratories. Today more than 30 papers have been published in refereed journals, showing the success of this adventure. The installation of a large and efficient gamma-ray array at a neutron beam facility is proven to be very effective in the study of neutron rich nuclei from cold neutron induced fission as well as in performing the detailed spectroscopy of close-to-stability nuclei in radiative capture reactions.

Acknowledgements
The EXILL campaign would not have been possible without the support of several services at the ILL and the LPSC. We are grateful to the EXOGAM collaboration for the loan of the detectors, to GANIL for assistance during installation and dismantling, and to the INFN Legnaro laboratory for the loan of the GASP detectors.

References

[1] M.A.C. Hotchkis et al., Phys. Rev. Lett. **64**, 3123 (1990)
[2] JEFF-3.1.1 database, http://www.oecd-nea.org/dbdata/jeff/
[3] M. Jentschel et al., JINST **12** P11003 (2017)
[4] J. Simpson and the EXOGAM collaboration, Acta Physica Hungarica, New Series, Heavy Ion Physics **11** 159 (2000)
[5] D. Bazzacco and the GASP Collaboration, Proc. Int. Conf. Nuclear Structure at High Angular Momentum, Ottawa, AECL-10613 376 (1992)
[6] http://nuclear.fis.ucm.es/fasttiming/
[7] H. Abele et al., Nucl. Instr. and Meth. A **562** 407 (2006)
[8] M.A.C. Hotchkis et al., Nucl. Phys. A **530** 111 (1991)
[9] I. Ahmad and W.R. Phillips, Rep. Prog. Phys. **58** 1415 (1995)
[10] K.E.G. Löbner The electromagnetic interaction in nuclear spectroscopy, (ed. W. D. Hamilton, North-Holland New York 1975) 173

[11] H. Mach, R.L. Gill, M. Moszyński Nucl. Instr. and Meth. A **280** 49 (1989)
[12] J.M. Régis et al., Nucl. Instr. and Meth. **A 622** 83 (2010)
[13] J.M. Régis et al., Nucl. Instr. and Meth. **A 763** 210 (2014)
[14] J.M. Régis et al., Nucl. Instr. and Meth. **A 684** 36 (2012)
[15] G. Bocchi et al., Phys. Lett. B **760**, 273 (2016)
[16] G. Colò et al., Phys. Rev. C **95**, 034303 (2017)
[17] W. Urban et al., Phys. Rev. C **62**, 027301 (2000); Phys. Rev. C **79**, 037304 (2009)
[18] J.A. Cizewski, R.F. Casten, G.J. Smith, M.L. Stelts, W.R. Krane, H.G. Börner, W.F. Davidson, Phys. Rev. Lett. **40** 167 (1978)
[19] H.G. Börner, J. Jolie, S.J. Robinson, R.F. Casten, J.A. Cizewski, Phys. Rev. C **42** R2271 (1990)
[20] J. Jolie et al., Nucl. Phys. A **934** 1 (2015)
[21] D.D. Warner, R.F. Casten, M.L. Stelts, H.G. Börner and G. Barreau, Phys. Rev. C **26** 1921 (1982)
[22] R.F. Casten, G.G. Colvin and K. Schreckenbach, J. Phys. G **13** 221 (1987)

EPJ Web of Conferences **178**, 01005 (2018) https://doi.org/10.1051/epjconf/201817801005
CGS16

Proposed BISOL Facility – a Conceptual Design

Yanlin Ye[1] (for BISOL Collaboration),*

[1]State Key Laboratory of Nuclear Physics and Technology, School of Physics, Peking University

Abstract. In China, a new large-scale nuclear-science research facility, namely the "Beijing Isotope-Separation-On-Line neutron-rich beam facility (BISOL)", has been proposed and reviewed by the governmental committees. This facility aims at both basic science and application goals, and is based on a double-driver concept. On the basic science side, the radioactive ion beams produced from the ISOL device, driven by a research reactor or by an intense deuteron-beam accelerator, will be used to study the new physics and technologies at the limit of the nuclear stability in the medium mass region. On the other side regarding to the applications, the facility will be devoted to the material research associated with the nuclear energy system, by using typically the intense neutron beams produced from the deuteron-accelerator driver. The initial design will be outlined in this report.

1 Brief history

Around the year 2010, China Institute of Atomic Energy (CIAE) and Peking University (PKU) had separately proposed two ISOL-type facilities, namely "China Advanced Rare Ion beam Facility (CARIF)" and "Intense multi-beam multi-PUrpose Facility (ImPUF)", respectively. In 2011, CIAE and PKU signed an agreement to jointly promoting an ISOL-type RIB facility in Beijing area. As a result the previously proposed CARIF and ImPUF were merged into a combined project "Beijing Isotope-Separation-On-Line neutron-rich beam facility (BISOL)".

In October 2012 an international advisory committee (IAC), led by Professors Shoji Nagamiya and Jiaer Chen, was formed and a review meeting was held at PKU to evaluate the "Initial Conceptual Design of the BISOL". The committee gave a high appraisal to the proposal and "considers the research potential of the proposed facility in both, basic as well as applied and interdisciplinary research, as excellent and highly competitive on the world level. It promises a unique science reach in several respects, in particular with regard to the most neutron rich exotic nuclei and the study of the astrophysical r-process."

In 2013, the advanced ISOL-type facility was adopted in "the national mid- and long-range plan (till 2030) of the major facilities for science and technology development".

In August 2014, a Xiang Shan Forum was successfully organized and a road map for major nuclear physics research facilities was established, including the BISOL as the next-generation major facility.

Then in May 2016, a domestic expert meeting was held at CIAE to evaluate the preparation works of the BISOL, aiming at a proposal to the 13th 5-year plan of the central government of China. Soon after, the project was included into a list of the large-scale-facility

*e-mail: yeyl@pku.edu.cn

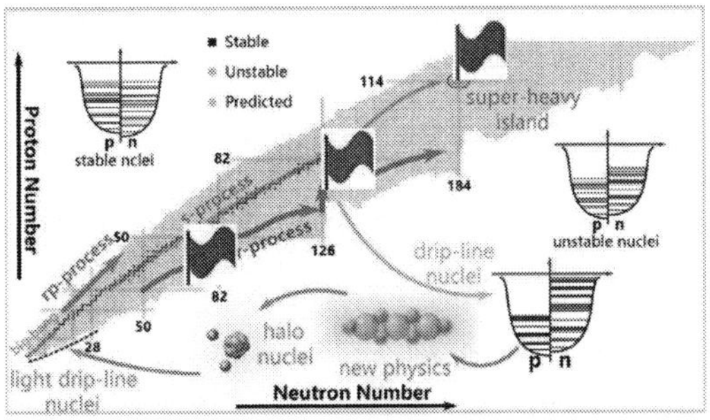

Figure 1. Great scientific questions at the expanded nuclear chart (see the text for details).

proposals and reviewed by the National Development and Reform Council in order to establish the priorities for the coming 5-year plan. As a result the BISOL proposal was successfully classed as one of the preparation facilities.

In January 2017, another topical IAC meeting was held at PKU-CIAE, which was dedicated to the accelerator based intense neutron source, in particular the high-power target systems. The first BISOL user meeting was also held at PKU in March with ~150 participants who enjoyed active discussions. Then in June, the conceptual design of BISOL (BISOL-CD-1) was finalized and evaluated by an domestic expert committee, and was ready for the next national review and the next IAC review.

2 Background in basic and applied sciences

The BISOL aims at both the basic nuclear science research as well as the nuclear technique applications. The basic scientific questions at the expanded nuclear chart, which will be addressed by BISOL, include: the new physics at the drip-lines; the nuclear-processes in creating heavy elements in the stars; the ways towards the stable island for super-heavy-elements. Fig. 1 gives an illustration of these major scientific goals. At the application side, BISOL will be dedicated to meet the urgent needs for applying the intense neutron beams, such as studying the mechanism of the neutron irradiation damages of the materials, the material evaluations for the nuclear-energy systems; the nuclear reaction data with fast neutrons; and so on. These applications are extremely important considering the planned tremendous growth of the nuclear power in China in the coming 30 years or so.

The requirement of RIBs for studies of unstable nuclei has been outlined previously [1], based on which we have defined the primary parameters and the basic configuration of the BISOL, as illustrated in Fig. 2. The idea is to combine the reactor driver (RD) and the intense deuteron-beam driver (IDD) to produce the isotopes, to apply both the isotope separation on line (ISOL) and the projectile fragmentation (PF) techniques (the latter follows the former) in order to obtain the most neutron-rich RIBs in the medium mass region, and to target both the basic science problems as well as the key application problems. This interwinding concept is quite in line with the traditional Chinese Dao-philosophy which emphasizes the combination of various aspects into one unit, as symbolized by the "Taiji picture" in the inset of Fig. 2. The reason to have two drivers is basically the complementarity. The deuteron-beam driver (IDD) should produce fast neutrons with energies up to 20 MeV while the reactor driver (RD) provides neutrons with much lower energies (~ 1 MeV). The higher-energy-neutron induced fission leads to more uniform isotope yields over the medium-mass region, whereas the low-energy neutron would induce fission-fragments focus on the two fission-fragment

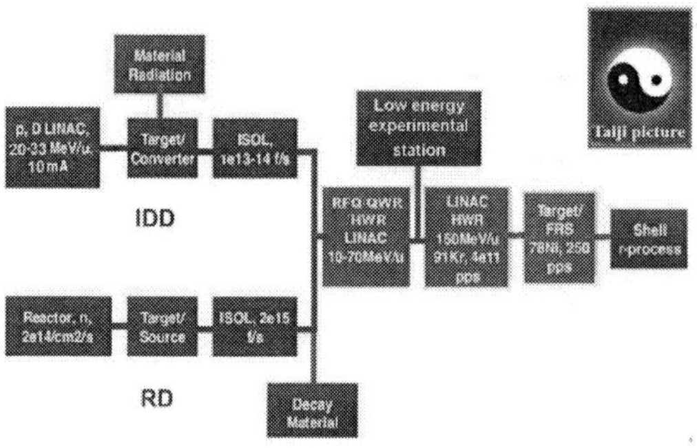

Figure 2. Primary parameters and the basic configuration of the BISOL. The inset at the upper-right corner shows a picture of the traditional Chinese Dao-philosophy, which symbolize the current design concept of combining various goals and technologies.

peaks. Therefore two drivers allow to produce high-intensity beams at different mass region. More importantly, these two drivers can be operated simultaneously to double the machine efficiency. For instance when the RD drives the ISOL target for the basic research purpose, the IDD may be used separately to irradiate materials for the application purpose. These dual-purposes are essential for the BISOL-design.

3 Introduction to the conceptual design

The overall design of the BISOL is displayed in Fig. 3. It is composed of the reactor driver (RD) using the existing China Advanced Research Reactor (CARR) and the intense deuteron linear accelerator driver (IDD, D-LINAC) together with the respective target systems, the low energy ISOL devices to separate the selected isotopes, the post-acceleration up to 150 MeV/u, the projectile fragmentation followed by the online separation of the very neutron-rich nuclei via a high-power separator (FRS).

The experimentation is configured into three areas shown in Fig. 3: the Separation Energy Experiments with beams at 20keV/q (SEE); the Low Energy Experiments with beams at 20 MeV/u (LEE); and the Intermediate Energy Experiments with beams at 150 MeV/u (IEE). Various instrumentations will be built in these areas.

Fig. 4 shows a more realistic placement of the facilities. Due to the limited land available to the facility, two levels (floors) arrangement is considered, with the high-intensity deuteron accelerator and the LE-ISOL together with the LEE placed in the underground floor, and the post-acceleration together with LEE and IEE placed at the ground floor.

The parameters for the isotope yields and the RIBs are specified in Figs 3 and 4. The estimated intensities for the typical ISOL-type beams are listed in Table 1. The combination of the applied technologies allows to produce the double-magic nucleus ^{78}Ni and the neutron-drip-line nucleus ^{120}Sr as high as 250 and 2×10^{-4} pps, respectively, based on the simulation calculations. These reported numbers are obtained based on the reactor-driver (RD) configuration. The neutron flux is about 8×10^{14} cm^{-2}s^{-1} and a 5g ^{235}U-target is assumed, resulting in a fission rate of about 2×10^{15} s^{-1}. Similar fission-rate is also estimated by using the deuteron-beam driver (IDD).

On the application side, BISOL will be used to study material properties under strong neutron irradiation. The deuteron-LINAC can be operated independently to generate high-flux neutrons. The liquid lithium will be used as target for neutron production. The target must sustain a beam power of 400 kW. The estimated neutron irradiation capabilities depend

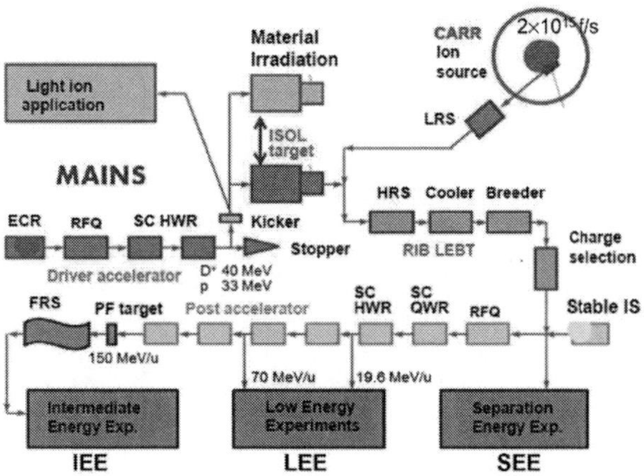

Figure 3. Specifications of various parts of the BISOL, together with the basic designed parameters.

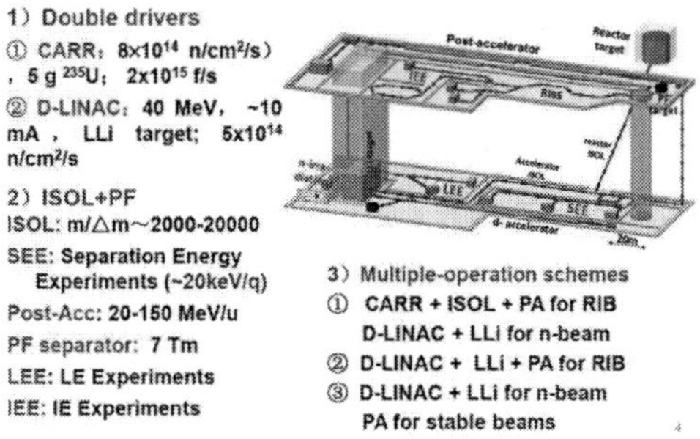

Figure 4. Schematic view of the arrangment of the BISOL facility. The basic parameters of the majors are listed.

Table 1. Estimated beam intensity

nuclei	Fis. yield	rate	Target + isol eff.	CB eff.	Linac eff.	intensity
^{91}Kr	3.2×10^{-2}	6.4×10^{13}	13.0%	10%	50%	4×10^{11}
^{142}Xe	4.3×10^{-3}	8.8×10^{12}	2.0%	10%	50%	9×10^{9}
^{132}Sn	5.7×10^{-3}	1.2×10^{13}	8.0%	10%	80%	7×10^{10}
^{81}Ga	7.6×10^{-5}	2.0×10^{11}	8.0%	10%	95%	1×10^{9}

on the distance from the target and are illustrated in Fig.5. In the most favorable case (central region) 8-15 displacement per atom (dpa) can be achieved for one full power year (fpy) of irradiation, which is already the highest record at the near-future accelerator facilities and is very useful for material studies. This part of the facility is also called "the Multi-purpose-Aimed Intense Neutron Source" (MAINS in Fig. 3).

The detailed design has been performed for each part of the facility, including: (1) the reactor target; (2) the deuteron-LINAC; (3) the high power target and material irradiation station; (4) the low energy ISOL; (5) the separation energy experiments (SEE); (6) the post-acceleration; (7) the low energy experiments (LEE); (8) the PF separator and the intermediate energy experiments (IEE). The infrastructure is also designed by the an expertized company.

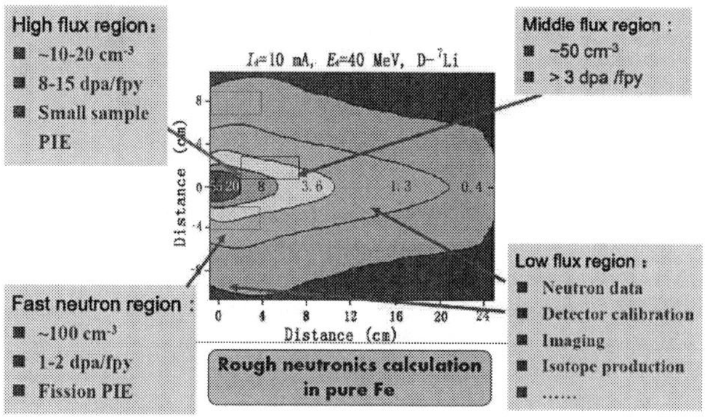

Figure 5. Irradiation performances at the target station down stream of the deuteron LINAC.

All details of the conceptual design version-1 are presented in an internal report BISOL-CD-1 (June 2017) [2].

The land for BISOL installation has already been allocated by CIAE, located at the central region in the CIAE campus near the CARR reactor. This is extremely important considering the safety and environment control for the strong-irradiation apparatus.

4 Summary

BISOL, in combination of the ISOL production through either the high-intensity deuteron beam driver or the high thermal neutron-flux from the reactor, together with re-acceleration of RIBs up to projectile-fragmentation energies (i.e. ~150 AMeV), promises the most intense RIB capabilities in the medium mass region. The availability of accelerator-driven high-flux neutrons, produced with an intense beam of 40 MeV deuteron, provides for a wide spectrum of applied research over a range of energies, in particular for unique material research relevant to the next generation fission and fusion reactors.

The facility cost is estimated to be about 3.5B RMB (~500M USD) plus the infrastructure and manpower investment. The timetable is not clear yet but we expect a completion time in 2025-2030.

ACKNOWLEDGEMENTS

The author wish to thank all members of the BISOL collaboration who have contributed to this conceptual design. Particularly the original figures presented here were drawn by Dr. Jian GAO, Weiping LIU, Zhiyu GUO, Yang Liu, Biao YANG, Yugang WANG et al., and should be acknowledged. This work is partially supported by the 973 Program of China under Grant Nos. 2013CB834400 and 2014CB845500.

References

[1] Baoqun Cui, Yuan Gao, Yucheng Ge, Zhiyu Guo, Zhihong Li, Weiping Liu, Shixiang Peng, Zhaohua Peng, Zhi Wang, Sha Yan, Yanlin Ye, Sheng Zeng, Guohui Zhang, Feng Zhu, *Nucl. Instr. Meth. in Phys. Res. B* **317**,257-262(2013).

[2] BISOL collaboration, *Internal report BISOL-CD-1* (June 2017).

The MAJORANA DEMONSTRATOR Status and Preliminary Results

C.-H. Yu[1], *S.I.* Alvis[2], *I.J.* Arnquist[3], *F.T.* Avignone III[1,4], *A.S.* Barabash[5], *C.J.* Barton[6], *F.E.* Bertrand[1], *T.* Bode[7], *V.* Brudanin[8], *M.* Busch[9,10], *M.* Buuck[2], *T.S.* Caldwell[9,11], *Y.-D.* Chan[12], *C.D.* Christofferson[13], *P.-H.* Chu[14], *C.* Cuesta[2,*] *J.A.* Detwiler[2], *C.* Dunagan[13], *Yu* Efremenko[1,15], *H.* Ejiri[16], *S.R.* Elliott[14], *T.* Gilliss[9,11], *G.K.* Giovanetti[17], *M.* Green[1,9,18], *J.* Gruszko[19], *I.S.* Guinn[2], *V.E.* Guiseppe[4], *C.R.* Haufe[9,11], *L.* Hehn[12], *R.* Henning[9,11], *E.W.* Hoppe[3], *M.A.* Howe[9,11], *K.J.* Keeter[20], *M.F.* Kidd[21], *S.I.* Konovalov[5], *R.T.* Kouzes[3], *A.M.* Lopez[15], *R.D.* Martin[22], *R.* Massarczyk[14], *S.J.* Meijer[9,11], *S.* Mertens[7,23], *J.* Myslik[12], *G.* Othman[9,11], *W.* Pettus[2], *A.W.P.* Poon[12], *D.C.* Radford[1], *J.* Rager[9,11], *A.L.* Reine[9,11], *K.* Rielage[14], *N.W.* Ruof[2], *B.* Shanks[1], *M.* Shirchenko[8], *A.M.* Suriano[13], *D.* Tedeschi[4], *R.L.* Varner[1], *S.* Vasilyev[8], *K.* Vetter[12,**], *K.* Vorren[9,11], *B.R.* White[14], *J.F.* Wilkerson[1,9,11], *C.* Wiseman[4], *W.* Xu[6], *E.* Yakushev[8], *V.* Yumatov[5], *I.* Zhitnikov[8], and *B.Z.* Zhu[14]

(MAJORANA Collaboration)

[1]Oak Ridge National Laboratory, Oak Ridge, TN, USA
[2]Center for Experimental Nuclear Physics and Astrophysics, and Department of Physics, University of Washington, WA, USA
[3]Pacific Northwest National Laboratory, Richland, WA, USA
[4]Department of Physics and Astronomy, University of South Carolina, Columbia, SC, USA
[5]National Research Center, "Kurchatov Institute" Institute for Theoretical and Experimental Physics, Moscow, Russia
[6]Department of Physics, University of South Dakota, Vermillion, SC, USA
[7]Max-Planc-Institute für Physik, München, Germany
[8]Joint Institute for Nuclear Research, Dubna, Russia
[9]Triangle Universities Nuclear Laboratory, Durham, NC, USA
[10]Department of Physics, Duke University, Durham, NC, USA
[11]Department of Physics and Astronomy, University of North Carolina, Chapel Hill, NC, USA
[12]Nuclear Science Division, Lawrence Berkeley National Laboratory, Berkeley, CA, USA
[13]South Dakota School of Mines and Technology, Rapid City, SD, USA
[14]Los Alamos National Laboratory, Los Alamos, NM, USA
[15]Department of Physics and Astronomy, University of Tennessee, Knoxville, TN, USA
[16]Research Center for Nuclear Physics, Osaka University, Ibaraki, Osaka, Japan
[17]Department of Physics, Princeton University, Princeton, NJ, USA
[18]Department of Physics, North Carolina State University, Raleigh, NC, USA
[19]Department of Physics, Massachusetts Institute of Technology, Cambridge, MA, USA
[20]Department of Physics, Black Hills State University, Spearfish, SD, USA
[21]Physics Department, Tennessee Tech University, Cookeville, TN, USA
[22]Department of Physics, Engineering and Astronomy, Queen's University, Kingston, ON, Canada
[23]Physik Department, Technische Universität, München, Germany

*Present Address: Centro de Investigaciones Energéticas, Medioambientales y Tecnológicas, CIEMAT, 28040, Madrid, Spain
**Alternative address: Department of Nuclear Engineering, University of California, Berkeley, CA, USA

Abstract. The MAJORANA Collaboration is using an array of high-purity Ge detectors to search for neutrinoless double-beta decay in ^{76}Ge. Searches for neutrinoless double-beta decay are understood to be the only viable experimental method for testing the Majorana nature of the neutrino. Observation of this decay would imply violation of lepton number, that neutrinos are Majorana in nature, and provide information on the neutrino mass. The MAJORANA DEMONSTRATOR comprises 44.1 kg of p-type point-contact Ge detectors (29.7 kg enriched in ^{76}Ge) surrounded by a low-background shield system. The experiment achieved a high efficiency of converting raw Ge material to detectors and an unprecedented detector energy resolution of 2.5 keV FWHM at $Q_{\beta\beta}$. The MAJORANA collaboration began taking physics data in 2016. This paper summarizes key construction aspects of the DEMONSTRATOR and shows preliminary results from initial data.

1 Introduction

Neutrinoless double-beta ($0\nu\beta\beta$) decay experiments play an important role in understanding the nature of neutrinos and their connection with the evolution of the universe. The observation of $0\nu\beta\beta$ decay would indicate that lepton number is not conserved – a key ingredient of thermal leptogenesis – and that neutrinos are Majorana particles [1, 2]. Measurements of the $0\nu\beta\beta$ decay rates would also yield information on the absolute neutrino mass. For more comprehensive reviews of $0\nu\beta\beta$ decays, see Refs. [3–11].

The MAJORANA DEMONSTRATOR is an experiment using enriched- and natural Ge as the detector and source to search for neutrinoless double-beta decay from ^{76}Ge. The primary goal of the DEMONSTRATOR is to build a scalable experiment achieving low enough backgrounds to justify the feasibility of a larger-scale experiment. The following sections give a short summary of the design, operation, data analysis and initial results from the MAJORANA DEMONSTRATOR.

2 The MAJORANA DEMONSTRATOR

By choosing the isotope ^{76}Ge for the $0\nu\beta\beta$ search, the MAJORANA DEMONSTRATOR is able to take advantage of the well-established Ge detector technology to achieve excellent energy resolution and event reconstruction (using pulse shape analysis) with intrinsically low backgrounds. For the DEMONSTRATOR , the average FWHM at the $Q_{\beta\beta}$ energy is 2.5 ± 0.1 keV, the best energy resolution achieved to date for a $0\nu\beta\beta$ decay experiment.

A total of 44.1 kg of Ge was used for the MAJORANA DEMONSTRATOR, 29.7 kg of which is enriched to 88% of isotope 76. The p-type point-contact (PPC) detector design [12–14] provides effective single-site/multi-site event discrimination, low noise, and low-energy thresholds that make it possible to search for new physics at low energies. To fabricate the Ge detectors, the enriched raw $^{76}GeO_2$ material was reduced, zone refined, grown into single crystals, and then made into 35 detectors. Overall, the MAJORANA DEMONSTRATOR reached an efficiency of 69.8% in converting the initial $^{76}GeO_2$ material into Ge detectors, the highest efficiency achieved to date [15]. The 35 detectors were assembled into 14 strings (3–5 detectors in each string) and installed in two cryostats with 7 strings in each. Care was taken to minimize the cosmogenic activation by limiting the total surface exposure time of these detectors. On average, the effective sea-level surface exposure time of the enriched detectors was approximately 23 days.

To reject external background, the detector cryostats are surrounded by multi-layer shields and installed in the clean lab at the Sanford Underground Research Facility (SURF) in Lead,

South Dakota, USA at 4850 feet below the surface. Figure 1 shows a cross sectional illustration of the Demonstrator. Starting from the center cavity, two cryostat modules constructed from electroformed copper house the detector strings. The cryostat cavity is surrounded by an electroformed copper inner shield and an outer copper shield constructed from clean commercial copper. Outside the copper shields is a thick layer of high-purity lead. A muon veto system consisting of 32 aluminum-clad plastic scintillator panels surrounds the lead and copper shields. The muon veto was used to measure the muon flux in the underground clean lab. The measured flux of $5 \times 10^{-9} \mu / s / cm^2$ is in good agreement with simulations[16]. To reduce neutron background, a polyethylene outer shield completes the shield system.

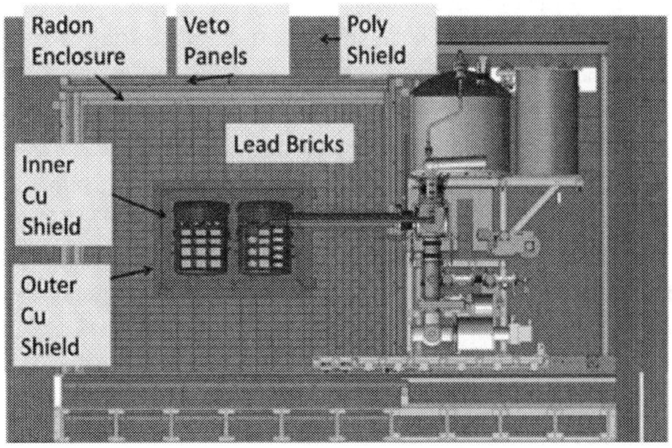

Figure 1. An illustration of the Majorana Demonstrator shield system in cross section. Detectors are denoted by blue disks stacked in strings installed in two electroformed copper cryostats. The cryostats are surrounded by electroformed inner- and clean-commercial outer copper shields. Lead bricks, a radon enclosure, muon veto panels, and polyethylene shields complete the shield system.

To reduce intrinsic background from sources near the detectors, careful designs enabled the least amount of material for detector assembly, including front-end circuit boards, connectors, cables, and string parts. Copper was chosen as the material for the Demonstrator cryostats and inner shielding layers due to its lack of naturally-occurring radioactive isotopes, as well as its suitability for machining. Electroforming further reduces radioactivity from contaminants. To minimize surface exposure of material, an electroforming lab, a machine shop, and part-cleaning facilities were established underground at SURF. As a result, most copper components installed in the Demonstrator have no cosmogenic activation from surface-level cosmic rays.

3 Data Accumulation and Analysis

The assembly of the first module (cryostat) of the Demonstrator was completed in 2015. This module consists of 16.8 kg of enriched Ge detectors and 5.6 kg of natural Ge detectors. Data taking of Module 1 alone began in June 2015. The assembly of Module 2 with 12.9 kg of enriched Ge and 8.8 kg natural Ge was completed in August 2016. The full operation of the Demonstrator started with both modules taking data in August 2016.

Since the start of operation up to this presentation, several data sets (DS) have been collected, corresponding to various experimental setup and running conditions. Before the installation of the inner copper shield, a serious of commissioning runs were carried out, and these were collected in DS0. After the inner copper shield was installed, data were collected in DS1. Subsequently, a series of tests were performed on signal multisampling of the digitized waveforms in order to extend the signal capture after an event, therefore making it more effective to discriminate against α surface-background events. These data were collected in DS2. After the implementation of signal multisampling, Module 1 and Module 2 data were respectively collected into DS3 and DS4 with separate data acquisition systems. Results presented at this conference are those from DS3 and DS4, which were the first two datasets with detector-, shield-, and data-acquisition configurations that are close to the final configuration.

The collected data were first scanned to remove non-physical waveforms. Then, events that are associated with multi-site energy deposits were removed by coincidence-event rejection and pulse-shape analysis[17]. Double-beta decay events are considered single-site events because the range of the electron in Ge is small compared to that of the Compton-scattered γ events, which are backgrounds. Finally, events associated with external α energy depositions on the passivated surface of detectors were rejected by analysis cuts. Extensive tests indicate that these α activities were most likely due to ^{210}Pb-supported ^{210}Po as daughters of Rn contamination on the Teflon components of the detector mount. For most of these α events, charge drifting takes place along the detector surface, whereas for most non-α events, charge drifting takes place inside the bulk of the detector. The difference of surface- and bulk-drifting times enables the rejection of α events by pulse shape analysis.

4 Preliminary Results

By the fall of 2017, the Demonstrator had collected more than 20 kg yr of data. Of these data, the first 1.39 kg yr of enriches-Ge data collected with both modules is presented in this paper. A summed spectrum of enriched-Ge detectors from DS3 and DS4 is shown in Figure 2. This spectrum is the result of applying all analysis cuts, including pulse-shape discrimination (PSD) and α-contamination cuts. After all cuts, 1 count remains in the 400-keV region of interest centered at the 2039-keV $0\nu\beta\beta$ Q value. This translates into a background rate of $5.0^{+8.8}_{-3.2}$ c/(ROI t yr) at 90% confidence level, where the ROI (region of interest) is a 2.8-keV window centered at the $0\nu\beta\beta$ Q value, or a background index of $1.8^{+3.1}_{-1.1} \times 10^{-3}$ c/(keV kg yr).

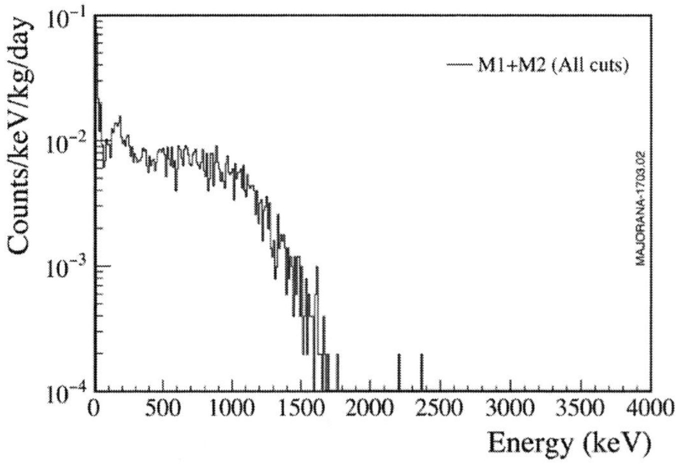

Figure 2. Summed background spectrum from DS3 and DS4 after all cuts (enriched Ge detectors only). One count remains in the 400-keV region of interest centered at the 2039-keV $0\nu\beta\beta$ Q value.

It should be noted that at the time of this proceeding, the MAJORANA DEMONSTRATOR has more than 7 times data collected and analyzed than what is shown in this presentation. For a current description of results from the MAJORANA DEMONSTRATOR, see Ref. [18].

Although data analyzed so far corresponds to limited exposure, the MAJORANA DEMONSTRATOR is approaching half-life limits comparable to the best effort to date, due to the very low background and excellent energy resolution of the experiment. Both the MAJORANA DEMONSTRATOR and GERDA are currently operating in the nearly-background free regime. The latest results from both experiments point to a ^{76}Ge half-life limit approaching 10^{26} yr[18, 19], and indicate that a future large-scale ^{76}Ge experiment, such as LEGEND [20], is indeed warranted.

Acknowledgments

This material is based upon work supported by the U.S. Department of Energy, Office of Science, Office of Nuclear Physics, the Particle Astrophysics and Nuclear Physics Programs of the National Science Foundation, the Russian Foundation for Basic Research, the Natural Sciences and Engineering Research Council of Canada, the Canada Foundation for Innovation John R. Evans Leaders Fund, the National Energy Research Scientific Computing Center, the Oak Ridge Leadership Computing Facility, and the Sanford Underground Research Facility. This research used resources of the Oak Ridge Leadership Computing Facility, which is a DOE Office of Science User Facility supported under Contract DE-AC05-00OR22725.

References

[1] M. Zralek, ACTA Phys. Pol. **B 28**, 2225 (1997)

[2] J. Schechter, J and J.W.F. Valle, Phys. Rev. **D 25**, 2951 (1982)

[3] F. T. Avignone III, S.R. Elliott and J. Engel , Rev. Mod. Phys. **80**, 481 (2008)

[4] A.S. Barabash, Phys. Part., Nucl. **42**, 613 (2011)

[5] W. Rodejohann, Int. J. Mod. Phys. **E 20**, 1833 (2011)

[6] S. R. Elliott, Mod. Phys. Lett. **A 27**, 1230009 (2012)

[7] J. D. Vergados, H. Ejiri and F. Simkovic, Rep. Prog. Phys. **75**, 106301 (2012)

[8] O. Cremonesi, Nucl. Phys. B Proc. Supp. **237**, 7 (2013) (Preprint arXiv: 1212.4885)

[9] B. Schwingenheuer, Ann. der Physik, **523** 269 (2013) (Preprint arXiv: 1210.7432)

[10] S.R. Elliott and M. Franz, Rev. Mod. Phys. **87**, 137 (2015)

[11] R. Henning, Rev. Phys. **1** 29 (2016)

[12] P.N. Luke, F.S. Goulding, N.W. Madden and R.H. Pehl, IEEE Trans. Nucl. Sci., **36**, 926 (1989)

[13] P.S. Barbeau, J.I. Collar and O. Tench, JCAP **09**, 009 (2007)

[14] E. Aguayo *et al.* sLAC eConf, C110809 (2011) (Preprint arXiv: 1109.6913)

[15] N. Abgrall *et al.*, Nucl. Instrum. Meth., **A 877**, 314 (2017) (Preprint arXiv: 1707.06225)

[16] N. Abgrall *et al.*, Astro. Phys., **93**,70 (2017) 70(Preprint arXiv: 1602.07742)

[17] C. Cuesta and MAJORANA, J.Phys.: Conf. Ser. **888**, 012240 (2017)

[18] C.E. Aalseth *et al.*, Phys. Rev. Lett. **120**, 132502 (2018)

[19] M. Agostini *et al.*, 2017 (preprint arXiv:1710.07776)

[20] N. Abgrall *et al.*, 2017 (Preprint arXiv:1709.01980)

Nuclear astrophysics at Gran Sasso Laboratory: the LUNA experiment

Francesca Cavanna[1,*](on behalf of the LUNA collaboration)

[1]Istituto Nazionale di Fisica Nucleare (INFN), Sezione di Genova

Abstract. LUNA is an experimental approach for the study of nuclear fusion reactions based on an underground accelerator laboratory. Aim of the experiment is the direct measurement of the cross section of nuclear reactions relevant for stellar and primordial nucleosynthesis. In the following the latest results and the future goals will be presented.

1 Introduction

The Laboratory for Underground Nuclear Astrophysics (LUNA) is located at Gran Sasso National Laboratories, Italy, where the 1400 meters of rocks dominating the laboratory guarantee a reduction of six orders of magnitude in the cosmic muon flux and a reduction of three orders of magnitude in the neutron flux. The 400 kV electrostatic accelerator provides an high intensity (~200 μA) proton or alpha beam. The beam energy spread at the exit of the accelerator was determined to be < 100 eV while the energy drift is < 5 eV/h. The uncertainty on the beam energy is 0.3 keV, mainly due to the uncertainty on the ^{12}C(p, γ)^{13}N reaction Q-value used in the energy calibration. The calibration has been rechecked several times and was found consistent with the new measurements. The beam can be delivered either to a solid or to a gas target. Different gamma-ray or particle detectors can be used, depending on the characteristics of the nuclear reaction to be studied.

Several experimental campaigns have been accomplished in the past; in particular reactions of hydrogen burning [16] e.g. ^{22}Ne(p,γ)^{23}Na [11–14], ^{17}O(p,α)^{14}N [9] and ^{23}Na(p,γ)^{24}Mg [17] were deeply studied and outstanding results were achieved. Also few reactions of Big Bang Nucleosynthesis were measured as the ^{2}H(α,γ)^{6}Li [15] and ^{2}H(p,γ)^{3}He reactions. This paper focuses on ^{17}O(p,α)^{14}N and ^{2}H(p,γ)^{3}He cross section measurements and, in the last part, on the future program.

2 ^{17}O(p,α)^{14}N

The ^{17}O(p,α)^{14}N reaction (Q_{val}=1.2 MeV) plays a key role in several astrophysical scenarios, and in AGB stars in particular [1–3]. Models predict that massive AGB stars should produce significant amounts of cosmic dust, and yet no pre-solar grain appears to match the HBB signature expected from these stars [4]. The most obvious candidates, Group II grains, have ^{17}O/^{16}O ratios that are a factor of two lower than expected. For this reason is crucial the measurement of the ^{17}O(p,α)^{14}N cross section.

*e-mail: francesca.cavanna@ge.infn.it

At energies of astrophysical interest its reaction rate is dominated [6] by a narrow and isolated resonance at E_p=70 keV. This resonance has been studied several times in the past, using both direct and indirect methods, as summarised in ref. [7]. However, the picture painted in the literature is still not completely satisfying. The uncertainty in the resonance strength is not negligible (≈20%). Furthermore, published strength values obtained with direct measurements have all been retracted or reanalysed [7].
An experimental campaign aimed at measuring the E_p=70 keV resonance in $^{17}O(p,\alpha)^{14}N$ was recently completed at the underground LUNA accelerator. We exploited the low background in the underground environment in order to carry out a direct investigation of this weak ($\omega\gamma \approx$ neV) resonance employing the thick-target yield technique. Protons were accelerated on a solid Ta_2O_5 target and alpha particles detected at backward angles using an array of silicon detectors. The setup was commissioned using the well-known E_p=193 keV resonance in $^{17}O(p,\alpha)^{14}N$ [8]. A clear peak was observed at E_p=71.5 keV at the expected energies. The alpha peak from the E_p=70 keV resonance appears where expected and has a signal significance higher than 5 sigma. Results of the analysis [9] indicate a factor of two increase in the rate of the $^{17}O(p,\alpha)^{14}N$ reaction compared to [10]. This translates directly into a factor of two reduction in the expected ^{17}O isotopic content, which now closely matches the observed values [4, 5].

3 $^2H(p,\gamma)^3He$

Big Bang Nucleosynthesis (BBN) describes the production of light nuclei in the first minutes of cosmic time. It started with deuterium formation through the $p(n,\gamma)^2H$ reaction, when the Universe was cold enough to allow ^2H nuclei to survive to photo-disintegration. The main source of uncertainty on standard BBN prediction for the deuterium abundance is actually due to the radiative capture process $^2H(p,\gamma)^3He$ that destroys deuterium, because of the poor knowledge of its S-factor at BBN energies. Moreover, the paucity of $^2H(p,\gamma)^3He$ data represents the main obstacle to improve the accuracy on the determination of the baryon density and to constrain of the density of relativistic species existing at BBN epoch (only photons and 3 families of neutrinos are foreseen in the standard model). The low energy limit of this reaction cross section is well known thanks to the data coming from the LUNA measurement performed under the solar Gamow peak [18], as shown in figure 1, in the energy range of interest for the solar processes (0 keV < E_{cm} < 50 keV), the measurements of Griffiths et al. [19], Schmid et al. [20] and Bystritsky et al. [21] are also available, using a proton beam on a D_2O ice solid target. Some discrepancies exist for the lower energies data, but in [20] the authors stress on the presence of a systematic error in Griffiths's and Bailey's data due to the wrong stopping power used for the heavy-ice target, affecting the S-factor value by about 15%. In the 200 keV < E_{cm} < 1 MeV energy range, Griffiths et al. [22] and Bailey et al. [23] have measured the $^2H(p,\gamma)^3He$ cross section using once again a solid target. Warren et al. [24], instead, has measured the cross section of ^3He photodisintegration on a gas target and reaching a good agreement with direct measurement data. Finally in the BBN energy range only a single data set is currently available with a systematic error of 9% [25]. A measurement of this reaction cross section with a 3% accuracy in the BBN energy range (30 keV $\lesssim E_{cm} \lesssim$ 300 keV) is thus desirable [26]. Adelberger et al. in [27] performed a fit of the reaction S-factor, using the experimental data [22], [18], [20] and [25]. The four data sets have been fitted by quadratic polynomial and plotted with the blue band in figure 1 that represents the 68% lower and upper bounds of the adopted best-fit. The fit performed by Angulo et al. in NACRE [28] is also shown. An updated version of NACRE fit has been developed by [29], adding the post-NACRE data set of LUNA [18], extending thus the energy range down to 2 keV. An evaluation of the $^{2}H(p,\gamma)^{3}He$ S-factor has been performed by Descouvemont et al. in [30]. Finally the result of ab-initio calculation are also reported [31].

The $^{2}H(p,\gamma)^{3}He$ experiment at LUNA consists of two main phases characterized by different setups. The former is a windowless gas target filled with deuterium surrounded by a 4π BGO detector. The high efficiency ($\sim$ 70%) of the BGO detector reduces the dependence of the reaction yield on the angular distribution of the emitted γ rays and thus allows to achieve a low systematic uncertainty. The detection efficiency has been determined by Monte Carlo simulations, as well as by measurements with radioactive sources (^{60}Co, ^{137}Cs, ^{88}Y) and $^{14}N(p,\gamma)^{15}O$ resonant reaction [12, 13]. The study of systematic uncertainties included also the determination of the beam heating effect obtained by varying the target pressure and beam intensity. The beam intensity value and uncertainty was determined by a proper calibration of the calorimeter. The $^{2}H(p,\gamma)^{3}He$ excitation function has been measured in the energy range $30\,\mathrm{keV} \lesssim E_{cm} \lesssim 200\,\mathrm{keV}$.

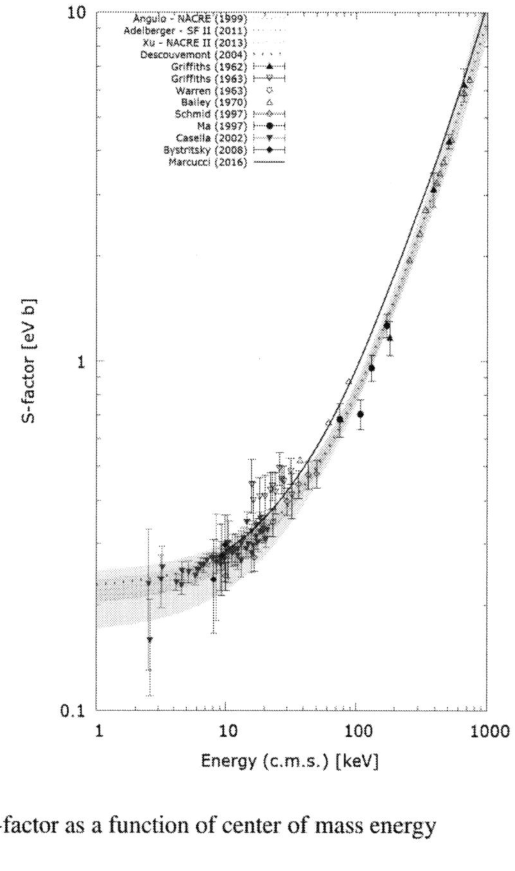

Figure 1. $^{2}H(p,\gamma)^{3}He$ S-factor as a function of center of mass energy

The latter phase, instead, focuses on the $70\,\mathrm{keV} \lesssim E_{cm} \lesssim 260\,\mathrm{keV}$ energy range. The set up consists of a 137% HPGe detector in close geometry with the interaction chamber. With this setup the angular distribution can be inferred by exploiting the high energy resolution of the detector and the Doppler effect that changes along the beam line the energy of the detected γ rays. This study provides an important experimental ground for theoretical nuclear physics. This measurement requires the detection efficiency to be known with high accuracy along all the target chamber. The procedure adopted couples the main HPGe detector (Ge1) to an auxiliary one (Ge2) and exploits the coincidence between two γ-rays emitted in cascade. The main decay channel of the E_r = 259 keV resonance (BR=57,8%) is particularly useful to determine the germanium efficiency, because it produces two gammas in cascade, with energy E_1= 1384 keV (close to the energy of ^{60}Co γ-rays) and E_1= 6172 keV (close to the energy of photons produced by the ^{2}H(p,γ)^{3}He reaction). Whenever the auxiliary counter detects an event 1, it enables the other counter that can thus detect the corresponding cascade emitted photon 2. The detection efficiency of Ge1 can be evaluated by the ratio between the number of events that has triggered the acquisition and the number of events actually acquired by the detector itself. The effect of the coincidence acquisition is shown in figure 2, where the spectrum of the ^{14}N(p,γ)^{15}O resonance acquired by Ge1 in inclusive configuration is shown on the left. Selecting the resonance emitted photon at 1.3 MeV with Ge2, the resulting coincidence spectrum of Ge1 (on the right in figure 2) shows the only peak of the corresponding cascade photon at 6.1 MeV.

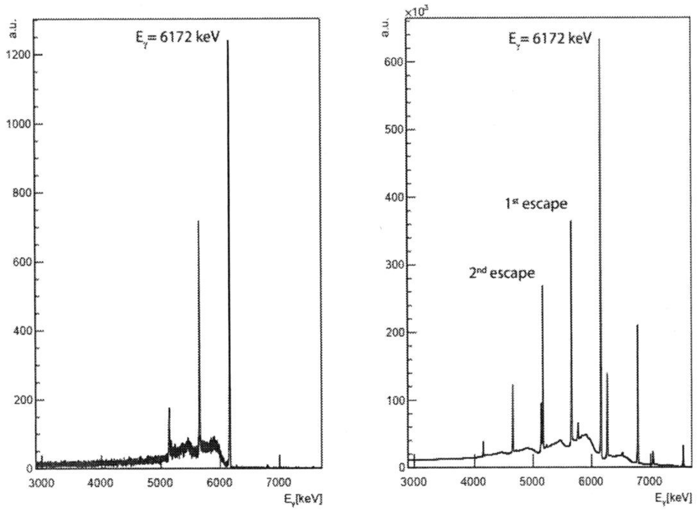

Figure 2. ^{14}N(p,γ)^{15}O resonance spectra: left the one acquired in coincidence configuration, right the one acquired in inclusive configuration

4 LUNA-MV

A new accelerator will be installed soon at LNGS to allow the study of nuclear reactions involved in helium and carbon burning. The LUNA MV accelerator will provide intense beams of H^+, $^4H^+$, $^{12}C^+$ e $^{12}C^{++}$ in the energy range: 350 keV - 3.5 MeV, and is presently under construction at High Voltage Engineering Europe (HVEE) in The Netherlands. It will be equipped with two beam lines, one for measurement at the solid target and the other one for gas or solid target, accordingly to the experimental need.

Delivery at LNGS is presently scheduled for December 2018. A six months commissioning phase is scheduled thereafter so that the first physics experiments are envisaged at middle 2019.

Since spring 2016, the design and construction of the new LUNA-MV laboratory inside Hall B has been started. The clearance of the LUNA-MV area in Hall B has been accomplished, the beginning of the construction works is forseen for the next months and the delivering of the full equipped LUNA-MV laboratory (accelerator room and technical building) by march 2018.

A full proposal (available on the LUNA web site https://luna.lngs.infn.it/) for the first five years of activity at LUNA-MV has been approved by the LUNA Collaboration in July 2016 and submitted both to LNGS and to INFN-CSN3. In particular, the first cross section to be measured will be the $^{14}N(p,\gamma)^{15}O$ in a much wider energy range compared to what was done at LUNA400. This way the tuning of LUNA MV will be done and we will more precisely extrapolate the reaction cross section within the Gamow peak of the Sun, i.e. the burning energy region. Then, the focus of the activity will be the study of $^{12}C+^{12}C$: the understanding of its cross section at low energy will be the main goal of the first 5 years of LUNA MV. Alternating in time with $^{12}C+^{12}C$, the study of $^{13}C(\alpha,n)^{16}O$ will be performed on the other beam line (the accelerator can feed only one line at a time). Finally, $^{22}Ne(\alpha,n)^{25}Mg$ will be the last reaction covered by this scientific plan. On the other hand, $^{12}C(\alpha,\gamma)^{16}O$ will be the main goal of the second scientific plan at LUNA MV, starting in the year 2023.

References

[1] K.M. Nollett *et al.*, Astrophys. J. **582**, 1036 (2003)
[2] M. Lugaro *et al.*, A&A **461**, 657 (2007)
[3] S. Palmerini *et al.*, ApJ. **764**, 128 (2013)
[4] M. Lugaro *et al.*, Nature Astronomy **1**, 0027 (2017)
[5] O. Straniero et al., Astronomy and Astrophysics **598**, A128 (2017)
[6] M.Q. Buckner *et al.*, Phys. Rev. C **91**, 015812 (2015)
[7] M.L. Sergi *et al.*, Phys. Rev. C **91**, 065803 (2015)
[8] C.G. Bruno *et al.*, Eur. Phys. J. A **51**, 94 (2015)
[9] C.G. Bruno *et al.*, Physical Review Letters **117**, 142502 (2016)
[10] J. Blackmon *et al.*, Phys. Rev. Lett. **74**, 2642 (1995)
[11] F. Cavanna *et al.*, Eur. Phys. J. A **50**, 179 (2014)
[12] F. Cavanna *et al.*, Phys. Rev. Lett. **115** 252501 (2015)
[13] R. Depalo *et al.*, Phys. Rev. C **94** 055804 (2016)
[14] A. Slemer *et al.* MNRAS **465** 4817-4837 (2017)
[15] D. Trezzi *et al.*, Astroparticle Physics **89**, 57 (2017)
[16] A. Boeltzig *et al.*, European Physical Journal A **52**, 75 (2016)
[17] A. Boeltzig *et al.*, Journal of Physics G, submitted (2017)
[18] C. Casella *et al.*, Nucl. Inst. Meth. A **489**, 160 (2002)
[19] G. Griffiths *et al.*, Can.J.Phys. **41**, 724 (1963)
[20] J. Schmid *et al.*, Phys. Rev. C **56**, 2565 (1997)
[21] V. M. Bystritsky *et al.*, Nucl. Inst. Meth. A **595**, 543 (2008)
[22] G. Griffiths *et al.*, Can.J.Phys. **40**, 402 (1962)
[23] G. Bailey *et al.*, Can. J. Phys. **48**, 3059 (1970)
[24] J. B. Warren *et al.*, Phys. Rev. 132, 1691 (1963)
[25] L. Ma *et al.*, Phys. Rev. C **55**, 588 (1997)
[26] E. Di Valentino *et al.*, Phys. Rev. D **90** 023543 (2014)
[27] E. G. Adelberger *et al.*, Rev. Mod. Phys **83**, 195 (2011)
[28] C. Angulo *et al.*, Nucl. Phys. A **656**, 3 (1999)
[29] Y. Xu *et al.*, Nucl. Phys. A **918**, 61 (2013)
[30] P. Descouvemont *et al.*, At. Data Nucl. Data Tab. 88, **203** (2004)
[31] L. E. Marcucci *et al.*, Phys. Rev. Lett. **116**, 102501 (2016)

Felsenkeller 5 MV underground accelerator: Towards the Holy Grail of Nuclear Astrophysics $^{12}C(\alpha,\gamma)^{16}O$

Daniel Bemmerer[1,*], *Thomas E.* Cowan[1,2], *Marcel* Grieger[1,2], *Sebastian* Hammer[1,2], *Thomas* Hensel[1,2], *Arnd R.* Junghans[1], *Martina* Koppitz[1,2], *Felix* Ludwig[1,2], *Stefan E.* Müller[1], *Bernd* Rimarzig[1], *Stefan* Reinicke[1,2], *Ronald* Schwengner[1], *Klaus* Stöckel[1,2], *Tamás* Szücs[1], *Marcell P.* Takács[1,2], *Steffen* Turkat[2], *Andreas* Wagner[1], *Louis* Wagner[1,2], and *Kai* Zuber[2]

[1]Helmholtz-Zentrum Dresden-Rossendorf, Bautzner Landstr. 400, 01328 Dresden, Germany
[2]Technische Universität Dresden, Institut für Kern- und Teilchenphysik, 01069 Dresden, Germany

Abstract. Low-background experiments with stable ion beams are an important tool for putting the model of stellar hydrogen, helium, and carbon burning on a solid experimental foundation. The pioneering work in this regard has been done by the LUNA collaboration at Gran Sasso, using a 0.4 MV accelerator. The present contribution reviews the status of the project for a higher-energy underground accelerator in Felsenkeller, Germany. Results from γ-ray, neutron, and muon background measurements in the Felsenkeller underground site in Dresden, Germany, show that the background conditions are satisfactory.
Two tunnels of the Felsenkeller site have recently been refurbished for the installation of a 5 MV high-current Pelletron accelerator. Civil construction work has completed in March 2018. The accelerator will provide intense, 50 μA, beams of $^{1}H^{+}$, $^{4}He^{+}$, and $^{12}C^{+}$ ions, enabling research on astrophysically relevant nuclear reactions with unprecedented sensitivity.

1 Introduction

Experimental nuclear astrophysics aims to understand the origin of the chemical elements by laboratory experiments [1].

One particularly challenging case is the $^{12}C(\alpha,\gamma)^{16}O$ reaction. Its rate affects the carbon to oxygen abundance ratio at the end of stellar helium burning and, hence, the abundances of a host of other nuclides that are produced during subsequent burning stages [2]. There are no direct cross section data in the astrophysically relevant energy range, so the present knowledge of its rate depends on indirect approaches and R-matrix fits of a complicated excitation function, resulting in different central values and widely different error estimates [3–6].

A second key aspect of nuclear astrophysics are the nuclear reactions of solar hydrogen burning. After the discovery of solar neutrino oscillations [7], the focus has shifted to precision studies of our sun. Neutrino fluxes are presently better known than the related nuclear reactions of the proton-proton (pp) chain. Recently, they have been used to determine the cross sections of two pp-chain nuclear reactions: $^{3}He(\alpha,\gamma)^{7}Be$ [8] and $^{7}Be(p,\gamma)^{8}B$ [9].

*e-mail: d.bemmerer@hzdr.de

The most precise technique hitherto developed to address stable-ion nuclear reactions such as these is underground nuclear astrophysics, which has been pioneered by the LUNA collaboration working deep underground in the Gran Sasso laboratory, Italy [10]. There, in recent years a number of nuclear reactions important for Big Bang nucleosynthesis [11, 12] and for stellar hydrogen burning [13–15, recent examples] have been studied using a 0.4 MV high-current accelerator.

The present contribution describes the progress of the higher-energy underground accelerator in Felsenkeller, Dresden, with a science program that is complementary to LUNA and also to the LUNA-MV project (F. Cavanna, these proceedings).

2 Background studies

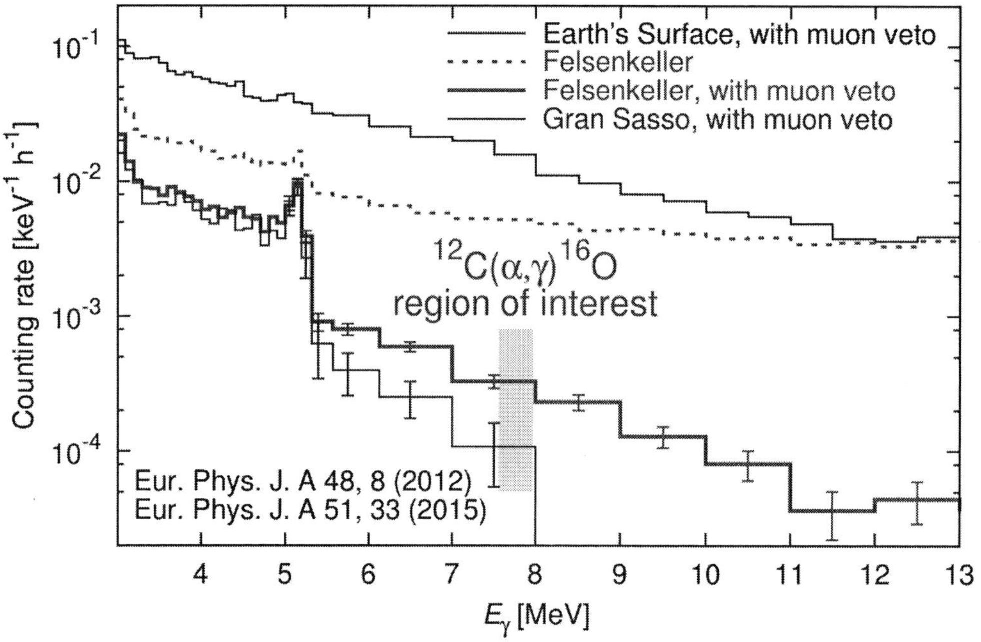

Figure 1. Background intercomparison using one and the same escape-suppressed HPGe detector, subsequently transported to different sites. The data [16, 17] show that using a muon veto, the Felsenkeller background is not far from the deep underground (Gran Sasso) case.

The Felsenkeller site is located 5 km from the city center of Dresden, Germany (population 500,000) and protected from cosmic rays by 45 m of hornblende monzonite rock. It consists of nine horizontal tunnels, two of which are used for the accelerator laboratory (see below). Detailed studies of the muon and neutron background have been carried out in the tunnels and will be described in forthcoming work.

The most important aspect for underground nuclear astrophysics, however, is the remaining high-energy background in a typical in-beam γ-ray detector. Detailed measurements at Felsenkeller [16, 17] have shown that the background in the crucial 7-8 MeV γ-energy range (affecting *inter alia* the $^{12}C(\alpha,\gamma)^{16}O$ reaction) is only a factor of 3 higher in Felsenkeller than deep underground at Gran Sasso (Figure 1).

https://doi.org/10.1051/epjconf/201817801008

3 Scientific program

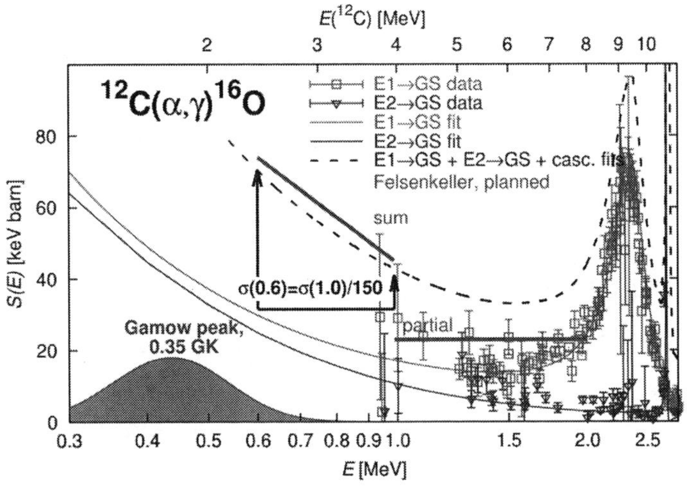

Figure 2. Excitation function of the $^{12}C(\alpha,\gamma)^{16}O$ reaction. Experimental data [18] and R-matrix fits for the dipole [18], quadrupole [4], and sum of all components [18] are plotted. The feasible energy range for direct total and partial cross section data at Felsenkeller, based on the measured background [16, 17] and assuming inverse kinematics, and the Gamow peak for 0.35 GK temperature are also shown.

The Felsenkeller scientific program draws on many years of experience in nuclear astrophysics experiments at the overground HZDR 3 MV Tandetron accelerator [19–24] and on a long-standing involvement in the LUNA collaboration [10, 11]. Complementary experiments studying the γ-widths of relevant levels [25, 26] have been performed at the local bremsstrahlung facility [27] and are underway at the local neutron time-of-flight facility [28].

For the in-house research by the two institutions HZDR and TU Dresden that together run the new Felsenkeller underground ion beam, two initial science cases have been selected: The $^{12}C(\alpha,\gamma)^{16}O$ reaction affecting helium burning and the $^{3}He(\alpha,\gamma)^{7}Be$ reaction affecting solar ^{7}Be and ^{8}B neutrinos.

In order to give an example of the science potential, the excitation function of the $^{12}C(\alpha,\gamma)^{16}O$ reaction is plotted in Figure 2. There are no significant data points below 1 MeV center-of-mass energy, yet in this range both the E1 and the E2 components of the cross section exhibit a steep rise to lower energies, due to two subthreshold resonances, one in each of the two channels. New data are clearly needed to constrain the fits, and actually with a summing detector it is possible to provide direct cross section values inside the stellar Gamow peak. This inverse-kinematics experiment will require intense ^{12}C beam, $\sim 50\,\mu A$.

In the $^{3}He(\alpha,\gamma)^{7}Be$ case, the main challenge lies in linking the ultra-low energy LUNA data [11] with a host of other experiments at much higher energy. Again, this can be accomplished at Felsenkeller, using $\sim 50\,\mu A$ ^{4}He beam, in the center-of-mass energy range of 0.2-2 MeV.

In addition to this in-house research, a significant amount of beam time will be made available, free of charge, as a facility to national and international users based on scientific proposals to be evaluated by an independent scientific advisory committee.

4 Project status

The ion accelerator, a 5 MV Pelletron with double charging chains and a nitrogen gas stripper, has been furnished with an additional terminal ion source that will provide intense ^{1}H and ^{4}He beams in single-ended mode. This terminal ion source is in addition to the remaining multi-cathode cesium sputter ion source that will supply intense ^{12}C beams in tandem mode. Taken together, with these two modes it will be possible to address the most challenging cases of low-energy nuclear astrophysics.

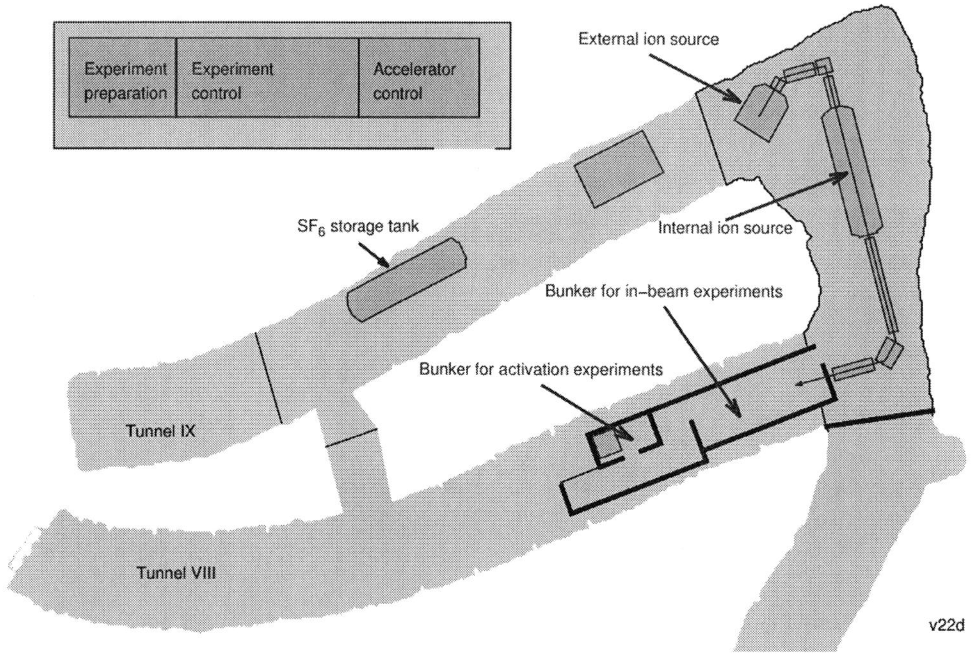

Figure 3. Sketch of the most important installations in Felsenkeller tunnels VIII and IX.

The original nine tunnels have been excavated in the 1850s to serve the former Felsenkeller brewery. Since 1982, tunnel IV hosts a low-activity γ-counting laboratory [29]. The new underground ion accelerator is hosted in tunnels VIII and IX (Figure 3).

As of the writing of this proceedings, the civil construction has completed. Most notably, a building has been erected inside the tunnels to house the accelerator and ion source, and two bunkers have been constructed from low-radioactivity concrete, one each for in-beam and activation experiments, respectively. A very large HPGe detector in a specially selected shield has been procured for the activity measurements. The ventilation and air conditioning system is already operating, and the accelerator and main electromagnets have been moved underground. The beam lines will be constructed in spring-summer 2018, with an aim to test both ion sources in the coming months.

It is hoped that first data from this new laboratory will be presented at the next CGS meeting.

References

[1] W.A. Fowler, Rev. Mod. Phys. **56**, 149 (1984)
[2] S.E. Woosley, A. Heger, Phys. Rep. **442**, 269 (2007)
[3] L. Buchmann, C. Barnes, Nucl. Phys. A **777**, 254 (2006)
[4] C.R. Brune, B. Davids, Annu. Rev. Nucl. Part. Sci. **65**, 87 (2015)
[5] Z.D. An, Z.P. Chen, Y.G. Ma, J.K. Yu, Y.Y. Sun, G.T. Fan, Y.J. Li, H.H. Xu, B.S. Huang, K. Wang, Phys. Rev. C **92**, 045802 (2015)
[6] R.J. deBoer, J. Görres, M. Wiescher, R.E. Azuma, A. Best, C.R. Brune, C.E. Fields, S. Jones, M. Pignatari, D. Sayre et al., Rev. Mod. Phys. **89**, 035007 (2017)
[7] A.B. McDonald, Rev. Mod. Phys. **88**, 030502 (2016)

[8] M.P. Takács, D. Bemmerer, T. Szücs, K. Zuber, Phys. Rev. D **91**, 123526 (2015)
[9] M.P. Takács, D. Bemmerer, A.R. Junghans, K. Zuber, Nucl. Phys. A **970**, 78 (2018)
[10] C. Broggini, D. Bemmerer, A. Caciolli, D. Trezzi, Prog. Part. Nucl. Phys. **98**, 55 (2018)
[11] D. Bemmerer, F. Confortola, H. Costantini, A. Formicola, G. Gyürky, R. Bonetti, C. Broggini, P. Corvisiero, Z. Elekes, Z. Fülöp et al., Phys. Rev. Lett. **97**, 122502 (2006)
[12] M. Anders, D. Trezzi, R. Menegazzo, M. Aliotta, A. Bellini, D. Bemmerer, C. Broggini, A. Caciolli, P. Corvisiero, H. Costantini et al., Phys. Rev. Lett. **113**, 042501 (2014)
[13] F. Cavanna, R. Depalo, M. Aliotta, M. Anders, D. Bemmerer, A. Best, A. Boeltzig, C. Broggini, C.G. Bruno, A. Caciolli et al., Phys. Rev. Lett. **115**, 252501 (2015)
[14] C.G. Bruno, D.A. Scott, M. Aliotta, A. Formicola, A. Best, A. Boeltzig, D. Bemmerer, C. Broggini, A. Caciolli, F. Cavanna et al., Phys. Rev. Lett. **117**, 142502 (2016)
[15] M. Lugaro, A.I. Karakas, C.G. Bruno, M. Aliotta, L.R. Nittler, D. Bemmerer, A. Best, A. Boeltzig, C. Broggini, A. Caciolli et al., Nature Astronomy **1**, 0027 (2017)
[16] T. Szücs, D. Bemmerer, T. Cowan, D. Degering, Z. Elekes, Z. Fülöp, G. Gyürky, A. Junghans, M. Köhler, M. Marta et al., Eur. Phys. J. A **48**, 8 (2012)
[17] T. Szücs, D. Bemmerer, T.P. Reinhardt, K. Schmidt, M.P. Takács, A. Wagner, L. Wagner, D. Weinberger, K. Zuber, Eur. Phys. J. A **51**, 33 (2015)
[18] Y. Xu, K. Takahashi, S. Goriely, M. Arnould, M. Ohta, H. Utsunomiya, Nucl. Phys. A **918**, 61 (2013)
[19] M. Marta, E. Trompler, D. Bemmerer, R. Beyer, C. Broggini, A. Caciolli, M. Erhard, Z. Fülöp, E. Grosse, G. Gyürky et al., Phys. Rev. C **81**, 055807 (2010)
[20] K. Schmidt, S. Akhmadaliev, M. Anders, D. Bemmerer, K. Boretzky, A. Caciolli, D. Degering, M. Dietz, R. Dressler, Z. Elekes et al., Phys. Rev. C **88**, 025803 (2013)
[21] K. Schmidt, S. Akhmadaliev, M. Anders, D. Bemmerer, A. Caciolli, M. Dietz, Z. Elekes, A.R. Junghans, M.L. Menzel, R. Schwengner et al., Phys. Rev. C **89**, 045802 (2014)
[22] R. Depalo, F. Cavanna, F. Ferraro, A. Slemer, T. Al-Abdullah, S. Akhmadaliev, M. Anders, D. Bemmerer, Z. Elekes, G. Mattei et al., Phys. Rev. C **92**, 045807 (2015)
[23] T.P. Reinhardt, S. Akhmadaliev, D. Bemmerer, K. Stöckel, L. Wagner, Nucl. Inst. Meth. B **381**, 58 (2016)
[24] L. Wagner, S. Akhmadaliev, M. Anders, D. Bemmerer, A. Caciolli, S. Gohl, M. Grieger, A. Junghans, M. Marta, F. Munnik et al., Phys. Rev. C **97**, 015801 (2018)
[25] R. Schwengner, A. Wagner, Y. Fujita, G. Rusev, M. Erhard, D. de Frenne, E. Grosse, A.R. Junghans, K. Kosev, K.D. Schilling, Phys. Rev. C **79**, 037303 (2009)
[26] T. Szücs, D. Bemmerer, A. Caciolli, Z. Fülöp, R. Massarczyk, C. Michelagnoli, T.P. Reinhardt, R. Schwengner, M.P. Takács, C.A. Ur et al., Phys. Rev. C **92**, 014315 (2015)
[27] R. Schwengner, A. Wagner, Nuclear Physics News **27**, 23 (2017)
[28] A. Frotscher, Master's thesis, Technische Universität Dresden (2017)
[29] M. Köhler, D. Degering, M. Laubenstein, P. Quirin, M.O. Lampert, M. Hult, D. Arnold, S. Neumaier, J.L. Reyss, Appl. Radiat. Isot. **67**, 736 (2009)

Covariant density functional theory: predictive power and first attempts of a microscopic derivation

Peter Ring[1,2,*]

[1]Physics Department, Technical University of Munich, 85748 Garching, Germany
[2]School of Physics, Peking University, Beijing 100871, People's Republic of China

Abstract. We discuss systematic global investigations with modern covariant density functionals. The number of their phenomenological parameters can be reduced considerable by using microscopic input from ab-initio calculations in nuclear matter. The size of the tensor force is still an open problem. Therefore we use the first full relativistic Brueckner-Hartree-Fock calculations in finite nuclear systems in order to study properties of such functionals, which cannot be obtained from nuclear matter calculations.

1 Introduction

In recent years important progress has been made in the ab-initio studies of nuclear properties. However, because of the tremendously growing dimensions of the underlying configuration spaces the majority of heavy nuclei are, so far, only accessible through nuclear density functional theory (DFT). Symmetries play an essential role in the construction of such functionals. One of the underlying symmetries of QCD is Lorentz invariance and therefore covariant density functional theory (CDFT) are of particular interest in nuclear physics.

As in the non-relativistic case, over the years the number of such phenomenological functionals has increased considerably. In recent years systematic investigations have been carried out to study the behaviour of the different types of functionals in regions far from stability with large neutron excess and in the area of super-heavy elements [1]. It has been found that the increase of the theoretical uncertainties in areas far from the valley of stability is connected in many cases with the single particle structure. It also well known that tensor terms have a considerable influence on such quantities [2] and that these terms are hard to determine from experimental data.

On the other hand, ab-initio calculations in nuclear matter cannot teach us much about tensor terms in density functionals, because they are usually based on spin-saturation. In order to obtain such information one has to carry out ab-initio calculations for finite nuclei. The Brueckner-Hartree-Fock (BHF) theory is one of the most promising theories for such a purpose. It provides in a fully microscopic way a density dependent effective force for mean field calculations in nuclei. Most of the investigations based on BHF-theory are devoted to infinite nuclear matter. In the seventies much effort has been used to carry out such calculations also in finite nuclei. However such non-relativistic calculations usually failed to reproduce saturation, because of the missing three-body forces. In the eighties it has been found that

*e-mail: ring@ph.tum.de

relativistic Brueckner-Hartree-Fock (RBHF) theory can describe saturation in nuclear matter without tree-body forces [3].

Recently the first full RHBF calculations have been carried out for finite nuclear systems [4, 5]. They provide a basis to determine in future the size of the tensor forces in covariant density functionals on a microscopic basis.

2 Phenomenological and Semi-microscopic CDFT's

In Coulombic system an essential input for the derivation of microscopic functionals is the exact numerical solution of the homogeneous electron gas at various densities. Starting from this energy functional $E[\rho]$ additional gradient terms and many other corrections have been added with great success. It seems to be reasonable to apply a similar concept in nuclear physics [6, 7]. Of course, at present there are no exact solutions of the nuclear matter problem available. One has to rely on approximate solutions, such as sophisticated variational calculations or modern Brueckner-Hartree-Fock methods.

The basic idea of microscopic covariant density functionals is therefore to start from RBHF-calculations in nuclear matter [8] of various densities and to map the results to a Walecka model with density dependent parameters [9, 10]. However, this mapping is not unique and therefore there are many open questions until today. In addition, none of these fully microscopic functionals is able reach the accuracy necessary for practical applications.

Much more successful are semi-microscopic functionals, which use microscopic calculations in nuclear matter in order to reduce the number of phenomenological parameters considerably. An example is the parameter set DD-MEδ [11]. Here the density dependence of the corresponding meson-nucleon couplings is adjusted to microscopic calculations in nuclear matter and only four phenomenological parameters $g_\sigma(\rho_{\rm sat})$, $g_\sigma(\rho_{\rm sat})$, $g_\rho(\rho_{\rm sat})$ ($\rho_{\rm sat}$ is the saturation density), and m_σ are fine tuned by a fit to masses and radii of finite nuclei, .

Extended applications of this functional on a global scale [1] have shown that the parameter set DD-MEδ it is comparable with all the modern relativistic high precision functionals containing considerably more phenomenological parameters. The deviations between theoretical and experimental binding energies are well below 0.5 %, with some exceptions for light nuclei. It should be emphasized, that there are also successful non-relativistic semi-microscopic functionals based on the same concept [7].

We can conclude that, at present, we are able to derive very successful semi-microscopic covariant density functionals [11] from ab-initio results for nuclear matter and neutron matter, containing only few remaining parameters for a fine tuning.

Table 2 shows the rms deviations for nuclear masses and radii between CDFT calculations and the experimental data for 835 even-even nuclei. Several types of covariant density functionals are used, NL3* for non-linear meson couplings, DD-ME2 and DD-MEδ for density dependent meson couplings, and DD-PC1 and DD-PK1 for density dependent point coupling models. The last line contains also rotational corrections [16] obtained in a microscopic way with the same parameters. Such beyond mean field corrections improve considerably the agreement with experiment.

3 The problem of tensor forces

We have to emphasize that all the microscopic input comes at this stage from ab-initio calculations in nuclear matter. On the other side, there are problems in finite nuclei, where simple density functionals of this type fail consistently. An example are single-particle energies and their distribution. This has been found in an indirect way by global investigations in the area

Table 1. The number of phenomenological parameters n together with the rms deviations for masses and radii for typical covariant density functionals [1]. See text for details.

CEDF	n	ΔE_{rms} (MeV)	Δr_c^{rms} (fm)
NL3* [12]	6	2.96	0.0283
DD-ME2 [13]	8	2.39	0.0230
DD-MEδ [11]	4	2.29	0.0329
DD-PC1 [14]	10	2.01	0.0253
PC-PK1 [15]	10	2.58	-
PC-PK1 + E$_{corr}$ [16]		1.24	-

of transitional nuclei [1], which depend in some cases in a sensitive way on the single particle structure, but also in experimental observations of systematic shifts of single particle energies in isotopic chains. A famous case is the splitting of $1h_{11/2}$ and $1g_{7/2}$ proton configurations in the Sn-region [17], which shows a very characteristic behavior as a function of the neutron number. None of the conventional density functionals can reproduce this behaviour on the Hartree level. Only relativistic Hartree-Fock calculations which contain tensor forces in the exchange terms can reproduce this behavior. However, the strength of the tensor has to be fitted in a phenomenological way [2]. Similar results have also been found in density dependent Relativistic Hartree-Fock (RHF) theory [18].

On the other side, it is an open question, whether these shifts in observed single particle energies are an indication of the necessity of effective tensor forces in relativistic density functionals. In fact, recently it has been shown by A. Afanasjev and E. Litvinova [19] that a large part of the shifts of the $1h_{11/2}$ and $1g_{7/2}$ proton configurations in the Sn-region [17] can be described with the simple functional NL3* going beyong mean field and including quasiparticle-vibrational coupling [20] .

4 Microscopic RBHF calculations in finite nuclei

The concept to derive density functionals from properties of infinite nuclear matter cannot teach us much when it comes to properties of the functionals, which do not show up in nuclear matter calculations as for instance the tensor term in spin non-saturated systems. Here one needs ab-initio calculations in finite nuclei. An obvious way to carry out such investigations is Brueckner theory, the mother of modern density functional theory in nuclei. The advantage of relativistic Brueckner theory is that one does not need three-body forces to get saturation in nuclear matter close to the experimental area. In most of these applications the local density approximation (LDA) has been used, i.e. in a first step the self-energies are calculated in nuclear matter at various densities. In a second step the relativistic Hartree-Fock equations for Walecka-type functionals are solved in finite nuclei with density dependent coupling constants adjusted to the results of nuclear matter calculations at the corresponding density. This provides a mapping of the microscopically obtained nuclear matter results onto RHF-models of Walecka type.

Although such calculations are successful, they cannot reach the accuracy of present day phenomenological covariant density functionals. In particular, the mapping is not unique and therefore the results of different groups deviate form each other considerably. Therefore we started a new project to solve the RBHF equations in a large finite basis. First we started in an oscillator basis [21], but then we found that a large Dirac-Woods-Saxon (DWS) basis is more appropriate [22]. There are basis states with positive and negative energies and cut-off-parameters are introduced for the basis, such that convergence is achieved. For details see Ref. [5].

Table 2. Ground state properties of ^{16}O obtained with RBHF theory are compared various other approximations and with experiment (from Ref. [4]).

	E (MeV)	r_c (fm)	r_m (fm)	$\Delta E^{ls}_{\pi 1p}$ (MeV)
EXP	−127.6	2.70	2.54(2)	6.3
RBHF, Bonn *A*	−113.5	2.56	2.42	5.4
RBHF (DWS) [5]	−120.7	2.52	2.38	6.0
DDRHF, PKO1 [18]	−128.3	2.68	2.54	6.4
DDRHF, PKA1 [23]	−127.0	2.80	2.67	6.0
BHF [24], AV18	−134.2		1.95	13.0
CC [25], N^{3}LO	−120.9		2.30	
NCSM [26], N^{3}LO	−119.7(6)			
NLEFT [27], N^{2}LO	−121.4(5)			

As an application we consider the nucleus ^{16}O. We use the realistic NN interaction Bonn A which has been adjusted to the NN scattering data in Ref. [28]. The ground-state properties are listed in Table 2. The results of our full RBHF calculation are compared with the corresponding experimental data and with several other calculations: BHF is a non-relativistic Brueckner calculation based on the interaction Bonn A. We also show results obtained in RHF-calculations with the phenomenological effective interaction PKO1, which has been fitted to binding energies and charge radii of a set of spherical nuclei. It is seen that the ground-state properties in RBHF theory are improved considerably as compared with the non-relativistic results. This energy of ^{16}O is also very close to the value of $E = -119.7$ MeV obtained in the No Core Shell Model (NCSM) calculation using the chiral NN interaction N^{3}LO [26]. Of course the results of the calculations with PKO1 which has been fitted to these data shows only a very small deviation from the experimental values.

5 Conclusions

In summary, the full relativistic Brueckner-Hartree-Fock (RBHF) equations have been solved in a Dirac-Woods-Saxon (DWS) basis. The relativistic structure of the two-body matrix elements as well as of the Pauli operator is fully taken into account. The only input is the bare NN-interaction Bonn A adjusted to the scattering phase shifts in Ref. [28]. No other parameter is used. Since nuclear matter calculations within the same framework produce results far away from the Coester line and close to the experimental values of saturation, we neglect at this stage three-body forces.

Despite the good agreement of these results, there is room for improvements. The RBHF-theory presented here is no exact solution of the nuclear many-body problem. So far, rearrangement terms are not taken into account and higher order diagrams in the hole-line expansion are not included.

On the other side, our method has the potential to investigate heavier nuclei, where exact solutions are impossible, in particular systems without spin saturation and with large neutron excess. In this case we hope to be able to gain a parameter-free, microscopic understanding of open questions in modern phenomenological density functional theories, such as their isospin dependence or the importance of the tensor terms. As an example it has been found that the spin-orbit splitting in neutron droplets shows a very characteristic behavior as a function of the neutron number, which can be reproduced in phenomenological functionals with tensor terms of an appropriate size [29].

6 Acknowledgements

This work was partly supported by the DFG cluster of excellence "Origin and Structure of the Universe" (www.universe-cluster.de).

References

[1] S.E. Agbemava, A.V. Afanasjev, D. Ray, P. Ring, Phys. Rev. C **89**, 054320 (2014)
[2] G.A. Lalazissis *et al*, Phys. Rev. C **80**, 041301 (2009)
[3] M.R. Anastasio, L.S. Celenza, C.M. Shakin, Phys. Rev. Lett. **45**, 2096 (1980)
[4] S.H. Shen, H.Z. Liang, J. Meng, P. Ring, S.Q. Zhang, Phys. Rev. C **96**, 014316 (2017)
[5] S.H. Shen *et al*, Chin. Phys. Lett. **33**, 102103 (2016)
[6] S. Fayans, JETP Lett. **68**, 169 (1998)
[7] M. Baldo, P. Schuck, X. Viñas, Phys. Lett. B **663**, 390 (2008)
[8] R. Brockmann, R. Machleidt, Phys. Rev. C **42**, 1965 (1990)
[9] R. Brockmann, H. Toki, Phys. Rev. Lett. **68**, 3408 (1992)
[10] F. Hofmann, C.M. Keil, H. Lenske, Phys. Rev. C **64**, 034314 (2001)
[11] X. Roca-Maza *et al*, Phys. Rev. C **84**, 054309 (2011)
[12] G.A. Lalazissis *et al*,Phys. Lett. B **671**, 36 (2009)
[13] G.A. Lalazissis, T. Nikšić, D. Vretenar, P. Ring, Phys. Rev. C **71**, 024312 (2005)
[14] T. Niksic, D. Vretenar, P. Ring, Phys. Rev. C **78**, 034318 (2008)
[15] P.W. Zhao, Z.P. Li, J.M. Yao, J. Meng, Phys. Rev. C **82**, 054319 (2010)
[16] Q.S. Zhang, Z.M. Niu, Z.P. Li, J.M. Yao, J. Meng, Front. Phys. p. 529 (2014)
[17] J.P. Schiffer *et al*, Phys. Rev. Lett. **92**, 162501 (2004)
[18] W.H. Long, N. Van Giai, J. Meng, Phys. Lett. B **640**, 150 (2006)
[19] A.V. Afanasjev, E. Litvinova, Phys. Rev. C **92**, 044317 (2015)
[20] E. Litvinova, P. Ring, Phys. Rev. C **73**, 044328 (2006)
[21] P. Ring, J. Phys. Conf. Ser. **580**, 012005 (2015)
[22] S.G. Zhou, J. Meng, P. Ring, Phys. Rev. C **68**, 034323 (2003)
[23] W.H. Long, H. Sagawa, N.V. Giai, J. Meng, Phys. Rev. C **76**, 034314 (2007)
[24] B.S. Hu, F.R. Xu, Y.Z. Ma, Q. Wu, Z.H. Sun, Phys. Rev. C **95**, 034321 (2017)
[25] G. Hagen *et al*, Phys. Rev. C **80**, 021306 (2009)
[26] R. Roth *et al*, Phys. Rev. Lett. **107**, 072501 (2011)
[27] T.A. Laehde *et al*, Phys. Lett. B **732**, 110 (2014)
[28] R. Machleidt, Adv. Nucl. Phys. **19**, 189 (1989)
[29] S.H. Shen, H.Z. Liang, J. Meng, P. Ring, S.Q. Zhang, Phys. Lett. B **778**, 344 (2018)

Pairing correlations and symmetries in odd-A nuclei.

J. Luis Egido[1,*] and *Marta* Borrajo[1]

[1]Departamento de Física Teórica, Universidad Autónoma de Madrid, E-28049 Madrid, Spain

Abstract. The pairing correlations in odd-A nuclei are analyzed in the mean field approximation and beyond. In particular the role of symmetry conservation is investigated. We find that particle number projection after the variation (PN-PAV) has little effect on the pairing correlations specially in the weak pairing regime. This is in contrast to the variation after particle number projection (PN-VAP) approach where a strong effect is found. The situation is specially critical in odd nuclei because the pairing correlations vanish due to the blocking effect and the Hartree-Fock-Bogoliubov wave function collapses to the Hartree-Fock one. The PN-VAP, however, handles perfectly the exact blocking providing highly correlated wave functions. The role of the angular momentum projection is studied only in the PAV approach. We find small changes of the pairing correlation, at least at small angular momentum. In the calculations we use the Gogny interaction well suited to this kind of studies.

1 Introduction

Pairing correlations play a predominant role in the description of nuclear phenomena [1]. Basic approaches like the mean field (MF) only provide a realistic description when the superfluidity is incorporated into the theory. The success of the BCS approach and the Hartree-Fock-Bogoliubov (HFB) theory are good examples of it. The consideration of pairing correlations provides a qualitative difference in the description of basic properties like odd-even mass differences, spectra of even-even and odd-even A nuclei and dynamical masses -e.g. moments of inertia- to mention a few. The great breakthrough of the BCS was to consider the *symmetry breaking mechanism* to include particle-particle correlations in a product (of quasiparticles) wave function. Another example of the fruitful idea is the angular momentum. In this case the breaking of the rotational symmetry allows to describe deformed nuclei by simple product wave functions.

BCS and HFB theories, however, have some drawbacks like the pairing collapse in the weak pairing regime and the non-conservation of the particle number (PN) and angular momentum (AM) symmetries. To overcome these problems and at the same time to benefit from the simplicity of product wave functions the symmetry conserving mean field (SCMF) approach has been developed [2–4]. In this theory use is made of the projectors technique to obtain eigenstates of the particle number operator and the angular momentum from symmetry breaking HFB wave functions [5]. At this point it is important to recall the two ways to recover a symmetry. They differ in the way the matrices (U, V) of the Bogoliubov transformation, see Eq. (1) below, are determined. The simplest way is the so-called projection after variation (PAV). In this approach the matrices (U, V) are determined by minimizing the HFB energy and only afterwards the HFB wave function is projected to calculate the different observables. In the more demanding variation after projection (VAP), (U, V) are determined by minimizing the projected energy. Obviously only the latter one provides a selfconsistent determination of (U, V) compatible with the symmetries of the system. An inherent failure of the PAV approach is that the pairing collapse in the weak pairing cases cannot be avoided.

The SCMF wave funtions are the building blocks of the modern beyond mean field (BMF) approaches recently developed with the Skyrme, Gogny and relativistic effective interactions [6–9]. In these theories a basis of SCMF wave funtions is provided by the generator coordinate method (GCM) using as coordinates the most relevant nuclear degrees of freedom [10–15]. The Hamiltonian eigenstates are developed in this basis and the mixing coefficients determined by the variational principle. These developments are called symmetry-conserving configuration mixing (SCCM) approaches and have been applied mostly to even-even nuclei. Attempts to extend these approaches to odd-even and odd-odd nuclei are underway [16–18]. This task is not easy because odd nuclei are far more complicated to deal with. Even at the mean-field level like in the HFB theories, odd nuclei are numerically cumbersome and to calculate ground states one must try several spins, parity, etc. Furthermore, the blocked structure of the wave function entails the breaking of the time-reversal symmetry and triaxial calculations must be performed. An important problem not to be overlooked in SCCM calculations is the fact that the blocked structure of the wave function favors the quenching of the pairing correlations. That means, if in some basis states of the GCM linear combination the pairing collapses, the dynamical correlations introduced by the diagonalization of the associated Griffins-Hill-Wheeler equation will be wrong. To investigate this fact in this paper we study the even-odd

*e-mail: j.luis.egido@uam.es

nucleus ^{27}Mg in the SCMF approach using as generator coordinates the quadrupole deformation parameters (β, γ) in the PAV and VAP approaches. In the calculations we use the Gogny interaction [19, 20], which is well suited to deal with the pairing correlations.

2 Theory

Our starting approach is the HFB theory [5]. As a mean-field approximation the HFB wave function $|\phi\rangle$ is a product of quasi-particles α_ρ defined by the general Bogoliubov transformation

$$\alpha_\rho^\dagger = \sum_\mu U_{\mu\rho} c_\mu^\dagger + V_{\mu\rho} c_\mu, \tag{1}$$

where $c_\mu^\dagger, c_\mu$ are the particle-creation and -annihilation operators in the reference basis, in our case the Harmonic Oscillator one. U and V are the Bogoliubov matrices to be determined by the Ritz variational principle. In our approach we have imposed the simplex, $\Pi_1 = \hat{P}e^{-i\pi J_x}$, which allows to characterize the blocking structure of odd and even nuclei [2, 21]. The index μ runs over the states $\{1, 2, ..., M, \bar{1}, \bar{2}, ..., \bar{M}\}$. The states $\alpha_m^\dagger$ have simplex $+i$ and $\alpha_{\bar{m}}^\dagger$ have $-i$.

The wave function of the ground state of an even-even nucleus is given by[1]

$$|\phi\rangle = \prod_{\mu=1}^{2M} \alpha_\mu |-\rangle = \prod_{\mu=1}^{M} \alpha_m \alpha_{\bar{m}} |-\rangle, \tag{2}$$

with $|-\rangle$ the particle vacuum. The quasiparticle vacuum $|\phi\rangle$ is defined by

$$\alpha_\mu |\phi\rangle = 0, \;\; \mu = 1, ..., 2M, \tag{3}$$

and has simplex +1. The quasiparticle excitations

$$|\tilde{\phi}\rangle = \alpha_{\rho_1}^\dagger |\phi\rangle \tag{4}$$

correspond to odd-even nuclei. They can be written as vacuum to the quasiparticle operators $\tilde{\alpha}_\rho$,

$$\tilde{\alpha}_\rho |\tilde{\phi}\rangle = 0, \;\; \rho = 1, ..., 2M. \tag{5}$$

The $2M$ operators $\{\tilde{\alpha}_\rho^\dagger\}$ are obtained from the set $\{\alpha_\mu^\dagger\}$ by replacing the creation operator $\alpha_{\rho_1}^\dagger$ by the annihilation operator α_{ρ_1}, the other $2M - 1$ operators remain unchanged. In this work we are interested in the nucleus ^{27}Mg. We therefore consider wave functions of the form

$$|\tilde{\phi}^\pi\rangle = \alpha_{\rho_1}^\dagger \prod_{\mu=1}^{2M} \alpha_\mu |-\rangle. \tag{6}$$

with ρ_1 a neutron state. Notice that in the running product of Eq. (6), orbitals with the same parity are occupied pairwise, therefore the parity, π, and the simplex of the state $|\tilde{\phi}^\pi\rangle$ are those of the blocked level $\alpha_{\rho_1}^\dagger$. One can, of course, block a state with positive or negative simplex, but since we do not break time reversal explicitly both possibilities are degenerated.

Though the state $|\tilde{\phi}\rangle$ has the right blocking structure, since the Bogoliubov transformation mixes creator and annihilator operators and states with different angular momenta, $|\tilde{\phi}\rangle$ is neither an eigenstate of the particle number nor of the angular momentum operators. As with even-even nuclei, to recover the symmetries one has to project to the right quantum numbers, see [5]. In the spirit of the SCMF approach, our ansatz is

$$|\Psi_{M,\sigma}^{N,I,\pi}\rangle = \sum_K g_{K\sigma}^I P^N P_{MK}^I |\tilde{\phi}^\pi\rangle, \tag{7}$$

the parameters $g_{K\sigma}^I$ and the matrices $\{\tilde{U}, \tilde{V}\}$ of the Bogoliubov transformation are determined by the variational principle

$$\delta E^{N,I,\pi} = \delta \frac{\langle \Psi_M^{N,I,\pi} | \hat{H} \hat{P}^N P_M^I | \Psi_M^{N,I,\pi} \rangle}{\langle \Psi_M^{N,I,\pi} | \hat{P}^N P_M^I | \Psi_M^{N,I,\pi} \rangle} = 0. \tag{8}$$

This way of solving the SCMF approach is called variation after projection of the particle number and angular momentum, PNAM-VAP. Unfortunately this equation is too complicated to be solved with effective interactions, as in our case, and for large configuration spaces. There are several ways to approximate the exact solution of Eq. (8). The simplest way would be to minimize the HFB energy, i.e., determine $(\tilde{U}, \tilde{V})$ and then perform the projections, i.e. the so-called projection-after-variation (PAV). This method does not provide a good approximation. A much better approach is the following: From even-even nuclei one knows that PN-VAP (that means, solving Eq. (8) projecting only on the particle number) is feasible while AM-VAP is very CPU-time consuming. The approach of solving the PN-VAP variational equation to find the self-consistent minimum and afterwards to perform an AM-PAV is not very good because the AMP is not able to exploit any degree of freedom of the HFB transformation and self-consistency with respect to the AMP is not guaranteed. An intermediate way is to perform an approximate AM-VAP approach by solving the variational PN-VAP equation for a large set of relevant wave functions as to cover the sensitive degrees of freedom. Afterwards an AM-PAV to this set of wave functions will determine the absolute minimum among these states for different angular momenta. We choose the deformation parameters (β, γ) as the additional degrees of freedom. Notice that this method guarantees, at least, AM-VAP self-consistency with respect to these relevant quantities. Therefore, in order to obtain a grid of wave functions we minimize the PN-VAP constrained energy

$$E'[\tilde{\phi}^\pi] = \frac{\langle \tilde{\phi}^\pi | \hat{H} \hat{P}^N | \tilde{\phi}^\pi \rangle}{\langle \tilde{\phi}^\pi | \hat{P}^N | \tilde{\phi}^\pi \rangle} - \langle \tilde{\phi}^\pi | \lambda_{q_0} \hat{Q}_{20} + \lambda_{q_2} \hat{Q}_{22} | \tilde{\phi}^\pi \rangle, \tag{9}$$

with the Lagrange multiplier λ_{q_0} and λ_{q_2} being determined by the constraints

$$\langle \tilde{\phi}^\pi | \hat{Q}_{20} | \tilde{\phi}^\pi \rangle = q_0, \;\; \langle \tilde{\phi}^\pi | \hat{Q}_{22} | \tilde{\phi}^\pi \rangle = q_2. \tag{10}$$

[1] The quasiparticle operators that annihilate trivially the particle vacuum are to be omitted from the product.

The relation between (β, γ) and (q_0, q_2) is given by $\beta = \sqrt{20\pi(q_0^2 + 2q_2^2)}/3r_0^2 A^{5/3}$, $\gamma = \arctan(\sqrt{2}q_2/q_0)$ with $r_0 = 1.2$ fm and A the mass number. The minimization of Eqs. (9-10) is performed with the conjugated-gradient method [22].

The next step is the simultaneous particle-number and angular-momentum projection (PNAMP) of each state $|\tilde{\phi}^\pi(\beta, \gamma)\rangle$ that conforms the (β, γ) grid,

$$|\Psi^{N,I,\pi}_{M,\sigma}(\beta, \gamma)\rangle = \sum_K g^I_{K\sigma} P^N P^I_{MK} |\tilde{\phi}^\pi(\beta, \gamma)\rangle = \sum_K g^I_{K\sigma} |IMK, \pi, N, (\beta, \gamma)\rangle, \quad (11)$$

where the coefficients $g^I_{K\sigma}$ are variational parameters. They are determined by the energy minimization which provides a reduced Hill-Wheeler-Griffin [23] equation

$$\sum_{K'} (\mathcal{H}^{N,I,\pi}_{K,K'} - E^{N,I,\pi}_\sigma N^{N,I,\pi}_{K,K'}) g^I_{K'\sigma} = 0. \quad (12)$$

where $\mathcal{H}^{N,I,\pi}_{KK'}$ and $N^{N,I,\pi}_{K,K'}$ are the Hamiltonian and norm overlaps defined by

$$\mathcal{H}^{N,I,\pi}_{K,K'} = \langle IMK, \pi, N, (\beta, \gamma)|H|IMK', \pi, N, (\beta, \gamma)\rangle \quad (13)$$

$$N^{N,I,\pi}_{K,K'} = \langle IMK, \pi, N, (\beta, \gamma)|IMK', \pi, N, (\beta, \gamma)\rangle \quad (14)$$

The presence of the norm matrix in Eq. (12) is due to the non-orthogonality of the states $|IMK, \pi, N, (\beta, \gamma)\rangle$. Eq. (12) is solved by standard techniques [5]. Notice that at each (β, γ) point one can have several eigenvalues $E^{N,I,\pi}_\sigma$ labeled by σ, the lowest solution corresponds to $\sigma = 0$.

The solution of Eqs. (12) in the (β, γ) grid for different angular momenta and parity provides $E^{N,I,\pi}_\sigma(\beta, \gamma)$ as a function of (β, γ), I, π and σ. This energy can be written as

$$E^{N,I,\pi}_\sigma(\beta, \gamma) = \frac{\langle \Psi^{N,I,\pi}_{M,\sigma}(\beta, \gamma)|H|\Psi^{N,I,\pi}_{M,\sigma}(\beta, \gamma)\rangle}{\langle \Psi^{N,I,\pi}_{M,\sigma}(\beta, \gamma)|\Psi^{N,I,\pi}_{M,\sigma}(\beta, \gamma)\rangle}, \quad (15)$$

which obviously represents the potential energy surface (PES) of the projected energy in the (β, γ) plane for the given quantum numbers. This projected PES differs from the usual mean field PES and is angular momentum (I), parity (π) and state (σ) dependent. The minimum value of $E^{N,I,\pi}_\sigma(\beta, \gamma)$ in the PES provides the energy and the deformation parameters $(\beta_{\min}, \gamma_{\min})$ of the state characterized by the quantum numbers (I, π, σ) in this approximation. Its wave function is given by $|\Psi^{N,I,\pi}_{M,\sigma}(\beta_{\min}, \gamma_{\min})\rangle$. Notice that by changing the values of I and π we can calculate the spectrum built by the lowest states of a given spin and parity.

In the precedent approach to generate the states $|\tilde{\phi}^\pi(\beta, \gamma)\rangle$ we have use the PNVAP approach solving Eq. (9). In this case the pairing degree of freedom is treated fully self consistently with the conservation of the particle number. This approach is rather time consuming and simpler approximations are often used in the literature to produce the states $|\tilde{\phi}^\pi(\beta, \gamma)\rangle$. The most popular is the constrained HFB approximation in which the HFB energy, given by an expression similar to Eq. (9) but without the projector $\hat{P}^N$, is minimized. Two additional constraints on the number of protons and neutrons have to be added to keep the right number of particles at least on the average. We will call this approach the HFB approach. Since the use of the HFB method only affects the wave functions $|\tilde{\phi}^\pi(\beta, \gamma)\rangle$ the rest of the procedure, i.e, the solution of Eqs. 12, is the same in the PNVAP and in the HFB approach. As mentioned in the Introduction is this paper we want to elucidate how good the HFB approach is as compared to the PNVAP in the SCMF and in SCCM calculations.

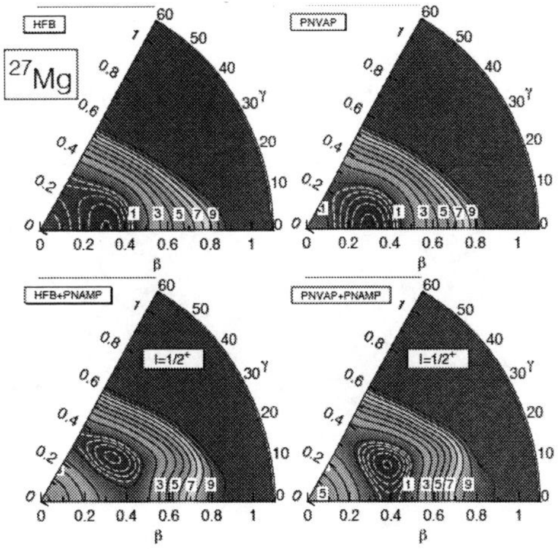

Figure 1. Potential energy surfaces in the (β, γ) plane for ^{27}Mg in several approaches.

3 Results

In the calculations the intrinsic many body wave functions $|\tilde{\phi}(\beta, \gamma)\rangle$ are expanded in a Cartesian harmonic oscillator basis and the number of spherical shells included in this basis is $N_{shells} = 6$ with an oscillator length of $b = 1.01A^{1/6}$. The (β, γ) grid of equilateral triangles contains 116 points. The angular momentum projection has been done with the set of integration points in the Euler angles ($N_\alpha = N_\beta = N_\gamma = 32$) in the intervals $\alpha \in [0, 2\pi], \beta \in [0, \pi], \gamma \in [0, 2\pi]$. The number of points to perform the integral of the particle-number projection is 11. In the calculations we use the Gogny interaction [19] with the D1S parameterization [20]. We consider all exchange terms of the interaction, the Coulomb force and the two-body correction of the kinetic energy to avoid problems with the PNP [24]. Concerning the density dependence of the force we adopt the projected density prescription for the PNP and the mixed one for the AMP, for further details see for example Refs. [8, 11, 25].

As an example we present in Fig. 1 contour plots of the potential energy surfaces for the nucleus ^{27}Mg calculated in several approaches. In the PN-VAP approach be means of Eq. (9) and in the PNVAP+PNAMP case by Eq. (15). In the HFB case we solve Eq. (9) but without the PN projector and in HFB+PNAMP case by Eq. (15) using the corresponding intrinsic wave functions. The wave functions used to calculate the HFB surface are neither eigenfunctions of the particle number nor of the angular momentum. Those of the PNVAP conserve the particle number but not the angular momentum. Finally, the ones of the HFB+PNAMP and PNVAP+PNAMP conserve both symmetries. The angular momentum and parity $I = 1/2^+$ corresponds to the ground state of this nucleus (see Ref. ([17]) for a detailed description). The HFB and PNVAP approaches predict an axially symmetric ground state ($\beta \approx 0.3$) though rather soft in γ. The angular momentum projection shifts the minimum to a triaxial shape and a larger deformation ($\beta \approx 0.4$). The HFB+PNAMP minimum is more triaxial ($\gamma = 30^o$) than the PNVAP+PNAMP one ($\gamma = 20^o$) and softer in the oblate direction. To analyze the

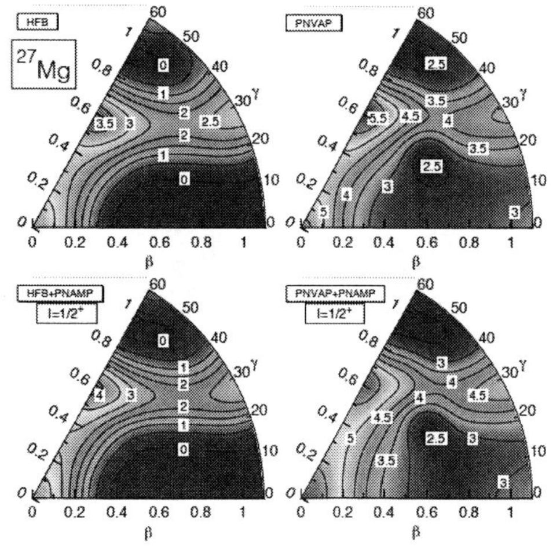

Figure 2. Contour plots of the absolute value of the proton pairing energies in the (β, *gamma*) plane in several approaches. The color code is panel dependent.

quality of the HFB approach in the (β, γ) plane we now study the pairing correlations in the different approaches, see Ref. [26] for a precise definition in BMF approaches. In Fig. 2 we show the absolute value of the pairing energies for protons, in the top panels in the HFB and the PNVAP approaches. The most striking aspect is that in the HFB case the pairing correlations vanish or are very small in large areas of the (β, γ) plane at variance with the PNVAP approach where the smallest value is 2.5 MeV. At first glance it may be surprising that, though quantitatively very different, the pairing energies in the HFB and the PNVAP approach look qualitatively similar. This is however expected, since pairing correlations are primarily determined by the level density and this is fixed by the (β, γ) deformation parameters. In the lower panels of Fig. 2 the projected versions are displayed. In the PNVAP+PNAMP by comparison with the PNVAP panel we can observe the effect of angular momentum projection. Though the PNVAP+PNAMP provides somewhat larger pairing than the PNVAP in some areas, both plots are very similar. This result is for $I = 1/2^+$. For larger values of the angular momentum greater deviations are expected. In the HFB+PNAMP case where we can learn about the effect of the PNP after the variation by comparison with the HFB panel we find that there are only small changes. This is to be expected since in the HFB case the pairing correlations are either zero or rather small.

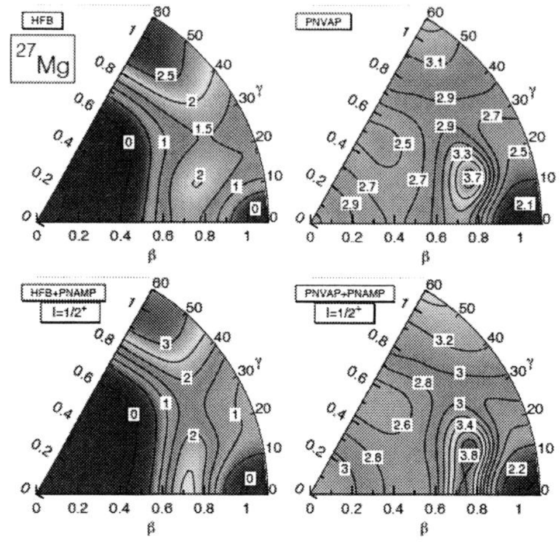

Figure 3. Same as Fig. 2 for neutrons.

In the neutron channel we anticipate a generalized lowering of the pairing correlations due to the blocked neutron. This can be observed in Fig. 3 where the neutron pairing energies are represented. We find that even in the PNVAP case the pairing correlations are reduced by about 30% as compared with the protons. In the HFB case we find again large areas where the pairing collapse takes place. We also observe that in this case the *qualitative* resemblance found in the proton channel between the HFB and the PNVAP pairing correlations is less pronounced. Apparently the VAP treatment is able to change the blocked structure as compared with the HFB one. We can appreciate the impact of the blocking effect by looking at the neutron energies in the even-even nucleus ^{28}Mg shown in Fig 4. Here we find smaller pairing energies in HFB as compared with the PNVAP approach, but there is a big difference as compared with ^{27}Mg, namely that the areas where the pairing collapse takes place are much smaller.

It is important to notice, by looking at the HFB+PNAMP panel of Fig. (1), that in large areas and in particular around the energy minimum the pairing correlations for protons and neutrons are either zero or small, at variance with the PNVAP+PNAMP approach. This will have a great impact on the SCCM approaches, mentioned in the Introduction, used for full spectroscopy [9]. In these approaches the wave function is given by

$$|\Psi^{N,I,\pi}_{M,\sigma}\rangle = \sum_{K\beta,\gamma} g^{I}_{K\sigma}(\beta,\gamma) P^{N} P^{I}_{MK} |\tilde{\phi}^{\pi}(\beta,\gamma)\rangle. \qquad (16)$$

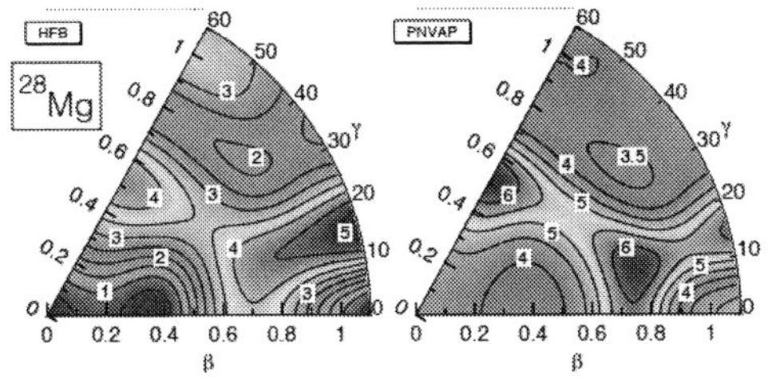

Figure 4. Neutron pairing energies for ^{28}Mg in the HFB and the PNVAP approches.

That means the previous states $|\tilde{\phi}^{\pi}(\beta,\gamma)\rangle$ are considered as basis states. The collapse of the pairing correlation for many (β,γ) values will have a strong impact in SCCM calculations not only locally, as in the SCMF calculations, but also on the dynamical correlations (i.e. the associated masses). To gauge this impact we have performed some calculations in the simpler SCMF approach. Though we know that in the SCMF theory additional correlations are missing for a good description of excited states, to investigate the impact of a proper treatment of the pairing correlations on the spectrum we now solve Eq. (12) for spin values $1/2, 3/2, 5/2, ...$ and positive and negative parity in the HFB+PNAMP and PNVAP+PNAMP cases. The corresponding spectrum is plotted in Fig. 5. Though the level ordering is the same in both approximations we find large differences in the spacing. In particular we find that the HFB+PNAMP provides a much more compressed spectrum as the PNVAP+PNAMP approach.

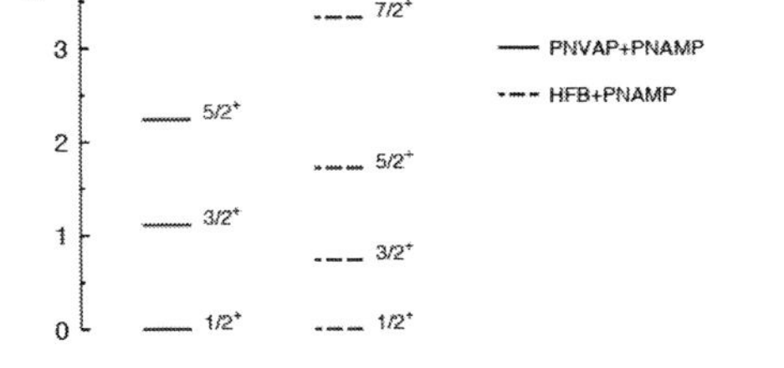

Figure 5. Spectrum of the nucleus ^{27}Mg in the HFB+PNAMP and the PNVAP+PNAMP.

In conclusion, we have presented a study of the pairing correlations in odd-A nuclei in the framework of the symmetry conserving mean field approximation with the effective Gogny force. We have analyzed for ^{27}Mg the pairing correlations in the (β,γ) plane with the striking result that in the HFB approach they collapse or are very small for many points of the grid, in contrast to the variation after projection approach. We illustrate the difference of both approaches in the case of the excitation spectrum.

4 Acknowledgements

This work was supported by the Spanish Ministerio de Economía y Competitividad under contract FPA2014-57196-C5-2-P.

References

[1] *Fifty Years of Nuclear BCS* R. A. Broglia and V. Zelevinsky Eds. World Scientific, Singapore, 2013.
[2] H.J. Mang, Phys. Rep. 18C (1975) 325.
[3] J.L. Egido, P. Ring, Nucl. Phys. **A388**, 19-36 (1982)
[4] E. Hammarén, K. W. Schmid, F. Grümmer, A. Faessler, and B. Fladt, Nucl. Phys. **A437**, 1 (1985).
[5] P. Ring and P. Schuck, *The Nuclear Many Body Problem* (Springer-Verlag, Berlin, 1980).
[6] M. Bender, P.-H. Heenen and P.-G. Reinhard, Rev. Mod. Phys. **75** (2003) 121.
[7] T. Nikšić, D. Vretenar and P. Ring, Prog. Part. Nucl. Phys. 66 (2011) 519.
[8] J.L. Egido, Phys. Scr. 91 (2016) 073003.
[9] J. L. Egido, M.Borrajo, T. R. Rodríguez, Phys. Rev. Lett. **116**, 052502 (2016)
[10] M. Bender and P.-H. Heenen, Phys. Rev. C 78 (2008) 024309.
[11] T.R. Rodríguez and J.L. Egido, *Phys. Rev.* **C81** (2010) 064323.
[12] J. M. Yao, J. Meng, P. Ring, and D. Vretenar, Phys. Rev. **C81** (2010) 044311.
[13] N. López Vaquero, T. R. Rodríguez and J. Luis Egido, *Phys. Lett.* **B704** (2011) 520.
[14] N. López Vaquero, J.L. Egido and T. R. Rodríguez *Phys. Rev. C* **88** (2013) 064311.
[15] N. López Vaquero, T. R. Rodríguez and J.L. Egido *Phys. Rev. Lett.* **111** (2013) 142501.
[16] B. Bally, B. Avez, M. Bender, and P.-H. Heenen, Phys. Rev. Lett. **113**, 162501 (2014).
[17] M. Borrajo and J.L. Egido, *Eur. Phys. J.* **A52** (2016) 277.
[18] M. Borrajo and J.L. Egido, Physics Letters **B 764** (2017) 328–334
[19] J. Dechargé and D. Gogny, Phys. Rev. **C 21** (1980) 1568.
[20] J. F. Berger, M. Girod and D. Gogny, Comp. Phys. Commun. **63** (1991) 365.
[21] J.L. Egido, H.J. Mang, P. Ring, Nucl. Phys. **A334** (1980)1
[22] J.L. Egido, J. Lessing, V. Martin, L.M. Robledo, *Nucl. Phys.* **A 594** (1995) 70.
[23] J.J. Griffin and J.A. Wheeler, Phys. Rev. **108**, 311 (1957)
[24] M. Anguiano, J.L. Egido and L. M. Robledo, *Nucl. Phys.* **A 683**(2001) 227.

[25] R. Rodriguez-Guzman, J. L. Egido, L. M. Robledo, Nucl. Phys. **A 709**, 201(2002)

[26] M. Anguiano, J.L. Egido, L.M. Robledo Phys. Lett. **B 545** 62–72 (2002)

Single-particle states vs. collective modes: friends or enemies ?

T. Otsuka[1,2,3,4,5], Y. Tsunoda[3], T. Togashi[3], N. Shimizu[3], and T. Abe[2]

[1]RIKEN Nishina Center, 2-1 Hirosawa, Wako, Saitama 351-0198, Japan
[2]Department of Physics, University of Tokyo, 7-3-1 Hongo, Bunkyo, Tokyo 113-0033, Japan
[3]Center for Nuclear Study, University of Tokyo, 7-3-1 Hongo, Bunkyo, Tokyo 113-0033, Japan
[4]Instituut voor Kern- en Stralingsfysica, K. U. Leuven, Celestijnenlaan 200D, B-3001 Leuven, Belgium
[5]National Superconducting Cyclotron Laboratory,Michigan State University, East Lansing, Michigan 48824, USA

Abstract. The quantum self-organization is introduced as one of the major underlying mechanisms of the quantum many-body systems. In the case of atomic nuclei as an example, two types of the motion of nucleons, single-particle states and collective modes, dominate the structure of the nucleus. The collective mode arises as the balance between the effect of the mode-driving force (*e.g.*, quadrupole force for the ellipsoidal deformation) and the resistance power against it. The single-particle energies are one of the sources to produce such resistance power: a coherent collective motion is more hindered by larger spacings between relevant single particle states. Thus, the single-particle state and the collective mode are "enemies" against each other. However, the nuclear forces are rich enough so as to enhance relevant collective mode by reducing the resistance power by changing single-particle energies for each eigenstate through monopole interactions. This will be verified with the concrete example taken from Zr isotopes. Thus, the quantum self-organization occurs: single-particle energies can be self-organized by (i) two quantum liquids, e.g., protons and neutrons, (ii) monopole interaction (to control resistance). In other words, atomic nuclei are not necessarily like simple rigid vases containing almost free nucleons, in contrast to the naïve Fermi liquid picture. Type II shell evolution is considered to be a simple visible case involving excitations across a (sub)magic gap. The quantum self-organization becomes more important in heavier nuclei where the number of active orbits and the number of active nucleons are larger.

1 Introduction

The underlying mechanisms of the many-body structure of atomic nuclei have been studied over decades as one of the most important objectives of nuclear physics. It has then been understood usually that there are two types of dominant motion of nucleons in the atomic nucleus : single-particle states and collective modes. Regarding the single-particle states, Mayer and Jensen introduced the shell structure and associated magic numbers [1–3]. The nuclear shell model developed on these concepts has been extremely successful in the description of the structure of many nuclei (see for example [4–6]). The collective modes include various cases. Among them, the deformation of the nuclear shape has been studied since Rainwater [7], and Bohr and Mottelson [8, 9]. The nuclear shapes have been one of the major focuses of the nuclear structure physics, including spherical, vibrational and rotational ones [10]. The relation between the single-particle states and the collective modes has naturally become of much interest, as described by Bohr and Mottelson in [10] as "the problem of reconciling the simultaneous occurrence of single-particle and collective degrees of freedom and exploring the variety of phenomena that arise from their interplay".

The atomic nucleus is a many-body quantum system comprised of protons and neutrons, which is often considered to be described in terms of Landau's Fermi Liquid picture. In a somewhat simplified expression of this picture, protons and neutrons of a nucleus are in a mean potential which is like a rigid "vase", and those nucleons are like free particles moving in this vase, interacting weakly among themselves through a "residual interaction". The single-particle energies (SPE) of such a system exhibit the shell structure, and are split in general. If the splitting is large enough, the many-body structure is dominated by the SPEs: nucleons occupy the lowest single particle states in the ground state, the next lowest configuration gives us the first excited state, and so forth. In such cases, the correlations due to the interaction between nucleons may contribute, but their effects are minor, more or less, compared to the effects of SPE splittings. However, if the energy gain from such correlations overcomes the relevant SPE splittings, a collective mode dominates the structure of the ground and low-lying states. Although the understanding of the relation between the single-particle states and the collective modes has been pursued in many ways, it seems to remain an open problem. For instance, G.E. Brown had kept, throughout his life, the question, "how single particle states can coexist with collective modes" as quoted from "Fermi liquid theory: A brief survey in memory of Gerald E. Brown" in [11]. We shall present a novel mechanism which is closely related to this problem.

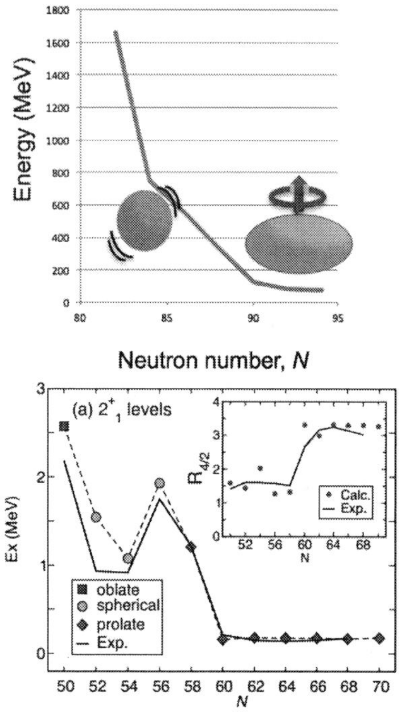

Figure 1. Systemtic changes of the 2^+_1 level in (upper) Sm and (lower) Zr isotopes, as functions of N. Data taken from [14] for Sm. The right panel is based on [12].

2 Nuclear shapes and quantum phase transition

We shall focus on the quadrupole deformation of the nuclear shape in this talk, while the scope can be extended. Figure 1 exhibits the excitation energy of 2^+_1 state, or the 2^+_1 level, for Sm and Zr isotopes as a function of the neutron number, N. In the Sm chain, the 2^+ level comes down rather gradually, similarly to many other isotopic chains. As shown in Fig. 1, a higher 2^+ level corresponds to a spherical shape and its surface oscillation, while a lower 2^+ level implies an ellipsoidal deformed shape and the rotation of the ellipsoid. On the other hand, in the Zr chain, the 2^+ level drops down abruptly in moving from N=58 to 60. Due to this abrupt change, this phenomenon can be referred to as a quantum phase transition [12]. Likewise, the ground-state structure of the Zr isotopes is changed drastically between N=58 and 60, also from the sphere to the strongly deformed ellipsoid. The Monte Carlo Shell Model (MCSM) describes both situations including the abrupt change with the same Hamiltonian [12, 13].

3 Monte Carlo Shell Model

We here sketch the MCSM briefly [15, 16]. The MCSM is a recent method to obtain eigensolutions of the shell-model calculation in nuclear physics. The shell-model calculation is quite similar to the configuration interaction (CI) calculation in other fields of science. The major differences are that (i) the two ingredients, protons and neutrons, are taken instead of electrons, and (ii) nuclear forces are considered instead of Coulomb or other forces. In conventional shell-model calculations, the matrix of the Hamiltonian with respect to many Slater determinants is diagonalized. Because a huge number of configurations are needed for the description of large systems, the dimension of the matrix can be enormous, making the calculation infeasible. On the other hand, many interesting and important problems lie beyond this limit. The MCSM provides a breakthrough in this regard. The MCSM is very different from the conventional shell-model calculation. A set of Slater determinants, called MCSM basis vectors, is introduced, and the diagonalization is performed in the Hilbert space spanned by the MCSM basis vectors. Each MCSM basis vector is a Slater determinant composed of single-particle states that are superpositions of the original single-particle states, and the amplitudes of these superpositions are determined through a combination of stochastic and variational methods. Even when the dimensions are in the order of 10^{23} for the conventional shell model, the problem can be solved, to a good approximation, with up to approximately 100 MCSM basis vectors.

4 Quantum Phase Transition in Zr isotopes

We shall look into the structure changes in Zr isotopes based on the MCSM calculation. The upper panel of Fig. 2 shows the occupation numbers of proton orbits for some states. The $g_{9/2}$ orbit is almost empty in the 0^+_1 state of ^{98}Zr, whereas it is occupied by about 3.5 protons in its 0^+_2 state. Note that this 0^+_1 (0^+_2) state is spherical (deformed). Such changes, including the numbers of proton holes in the pf orbits, result in substantial shifts of the neutron (effective) SPEs as shown schematically in the middle panel of Fig. 2. The proton-neutron monopole interaction (wavy line in the figure) generates those shifts. The right panel depicts the actual neutron effective SPEs. One notices a substantial change in SPE for the different states considered. One sees that the spacing between the $d_{5/2}$ and $g_{7/2}$ orbits is nearly 5 MeV for the 0^+_1 state of ^{98}Zr, but it is reduced to about 2 MeV in 0^+_2 state. Such a reduced splitting is found also in the 0^+_1 state of ^{100}Zr which is also strongly deformed.

We now discuss why the SPEs are so different between spherical and deformed states. We first point out that the nuclear deformation at low excitation energy is a Jahn-Teller effect [17], meaning that the collective motion causing the deformation occurs as a consequence of coherent contributions from some relevant orbits near the Fermi energy. For such coherent effects, larger splittings of SPEs weaken the coherence, leading to less collectivity. On the other hand, the monopole interaction can change the effective SPEs depending on the occupancy of the other nucleons. If the monopole interaction were uniform, no configuration dependence would appear, and this change should

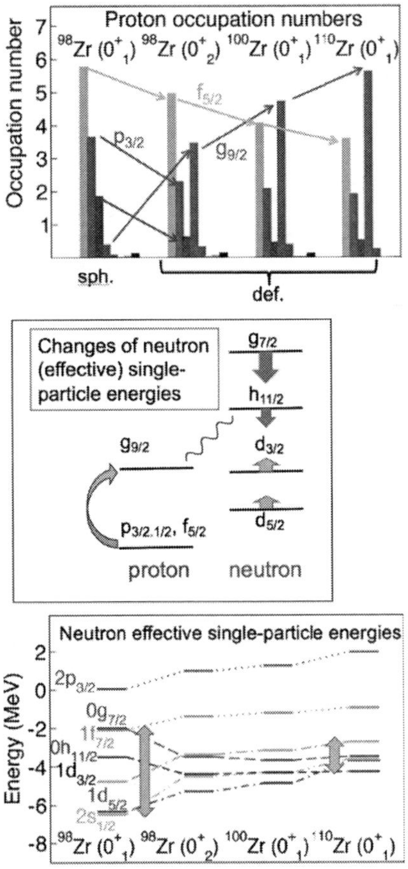

Figure 2. (upper) Occupation numbers of proton orbits of Zr isotopes. (middle) Schematic illustration of the changes of neutron (effective) single-particle energies in Zr isotopes. (lower) Actual values of neutron (effective) single-particle energies obtained in the calculation of [12]. Upper and lower panels are based on [12].

be absent. The tensor-force component of the nuclear force makes the monopole interaction attractive or repulsive, depending on the combination of the orbits [18, 19]. This is certainly against the uniformity, and its effect can be crucial. The central-force component changes its magnitude also depending on the combination of the orbits mainly due to varying overlaps of radial wave functions of single-particle states [19]. Thus, the monopole interaction is indeed far from being uniform, and the selection of favored configurations can move the SPEs of relevant orbits substantially. If relevant SPEs can be made closer to being degenerate, it helps the deformation. We shall formulate this novel mechanism in the next section.

5 Quantum self-organization

The nuclear deformation is determined by the balance between the effect of the collective-mode driving force and the resistance power against this collective mode. A schematic expression of this property is,

$$\text{deformation} = \frac{\text{quadrupole force}}{\text{resistance power}}. \tag{1}$$

The collective-mode driving force is the quadrupole (or quadrupole-quadrupole) interaction in the case of the ellipsoidal shape. This interaction is one of the major components of the proton-neutron realistic force. A typical example of the resistance power is the pairing interaction, which tends to make the shape more spherical because all time-reversal pairs are equally favored. Keeping the pairing interaction aside, we shall consider another source of the resistance power. That is the SPEs.

We here propose a novel mechanism called, **Quantum Self-Organization.** This mechanism means the following property: Atomic nuclei can "organize" their single-particle energies by taking particular configurations of protons and neutrons, optimized for each eigenstate, thanks to orbit-dependences of monopole components of nuclear forces (*e.g.*, tensor and central forces). This results in an enhancement of Jahn-Teller effect, *i.e.*, an enhancement of the collective mode. The deformation andk quantum self-organization can be linked in a non-linear way with a positive feedback : once some nucleons are excited to particular orbits, the SPEs are shifted in favor of a larger deformation. A larger deformation can promote such excitations with more nucleons. This cycle continues until a self-consistency is achieved, whereas intermediate situations are skipped. In many cases, massive excitations are involved, and the particle-hole hierarchy is broken, for instance, a 6p-6h deformed state comes right after a 2p-2h near-spherical state, skipping 4p-4h state [20].

The property shown in eq. (1) is somewhat analogous to the relation,

$$\text{electric current} = \frac{\text{voltage}}{\text{resistance}}, \tag{2}$$

where the electric current, voltage and resistance mean the usual quantities regarding the electricity. The higher voltage produces a higher current, but the current can be increased also by decreasing the resistance. The quantum self-organization implies that the atomic nucleus finds particular configurations which decrease the resistance power.

The most favorable configurations and associated SPEs vary for individual eigenstate even within the same type of the collective mode. For instance, prolate, oblate or triaxial shapes belong to the quadrupole deformation, but can appear with different patterns of the SPEs within the same nucleus. The oblate shape is less affected by the quantum self-organization, because smaller numbers of nucleons on unique-parity orbits are the key element of the oblate shape in most cases. In those cases, the organization of many orbits are rather irrelevant, and the quantum self-organization may not occur to a sizable extent. This feature has been verified with concrete cases. On the other hand, many orbits contribute coherently to the prolate deformation, and the quantum self-organization can produce crucial effects. This has been confirmed by changing the

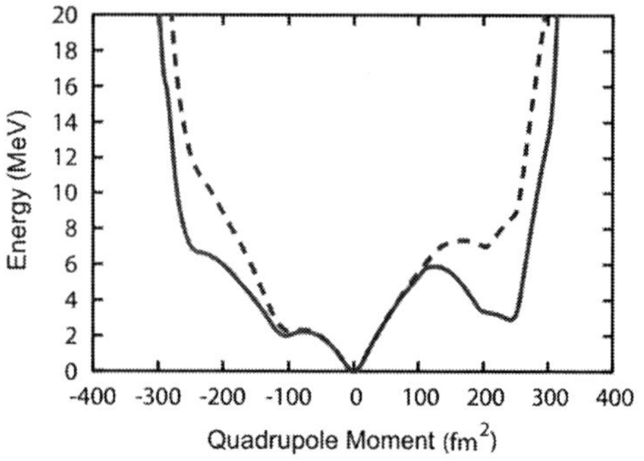

Figure 3. Potential Energy Surface with the axially symmetric deformation for ^{68}Ni. The red solid line denotes the energy of the constrained Hartree-Fock calculation with the original Hamiltonian. The blue dashed line implies the same calculation except that the quantum self-organization is suppressed (see the text). Figure taken from [20].

monopole interactions, for instance, closer to the uniform one.

We present a concrete example by taking the case of the prolate band in ^{68}Ni [20]. The monopole interaction between the neutron $1g_{9/2}$ orbit and the proton $1f_{5/2}$ orbit is more attractive than that between the neutron $1g_{9/2}$ orbit and the proton $1f_{7/2}$ orbit mainly due to the robust property of the monopole interaction of the tensor force, and this difference serves as the major origin of the quantum self-organization in this particular case: more neutrons in the $1g_{9/2}$ orbit reduces the $1f_{7/2}$-$1f_{5/2}$ spin-orbit splitting for protons [18, 19, 21]. The effect of this difference on the deformation can be seen quantitatively by replacing the strengths of these monopole interactions with the average of their original values, *i.e.*, the same value. Likewise, we reset the monopole interaction between the neutron $1f_{5/2}$ and the proton $1f_{7/2}$ orbits and that between the neutron $1f_{5/2}$ and the proton $1f_{5/2}$ orbits. These modifications correspond basically to the removal of the tensor-force monopole contributions, and are nothing but the suppression of the present effects of the quantum self-organization. The resulting Potential Energy Surface is shown in Fig. 3 for the axially symmetric deformation compared with that obtained from the original Hamiltonian. Around the spherical minimum the energy curves of the two calculations are similar, however, when going to stronger deformation values, the two approaches differ substantially. In particular, the prolate profound local minimum, seen in the original calculation (red solid line), is pushed up by about 4 MeV, if the quantum self-organization is suppressed as described above (blue dashed line). Thus, the quantum self-organization is a part of the crucial mechanisms producing the nuclear deformation.

At this point, we mention that the SPE being discussed corresponds somehow to the spherical terms in the Nilsson model [10] which are comprised of the $\ell\ell$ and ℓs terms as well as the harmonic-oscillator-quanta term. As their strengths are independent of the deformation, the present effect is not included in the Nilsson model.

Type-II shell evolution [20] has been discussed, for instance, in Co/Ni region [21–23], where neutrons are excited from the pf shell to $g_{9/2}$ across the N=40 sub-magic gap. The neutrons in $g_{9/2}$ and neutron holes in $f_{5/2}$ provide similar sizable monopole effects, as discussed above. A smaller $1f_{7/2}$-$1f_{5/2}$ spin-orbit splitting for protons reduces the resistance power against deformation, pulling down the prolate band as seen in Fig. 3. Type II shell evolution was introduced as the particle-hole excitation over a magic or sub-magic gap. Clearly, this kind of mechanism is a very simple and visible case of the quantum self-organization. On the other hand, the quantum self-organization can occur certainly in more complex ways. Such a complex way may be found in the shape transition of Sm isotopes (see Fig. 1), where no magic or sub-magic gap is involved. We can see the spherical-vibrational-rotational shape evolution in MCSM calculations, as will be reported in detail elsewhere.

Likewise, the shape coexistence in Hg/Pb isotopes have been studied. In those cases, the quantum self-organization gives intriguing contributions on the pattern of the shape coexistence, as reported also elsewhere.

6 Summary and Perspectives

We presented a novel mechanism on the relation between single-particle states and collective modes. A summary is given below.

- The atomic nuclei are not like simple rigid vases containing almost free nucleons interacting only weakly. This is in contrast to the naïve Fermi liquid picture.
- Nuclear forces are rich enough to change single-particle energies for each eigenstate, and can lead to the quantum self-organization.
- Single-particle energies can be self-organized, being enhanced by
  (i) two quantum liquids (*e.g.*, protons and neutrons)
  (ii) two major force components
  e.g., quadrupole interaction: to drive collective mode
  monopole interaction: to control resistance
- Type II shell evolution is a simple visible case involving excitations across (sub)magic gap.
- Actual cases such as shape coexistence, quantum phase transition, octupole vibration/deformation, super deformation, *etc.* can be studied with this scope.
- The quantum self-organization becomes more important in heavier nuclei where the number of active orbits and the number of active nucleons are larger. With larger numbers of them, the effects of the organization can be more significant. This feature may be linked to fission and superheavy elements. On the other hand, the quantum self-organization may not be so visible in light nuclei except for particular cases like intruder bands or cluster (or multiple particle-hole excited) states.

- Time-dependent version of quantum self-organization may be of another interest for reactions and fission.

The MCSM calculations were performed on the K computer at RIKEN AICS (hp140210, hp150224, hp160211,hp170230). This work was also supported in part by the HPCI Strategic Program (The origin of matter and the universe) and "Priority Issue on Post-K computer" (Elucidation of the Fundamental Laws and Evolution of the Universe) from MEXT and JICFuS.

References

[1] M. G. Mayer, Phys. Rev., **75**, 1969 (1949).
[2] O. Haxel,J. H. D. Jensen and H. E. Suess, Phys. Rev., **75**, 1766 (1949).
[3] M. G. Mayer and J. H. D. Jensen, Elementary Theory of Nuclear Shell Structure (Wiley, New York, 1955).
[4] de Shalit A. and I. Talmi, Nuclear Shell Theory (Dover,2004).
[5] K. L. G. Heyde, The Nuclear Shell Model (Springer,1990).
[6] E. Caurier, G. Martinez-Pinedo, F. Nowacki, A. Poves and A. P. Zuker, Rev. Mod. Phys., **77**, 427 (2005).
[7] J. Rainwater, Phys. Rev., **79**, 432 (1950).
[8] A. Bohr, Mat. Fys. Medd. Dan. Vid. Selsk., **26** (1952).
[9] A. Bohr and B. R. Mottelson, Mat. Fys. Medd. Dan. Vid. Selsk., **27** (1952).
[10] A. Bohr and B. R. Mottelson, Nuclear Structure, vol. 2 (Benjamin, New York, 1975).
[11] T. Schaefer, Nucl. Phys. A, **928**, 180 (2014).
[12] T. Togashi, Y. Tsunoda, T. Otsuka, and N. Shimizu, Phys. Rev. Lett., **117**, 172502 (2016).
[13] C. Kremer, S. Aslanidou, S. Bassauer, M. Hilcker, A. Krugmann, P. von Neumann-Cosel, T. Otsuka,N. Pietralla, V. Y. Ponomarev, N. Shimizu, et al., Phys. Rev. Lett., **117**, 172503 (2016).
[14] NNDC, Nudat 2, http://www.nndc.bnl.gov/nudat2/ (2016).
[15] T. Otsuka et al., Prog. Part. Nucl. Phys., **47**, 319 (2001).
[16] N. Shimizu et al., Prog. Theor. Exp. Phys., **2012**, 01A205 (2012).
[17] H. A. Jahn and E. Teller, Proc. R. Soc. A, **161**, 220 (1937).
[18] T. Otsuka, T. Suzuki, R. Fujimoto, H. Grawe, and Y. Akaishi, Phys. Rev. Lett., **95**, 232502 (2005).
[19] T. Otsuka, T. Suzuki, M. Honma, Y. Utsuno, N. Tsun-oda, K. Tsukiyama, and M. Hjorth-Jensen, Phys. Rev. Lett., **104**, 012501 (2010).
[20] T. Otsuka and Y. Tsunoda, J. Phys. G, **43**, 024009 (2016).
[21] Y. Tsunoda, T. Otsuka, N. Shimizu, M. Honma, and Y. Utsuno, Phys. Rev. C, **89**, 031301(R) (2014).
[22] A. Morales, G. Benzoni, et al., Phys. Lett. B, **765**, 328 (2017).
[23] S. Leoni, B. Fornal, et al., Phys. Rev. Lett., **118**, 162502 (2017).

Recent progress of the sign problem in the Onishi formula

Noritaka Shimizu[1,⋆], *Takahiro* Mizusaki[2], and *Makito* Oi[2]

[1] *Center for Nuclear Study, The University of Tokyo, 7-3-1 Hongo, Bunkyo-ku, Tokyo 113-0033, Japan*
[2] *Institute of Natural Sciences, Senshu University, 3-8-1 Kanda-Jinbocho, Chiyoda-ku, Tokyo 101-8425, Japan*

Abstract. The Hartree-Fock-Bogoliubov (HFB) method is one of the major methods to study a nuclear wave function. We discuss the sign problem in the Onishi formula, which emerges in calculating the overlap between two HFB wave functions. The origin of this problem is elucidated by expanding the overlap as a polynomial function. This article is based on Ref.[1].

1 Introduction

The Hartree-Fock-Bogoliubov (HFB) wave function is one of the most efficient and essential forms to describe a quantum many-body wave function which has strong pairing correlations. While the HFB wave function is usually defined as a quasi-particle vacuum, in the present work we begin with the following form of the HFB wave function:

$$|\phi^{(k)}\rangle = e^{\sum_{i<j} Z^{(k)}_{ij} c^\dagger_i c^\dagger_j}|-\rangle = e^{\hat{Z}^{(k)}}|-\rangle, \qquad \hat{Z}^{(k)} = \sum_{i<j} Z^{(k)}_{ij} c^\dagger_i c^\dagger_j, \tag{1}$$

where $c^\dagger_i$ and $|-\rangle$ are the creation operator of the single-particle state i and the vacuum, respectively. This wave function is parametrized by a skew-symmetric matrix $Z^{(k)}$.

The overlap of two HFB wave functions is obtained by the Onishi formula [2] as

$$\langle\phi^{(0)}|\phi^{(1)}\rangle = \sqrt{\mathrm{Det}(I - Z^{(0)*}Z^{(1)})}, \tag{2}$$

where I is the unit matrix. However, since a square root appears in Eq.(2), the sign of this equation is not determined. Hereafter we call this sign uncertainty "sign problem" in the Onishi formula. This problem can be serious when we perform angular-momentum projection.

2 Remedies for the sign problem

A lot of effort had been paid to solve the sign problem till now (e.g. [3]). Among them, in 1980's, two remedies were suggested. One is called "continuity method" [4]. In this method, the continuity of the overlap kernel as a function of the Euler angles is utilized to determine the sign. This procedure,

⋆e-mail: shimizu@cns.s.u-tokyo.ac.jp

however, becomes very complicated, especially in high-spin states. The other method was suggested by Neergård and Wüst. This method is based on the fact that the matrix $Z^{0*}Z^{1}$ has pair-wisely degenerate eigenvalues [5]. In this method, however, since this matrix has no particular symmetry such as hermiticity, its diagonalization might cause numerical instability. Thus, these two methods are useful in numerical calculations, while neither of them is a perfect solution.

In 2009, Robledo proposed a sophisticated formula utilizing Pfaffian [6] as

$$\langle \phi^{(0)} | \phi^{(1)} \rangle = (-1)^{N(N+1)/2} \mathrm{Pf} \begin{pmatrix} Z^{(1)} & -I \\ I & -Z^{(0)*} \end{pmatrix}, \tag{3}$$

where N is the number of the single-particle states, namely the dimension of $Z^{(k)}$. In Eq.(3), the overlap is provided by an $N/2$-th order polynomial of the matrix elements of $Z^{(1)}$ and $Z^{(0)*}$. Thus, this formula apparently does not have the sign problem and is simple without numerical instability.

By comparing Eqs.(2) and (3), one question can arise: why does the transcendental function (square root) appear in Eq.(2) while it can be represented in a polynomial function in Eq. (3)? Since the square root is also the origin of the sign problem, it is worth discussing the derivation of the Onishi formula to elucidate how the square root appears in the derivation.

3 Onishi formula and polynomial expansion

While the derivation of the Onishi formula is not shown in Ref.[2], we show the derivation utilizing the linked cluster theorem [7] in this section. Further details of the derivation can be found in Ref.[1].

According to the linked cluster theorem, the overlap is provided by the exponential function of the summation of the connected terms. Some lengthy calculations lead to the connected term given as

$$\ln \left(\langle \phi^{0} | \phi^{1} \rangle \right) = \langle \phi^{0} | \phi^{1} \rangle_{c} = \sum_{k=1}^{\infty} -\frac{1}{2k} \mathrm{Tr}(Y^{k}) \tag{4}$$

where $Y = Z^{(0)*}Z^{(1)}$ and the subscript c denotes the connected term. Note that the summation up to the infinite number of terms appears in Eq.(4) although only finite number of terms appear in Eq.(3). The Onishi formula is obtained by the linked-cluster theorem as

$$\langle \phi^{0} | \phi^{1} \rangle = \exp\left(\langle \phi^{0} | \phi^{1} \rangle_{c} \right) = \exp\left(-\frac{1}{2} \sum_{k=1}^{\infty} \frac{1}{k} \mathrm{Tr}(Y^{k}) \right) = \exp\left(\frac{1}{2} \mathrm{Tr} \ln(I - Y) \right) = \sqrt{\mathrm{Det}(I - Y)}. \tag{5}$$

The rightmost term is nothing but the Onishi formula. The coefficient $\frac{1}{2}$ in the middle of Eq.(5) brings about the square root of the Onishi formula. In the Neergård and Wüst method, the pairwise degeneracy of Y cancels this coefficient.

On the other hand, the overlap is obtained by expanding the exponential function as

$$\langle \phi^{0} | \phi^{1} \rangle = \langle - | \exp(\hat{Z}^{(0)\dagger}) \exp(\hat{Z}^{(1)}) | - \rangle = \sum_{k=0}^{N/2} \frac{1}{(k!)^{2}} \langle - | (\hat{Z}^{(0)\dagger})^{k} (\hat{Z}^{(1)})^{k} | - \rangle \tag{6}$$

where the Pauli exclusion principle terminates the summation concerning k at the $N/2$-th order. The $(N/2+1)$-th and the higher-order terms become zero thanks to the Pauli exclusion principle. It is consistent with Robledo's formula. More explicitly, each term of this polynomial function is shown in Fig.1.

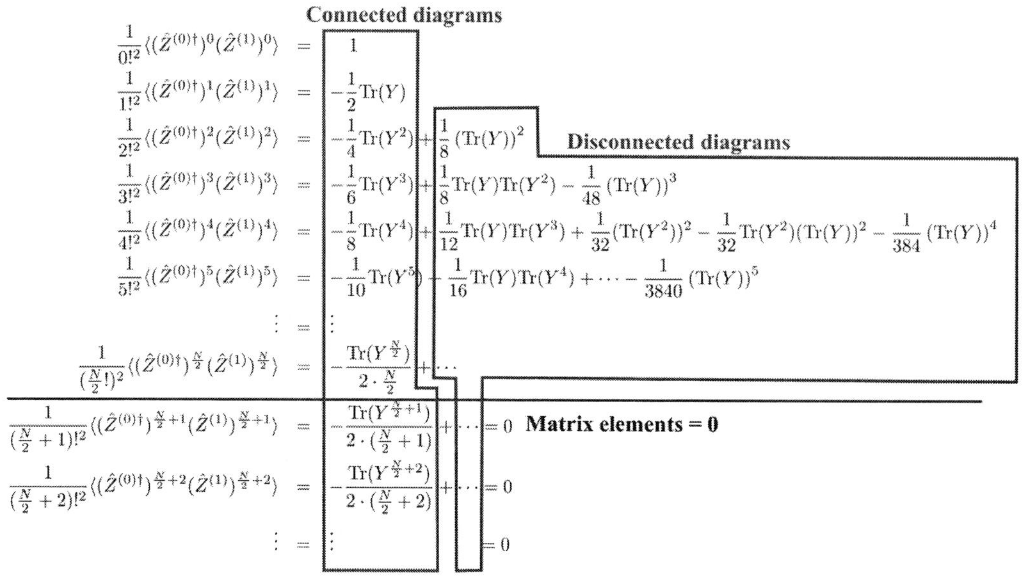

Figure 1. Polynomial expansion of the overlap of two HFB wave functions.

In each term, the corresponding connected term is enclosed by the red line. Even though the value of the higher terms in the left-hand-side equation is zero, the connected term is finite. In the Onishi formula, the connected terms enclosed by the red line is summed up infinitely. This causes the transcendental function, although the higher-order term becomes zero due to the cancellation of the connected term and the disconnected terms.

4 Summary

By comparing the Onishi formula to the polynomial expansion of the overlap of two HFB wave functions explicitly, we clarified that the sign problem is caused by the apparent infinite summation of the polynomial terms, which can be reduced to the finite summation in Robledo's formula. This consideration brought about a new idea of the efficient way of the computation of the HFB overlap, which will be published in near future [8].

NS acknowledges KAKENHI Grant 17K05433 and the priority issue 9 to be tackled by using Post-K computer.

References

[1] T. Mizusaki, M. Oi, and N. Shimizu, Phys. Lett. B **779**, 237 (2018).
[2] N. Onishi and S. Yoshida, Nucl. Phys. **80**, 367 (1966).
[3] B. Bally and T. Duguet, Phys. Rev. C **97**, 024304 (2018).
[4] K. Hara, Y. Hayashi, and P. Ring, Nucl. Phys. A **358**, 14 (1982).
[5] K. Neergård and E. Wüst, Nucl. Phys. A **402**, 311, (1983).
[6] L.M. Robledo, Phys. Rev. C **79**, 021302(R) (2009).
[7] C. Bloch, Nucl. Phys. **7**, 451 (1958).
[8] M. Oi, T. Mizusaki, N. Shimizu, and Y. Sun, in preparation.

Pushing the limits of excited-state g-factor measurements

Andrew E. Stuchbery[1,*], *Brendan P.* McCormick[1], *Timothy J.* Gray[1], and *Ben J.* Coombes[1]

[1]Department of Nuclear Physics, Research School of Physics and Engineering, The Australian National University, Canberra ACT 2601, Australia

Abstract. Current developments in excited-state g-factor measurements are discussed with an emphasis on cases where the experimental methodology is being extended into new regimes. The transient-field technique, the recoil in vacuum method, and moment measurements with $LaBr_3$ detectors are discussed.

1 Introduction

The magnetic moments of nuclear ground states provided important empirical evidence for the development of the nuclear shell model [1]. Today magnetic moment measurements on ground and excited nuclear states remain important observables to gain insights into nuclear structure – they are sensitive to the single-particle structure of the quantum state, give insight into how the nucleus carries its angular momentum, and can distinguish single-particle versus collective contributions to the wavefunction.

This paper reviews some current developments in excited-state g-factor measurements. The transient-field technique (sect. 2), the recoil in vacuum method (sect. 3), and moment measurements with $LaBr_3$ detectors (sect. 4) are discussed. As the gyromagnetic precession of the nucleus is the experimentally measured quantity, the following discussion generally refers to g factors rather than magnetic moments: $g = \mu/I$ where I is the nuclear spin and the moment μ is given in nuclear magnetons.

2 Transient field measurements

2.1 50 years of transient fields

An intense hyperfine magnetic field called the transient field (TF) acts on the nuclei of ions moving swiftly within a magnetized ferromagnetic medium. The discovery of the TF [2] will reach its 50th anniversary in 2018. While some g-factor measurements performed between 1968 and 1975 made use of the TF (e.g. [3]), it was not until after the 1975 discovery that the TF increases with the velocity of the moving ion [4] that the method became widely used [5]. It has continued in regular use ever since [6]. The TF method gives the sign of the g factor and is best suited for relative g-factor measurements on excited states with lifetimes in the picosecond range. The following subsection describes a contemporary measurement of $g(^{26}\mathrm{Mg}; 2^+)/g(^{24}\mathrm{Mg}; 2^+)$ by the high-velocity TF method [7].

*e-mail: andrew.stuchbery@anu.edu.au

2.2 High-velocity transient-field method: $N = 14$ subshell closure in ^{26}Mg

The $E(2_1^+)$ and $B(E2)$ systematics for $Z = 12$ indicate a subshell closure at $N = 14$, i.e. ^{26}Mg: as N increases from $^{22}_{12}\mathrm{Mg}_{10}$ the $E(2^+)$ value spikes at $N = 14$ and the $B(E2)$ value dips, indicative of a subshell closure. The expectation, then, is that the 2_1^+ state of ^{26}Mg should be dominated by proton excitations, giving $g(2_1^+) \sim +1$. Indeed, shell model calculations, using NuShellX [8] and the USDB interactions [9, 10], predict $g(2_1^+) = +0.959$. Surprisingly, the currently adopted value is $g(2_1^+) = +0.50(13)$ [11, 12], half the expected value.

The first $g(2_1^+, {}^{26}\mathrm{Mg})$ measurement by Eberhardt *et al.* in 1974, using the thick foil transient-field method in which the excited ^{26}Mg ions slowed and stopped in a magnetized iron host, found $g = +0.97(18)$ [3, 13]. Later, in 1981, Speidel *et al.* [12] argued that Eberhardt *et al.* had incorrectly accounted for the static-field contribution, which came into effect after the ions came to rest in the iron host. Speidel *et al.* made a new measurement using the thin-foil transient-field method, thereby excluding the static-field contribution, and obtained $g = +0.50(13)$, in agreement with Hartree-Fock calculations available at the time. This result, which implies near equal contributions from protons and neutrons, is currently listed as the adopted value in Nuclear Data Sheets [11]. As noted above, modern shell model calculations and single-particle arguments contend that the $N = 14$ subshell closure should result in $g(2_1^+)$ being much more heavily influenced by the proton contribution than the currently adopted measurement indicates. Both Eberhardt *et al.* and Speidel *et al.* used (α, α') reactions to excite and recoil ^{26}Mg ions into an iron host. The recoil velocity was relatively low, $v/c \sim 1\%$, and precession angles due to the transient field were very small, $\sim$1 mrad. These were challenging experiments.

A high-velocity transient-field measurement [14, 15] on beams of ^{24,26}Mg ions was performed. Beams of 120-MeV $^{24}\mathrm{Mg}^{8+}$ and $^{26}\mathrm{Mg}^{8+}$ were provided by the ANU Heavy Ion Accelerator Facility. The beams were Coulomb excited on a 9.9 mg/cm^2 natural gadolinium target, which also served as the ferromagnetic layer for the transient-field precession effect. Precession angles an order of magnitude larger than the earlier works [3, 12] were observed. Moreover, the same target was used with beam excitation to measure the ratio of 2^+_1-state g factors in ^{24}Mg and ^{26}Mg. Taking $g(2^+_1;^{24}\mathrm{Mg}) = +0.538(13)$ [16] gave $g(2^+_1;^{26}\mathrm{Mg}) = +0.86(10)$. The present g-factor measurement agrees with that of Eberhardt *et al.* [3], but with a reduced uncertainty. More details of the experiment and results are given in Ref. [7].

The $E(2^+_1)$, $B(E2)$ and $g(2^+_1)$ systematics of the even-A magnesium isotopes from ^{22}Mg to ^{32}Mg are shown in Fig. 1. The peak in $E(2^+_1)$ and the dip in $B(E2)$ value at ^{26}Mg together are indicative of the $\nu d_{5/2}$ subshell being filled. Shell-model calculations performed with NuShellX [8] and the USDB interaction [9, 10] are in agreement with most of the experimental data in Fig. 1. Of course, as the Island of Inversion is approached toward ^{32}Mg, the *sd*-shell model breaks down. More realistic predictions of g factors in the *sdpf* model space [17] are indicated in the lowest panel of Fig. 1. To measure the magnitudes of these g factors requires measurements on radioactive beams, for which the recoil in vacuum method has advantages over the TF method.

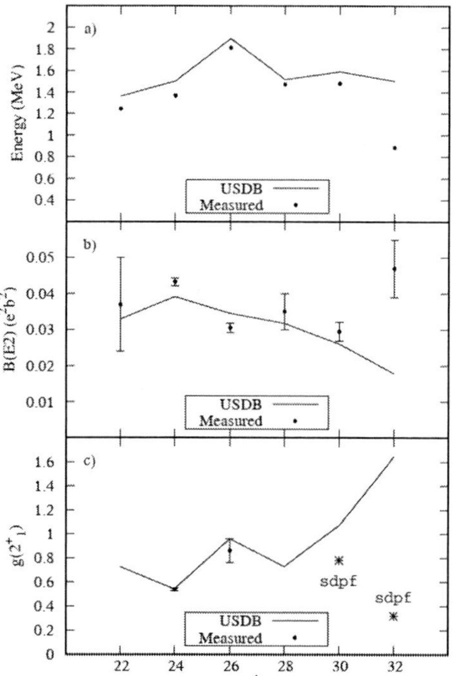

Figure 1. Comparison of USDB shell model calculations and experiment for the magnesium isotopes from $A = 22$ to 32 a) $E(2^+_1)$ energies, b) $B(E2)$ rates, and c) g factors. The theoretical g factors for ^{30}Mg and ^{32}Mg in a more realistic *sdpf* model space are shown as stars [17].

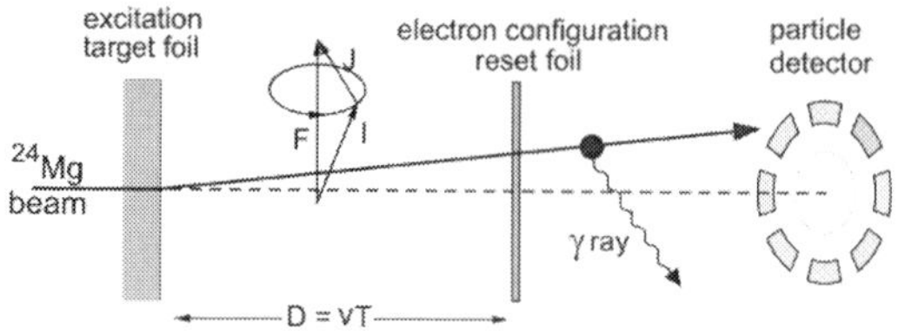

Figure 2. Sketch of TDRIV experiment. The 'stopper' of the traditional plunger technique is replaced by a thin foil that resets the electron configuration of H-like ions. The particle detector, with segmentation around the beam axis, is located downstream of the γ-ray detectors.

3 Recoil in Vacuum

In general, although the RIV method gives only the magnitude of the g factor, it has proven to give it more precisely than the transient-field method, particularly in the case of radioactive beam measurements where statistical precision is limited; compare Refs. [18, 19]. The primary reason is that the transient-field method requires γ-ray detection at a few specific angles in the plane perpendicular to the direction of the applied magnetic field whereas the RIV method can take advantage of γ-ray detection over a much broader angular range. A second reason, applicable for the time-dependent recoil in vacuum (TDRIV) method as applied recently to hydrogen-like ^{24}Mg 2^+_1 ions [16], is that the hyperfine interaction of the free ion in vacuum can be calculated from first principles with very high accuracy. The case of simple atomic systems such as H-like and Li-like ions will now be discussed, followed by some remarks on the RIV measurements on complex ions, which still have to be calibrated empirically.

3.1 RIV with H-like, Li-like and Na-like ions

The experimental method [20] used to measure $g(2^+_1)$ in ^{24}Mg [16] is illustrated in Fig. 2. Excited nuclei emerge from a target foil as ions carrying one electron. The nuclear spin $\boldsymbol{I}$ is aligned by the Coulomb-excitation reaction whereas the atomic spin $\boldsymbol{J}$ is oriented randomly. The hyperfine interaction couples the atomic spin to the nuclear spin, and together they precess about the total $\boldsymbol{F} = \boldsymbol{I} + \boldsymbol{J}$ with a frequency proportional to the nuclear g factor. Thus the orientation of the nuclear spin is periodically reduced and restored during the flight through vacuum. As a consequence, the angular intensity pattern of the γ rays emitted by the nuclei varies periodically, in step with the orientation of the nuclear spin. The nuclear precession frequency is determined by observing changes in the radiation pattern as the flight time is varied by changing the distance between the target and reset foils.

Experimental data obtained at the ALTO facility at IPN Orsay showing the time dependence of the radiation pattern are shown in Fig. 3 [16]. Results of the fits to these $R(T)$ data having strong, intermediate and weak amplitude oscillations give $g = 0.538(13)$, 0.539(24) and 0.54(3), respectively, where the uncertainties are statistical only.

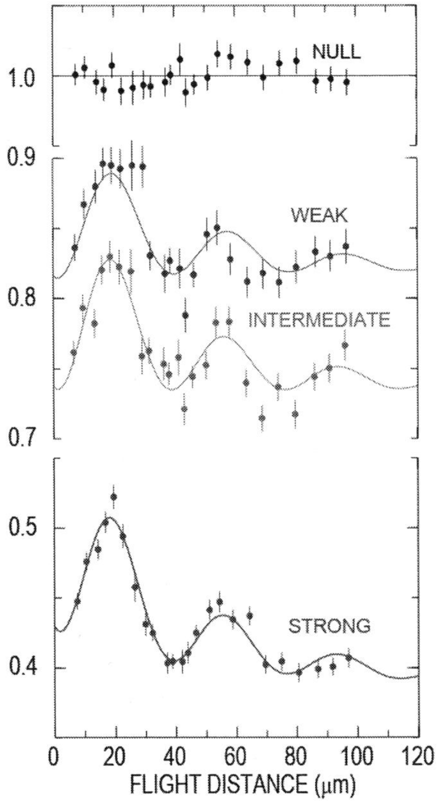

Figure 3. $R(T)$ ratio data showing the time-dependence of the radiation pattern and fits based on detailed modeling of the TDRIV experiment on ^{24}Mg [16]. The distance is the separation of target and reset foils (22.4 μm = 1 ps flight time). $R(T)$ is a ratio of γ-ray intensities observed at different angles versus the flight time T (or plunger separation). See [16] for details. The frequency of the oscillation determines the g factor whereas the amplitude affects the precision achieved. The rate of damping is determined by the nuclear level lifetime.

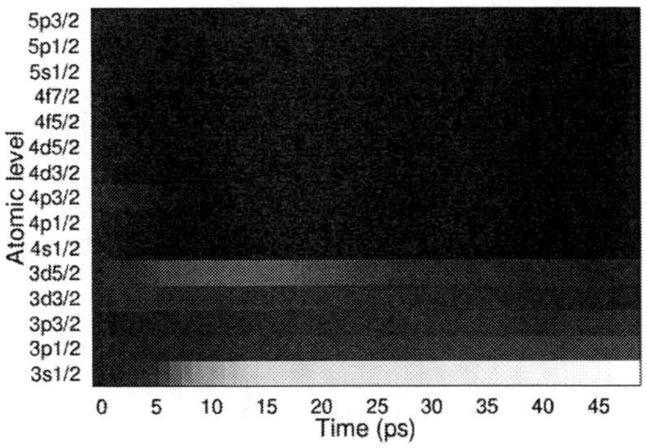

Figure 4. Evolution of the population of atomic states for Na-like Fe assuming an initial Gaussian distribution centred around 200 eV excitation energy with σ = 100 eV. Black represents least (no) population and white most.

The weighted average is $g = 0.538(11)$ (statistical error). Even the weakest amplitude oscillations yield a g factor with a statistical uncertainty of better than 6%. These data demonstrate that if the precession frequency can be observed in a TDRIV measurement, then the statistical precision is likely to exceed that of a transient-field g-factor measurement on the same state.

Unfortunately the TDRIV method can be applied to H-like ions only up to $Z \sim 20$ because the hyperfine frequency scales with Z^3 and the period becomes too short to measure; for $g = 0.5$, the period of the oscillations is 3.1 ps at $Z = 10$ and 0.38 ps at $Z = 20$.

To apply this type of TDRIV method to higher-Z nuclei requires the use of the weaker fields of shielded electrons, such as Li-like ions (2s electron) or Na-like ions (3s electron). The use of the method with Li-like or Na-like ions is under investigation. There are indications from measurements of charge-state distributions and integral recoil in vacuum measurements on relatively low-Z ions that atomic ground states and low-excitation atomic states dominate the free-ion interactions [21]. However a Na-like ion, for example, has many more excited states than an H-like ion, with the potential to wash out the unique frequency of the atomic ground-state configuration. Along with the $^2S_{1/2}$ ground state the low-excitation $^2P_{1/2}$ and $^2P_{3/2}$ atomic states are likely to contribute. Even so, Lin et al. [22] have reported evidence of population of Na-like ^{41}Ca ions in the $^2S_{1/2}$ state via the observation of a spin precession at the expected frequency in transverse decoupling experiments on the $15/2^+$ level with $\tau = 4.7$ ns.

We have also begun to perform calculations with the GRASP2K atomic structure codes [23] to explore how the population of atomic states might evolve as ions recoil in vacuum. The time evolution of excited atomic states of Na-like Fe assuming an initial Gaussian distribution of excited states is shown in Fig. 4. Note that there is a prominent decay to the lower-excited states and the ground state on the time-scale of several picoseconds. This calculation is schematic; at present it is unclear what initial population distribution is appropriate. However it does tend to confirm the empirical observations that atomic ground states and low-excitation atomic states dominate the free-ion interactions on the time scale of several picoseconds.

Attempts are underway to apply the TDRIV method to Na-like ions of fp shell nuclei. Whether oscillations associated with the Na-like $^2S_{1/2}$ atomic ground state are observed or not, important information will be obtained to understand the free-ion hyperfine fields and develop their application to future g-factor measurements. If an identifiable oscillation frequency is clearly observed, it may provide an independent and reliable measure of the absolute g factors in the fp shell. Such measurements are very important because transient-field strengths in the fp shell are somewhat uncertain due to a dearth of independently determined g factors that can serve to calibrate the transient-field strength - see Ref. [24].

3.2 RIV with complex ions

The time integral RIV technique on complex ions with ~ 20 - 30 electrons has proved to be a powerful method to measure the g factors of excited states of neutron-rich nuclei produced as radioactive beams, particularly in the Sn and Te isotopes near the neutron-rich doubly magic nuclide ^{132}Sn [25–31]. One of the method's advantages is that the g factor of the 2_1^+ state can be measured simultaneously with $B(E2; 0^+ \rightarrow 2^+)$ and $Q(2^+)$ [27–29, 32]. Although the hyperfine interaction must be calibrated empirically, the RIV method has proven to give the magnitude of the g factor more precisely [25, 26, 28] than the transient-field method [19, 33] A detailed description of calibration and analysis procedures has been included in the report on the case of ^{136}Te [31].

We are pushing the limits by developing this methodology for simultaneous measurements of $B(E2)$, $Q(2^+)$ and $g(2^+)$ in other regions of the nuclear chart. In particular, applications to few-nucleon 2_1^+ states around ^{208}Pb are of considerable interest. The shell structure in the neutron-rich ^{132}Sn region can be compared with that in the vicinity of stable ^{208}Pb [34, 35]. While the high-spin structure has been quite thoroughly studied experimentally around ^{208}Pb, the electromagnetic properties of low-excitation, low-spin states associated with a few pairs of valence nucleons outside ^{208}Pb have not. Thus direct comparisons of the related few-particle states around ^{132}Sn and ^{208}Pb are currently limited by the lack of experimental data on electromagnetic properties near ^{208}Pb rather than near ^{132}Sn. Measurements of the type described here on ^{210}Pb, ^{210}Po and ^{212}Po would enable comparison with their equivalents, ^{134}Sn, ^{134}Te and ^{136}Te. At present little is known about the strengths of the relevant free-ion hyperfine interactions near $Z = 82$; it is important to determine the strength of the RIV interaction for ions with $v/c \sim 0.08$ and ~ 30 - 40 electrons recoiling in vacuum. The effective fields are expected to be much stronger than in the ^{132}Sn region, so the g-factor measurements in the vicinity of ^{208}Pb will need to control the interaction time of the nuclear moment with the electronic configuration by use of a plunger.

4 Applications of $LaBr_3$ detectors

The development of Lanthanum Bromide ($LaBr_3$) detectors, which have excellent time resolution (typically ~300 ps) and energy resolution much superior to scintillators such as NaI and BaF_2, provides an opportunity to perform perturbed angular distribution g-factor measurements under new experimental conditions. We have investigated the application of $LaBr_3$ detectors in beam to measure g factors of nuclear states with nanosecond lifetimes using static hyperfine magnetic fields in magnetic hosts.

A preliminary experiment on ^{54}Fe implanted into nickel was performed. The target consisted of 0.76 mg/cm^2 of ^{45}Sc evaporated onto a nickel foil of thickness 2.44 mg/cm^2 that had previously been annealed at

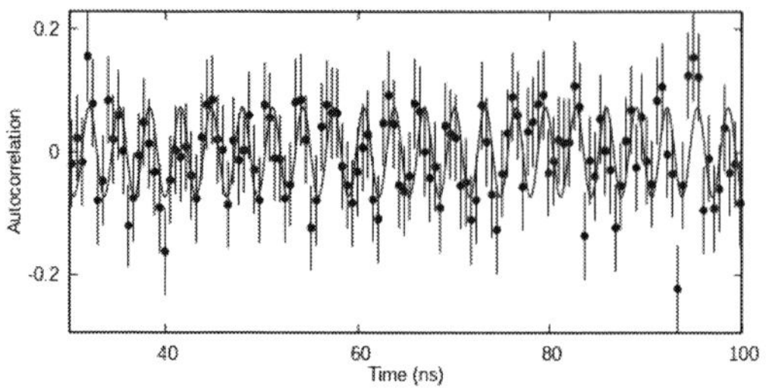

Figure 5. Autocorrelation function for the 3432-keV transition below the 10^+ isomer of ^{54}Fe after implantation into nickel and showing the expected 3.3 ns oscillation period. The autocorrelation procedure was applied to the time-dependent perturbed angular distribution data, as described in Ref. [36], to make the oscillations visually more apparent.

~ 790°C. The 10^+ isomer with $\tau = 525(10)$ ns and $g = +0.728(1)$ was populated by the ^{45}Sc(^{12}C, p2n)^{54}Fe reaction with 40-MeV ^{12}C beams from the ANU 14UD Pelletron. For the known g factor and hyperfine field $B_{hf} = -27.00(3)$ T, the expected precession period is ~ 3 ns. This period approached the limit of the experimental setup because the time width of the beam-pulse was about 2 ns. The expected frequency was observed in the measured autocorrelation function for the 3432 transition as shown in Fig. 5. These results demonstrate that cases where the precession period is of the order of a few nanoseconds are accessible for in-beam measurement.

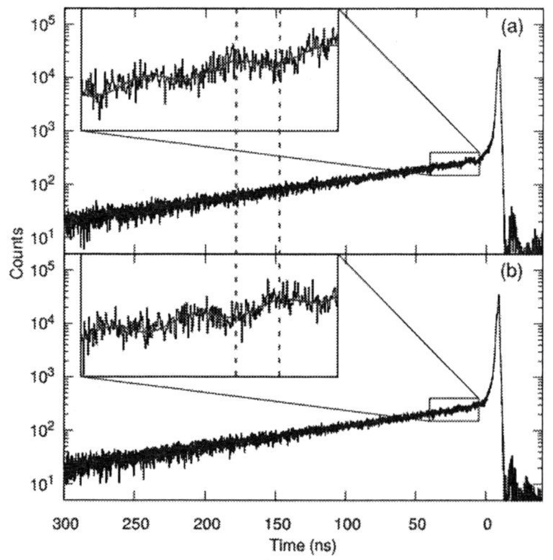

Figure 6. Time spectra for the 640-keV transition depopulating the $I^\pi = \frac{11}{2}^-$ isomer in ^{107}Cd. Time as displayed starts with the beam pulse and stops when a γ-ray is detected. Fits to guide the eye show the oscillations in the two spectra out of phase. The prompt peak is due to a prompt 632-keV transition in ^{106}Cd that cannot be resolved from the 640-keV line by the $LaBr_3$ detectors.

The hyperfine field of Cd implanted into gadolinium was then investigated. The motivation was to reassess the g-factor measurement on the $I^{\pi} = 10^{+}$ state in ^{110}Cd which reported $g(10^{+}) = -0.09(3)$, at least a factor of two smaller than $g \approx -0.2$ to -0.3 that would be expected for a rather pure $(h_{\frac{11}{2}})^2$ neutron configuration [37].

The hyperfine fields at ^{107}Cd ions implanted into gadolinium following the ^{98}Mo(^{12}C, $3n$)^{107}Cd reaction were measured under similar conditions to the ^{110}Cd g-factor measurement. Examples of time spectra for the 640-keV transition depopulating the $11/2^{-}$ isomer in ^{107}Cd are shown in Fig. 6. The oscillations show close to the expected frequency, but their amplitude indicates that the fraction of nuclei on field-free sites is significant. The consequence is that the effective hyperfine field for the integral perturbed angular distribution measurement of $g(10^{+})$ in ^{110}Cd was much smaller than adopted in Ref. [37]. With a corrected effective hyperfine field a re-analysis of the data from Ref. [37] gives $g(10^{+}) = -0.29(16)$, consistent with that of the expected seniority-two $\nu h_{11/2}$ configuration. A full account of this work has been published [38].

5 Conclusions

Some current developments in excited-state g-factor measurements have been described. The transient-field method continues to give important results a half century after the transient field itself was discovered. The recoil in vacuum method has advanced greatly over the past decade, yet it has much potential for further development, particularly through time-dependent measurements. Moment measurements with $LaBr_3$ detectors and fast timing have opened up new opportunities for in-beam g-factor measurements on states that live a few nanoseconds.

Acknowledgements

We thank our many colleagues who contributed to the work discussed here. This research was supported in part by the Australian Research Council grant numbers DP140102986, DP140103317 and DP70101673. B.P.M. T.J.G. and B.J.C. acknowledge the support of the Australian Government Research Training Program. Support for the Heavy Ion Accelerator Facility operations through the Australian National Collaborative Research Infrastructure Strategy (NCRIS) program is acknowledged.

References

[1] M.G. Mayer, Phys. Rev. **78**, 16 (1950)
[2] R.R. Borchers *et al.*, Phys. Rev. Lett. **20**, 424 (1968)
[3] J.L. Eberhardt, R.E. Horstman, H.W. Heeman, G.V. Middelkoop, Nuclear Physics A **229**, 162 (1974)
[4] J. Eberhardt, G.V. Middelkoop, R. Horstman, H. Doubt, Phys. Lett. B **56**, 329 (1975)
[5] N. Benczer-Koller, M. Hass, J. Sak, Annual Review of Nuclear and Particle Science **30**, 53 (1980)
[6] N. Benczer-Koller, G.J. Kumbartzki, J. Phys. G **34**, R321 (2007)
[7] B.P. McCormick *et al.*, Phys. Lett. B **779C**, 445 (2018)
[8] B.A. Brown, W.D.M. Rae, Nuclear Data Sheets **120**, 115 (2014)
[9] B.A. Brown, W.A. Richter, Phys. Rev. C **74**, 034315 (2006)
[10] W.A. Richter, S. Mkhize, B.A. Brown, Phys. Rev. C **78**, 064302 (2008)
[11] M. Basunia, A. Hurst, Nuclear Data Sheets **134**, 1 (2016)
[12] K.H. Speidel *et al.*, Physics Letters B **102**, 6 (1981)
[13] P.C. Zalm *et al.*, Hyperfine Interact. **5**, 347 (1977)
[14] A.E. Stuchbery *et al.*, Phys. Lett. B **611**, 81 (2005)
[15] E. Fiori *et al.*, Phys. Rev. C **85**, 034334 (2012)
[16] A. Kusoglu *et al.*, Phys. Rev. Lett. **114**, 062501 (2015)
[17] T. Otsuka, M. Honma, T. Mizusaki, N. Shimizu, Y. Utsuno, Prog. Part. Nucl. Phys. **47**, 319 (2001)
[18] J.M. Allmond *et al.*, Phys. Rev. C **87**, 054325 (2013)
[19] G.J. Kumbartzki *et al.*, Phys. Rev. C **86**, 034319 (2012)
[20] A.E. Stuchbery, P.F. Mantica, A.N. Wilson, Phys. Rev. C **71**, 047302 (2005)
[21] A.E. Stuchbery, Hyperfine Interact. **220**, 29 (2013)
[22] Y.C. Lin, W.A. Little, G.D. Sprouse, P.D. Bond, Hyperfine Interact. **4**, 105 (1978)
[23] P. Jönsson, G. Gaigalas, J. Bieroń, C.F. Fischer, I. Grant, Comp. Phys. Comm. **184**, 2197 (2013)
[24] M.C. East *et al.*, Phys. Rev. C **79**, 024303 (2009)
[25] N.J. Stone *et al.*, Phys. Rev. Lett. **94**, 192501 (2005)
[26] A.E. Stuchbery, N.J. Stone, Phys. Rev. C **76**, 034307 (2007)
[27] A.E. Stuchbery *et al.*, Phys. Rev. C **88**, 051304 (2013)
[28] J.M. Allmond *et al.*, Phys. Rev. C **87**, 054325 (2013)
[29] J.M. Allmond *et al.*, Phys. Rev. C **92**, 041303 (2015)
[30] J.M. Allmond *et al.*, Phys. Rev. Lett. **118**, 092503 (2017)
[31] A.E. Stuchbery *et al.*, Phys. Rev. C **96**, 014321 (2017)
[32] J.M. Allmond *et al.*, Phys. Rev. C **84**, 061303 (2011)
[33] N. Benczer-Koller *et al.*, Phys. Lett. B **664**, 241 (2008)
[34] L. Coraggio, A. Covello, A. Gargano, N. Itaco, Phys. Rev. C **80**, 021305 (2009)
[35] A. Gargano, L. Coraggio, A. Covello, N. Itaco, Journal of Physics: Conf. Ser. **168**, 012013 (2009)
[36] G. Georgiev *et al.*, J. Phys. G **28**, 2993 (2002)
[37] P. Regan, A. Stuchbery, S. Anderssen, Nucl. Phys. A **591**, 533 (1995)
[38] T.J. Gray *et al.*, Phys. Rev. C **96**, 054332 (2017)

g-factor calculations from the generalized seniority approach

Bhoomika Maheshwari[1,*], and *Ashok Kumar* Jain[2]

[1]Department of Physics, Banasthali Vidyapith, 304022 Banasthali, India
[2]Department of Physics, Indian Institute of Technology Roorkee, 247667 Roorkee, India

Abstract. The generalized seniority approach proposed by us to understand the B(E1)/B(E2)/B(E3) properties of semi-magic nuclei has been widely successful in the explanation of the same and has led to an expansion in the scope of seniority isomers. In the present paper, we apply the generalized seniority scheme to understand the behavior of g-factors in semi-magic nuclei. We find that the magnetic moment and the g-factors do show a particle number independent behavior as expected and the understanding is consistent with the explanation of transition probabilities.

1 Introduction

Recent experimental advances towards the neutron drip line provide us opportunities to test the known theoretical ideas and understand the systematics and trends in newer contexts. We have recently developed a simple generalized seniority scheme for multi-j configurations. A new kind of seniority isomers and their related seniority selection rules have been established for the first time[1]. This generalized seniority scheme is quite successful in explaining the reduced transition probabilities and the corresponding half-lives of the semi-magic isomers in different mass regions[1-5]. However, some contribution from core-excitation across Z=50 shell gap is not ruled out [6].

The magnetic dipole moments (g-factors) of a nuclear state depend on the orbital and spin angular momentum contributions of protons and neutrons involved. So the configuration mixing and the single particle structure near the Fermi surface affect the g-factor significantly. In the present work, we extend the generalized seniority calculations to obtain the g-factors, particularly in the neutron-rich Sn isotopes.

Several theoretical attempts have predicted the g-factors (magnetic moments) of the first excited 2^+ states in Sn isotopes [7-10]. However, the experimental information still remains incomplete towards ^{100}Sn and ^{132}Sn, the two extreme doubly magic nuclei. The g-factor of the first excited 2^+ states in the stable and even-even Sn isotopes were measured by Hass et al. [11]; they reported a positive and almost zero value of the g-factor in the case of ^{118}Sn and negative values for the neighboring nuclei ^{116,120}Sn on either side. In this paper, we present the g-factor calculations for the first excited 2^+ states in Sn isotopes by using our generalized seniority approach.

Besides this, we also discuss the g-factor calculations for the $11/2^-$ states and the 10^+ high spin isomers in Sn isotopes. The calculated results are in line with the experimental data. Predictions for the $27/2^-$ isomers have also been made.

2 Theoretical Framework

The magnetic moment of identical nucleons in pure-j^n configuration is given by

$$\mu = \sum_{i=1}^{n} g j_i = g \sum_{i=1}^{n} j_i = gJ \qquad (1)$$

It is well known that the magnetic moments and g-factors show a particle number independent behavior for identical nucleons in single-j shell of a pure-seniority scheme [12]. Hence, g-factors of all the states arising from a pure j^n configuration of identical nucleons should be equal to the g-factor of a single-j nucleon in the j^n configuration.

As per our previous extension of single-j seniority scheme to the multi-j generalized seniority scheme [1], we can write themagnetic moment of identical nucleons in the mixed configuration $\tilde{j}^n$as follows

$$\mu = g \sum_{i=1}^{n} \tilde{j}_i = gJ \qquad (2)$$

where $\tilde{j} = j_1 \otimes j_2 \otimes \ldots$ is the multi-j, mixed configuration. This means that the g-factors of a mixed configuration also show a particle number independent behavior in the generalized seniority scheme. Hence, g-factors of all the states arising from a given mixed configuration having identical nucleons must be equal to the g-factor of a single nucleon arising from the same mixed configuration.

3 Calculations and results

We present in Fig. 1, the experimental and the calculated g-factor trends for the first excited 2^+ states in Sn

*Corresponding author: bhoomika.physics@gmail.com

isotopes. Wefit one of the experimental data of magnetic moments and calculate the g-factor using the formula in Equation (2) for the full chain by using the two mixed configurations $\tilde{j} = g_{7/2} \otimes d_{5/2} \otimes s_{1/2} \otimes d_{3/2}$, and $\tilde{j} = d_{5/2} \otimes s_{1/2} \otimes d_{3/2} \otimes h_{11/2}$, respectively (before and after the middle of the shell). The calculated g-factor results and configurations are consistent withour previous results on two B(E2) parabolas[2]. All the experimental data have been taken from Stone's table [13].Two parabolas in B(E2) curve highlight the difference in the configuration mixing, before and after the middle for the generation of the 2^+ states in Sn isotopes, in contrast to a single peak at ^{116}Sn expected from the pure seniority scheme[2], and references therein, which is line with the interpretations of Irving et al. [14]. Therefore, g-factors are found to be of different signs before and after the middle. The positive g-factor can be attributed to the $g_{7/2}$ before the middle, whereas the negative value can be understood as the dominance of $h_{11/2}$ orbital after the middle.

We then present in Fig. 2 and 3, the g-factor calculations for the $11/2^-$ states and 10^+ isomers in Sn isotopes, respectively. Note that the $11/2^-$ states correspond to the unique-parity $h_{11/2}$ orbital of the 50-82 neutron valence space. On the other hand, we use the configuration mixing $\tilde{j} = s_{1/2} \otimes d_{3/2} \otimes h_{11/2}$, as reported in our previous papers [1, 3], for the g-factor calculations of 10^+ isomers in Sn isotopes. We get the expected trend (particle number independent), which matches with the experimental data well.

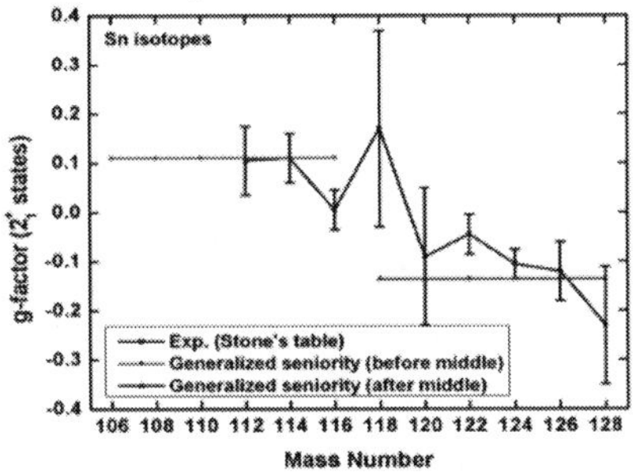

Fig. 1. Experimental [12] and calculated g-factor trend for the first excited 2^+ states in Sn isotopes.

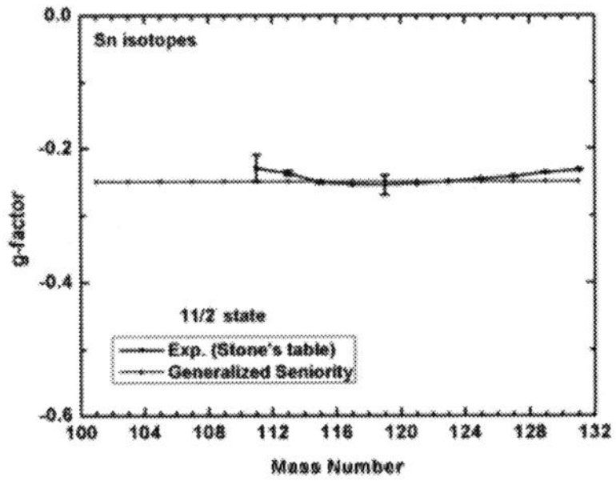

Fig. 2. Experimental [12] and calculated g-factor trend for the $11/2^-$states in Sn isotopes.

However, the data on $27/2^-$ isomers is not knowntill date. We expect the g-factor of this state to be of the same order as the 10^+ isomers, since both follow the same configuration mixing [8].

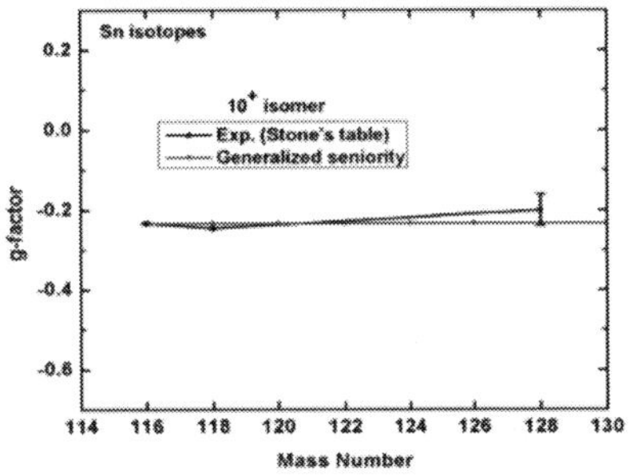

Fig. 3. Experimental [12] and calculated g-factor trend for the 10^+ isomers in Sn isotopes.

4 Conclusion

In the present work, we present the g-factor calculations for the first excited 2^+ states, $11/2^-$ states and 10^+ isomers in Sn isotopes by extending the usage of our scheme. We find that the states show a particle number independent trend, as per the seniority and generalized seniority scheme. The calculated results explain the experimental data quite well.Prediction for the g-factors of the $27/2^-$ isomers in Sn isotopes has been made.

References

1. B. Maheshwari, and A. K. Jain, Phys. Lett. B **753**, 122 (2016)
2. B. Maheshwari, A. K. Jain and B. Singh, Nucl. Phys. A **952**, 62 (2016)
3. A. K. Jain, and B. Maheshwari, Nuclear Phys. Review **34**, 73 (2017)
4. A. K. Jain, and B. Maheshwari, Physica Scripta **92**, 074004 (2017)
5. B. Maheshwari, S. Garg, and A. K. Jain, Pramana-Journal of Physics **89**, 75 (2017)
6. T. Otsuka, Private. Comm.
7. J. Terasaki et al., Phys. Rev. C **66**, 054313 (2002)
8. B. A. Brown et al., Phys. Rev. C **71**, 044317 (2005)
9. A. Ansari, and P. Ring, Phys. Lett. B **649**, 128 (2007)
10. L. Y. Jia, H. Zhang, and Y. M. Zhao, Phys. Rev. C **75**, 034307 (2007)
11. M. Hass et al., Phys. Rev. C **22**, 97 (1980)
12. I. Talmi, Simple Models of Complex Nuclei, Harwood Academic (1993)
13. N.J. Stone, www-nds.iaea.org/publications, INDC(NDS)-0658, Feb. (2014)
14. I.O. Morales, P. Van Isacker, I. Talmi, Phys. Lett. B **703**, 606 (2011)

EPJ Web of Conferences **178**, 02007 (2018) https://doi.org/10.1051/epjconf/201817802007
CGS16

Shell Evolution and $E2$ Collectivity: New Spectroscopic Information

N. Pietralla[1,*], *C.* Kremer[1], *T.* Beck[1], *W.* Witt[1], *U.* Gayer[1], *P. von* Neumann-Cosel[1], and *V.* Werner[1]

[1]Institut für Kernphysik, Technische Universität Darmstadt, Darmstadt, Germany

Abstract. New spectroscopic information on electric quadrupole collectivity has recently been obtained. We report on our identification of a coexisting deformed structure in the quasi-doubly closed-shell nucleus ^{96}Zr from our measurement of a corresponding absolute $E2$ intraband transition strength in electron scattering reactions and its small mixing matrix element with the spherical ground state structure. The even-mass zirconium isotopes exhibit a first order quantum shape phase transition between neutron numbers 58 and 60. We have used photon scattering reactions for a first measurement of the $E2$ decay strength of the nuclear 1^+ scissors mode. Evidence for its 2^+ rotational excitation and its extraordinarily large rotational moment of inertia are presented.

1 Introduction

Nuclear collective motion and evolution of nuclear structure with the numbers, Z and N, of constituting nucleons have traditionally been at the core of the discussions at the International Symposia on Capture Gamma-Ray Spectroscopy and Related Topics. While our understanding has progressed considerably over the years, we are still witnessing the unveiling of hitherto unknown phenomena or the identification of new mechanisms that govern the evolution of nuclear structure.

One example is the occurrence of rapidly changing nuclear shapes as a function of nucleon number in a sequence of nuclides in the nuclear chart. While these so-called shape phase transitions have been known for a long time, high-quality data are needed in order to clarify the questions whether a particular shape phase transition exists due to a gradual evolution of one minimum in the nuclear potential energy surface to another, or whether the energies of separate minima exchange their order. The former corresponds to a situation where the nuclear potential energy as a function of deformation parameters evolves through a rather flat potential with large deformation fluctuations in the nuclear wave function at the transitional point [1]. The latter corresponds to a scenario where near the shape phase transitional point nuclear states with distinct deformation coexist at low excitation energies, a phenomenon called shape coexistence [2].

The evolution of nuclear single-particle energies due to the varying interaction between protons and neutrons has been identified [3] as a microscopic cause for the change in shell structure and for the occurrence of nuclear deformation along a sequence of nuclides. Recently, a quantitative mechanism has been formulated which relates the evolution of ground

*e-mail: pietralla@ikp.tu-darmstadt.de

state deformation as a function of particle numbers to the structure of excited states in a single nucleus. This, so-called, *Type II Shell Evolution* (T2SE) has been held responsible for the occurrence of shape coexistence [4] in particular quasi-doubly magic nuclei such as ^{68}Ni or ^{96}Zr. The validation of the mechanism of T2SE based for the first time on the absolute measurement of the amount of quadrupole collectivity in both coexisting structures from our recent electron scattering experiment on ^{96}Zr [5] will be discussed. The phenomenon of T2SE is closely related to the proton-neutron interaction in the valence shells.

Another nuclear mode which is intimately linked to the proton-neutron interaction in the nuclear valence shell in an already deformed nucleus is the *Scissors Mode* [6]. While the scissors mode occurs due to nuclear quadrupole deformation, its $E2$ decay strength has still been unknown. We will report on our recent measurement [7] of the absolute $E2$ transition rate of the scissors mode into the ground state band using photon scattering reactions.

2 Shape coexistence in ^{96}Zr inferred from electron scattering

We report here on the first evidence for shape coexistence caused by T2SE which is firmly based on the measurement of absolute $E2$ transition rates. The data [5] have been obtained in high-resolution inelastic electron scattering spectroscopy of the nucleus ^{96}Zr at the Superconducting Darmstadt Linear electron Accelerator (S-DALINAC).

Figure 1 shows a spectrum of electrons scattered off a self-supporting metal target enriched to 57.4% in the isotope ^{96}Zr which otherwise has a natural abundance of only 2.8%. The target also contained other isotopes of zirconium including 9.2% of the isotope ^{90}Zr. The spectrum around 2.2 MeV contains a signal from the 2_2^+ state of ^{96}Zr at an excitation energy of 2226 keV and a signal from the 2_1^+ state of ^{90}Zr at 2186 keV with a known excitation strength of $B(E2; 0_1^+ \to 2_1^+) = 660\ e^2\text{fm}^4$. Data were taken at quadratic momentum transfers $q^2 \in [0.09, 0.35]\ \text{fm}^{-2}$. An analysis of the peak areas for the excitation of the 2_2^+ state of ^{96}Zr was done relative to the peak areas of other 2^+ excitations with previously known $E2$ excitation strengths, such as the 2_1^+ states of ^{90}Zr and ^{96}Zr.

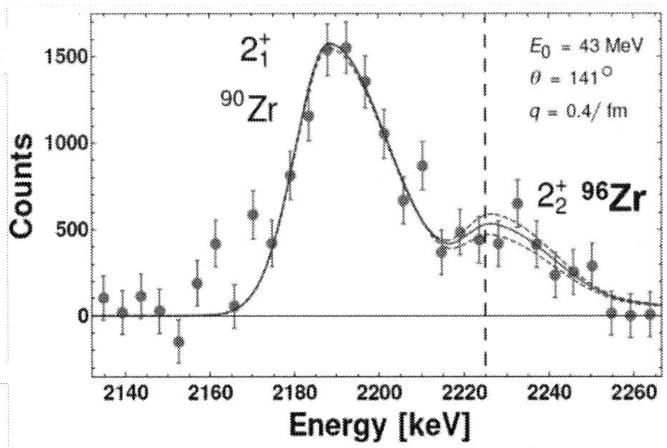

Figure 1. Region of interest of an energy-loss spectrum of electrons scattered off a metal zirconium target at the S-DALINAC. The incident electron energy amounts to 43 MeV. The polar scattering angle of the electrons is 141°. The scattering reaction yields a momentum transfer of 0.4 fm^{-1}. Spectra were taken with a total energy resolution of about 25 keV. Adapted from Ref. [5].

This resulted in the assignment of the $E2$ transition strength from the 2_2^+ state of ^{96}Zr to its ground state with a value of $B(E2; 2_2^+ \to 0_1^+) = 7(2)\ e^2\text{fm}^4 = 0.26^{+9}_{-7}$ W.u. [5]. This $E2$ strength is an order of magnitude smaller than the already small value of 2.3 W.u. for the $2_1^+ \to 0_1^+$ transition [8], and it is 140 times smaller than the $E2$ decay strength of the 2_2^+ state to the first excited 0_2^+ state. The latter has a collective character with a transition strength of $B(E2; 2_2^+ \to 0_2^+) = 36$ W.u. as can be inferred from the excitation cross section measured here in combination with previously known γ-decay branching ratios in ^{96}Zr [8]. The absolute transition rates clearly indicate the 0_1^+ and 2_1^+ states of ^{96}Zr belonging to a non-collective spherical structure and the 0_2^+ and 2_2^+ states forming a deformed rotational sequence of levels. The tiny $E2$ rate between them indicates very small mixing of these structures coexisting at low excitation energies in the level scheme of ^{96}Zr with considerably different shapes.

3 Type II Shell Evolution in ^{96}Zr

Large-scale shell model calculations performed by the Tokyo group [9] are able to reproduce the experimental findings very well. The occurrence of coexisting weakly collective and deformed states is understood in terms of T2SE. The ground state of ^{96}Zr reflects the quasi-doubly magic character at proton number $Z = 40$ and neutron number $N = 56$ with considerable energy gaps between the occupied pf-shell and the $1g_{9/2}$ intruder orbital for protons and between the occupied $n(2d_{5/2})$ orbital for neutrons and the upper part of the 50-82 neutron major shell. The 2_1^+ state's wave function is dominated by a $n(2d_{5/2} \to 3s_{1/2})$ single neutron excitation.

The wave functions of the deformed 0_2^+ and 2_2^+ states are characterized by a considerable population of the $p(1g_{9/2})$ proton orbital. This results in a vanishing of the $N = 56$ sub-shell gap because of the increased binding of the $n(2d_{3/2}, 1g_{7/2})$ neutron orbitals with $j = l - 1/2$ due to the monopole part of the nucleon-nucleon tensor interaction as well as of the $n(1h_{11/2})$ high-l orbital. Consequently, more neutrons are occupying the $j = l - 1/2$ neutron orbitals that induce an increased binding of the $p(1g_{9/2})$ $j = l + 1/2$ proton orbital and hence a reduction of the $Z = 40$ sub-shell gap for protons in turn facilitating its occupation. This mechanism is self-reinforcing and, thus, robustly results in a nuclear wave function which differs significantly from that of the closed-shell ground state. It is characterized by a more even spread of neutrons and protons over the $Z = 28 - 50$ proton orbitals and the $N = 50 - 82$ neutron orbitals indicative of nuclear deformation. The nucleus ^{96}Zr features quantum states that either do populate significantly the $p(1g_{9/2})$ proton orbital and are deformed or that do not and are spherical [5].

The barrier between these configurations is high and prevents significant mixing between them. A simple two-structure mixing scheme provides a quantitative estimate for the rather small mixing matrix element, $V_{\rm mix} = 76$ keV, between the spherical and deformed configurations [5]. They form the 0_1^+ ground state and the first excited 0_2^+ state, respectively, with a mixture between them of less then 1%. This can be concluded quantitatively from the fact that the inter-structural $B(E2; 2_2^+ \to 0_1^+)$ value measured in this work is two orders of magnitude smaller than the $B(E2; 2_2^+ \to 0_2^+)$ value within the collective structure. A small mixing matrix element between configurations with different deformations is indicative of a high barrier between their corresponding minima in the potential energy surfaces. The situation encountered in ^{96}Zr with a spherical ground state, a deformed excited 0_2^+ state at low excitation energy, and a high potential barrier between them, resembles to a considerable extent the situation studied previously by Leviatan in the theoretical context of partial dynamical symmetries [10]. The high barrier between the spherical and deformed structures and their energetic evolution as a function of neutron number provides the main ingredients for the occurrence of a first order quantum shape phase transition in the sequence of zirconium isotopes beyond $N = 56$.

The addition of neutrons beyond $N = 56$ results in an increasing occupation of $j = l - 1/2$ neutron orbitals already in the ground state, and a corresponding reduction of the $Z = 40$ proton shell gap. The deformed configuration decreases in energy and eventually becomes the ground state. This mechanism is responsible for the extraordinarily sharp shape phase transition between the isotopes ^{98}Zr with still spherical ground state and ^{100}Zr with a collective quadrupole deformed ground state. The latter can be concluded from the large $B(E2; 2_1^+ \to 0_1^+) = 75(4)$ W.u. value known for ^{100}Zr [11]. The spherical character of ^{98}Zr

https://doi.org/10.1051/epjconf/201817802007

is inferred from the high excitation energy of its 2_1^+ state at 1.2 MeV [12]. Unfortunately, an absolute value for a vibrational small $B(E2; 2_1^+ \rightarrow 0_1^+)$ transition strength for ^{98}Zr is still missing in the literature. To that end we have recently studied Coulomb excitation of a beam of ^{252}Cf-fission fragments of mass 98 from which we can determine a decisive upper limit for the $B(E2; 2_1^+ \rightarrow 0_1^+) < 11$ W.u. for the $E2$ transition strength of ^{98}Zr. The data will be published elsewhere [13]. They indicate the transitional character of ^{98}Zr at neutron number 58 in contrast to its deformed rotational even-even neighbor ^{100}Zr with 60 neutrons.

The prime example for a first order nuclear shape phase transition in the sequence of zirconium nuclei between neutron numbers $N = 58$ and 60 is understood microscopically in terms of the details of the proton-neutron interaction leading to quadrupole deformation. An example of an excitation mode of deformed nuclei based on the proton-neutron degree of freedom is the $J^\pi = 1^+$ nuclear scissors mode. In the last section of this contribution we present our measurement [7] of its $E2$ decay strength.

4 $E2$ strength of the Scissors Mode inferred from photon scattering

While the formation of the nuclear scissors mode originates from the nuclear quadrupole deformation its dominant decay mode is enhanced $M1$ transitions due to its isovector character. In fact, the $E2$ decay strength of the scissors mode of deformed nuclei has been unknown until recently. We have measured now the (small) $E2$ decay strength of the $J^\pi = 1^+$ band head of the scissors mode to the ground state band by γ-ray polarimetry in Nuclear Resonance Fluorescence [7]. Photon beams of the HIγS facility were scattered off a Gd_2O_3 target

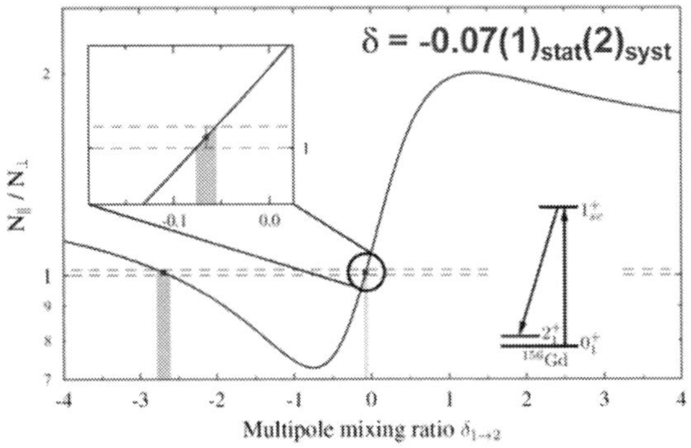

Figure 2. Measurement of the $E2/M1$ mixing ratio δ of the $1^+ \rightarrow 2_1^+$ transition of the main fragment of the scissors mode of ^{156}Gd from azimuthal angular correlations in photon scattering reactions at the High Intensity γ-ray Source at the Duke Free Electron Laser Laboratory. Adapted from Ref. [7].

which contained 10 g of Gadolinium with an enrichment of 93.79(3)% in the isotope ^{156}Gd. The azimuthal intensity distribution of the scattered γ radiation about the axis of the incident horizontally polarized γ-ray beam was measured at a polar angle of 135°. The ratio of count rates in the horizontal polarization plane and perpendicular to it was measured such that a finite $E2/M1$ mixing ratio δ, the square of which is equal to the ratio of $E2$ and $M1$ decay intensities, could be determined [7] for the $1_{sc}^+ \rightarrow 2_1^+$ transition. The measured $E2$ intensity corresponds to an electric quadrupole transition strength of $B(E2; 1_{sc}^+ \rightarrow 2_1^+) = 0.037(26)$ W.u. Within the framework of the interacting boson model, the $1_{sc}^+ \rightarrow 2_1^+$ transition corresponds to a transition between a mixed-symmetry state with F-spin quantum number $F = F_{max} - 1$ and the ground state band of proton-neutron symmetric states with F-spin quantum number $F = F_{max} = N/2$ where $N = N_\pi + N_\nu$ is the total number of proton and neutron bosons. This first measurement of an F-vector $E2$ transition rate in deformed nuclei provides direct evidence for the small F-vector quadrupole boson charge in deformed nuclei [7] conjectured already 30 years ago [14].

The data also provide evidence for the rotational excitation of the $K = 1$ scissors mode with spin quantum number $J = 2$. That 2_{sc}^+ state has been identified [7] on the basis of a reduced excitation strength $B(E2; 0_1^+ \to 2_{sc}^+)$ corresponding within uncertainties to the intrinsic $E2$ transition matrix element between the $J^\pi = 1^+$ scissors mode state and the ground state band discussed above. The energy difference $E(2_{sc}^+) - E(1_{sc}^+) = 19(1)$ keV provides first experimental evidence for the rotational moment of inertia of the scissors mode. It turns out to be very large even exceeding the rigid-body value for the ground state band of ^{156}Gd by about 50% if the alternative of an extraordinarily large signature splitting is neglected. Whether the scissors mode band in deformed nuclei is subject to extraordinarily large signature splitting or whether its moment of inertia is extraordinarily large can only be answered by a future observation of the next higher rotational excitation of the scissors mode band, the 3_{sc}^+ state. Its observation might be possible in high-resolution electron scattering spectroscopy by direct $M3$ excitation in inelastic (e, e') reactions.

5 Acknowledgments

We thank all those who have contributed to our experiments, in particular the accelerator crew, students, and doctoral researchers at the S-DALINAC and in the spectrometer group at TU Darmstadt, and those at HIγS, TUNL. We thank T. Otsuka for numerous discussions. We gratefully acknowledge the financial support by the German Research Council (DFG) under grant No. SFB 1245 and by the German Ministry for Education and Research (BMBF) under grant No. 05P15RDEN9.

References

[1] P. Cejnar, J. Jolie, and R.F. Casten, Rev. Mod. Phys. **82**, 2155 (2010).

[2] K. Heyde and J.L. Wood, Rev. Mod. Phys. **83**, 1467 (2011); *erratum ibid.* **83**, 1655 (2011).

[3] T. Otsuka, T. Suzuki, R. Fujimoto, H. Grawe, and Y. Akaishi, Phys. Rev. Lett. **95**, 232502 (2005).

[4] T. Otsuka and Y. Tsunoda, J. Phys. G **43**, 024009 (2016).

[5] C. Kremer, S. Aslanidou, S. Bassauer, M. Hilcker, A. Krugmann, P. von Neumann-Cosel, T. Otsuka, N. Pietralla, V.Yu. Ponomarev, N. Shimizu, M. Singer, G. Steinhilber, T.Togashi, Y. Tsunoda, V. Werner, M. Zweidinger, Phys. Rev. Lett. **117**, 172503 (2016).

[6] D. Bohle, A. Richter, W. Steffen, A. E. L. Dieperink, N. Lo Iudice, F. Palumbo, and O. Scholten, Phys. Lett. **137B**, 27 (1984).

[7] T. Beck, J. Beller, N. Pietralla, M. Bhike, J. Birkhan, V. Derya, U. Gayer, A. Hennig, J. Isaak, B. Löher, V. Yu. Ponomarev, A. Richter, C. Romig, D. Savran, M. Scheck, W. Tornow, V. Werner, A. Zilges, and M. Zweidinger, Phys. Rev. Lett. **118**, 212502 (2017).

[8] D. Abriola, A.A. Sonzogni, Nucl. Data Sheets **109**, 2501 (2008).

[9] T. Togashi, Y. Tsunoda, T. Otsuka, and N. Shimizu, Phys. Rev. Lett. **117**, 172502 (2016).

[10] A. Leviatan, Phys. Rev. C **74**, 051301(R) (2006); Phys. Rev. Lett. **98**, 242502 (2007).

[11] B. Singh, Nucl. Data Sheets **109**, 297 (2008).

[12] B. Singh, Z. Hu, Nucl. Data Sheets **98**, 335 (2003).

[13] W. Witt, V. Werner, N. Pietralla, M. Albers, A.D. Ayangeakaa, B.P. Bucher, M.P. Carpenter, D. Cline, H.M. David, A. Hayes, C. Hoffman, R.V.F. Janssens, B.P. Kay, F.G. Kondev, W. Korten, T. Lauritsen, O. Möller, G. Rainovski, G. Savard, D. Seweryniak, J. Smith, R. Stegmann, S. Zhu, C.Y. Wu, Phys. Rev. C, submitted for publication (2018).

[14] J.N. Ginocchio and P. Van Isacker, Phys. Rev. C **33**, 365 (1986).

Studying collective nuclear excitations with fast neutrons: a comparison of ^{76}Ge and ^{76}Se

Steven W. Yates[1,*], *Sharmistha* Mukhopadhyay[1], *Benjamin P.* Crider[2], *and Erin E.* Peters[1]

[1]Departments of Chemistry and Physics & Astronomy, University of Kentucky, Lexington, Kentucky, 40506-0055, USA
[2]Department of Physics and Astronomy, Mississippi State University, Mississippi State, Mississippi, 39762, USA

Abstract. The low-lying, low-spin level structures of ^{76}Ge and ^{76}Se were studied with the (n,n'γ) reaction. Gamma-ray excitation function and angular distribution measurements were performed, and level spins, level lifetimes, γ-ray intensities, and multipole mixing ratios were determined. Large observed B(E2) values support the identification of low-lying band structures.

1 Introduction

There are many excellent reasons for studying the stable A = 76 nuclei in detail. In recent years, the most frequently expressed motivation is related to the possibility of observing neutrinoless double-beta decay (0νββ). Indeed, the two-neutrino double-beta decay (2νββ) of ^{76}Ge, the parent, into ^{76}Se, the daughter, has been observed, and ^{76}Ge is considered one of the best 0νββ candidates [1]. The observation of 0νββ would have profound implications as the neutrino would be exposed as its own antiparticle, lepton number non-conservation would be verified for the first time, and a new window would be opened for determining the mass of the neutrino. Three factors determine the 0νββ decay rate: a phase space factor, which can be reliably calculated; the effective Majorana neutrino mass, for which only an upper limit is known; and the nuclear matrix element, NME [2]. The NME cannot be experimentally determined but must be obtained from nuclear structure theory. Unfortunately, comparisons of NMEs calculated with various theories significantly disagree, frequently by factors of two or three [2]. However, measurements of the structural characteristics and properties of ^{76}Ge and ^{76}Se can be used to constrain the theoretical calculations.

In addition to the reasons related to 0νββ, there are other goals for our studies. The A = 76 shape-transitional region is structurally interesting. Shape coexistence was observed many years ago [3, 4], and recent arguments about existence of rigid triaxiality have emerged [5, 6].

* Corresponding author: yates@uky.edu

2 Experimental methods

1.1 Inelastic neutron scattering

Inelastic scattering of fast neutrons in the few MeV range followed by γ-ray detection can yield a variety of valuable spectroscopic quantities. At the University of Kentucky Accelerator Laboratory (UKAL), neutrons are produced by the $^{3}H(p,n)^{3}He$ or $^{2}H(d,n)^{3}He$ reactions, and these fast, nearly monoenergetic ($\Delta E \approx 50$ keV) neutrons are used to excite the target nuclei.

There are a number of advantages of the (n,n'γ) reaction.

- As neutrons are uncharged and experience no Coulomb barrier, the nucleus can be gently probed. Moreover, the degree of excitation can be selected because the incident neutron energy can be controlled by changing the energy of the incident charged particles.
- As γ rays are detected, the energy resolution is excellent.
- At the neutron energies used in these measurements (< 4 MeV), the direct reaction contributions are generally regarded as small, and the compound nucleus assumption permits us to treat the population of levels statistically (*i.e.*, non-selectively, but limited by angular momentum considerations).
- Lifetimes can be determined by the Doppler-shift attenuation method (DSAM) [7, 8].
- When required, γ-γ coincidence measurements are possible [9, 10].

As with any method, there are limitations.

- Studies are generally limited to stable nuclei or those with long lifetimes, and large scattering samples are typically required. These reactions occur with secondary particles, *i.e.*, the fast neutrons, and incident fluxes are low.
- Large amounts (> 10 grams) of enriched isotopes are required in order to assure that the γ rays are attributed to the correct nuclear state and that its properties are correctly determined. Large scattering samples are generally available on a lease arrangement from Oak Ridge National Laboratory, but the costs are high.

1.1.1 Excitation functions

By keeping the detection angle constant and varying the incident neutron energy, it is possible to build a level scheme. The γ-ray thresholds permit the placement of the emitting levels. These data complement the information obtained from other probes and from γ-γ coincidence measurements. In addition, cross section data provide information about the angular momentum of the emitting levels.

1.1.2 Angular distributions

Gamma-ray angular distributions are produced by keeping the incident neutron energy constant and changing the detection angle and yield multipole mixing ratios, δ(E2/M1), and level lifetimes from DSAM measurements. From these quantities, reduced transition probabilities are obtained.

1.2 Experimental measurements

To minimize the spectral contributions from other naturally occurring isotopes of Ge and Se and increase the sensitivity of the measurements, enriched targets (19.56 g of $^{76}GeO_2$ enriched to 84% and 11.96 g of elemental ^{76}Se powder enriched to 96.95%) contained in polyethylene vials were used in the (n,n'γ) measurements at UKAL. Gamma rays were detected in a high-purity germanium (HPGe) detector of 50% relative efficiency and an energy resolution of less than 2.2 keV(FWHM) at 1333 keV surrounded by a bismuth germanate (BGO) annulus. Neutrons with energies up to 3.7 MeV were used in the excitation function measurements, and angular distributions were measured at E_n = 3.0 and 3.5 MeV for ^{76}Ge and at 2.4, 3.0, and 3.7 MeV for ^{76}Se. For each nucleus, level spins, level lifetimes, γ-ray branchings, and multipole mixing ratios were determined. In ^{76}Ge, many previously suggested levels were found to have been placed in error and were eliminated. This issue proved to be less problematic in ^{76}Se. The detailed results for ^{76}Ge have been published [11] and the results of the study of ^{76}Se will soon be submitted [12].

2 Results and discussion

A brief examination of the level structures of the ground bands (see Fig. 1) in these nuclei suggests that ^{76}Se is slightly more deformed and more collective than ^{76}Ge. As noted earlier, nuclei in this region exhibit an array of structural features, including shape coexistence [4]. While shell model calculations reproduced the low-lying level structure of ^{76}Ge very well, they have not met with similar success in ^{76}Se and are ongoing [12].

An obvious structural feature in the data (see Figs. 2 and 3) is the collective band built on the lowest 0^+ excitation in ^{76}Se (see Fig. 3), which occurs nearly 1 MeV lower in energy than in ^{76}Ge (see Fig. 2). Interestingly, the calculations by Nomura *et al.* [13] within mapped interacting boson model calculations based on the Gogny energy density functional predicts this state to lie even lower in excitation energy.

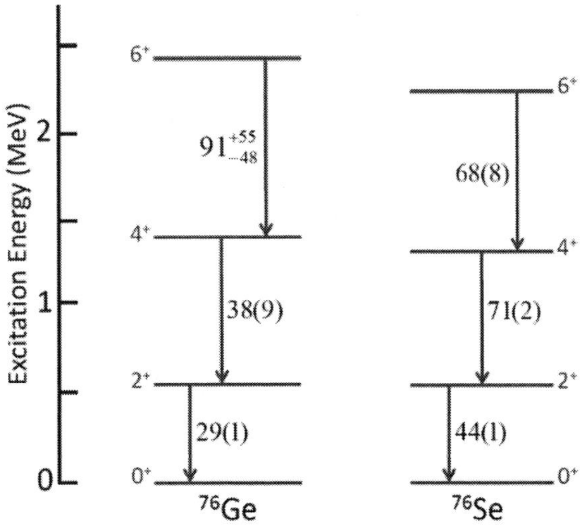

Fig. 1. Ground bands in ^{76}Ge and ^{76}Se. B(E2) values on the transition arrows, in W.u., are from ENSDF [14], except for the B(E2;$6^+ \rightarrow 4^+$) in ^{76}Ge, which is from the current measurements.

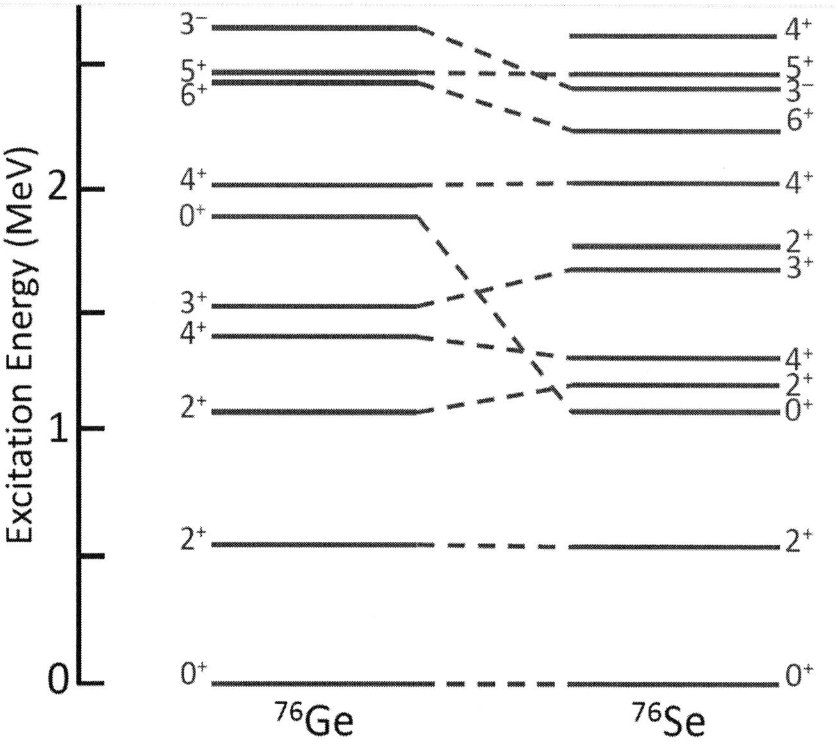

Fig. 2. Comparison of the level structures of ^{76}Ge and ^{76}Se. Band members are shown in the same color.

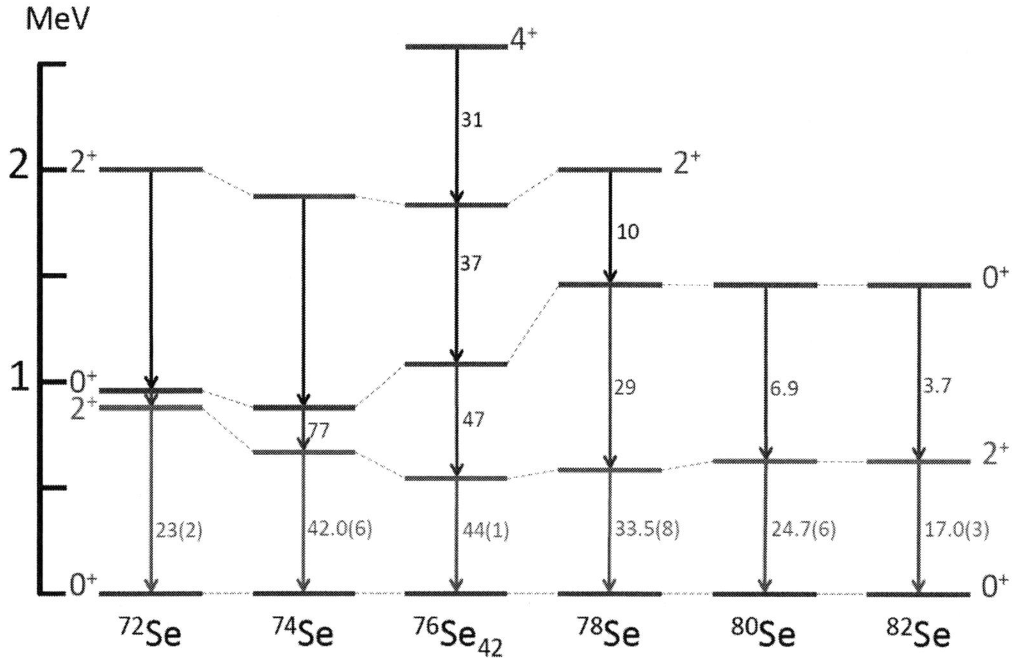

Fig. 3. Level systematics of the even-A Se nuclei. Except for the low-lying 0⁺ band in ^{76}Se, B(E2) values for the indicated transitions, given in W.u., were obtained from ENSDF [14]. The 2⁺ and 4⁺ states of this band were identified in the current measurements.

As in ^{76}Ge, the low energy of the second excited 2^+ state, the band head of a quasi-γ band, suggests triaxiality in ^{76}Se. The calculations of Nomura *et al.* [13] describe the quasi-γ band in ^{76}Se reasonably well and the 3^+ and 4^+ levels are close in energy, following the systematics of the γ-unstable-rotor model of Wilets and Jean [15].

3 Summary

The stable A = 76 nuclei, ^{76}Ge and ^{76}Se, have been studied with the inelastic neutron scattering reaction. While there are clearly similarities between these nuclei, it may be that their deformations are considerably different, as suggested by Nomura *et al.* [13], and a recent theoretical calculation indicates that a transition from prolate ^{76}Ge to oblate ^{76}Se occurs [13]. The impact of this finding on the exotic decay process, $0\nu\beta\beta$, remains to be evaluated.

We sincerely thank H.E. Baber and other colleagues at the UKAL for their contributions to these experiments. This material is based upon work supported by the U.S. National Science Foundation under Grant. No. PHY-1606890.

References

1. F.T. Avignone, S.R. Elliott, J. Engel, Rev. Mod. Phys. **80**, 481 (2008)
2. J. Engel, J. Menéndez, Rep. Prog. Phys. **80**, 046301 (2017)
3. Z. Podolyák, *et al.*, Int. J. Mod. Phys. E **13**, 123 (2004)
4. K. Heyde and J.L. Wood, Rev. Mod. Phys. **83**, 1467 (2011)
5. Y. Toh, *et al.*, Phys. Rev. C **87**, 041304(R), (2013)
6. J.J. Sun, *et al.*, Phys. Lett. B **734**, 308 (2014)
7. T. Belgya, G. Molnár, S.W. Yates, Nucl. Phys. **A607**, 43 (1996)
8. E.E. Peters, *et al.*, Phys. Rev. C **88**, 024317 (2013)
9. C.A. McGrath, *et al.*, Nucl. Instrum. Meth. **A421**, 458 (1999)
10. E. Elhami, *et al.*, Phys. Rev. C **78**, 064303 (2008)
11. S. Mukhopadhyay, *et al.*, Phys. Rev. C **95**, 014327 (2017)
12. S. Mukhopadhyay, *et al.*, Phys. Rev. C, to be submitted (2018)
13. R. Nomura, K. Rodríguez-Guzmán, L.M. Robledo, Phys. Rev. C **95**, 064310 (2017)
14. ENSDF, NNDC Online Data Service, ENSDF data base, http://www.nndc.bnl.gov/ensdf/ (2017)
15. L. Wilets and M. Jean, Phys. Rev. **102**, 788 (1956).

EPJ Web of Conferences **178**, 02009 (2018) https://doi.org/10.1051/epjconf/201817802009
CGS16

Low lying oscillations of deformed nuclei

Ani Aprahamian[1,*] and *Shelly R. Lesher*[2]

[1]Physics Department, University of Notre Dame, Notre Dame, IN 46556, USA
[2]Physics Department, University of Wisconsin-La Crosse, La Crosse, WI, 54601, USA

Abstract. Low-lying oscillations of the intrinsic deformed shape of a nucleus remain an open challenge in nuclear structure. The question or challenge revolves around the viability of single or multiple quanta of vibrational excitations superimposed on the equilibrium, deformed shape of a nucleus. The K=2 or "γ" vibrations are fairly widespread and nominally conform to expectations whereas the existence of the K=0 or "β" vibrational excitation is yet to be distinguished from other possible origins including the coexistence of other potential minima.

1 Introduction

The 1975 Nobel prize in Physics was awarded to Bohr, Mottelson, and Rainwater with the citation that read "for the discovery of the connection between collective motion and particle motion in atomic nuclei and the development of the theory of the structure of the atomic nucleus based on this connection". They were recognized for a geometric description of the nucleus from spherical, to deformed, and triaxial shapes with superimposed oscillations or vibrations of those shapes. There have been many advances since in describing nuclei in terms of sophisticated approaches and algebraic models but the question remains about the nature of the low lying oscillations in deformed nuclei. The lowest shape affecting oscillations are expected to be quadrupole in nature resulting in two types of vibrations, a "β" and a "γ" type of vibrations where one type of vibration is along the symmetry axis and the other is against it with a projection of $K=2^+$ on the symmetry axis [1]. The "γ" type of vibration seems to be well characterized as the first excited $K=2^+$ bands in deformed nuclei with typical B(E2) values of several Weisskopf units (W.u.) connecting states to the ground state (g.s.) band. Today, however, some forty years later, the nature of the "β" or the $K=0^+$ excitation is still unsettled. The topic has been discussed, questioned, challenged and redefined in numerous publications. A part of the challenge lies in the paucity of data and also to a significant extent on the interpretation or characterization of what a "β" vibration is expected to be. Fig. 1 [2] shows the experimental situation in the rare-earth region of the chart of nuclides. The situation is accepted and fairly predictable for the first excited $K=2^+$ bands but not so for the first excited $K^{\pi}=0^+$ band. Perhaps the most resounding criticism of labeling the first excited $K=0^+$ excitations as a "β"

* Corresponding author : aapraham@nd.edu

vibration built on the intrinsic ground state deformation, come from the coexistence approaches claiming that the first excited K=0^+ band is in fact the minimum for a competing shape and therefore coexisting minimum in the potential. Other approaches include calculations using a five dimensional collective Hamiltonian for the quadrupole vibrational and rotational degrees of freedom where instead of solving a potential in β and γ, potential energy surfaces are calculated using Relativistic Mean Field theory [3-6]. In this approach, the first excited $K^{\pi}=0^+$ excitation often ends up being a mixture of many coexisting shapes. There is also the constrained-Hartree-Fock-Bogoliubov (CHFB) theory together with a mapping to the five dimensional collective Hamiltonian, using the Cogny D1S interaction in the nuclear Hamiltonian [7,8]. The aim ofthis approach is the development of a theory of nuclear structure that is universal and includes spectroscopic calculations for nearly two thousand nuclei. They find that many of the properties are strongly dependent on the intrinsic deformation. On average, the predicted energies of the γ band or excited K=2^+ band are 12% higher than experiment whereas the predicted energies for the first excited 0^+ state are systematically 50% higher [7]. The character of the first excited 0^+_2 is interpreted either as a β-vibrational excitation or as shape coexistence depending on the relative quadrupole transition strengths. The result is a prediction of 0^+ bands with β vibrational character for many well-deformed nuclei with $4^+/2^+$ ratios approaching that of the axial rotor at 10/3 although they find coexistence to be more prevalent.

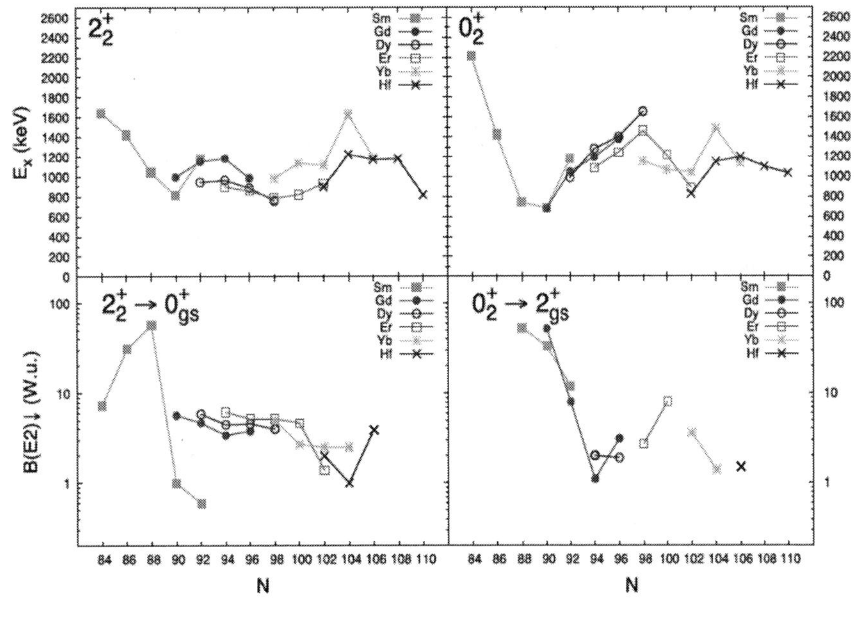

Figure 1: Systematics of the first excited $K^{\pi}=2^+$ "γ" and $K^{\pi}=0^+$ bands in several rare earth nuclear isotopes as a function of the neutron number N along with their corresponding B(E2) values connecting the shown excited bands to members of the ground state band [2].

Our goal in this work is to discern the nature of the low-lying excited 0^+ states in deformed nuclei. The Gd nuclei lie in the well-deformed rare earth region of the chart of nuclides with $4^+/2^+$ energy ratios that vary between 3.0 and 3.33. In an earlier study [9], we had uncovered the existence of thirteen 0^+ states in ^{158}Gd with experiments at the Munich Q3D spectrometer. Here, we report on lifetime measurements of states in the ^{156}Gd nucleus. The

$4^+/2^+$ energy ratio is 3.27 and there are three low-lying 0^+ bands. We report on measurements of lifetimes and the extracted quadrupole transition strengths for 8 members of the three excited $K=0^+$ bands using the Gamma-Ray-Induced Doppler (GRID) broadening technique [10] following neutron capture at the Institut Laue Langevin in Grenoble, France. The results point to an excited $K=0^+_2$ collective excitation built on the g.s. band in comparison to a nearby and much weaker $K=0^+_3$ excitation. An additional surprise of a third excited $K=0^+_4$ band emerged as a collective excitation built on the $K=2^+$ band. We evaluate the various excitation bands in terms of quadrupole transition probabilities as well as dynamic moments of inertia.

2 Experiments

The experiment to measure lifetimes was performed at the High Flux Reactor at the Institute Laue-Langevin in Grenoble, France. GRID allows the measurement of the broadening of a given transition affected by the recoil of a previously emitted γ-ray. The technique uses nearly perfect crystals of Si and Ge to measure line shapes of the transitions of interest. The recoil velocities are very small (10^{-4} to 10^{-6} c) and the measured broadenings are in the order of a few keV. Typically, there are very few known lifetimes of the 0^+ band heads in nuclei and in this case, we were able to measure all three. Additionally we re-measured some previously known lifetimes in the $K=2^+$ band.

3 Results

The resulting transition probabilities for the ^{156}Gd from the lifetime measurements are shown in Fig. 2 along with the systematics of the neighboring isotopes of ^{154,158,160}Gd. There are numerous 0^+ states identified in ^{154}Gd but lifetimes are not yet available. We report on the results for ^{156}Gd here. Lifetimes of levels in ^{158,160}Gd were measured using (n,n'γ) method at the University of Kentucky.

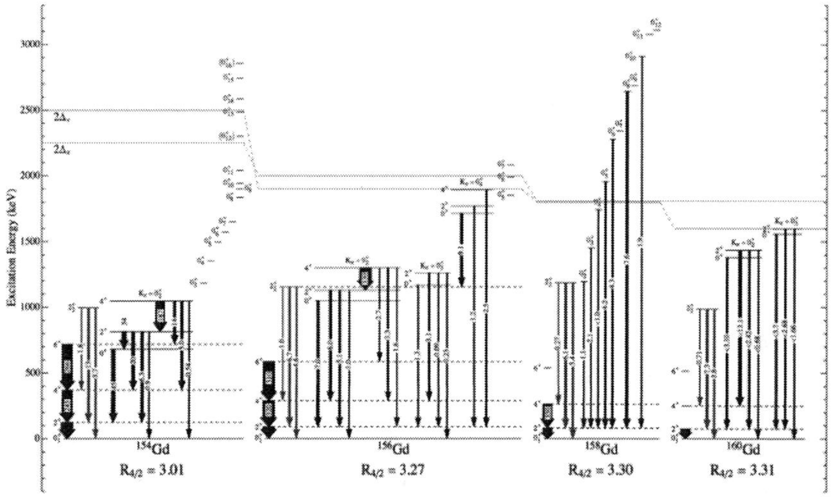

Figure 2: Partial level schemes are shown for the isotopes of ^{154,156,158,160}Gd. The B(E2) values from the lowest $K=2^+$ bands are included as well as the $K=0^+$ bands where lifetimes are available. The widths of the arrows are proportional to the B(E2) strengths and given in W.u.

Below the isotope label are the energy ratios of the first $4^+/2^+$ values. The horizontal lines represent the two proton and two neutron pairing gaps.

4 Discussion

The B(E2) values for transitions depopulating the first excited $K=0^+$ band in ^{156}Gd are of the same order as those depopulating the first excited $K=2^+$ band. Indicating that the first excited $K=0^+_2$ band may in fact be a "β" vibrational excitation built on the ground state in this nucleus. We have further evidence that the second $K=0^+_3$ band is much more weakly connected to the ground state levels while the third excited $K=0^+_4$ band is strongly connected to the $K=2^+$ band. This latter band behaves as the two-phonon excitation built on the single phonon "γ" vibrational excitation. In ^{156}Gd, there had been an earlier measurement of the lifetime of the $K^\pi=4^+$ band head by coulomb excitation. B(E2) values depopulating this state to the $K^\pi=2^+$ band are strong however decay of the K^π 4^+ band to the g.s. band is K forbidden having no other open channels for decay. Also, the energy ratio of the $K^\pi=4^+/K=2^+$ bands is strongly anharmonic with a ratio of 1.31 instead of the expected harmonic value of 2. The $K^\pi=0^+_4$ band starting at 1715.181 however is strongly connected to the K=2+ band as well and shows an energy ratio of approximately 1.5. This is also strongly anharmonic but all evidence points to the band at 1715.181 keV to be the two-phonon γγ vibrational excitation built on the $K^\pi=2^+$ band. Earlier works [11, 12] had pointed to the existence of identical dynamic moments of inertia for bands of similar character including the case for the ^{156}Gd, $K^\pi=2^+$ and $K^\pi=4^+$ bands. Here we include the new $K^\pi=0^+_4$ band shown to be the two-phonon "γγ" vibrational excitation built on the "γ" band by the collective B(E2) values, in addition to the $K^\pi=4^+$ band in Fig. 3. The dynamic moments of inertia for single $K^\pi=2^+$ and two phonon vibrational bands ($K^\pi=0^+$, $K^\pi=4^+$). The somewhat puzzling point is that the $K^\pi=0^+$ band does not have the same dynamic moment of inertia as the $K^\pi=4^+$ band in spite of the collective B(E2) values connecting them both to the $K^\pi=2^+$ band. The 5DCH calculations within RMF approach show identical moments of inertia for the g.s. band and the gamma band but no information regarding the two-phonon excitation bands.

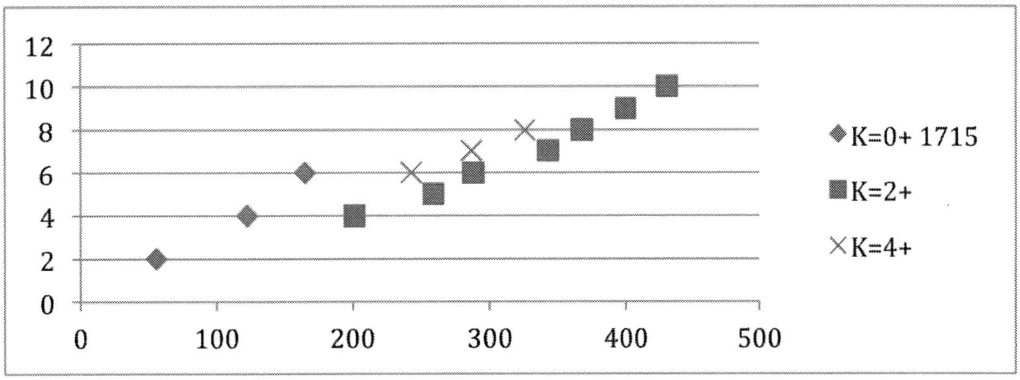

Figure 3: A plot of the initial spins against the level energy differences for the individual levels within a given band. Only ΔJ=2 transitions are considered here. The dynamic moment of inertia for each band in ^{156}Gd is the slope of the band. This figures shows that the $K^\pi=2^+$ and K^π

$=4^+$ bands have identical slopes but the $K^\pi=0^+$ band at 1715 keV has a slightly different slope if the uncertain (6^+) state is also included in spite of showing strong B(E2) transitions connecting the band to the $K=2^+$ band.

Following on the dynamic moments of inertia for the three K=0+ bands, Fig. 4 below shows the relation of the three excited $K=0^+$ bands. The $K=0^+$ band at 1049.5 keV connecting with collective transitions to the ground state band has the same slope as the $K=0^+$ band starting at 1715 keV which is connecting to the $K=2^+$ band with collective transitions. Furthermore, the g.s. band dynamic moment of inertia is almost exactly identical to the $K=0^+$ at band starting at 1168 keV however the quadrupole transitions connecting them is considerably weaker.

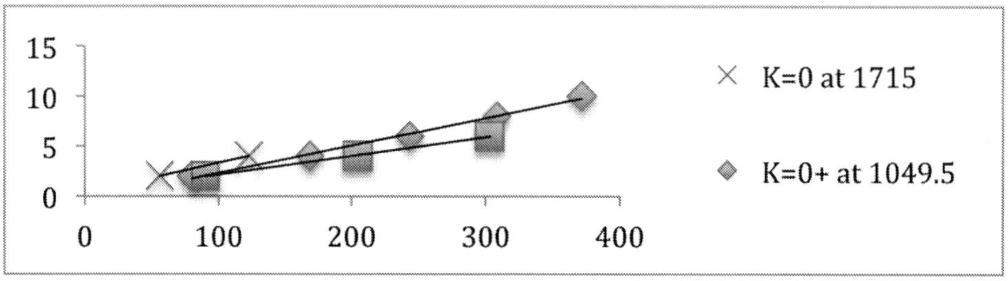

Figure 4: A plot of the initial spins (J) plotted against the $\Delta J=2$ of the energy levels within a given band for the three excited $K^\pi=0^+$ bands in ^{156}Gd. All indications here are that the $K^\pi=0^+$ at 1168 keV is a band of a very different nature than the other two $K^\pi=0^+$ bands at 1049.5 and 1715 keV.

5 Conclusions

We have reported on the lifetime measurements of three $K^\pi=0^+$ bands in the well-deformed ^{156}Gd nucleus. B(E2) transitions indicate that the first excited $K^\pi=0^+$ band is more strongly connected to the g.s. band than the nearby but higher $K^\pi=0^+$ band starting at 1168 keV. The $K^\pi=0^+$ band starting at 1715 keV shows strong B(E2) transitions to the first excited $K^\pi=2^+$ band. The dynamic moments of inertia for the bands in consideration show somewhat puzzling results.

This work was funded by the NSF-USA under contracts PHY-1419765 and PHY-1713857.

References:

1. A. Bohr and B.R. Mottelson, Nuclear Structure Vol. II ,(1975)
2. A. Aprahamian, S.R. Lesher, and A. Stratman, Bulg. J. Phys. 44, 372 (2017)
3. A. Aprahamian et al., Phys. Rev. C 95, 024329 (2017).
4. T. Nikšić et al.,Phys. Rev. C **79**, 034303 (2009)
5. Z. P. Li et al., Phys. Rev. C **79**, 054301 (2009)
6. J.P. Delaroche et al, ,Phys. Rev. C 81, 014303 (2010)
7. J. Decharge and D. Gogny, Phys. Rev. C21, 1568 (1980)
8. G. F Bertsch et al, , Phys. Rev. Lett. 99, 032502 (2007)
9. S.R. Lesher et al, Phys. Rev. C 66, 051305 (R) (2002)
10. H.G. Boerner et al., Phys. Lett. B 215, 45 (1988)
11. X. Wu, A. Aprahamian, J. Castro-Ceron, Phys. Lett. B 316, 235 (1993)
12. A. Aprahamian et al., Phys. Rev. C 65, 031301 (R) (2002)

Lifetime measurements in A~100 nuclei using $LaBr_3$(Ce) arrays.

A.M. Bruce[1,*]

[1]School of Computing, Engineering and Mathematics, University of Brighton, Brighton, BN2 4GJ, UK.

Abstract. The region of the nuclear chart around neutron-rich A~100 nuclei is one where prolate and oblate nuclear shapes are predicted to be in close competition. An indirect measurement of the shape of the nucleus can be obtained from measuring level lifetimes which relate, via transition rates, to β_2 deformation. In order to make measurements of level lifetimes in the sub nanosecond range an array of 36 $LaBr_3$(Ce) detectors has been constructed for use at the FAIR facility in Darmstadt, Germany. This presentation will give an overview of the array and examples of its use in commissioning experiments at the RIKEN Nishina Center in Japan and the Argonne National Laboratory in the USA.

1 Introduction

An array of 36 $LaBr_3$(Ce) detectors [1, 2] has been designed and procured for use at the FAIR facility in Germany. It uses detectors of diameter 38.1 mm and length 50.8 mm, a size selected as a balance between timing resolution (favouring smaller detectors) and efficiency (favouring larger detectors). The array is predicted to have an efficiency of ~10% at 1.5 MeV [1] and is planned to operate using a digital data acquisition system. In order to ensure correct operation of the array, it has been commissioned in a range of configurations, more details of which are given below.

2 Commissioning experiments with FATIMA

2.1 Lifetimes in $^{102-106}$Zr using FATIMA in coincidence with EURICA

Fission fragments were produced at the Radioactive Isotope Beam Factory, operated by the RIKEN Nishina Center and the Center of Nuclear Study, University of Tokyo using the in-flight abrasion fission of a $^{238}U^{86+}$ beam of average intensity 6.24×10^{10} particles/s accelerated to an energy of 345 MeV/nucleon. The fission fragments were selected and separated by the Big RIKEN Projectile Fragment Separator (BigRIPS) and implanted into the Wide-range Active Silicon Strip Stopper Array for β and Ion detection (WAS3ABi) [3], which detected ion implantations and their subsequent β-decay electrons. For this experiment, WAS3ABi comprised 5 layers of double-sided silicon-strip detectors placed 0.5 mm apart, each with 60 vertical and 40 horizontal strips of width and depth 1 mm.

*e-mail: alison.bruce@brighton.ac.uk

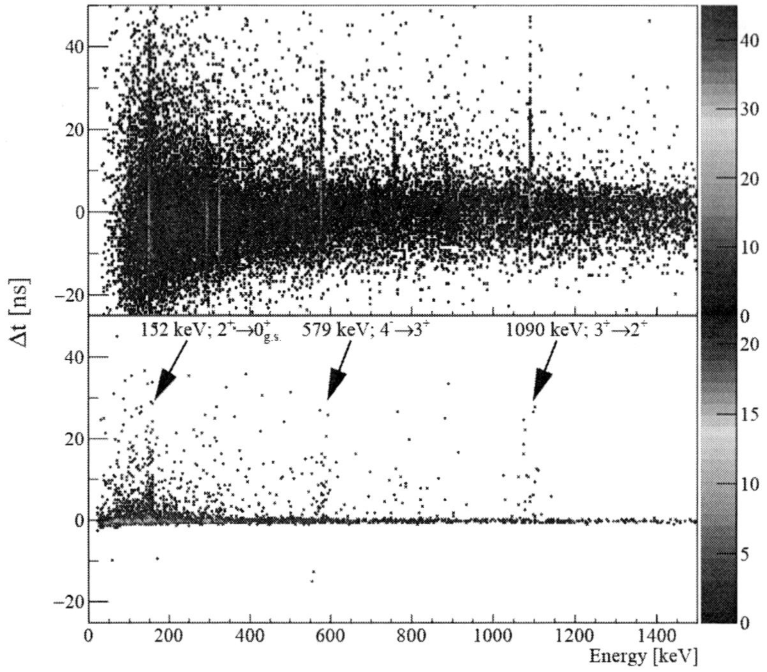

Figure 1. Energies and relative times of γ rays observed in EURICA (top) and the $LaBr_3(Ce)$ array (bottom) following the β-decay of ^{102}Y. This figure shows very clearly the power of using $LaBr_3(Ce)$ detectors rather than germanium detectors for measuring lifetimes in the sub-nanosecond range.

Fast-timing measurements were carried out by adding an array of 18 $LaBr_3(Ce)$ detectors to the EUROBALL-RIKEN cluster array (EURICA) [4]. Figure 1 shows energy-time matrices for transitions in ^{102}Zr produced in the β decay of implantated ^{102}Y ions [5]. For the Δt measurement, the time of the β decay was measured using plastic scintillators placed up-stream and down-stream of WAS3ABi. While the delayed components of the 152-, 579- and 1090 keV peaks are clearly evident in the $LaBr_3(Ce)$ spectrum (bottom), they are masked by noise in the spectrum measured in the germanium detector (top). The 152-, 579- and 1090 keV transitions in ^{102}Zr are produced in the decay of the $(\nu\frac{5}{2}[532]\otimes\frac{3}{2}[411])\,4^-$ state and figure 2 shows the sum of the background-subtracted time-difference projections of the 579- and 1090 keV transitions which have been fitted to give a level lifetime of $\tau = 9.5(7)$ ns [5]. The hindrances of the de-exciting γ-ray transitions are consistent with the systematics of the level being a K-isomer and with the decay of the known K-isomer in the isotone ^{100}Sr [6].

In the same experiment, the lifetimes of the 2^+_1 states in $^{104,106}Zr$ have been measured, following the β decay of implanted $^{104,106}Y$ ions, to be 2.90^{+25}_{-20} ns and 2.60^{+20}_{-15} ns respectively [7] corresponding to ground-state deformations of $|\beta_2| = 0.39(1)$ and $0.31(1)$ e^2b^2. The magnitude of these deformations is inconsistent with the oblate solutions of Hartree-Fock-Bogolyubov (HFB) calculations employing the Gogny-D1S interaction which indicate $|\beta_2|$ deformations of ~ 0.1 [8], but supports the conclusion that the neutron-rich Zr nuclei retain a prolate shape.

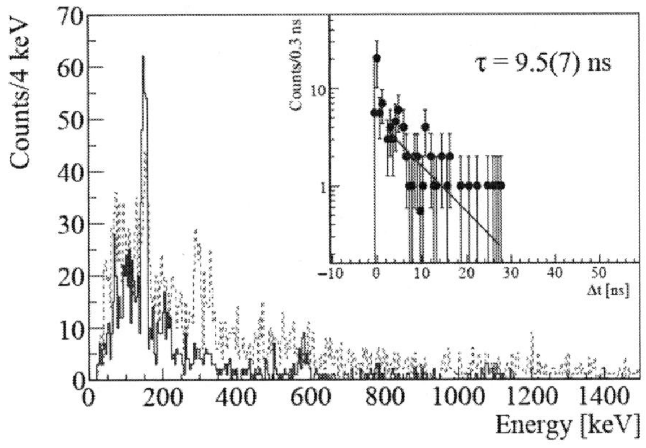

Figure 2. The gamma-ray energy spectra measured in the $LaBr_3$(Ce) detectors in prompt ($\Delta t < \pm 0.5$ ns, red) and delayed ($0.5 < \Delta t < 50$ ns, blue) coincidence with an electron emitted in the β decay of implantated ^{102}Y ions. The inset shows the sum of the background-subtracted time spectra for the 579- and 1090 keV transitions and the associated fit which yields a lifetime of $\tau = 9.5(7)$ ns for the 4^- state at 1821 keV in ^{102}Zr.

2.2 Lifetimes in ^{100}Zr using FATIMA in coincidence with Gammasphere

In a separate experiment at Argonne National Laboratory, fission fragments produced from the decay of a 34.4 μCi ^{252}Cf source, placed at the centre of a hybrid array, were studied. The array comprised 25 $LaBr_3$(Ce) detectors arranged in a hemisphere coupled with half of the Gammasphere [9] germanium detector array (51 detectors). Lead shields were used to avoid crosstalk between neighbouring $LaBr_3$(Ce) detectors. Two separate data sets were collected, one, to be used for spectroscopy, contained fold ≥ 3 events registered in Gammasphere while the other, to be used for lifetime measurements, collected fold ≥ 2 events in the $LaBr_3$(Ce) array which were in coincidence with fold ≥ 1 events in Gammasphere. The coincidence window between Gammasphere and the $LaBr_3$(Ce) array was 500 ns. In addition to the fact that the $LaBr_3$(Ce) detectors have poor energy resolution, γ-ray spectra from fission sources are always very messy. Therefore gates need to be set on transitions in Gammasphere in order to clean the $LaBr_3$(Ce)-$LaBr_3$(Ce) data sufficiently to allow timing measurements to be made. Figure 3 shows an example of one such procedure where a $LaBr_3$(Ce)-$LaBr_3$(Ce) E_γ-E_γ-Δt cube has been constructed gated on the 212- and 626 keV $2^+ \rightarrow 0^+$ and $8^+ \rightarrow 6^+$ transitions in ^{100}Zr. The coincidence between the 351 keV $4^+ \rightarrow 2^+$ and 497 keV $6^+ \rightarrow 4^+$ transitions in ^{100}Zr is clearly evident in the figure. The time spectra shown in figure 4 have then been obtained by subtracting peak/background and background/peak gated (enclosed by 2 black and 2 red lines) time spectra from the peak/peak gated (four black lines) time spectra and adding the background/background gated (four red lines) time spectra to avoid double counting.

Figure 4 shows the delayed and anti-delayed time-difference spectra for the 351 keV $4^+ \rightarrow 2^+$ and 497 keV $6^+ \rightarrow 4^+$ transitions in ^{100}Zr. The difference between the centroids of the distributions ΔC is measured to be 225(15) ps. The prompt response difference (PRD), obtained for this combination of energies from ^{152}Eu calibration data, is 124(7) ps. The lifetime of the level calculated from $2\tau = \Delta C - \text{PRD}$ [10] is then 51(9) ps. This value is consistent with the literature value of $\tau = 53(4)$ ps [11] but slightly larger than the value

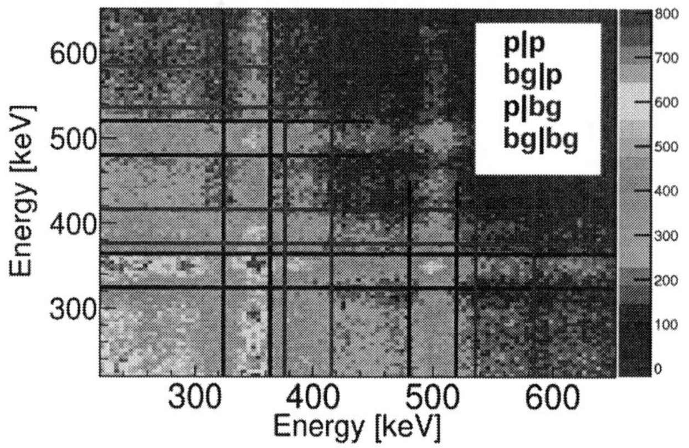

Figure 3. The projection of the $LaBr_3$(Ce)-$LaBr_3$(Ce) E_γ-E_γ-Δt cube measured in coincidence with the $2^+ \rightarrow 0^+$ and $8^+ \rightarrow 6^+$ transitions in ^{100}Zr observed in Gammasphere.

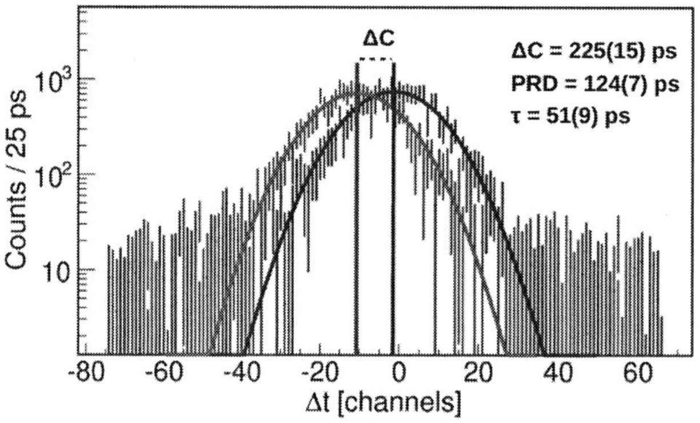

Figure 4. The background-subtracted delayed and anti-delayed time-difference spectra for the 351 keV $4^+ \rightarrow 2^+$ and 497 keV $6^+ \rightarrow 4^+$ transitions in ^{100}Zr.

of τ = 37(4) ps [12] obtained in the recent EXILL-FATIMA campaign at the Institut Laue-Langevin in Grenoble, France.

3 Conclusions

The data presented show that subsets of the 36 element $LaBr_3$(Ce) detector array have been installed, commissioned and tested successfully in coincidence with EURICA and with Gammasphere. An additional configuration has been used successfully during the AGATA campaign at GANIL in 2017 [13]. The full array will now be installed at GSI/FAIR and used for FAIR phase-0 experiments in Sept/Oct 2018.

4 Acknowledgements

This work was carried out in collaboration with the FATIMA collaboration. F.Browne was responsible for the analysis of the data from RIKEN and E.R.Gamba for the data from Gammasphere.

References

[1] O.J.Roberts *et al.*, Nucl. Instrum. Methods **A748** 91 (2014).
[2] L.M. Fraile *et al.*, Technical design report for the DESPEC fast timing array (2015). http://www.fair-center.eu/fileadmin/fair/publications_exp/TDR_HISPEC_DESPEC_FATIMA_public.pdf
[3] S.Nishimura *et al.*, Prog. Theor. Exp. Phys. **2012** 03C006 (2012).
[4] P.-A.Söderström *et al.*, Nucl. Instrm. Methods **B317** 649 (2013).
[5] F.Browne *et al.*, Phys. Rev. **C96** 024309 (2017).
[6] B.Pfeiffer *et al.*, Z. Phys. **A353** 1 (1995).
[7] F.Browne *et al.*, Phys. Letts. **B750** 448 (2015).
[8] R.Rodríguez-Guzmán *et al.*, Phys. Letts. **B691** 202 (2010).
[9] I.-Yang Lee, Nucl. Phys. **A520** 641 (1990).
[10] J.-M.Régis, PhD thesis, University of Köln (2011).
[11] B.Singh, Nucl. Data Sheets **109** 297 (2008).
[12] S.Ansari *et al.*, Phys. Rev. **C96** 054323 (2017).
[13] P.R.John private communication.

Collective states and shape coexistence in the Z = 50 region

Paul E. Garrett[1,2,*]

[1]Department of Physics, University of Guelph, Guelph, ON N1G2W1, Canada
[2]Department of Physics and Astronomy, University of the Western Cape, Bellville 7535, South Africa

Abstract.
Shape coexistence was established in the Sn and Cd nuclei, and in the odd mass In and Sb nuclei, some time ago but firm evidence has been lacking for its presence at low spin in the Te nuclei. However, strong circumstantial evidence exists, and recent highly-sensitive γ-ray spectroscopic studies have observed transitions for candidate rotational 2^+ band members in ^{122}Te and ^{124}Te.

1 Introduction

Shape coexistence in the $Z = 50$ region was first proposed nearly 50 years ago, in part of explain the appearance of "extra" states in the low-excitation part of the level spectrum of the mid-shell Cd isotopes [1, 2]. Since that time, there has been tremendous progress in characterising the shape-coexisting states in Sn and Cd isotopes, as well as the odd-mass In and Sb. Shown in Fig. 1 are the energy systematics of the low-lying states in the Pb and Hg isotopes, with $Z = 82$ and 80, respectively, and the states in the Sn and Cd isotopes, with $Z = 50$ and 48. The similarity of the pattern of the shape-coexisting states as a function of the neutron number is striking and points to a common underlying cause. As has been discussed for many years, this cause is the promotion of a pair of protons across the closed-shell gap into unoccupied orbitals increasing the correlation energy which maximizes at the neutron mid-shell, giving rise to the characteristic "V"-shaped pattern of the shape-coexisting states [3, 4]. For a review of the presence of shape coexistence in the $Z = 50$ region, and the data associated with its assignment, see Refs. [5, 6]. While the evidence in support of shape coexisting states, based on $2p - 2h$ proton "intruder" excitations, is quite firm for the Cd and Sn isotopes (and $1p - 1h$ in the odd-mass In and Sb), it is much less so for other nuclei in the region [6]. In the Te isotopes, for example, while it can be expected that they should possess shape-coexisting states based on $4p - 2h$ proton configurations, analogous to the $2p - 4h$ states for the Cd isotopes, there has been considerable debate regarding their appearance in the low-excitation energy region; the energy systematics, shown in Fig. 2 could be nearly equally well explained by the evolution of structure from a predominately $U(5)$, or spherical vibrator, towards an $O(6)$, or γ-soft, behaviour (see, for example, Refs. [7, 8]), and even the possibility of $E(5)$ [9] behaviour in some cases [10–12].

The occurrence of shape coexistence in the low-excitation energy region of nuclei can greatly complicate the elucidation of their underlying structure. This is particularly the case when the shape-coexisting states strongly mix; signatures of structure can then be somewhat smeared. For a long time, this was suggested to be the case for the Cd isotopes [13–17],

*e-mail: pgarrett@physics.uoguelph.ca

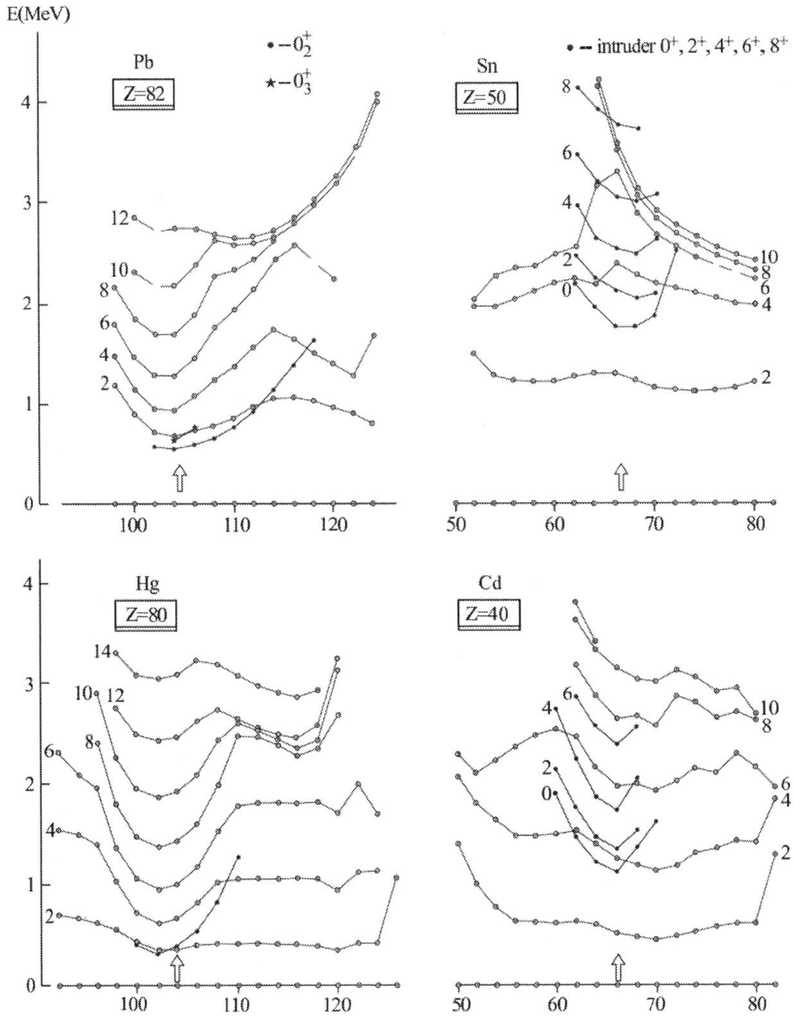

Figure 1. Systematics of the energies of the yrast, and near-yrast, states in the closed shell isotopes of Pb and Sn (top panels), and the 2-proton-hole nuclei Hg and Cd (bottom) as a function of neutron number. The arrows indicate the neutron mid-shell. Red indicates non-yrast states. Figure courtesy of J.L. Wood.

although there were issues associated with such a strong-mixing interpretation [18, 19]. It was pointed out in ^{114}Cd that the available $B(E2)$ data could be interpreted with either a strong or weak mixing scenario [20]. More recently, evidence in favour of weak mixing between the configurations was presented for ^{110}Cd [21], and the analysis of $E0$ transitions in ^{114}Cd also strongly support weak mixing [6, 22]. While the mixing may be weak, it can still have important consequences and lead to enhanced $B(E2)$ values, especially for the $0_2^+ \to 2_1^+$ transitions that arise from the contribution of the large intruder in-band $B(E2; 2^+ \to 0^+)$ value and the $2I + 1$ spin factor involved [23].

2 Evidence for intruder states in Te

While the energy systematics presented in Fig. 2 for the Te isotopes has a characteristic "V"-shaped pattern for the 0_2^+ states (and even the 0_3^+ states), this does not prove that they are of intruder origin. Shown in Fig. 3 is a detailed comparison of the available data for assigned intruder band heads in the Cd isotopes (top) and the suggested analogous states in the Te isotopes (bottom). Concentrating first on the Cd isotopes, it is seen that, especially in ^{110,112}Cd, the 0_2^+ states have a very strong population in the Pd($^3He, n$) reactions [24], of 42% and 55% relative to the ground state population, respectively. The Pd target nuclei have 4 proton holes

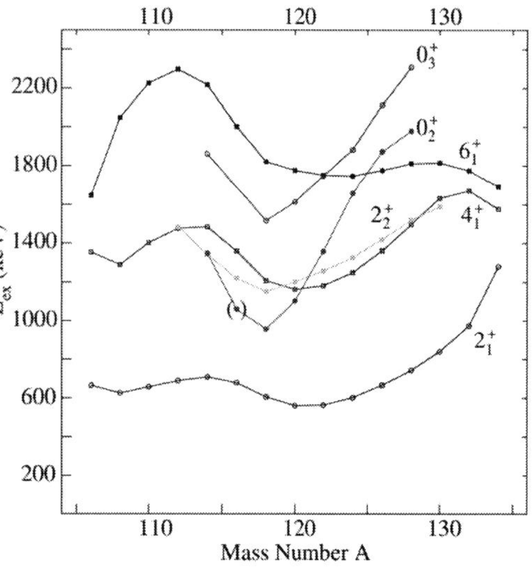

Figure 2. Systematics of the low-lying states in the even Te isotopes. The energy pattern of the 0_2^+ and 0_3^+ states suggests that there may be a crossing in the vicinity of ^{126}Te. Figure from Ref. [23].

with respect to the $Z = 50$ closed shell, and thus the two-proton-stripping reaction would be expected to populate either the ground state ($2h$) or the $2p - 4h$ configurations. These data

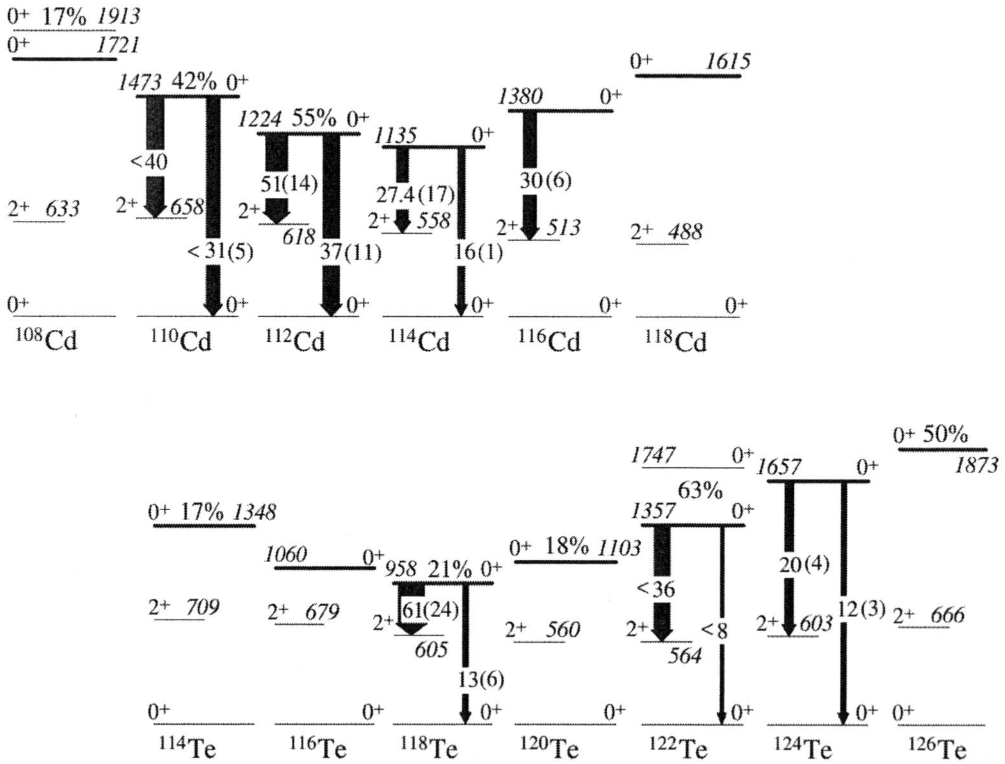

Figure 3. A comparison of the key data for the assigned deformed intruder band heads in the $^{108-118}$Cd isotopes (top) and those for the suggested 0^+ intruder states in $^{114-126}$Te (bottom). The levels with bold horizontal lines are the assigned or suggested intruder band heads. The numbers in percentages are the ($^3He, n$) cross sections relative to the ground state cross sections. The arrows represent the $B(E2)$ values for the $0^+ \rightarrow 2_1^+$ transitions, with their values in W.u., or $\rho^2(E0) \cdot 10^3$ values for $0^+ \rightarrow 0^+$ transitions. The nuclei are aligned vertically to a common neutron number. Figure adapted from Ref. [6].

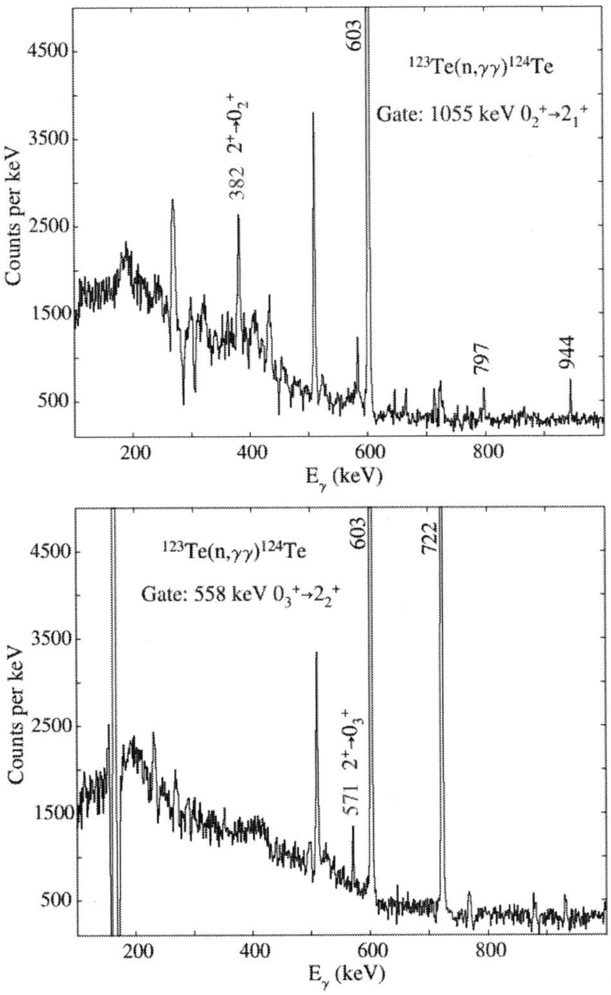

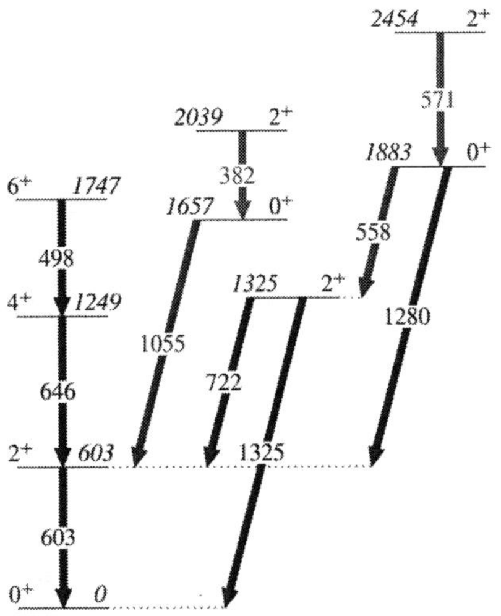

Figure 4. Partial results from the ^{123}Te$(n, \gamma\gamma)^{124}$Te reaction performed at the FIPPS facility. The spectra on the left are obtained by selecting coincidence conditions for the 1055-keV $0_2^+ \rightarrow 2_1^+$ (top) and 558-keV $0_3^+ \rightarrow 2_2^+$ (bottom) transitions, as indicated in blue on the partial level scheme shown above. The transitions marked in red are the suggested $2^+ \rightarrow 0^+$ "in-band" transitions, and are either newly observed (571 keV) or have a previously unmeasured branching ratio (382 keV).

were, in fact, key in establishing the nature of the intruder configurations. The intruder band heads also, generally, possess enhanced $B(E2)$ values for decay to the 2_1^+ state, with values of 51 ± 14 W.u., 27.4 ± 1.7 W.u., and 30 ± 6 W.u. for ^{112,114,116}Cd, respectively. Furthermore, moderately enhanced $E0$ transitions for decay to the ground state are also observed.

Turning to the Te isotopes, a very similar pattern is observed. Where the data are available, it is seen that the Sn($^3He, n$) reactions strongly populates the 0_2^+ states in ^{114,118,120}Te, and apparently both the 0_2^+ and 0_3^+ levels in ^{122}Te. In ^{118,120,122,124}Te the 0_2^+ states have enhanced $B(E2)$ values for decay to the 2_1^+ levels, and also moderately enhanced $E0$ transitions to the ground state. This striking similarity to the Cd isotopes is strongly suggestive that the states identified in Fig. 3 for the Te isotopes are the analogous intruder $4p - 2h$ excitations.

Until very recently, one of main issues in preventing a firm identification of the intruder excitations at low spin in the Te isotopes was the non-observation of enhanced in-band $E2$ transitions. In ^{122}Te, Hicks *et al.* [25] observed a 395-keV γ ray with $B(E2; 2_I^+ \rightarrow 0_I^+) = 194^{+26}_{-24}$ W.u. The 4^+ band member was not identified. In order to remedy this situation for other Te isotopes, a program of detailed spectroscopy using the newly constructed fission product prompt γ-ray spectrometer (FIPPS) [26] at the Institute Laue Langevin has been initiated. For the first experiment of our program, the ^{123}Te(n, γ) reaction, FIPPS consisted of 10 clover HPGe detectors. The flux of neutrons available is $\approx 1 \times 10^8$ n/cm^2 s^{-1}. An ≈ 4

mg sample of enriched ^{123}Te was irradiated in the neutron beam. Figure 4 displays partial results from the ^{123}Te(n,γ) reaction. The top-left panel shows the γ-ray spectrum obtained using a coincidence gate on the 1055-keV $0_2^+ \rightarrow 2_1^+$ γ ray (top); a 382-keV γ ray, placed as the $2_3^+ \rightarrow 0_2^+$ transition, is clearly seen. The bottom-left panel displays the γ-ray spectrum with a gate on the 558-keV $0_3^+ \rightarrow 2_2^+$ γ ray, showing clearly the presence of a new 571-keV γ ray. This is assigned as the $2_7^+ \rightarrow 0_3^+$ transition as shown in the partial level scheme on the right. The analysis is ongoing, but with the extraction of the branching ratio the $B(E2)$ values will be determined since the level lifetimes are known [12].

3 Summary

Using the properties of the well-established intruder structures in the Cd nuclei as a guide, the 0_2^+ states in the mid-shell Te nuclei are suggested to be of intruder origin based on $4p-2h$ proton excitations. To firmly establish their nature, evidence for the associated rotational-like band must be obtained. An experiment using the (n,γ) reaction has permitted the observation of a γ-ray transition linking the 2_3^+ state to the 0_2^+ state in ^{124}Te, and it is suggested that the 2_3^+ state is the 2^+ band member of an intruder band based on the 0_2^+ level.

References

[1] L.K. Peker and M.E. Voichansky, Izv. Akad. Nauk SSSR Ser Fiz. **32**, 892 (1968).
[2] G. Gneuss and W. Greiner, Nucl. Phys. **A171**, 449 (1971).
[3] P. Federman and S. Pittel, Phys. Lett. **B69**, 385 (1977).
[4] K. Heyde, P. Van Isacker, R.F. Casten, and J.L. Wood, Phys. Lett. **B155**, 303 (1985).
[5] K. Heyde and J.L. Wood, Rev. Mod. Phys. **83**, 1467 (2011).
[6] P.E. Garrett, J. Phys. G: Nucl. Part. Phys. **43**, 084002 (2016).
[7] N. Warr *et al.*, Nucl. Phys. **A636**, 379 (1998).
[8] J.R. Vanhoy *et al.*, Phys. Rev. C **69**, 064323 (2004).
[9] F. Iachello, Phys. Rev. Lett. **85**, 3580 (2000).
[10] R.M. Clark *et al.*, Phys. Rev. C **69**, 064322 (2004).
[11] D.G. Ghiţă *et al.*, Int. J. Mod. Phys. E: Nucl. Phys. **17**, 1453 (2008).
[12] S.F. Hicks *et al.*, Phys. Rev. C **95**, 034322 (2017).
[13] K. Heyde *et al.*, Phys. Rev. C **25**, 3160 (1982).
[14] J. Kern *et al.*, Nucl. Phys. **A512**, 1 (1990).
[15] M. Délèze *et al.*, Nucl. Phys. **A551**, 269 (1993).
[16] F. Corminboeuf *et al.*, Phys. Rev. Lett. **84**, 4060 (2000).
[17] F. Corminboeuf *et al.*, Phys. Rev. C **63**, 014305 (2000).
[18] R. Julin *et al.*, Z. Phys. **A296**, 315 (1980)
[19] C. Fahlander *et al.*, Nucl. Phys. **A485**, 318 (1988).
[20] R.F. Casten *et al.*, Phys. Lett. **297B**, 19 (1992).
[21] P.E. Garrett *et al.*, Phys. Rev. C **86**, 044304 (2012).
[22] P.E. Garrett *et al.*, E. Phys. J. Web of Conf. **123**, 02005 (2016).
[23] P.E. Garrett, J.L. Wood, and S.W. Yates, Phys. Scr. *in press* (2018).
[24] H.W. Fielding *et al.*, Nucl. Phys. **A281**, 389 (1977).
[25] S.F. Hicks *et al.*, Phys. Rev. C **71**, 034307 (2005).
[26] C. Michelagnoli *et al.*, *Proc. 6th Workshop on Nuclear Fission and Spectroscopy of Neutron-Rich Nuclei*, Chamrousse, Mar. 2017, E. Phys. J. Web of Conf. *in press* (2018).

Studies of positive-parity low-spin states in the A = 150 region

Robert Bark[1,*], *Zhipan* Li[2], *Siyabonga* Majola[3], *John* Sharpey-Schafer[4], *Zhi* Shi[5] and *Shuangquan* Zhang[6]

[1]iThemba LABS, P.O. Box 722, Somerset West 7129, South Africa
[2]School of Physical Science and Technology, Southwest University, Chongqing, 400715 China
[3]Department of Physics, University of Zululand, Private Bag X1001, KwaDlangezwa, 3886, South Africa
[4]Department of Physics, University of Western Cape, P/B X17, Bellville 7535 South Africa
[5]School of Physics and Nuclear Energy Engineering, Beihang University, Beijing 100191, China
[6]State Key Laboratory of Nuclear Physics and Technology, School of Physics, Peking University, Beijing 100871, China

Abstract. A systematic investigation of low-lying levels of nuclides in the mass 150 region has been undertaken at iThemba LABS. An extensive set of data on the low-lying, positive-parity bands in the nuclides between N = 88 and 92 and Sm to Yb has been obtained from γ-γ coincidence measurements following fusion-evaporation reactions optimized of the population of low-spin states. The energies and electromagnetic properties of the so-called β- and γ-bands of nuclei in this region have been compared with the solutions of a five dimensional collective Hamiltonian for quadrupole vibrational and rotational degrees of freedom, with moments-of-inertia and mass parameters determined by constrained self-consistent relativistic mean-field calculations using the PC-F1 relativistic functional. Some of the results of this comparison are presented here.

1 Introduction

The Bohr Hamiltonian [1], one of the earliest models describing the lowest-lying excitations of deformed nuclei, predicted the existence of the so-called β and γ vibrational bands, based on $K=0$ and $K=2$ excitations.

While numerous examples of $K=0$ and $K=2$ bands have been identified, the identification of the lowest lying $K=0$ band with the β-band of the Bohr Hamiltonian is under question [2]. The mass 150 region elicits experimental interest as a testing ground for alternate descriptions of low-lying 0^+ bands. The phenomenon of shape-coexistence (see e.g. [3,4]), pairing isomers [5] and X(5) symmetry [6] have been invoked to interpret the low-lying structures around ^{152}Sm at $N=90$. However, the original Bohr-Hamiltonian approximated the nuclear potential with a harmonic oscillator in β and γ. A more realistic potential would be expected to be more complex, possibly including co-existing wells in

* Corresponding author: bark@tlabs.ac.za

the β - γ plane. Such a model would then automatically include shape-coexistence and vibrational bands under the one umbrella. Indeed a more realistic "Bohr Hamiltonian" has been developed - a five-dimensional Hamiltonian for quadrupole vibrational and rotational degrees of freedom, with collective parameters determined by constrained self-consistent covariant density functional calculations for triaxial shapes [7,8]. Rather than use a harmonic oscillator potential in β and γ, a realistic potential $V_{coll}(\beta,\gamma)$ is determined from the potential energy surfaces calculated using relativistic mean field (RMF) theory.

Experimentally, a large body of data from in-beam $\gamma\gamma$-coincidence measurements, following fusion evaporation reactions, has been collected in experiments performed at iThemba LABS, Argonne National Laboratory and the University of Jyväskylä. In most even-even nuclides between N = 88 and 92 and between Sm and Yb, the 0_2^+, and gamma bands have been delineated up to at least spin $12\hbar$. Here, we will combine a subset of our results with data from the literature and compare the data with the predictions of the "realistic" five-dimensional Hamiltonian.

2 Theory

Detailed formalism of the five-dimensional collective Hamiltonian (5DCH) can be found in a number of articles, see, e.g. [7-10]. For completeness, a brief introduction is presented here. The five-dimensional collective Hamiltonian, which can simultaneously treat the quadrupole vibrational and rotational excitations, is expressed in terms of the two deformation parameters β and γ, and three Euler angles Ω that define the orientation of the intrinsic principal axes in the laboratory frame,

$$H_{coll}(\beta,\gamma) = T_{vib}(\beta,\gamma) + T_{rot}(\beta,\gamma,\Omega) + V_{coll}(\beta,\gamma) \tag{1}$$

The three terms in collective Hamiltonian are respectively the vibrational kinetic energy, rotational kinetic energy, and collective potential. There are seven collective parameters, including the mass parameters $B_{\beta\beta}$, $B_{\beta\gamma}$, $B_{\gamma\gamma}$, the moments of inertia I_k (k=1, 2, 3) and the collective potential V_{coll}, which are all functions of the intrinsic deformations β and γ. They in total govern the dynamics of the collective Hamiltonian. The eigenvalue problem of the Hamiltonian (1) is solved using an expansion of eigenfunctions in terms of a complete set of basis functions that depend on the five collective coordinates β, γ, and Ω. Using the collective wave functions thus obtained, various observables such as transition probabilities can be calculated.

In recent years, the five-dimensional collective Hamiltonian based on the covariant density functional theory (5DCH-CDFT) has been developed and extensively applied to describe the nuclear collective properties, such as the phase transitions, shape evolutions as well as the low-lying spectra along with the isotopic and isotonic chains in different mass regions [7,8,11-13]. In the framework of 5DCH-CDFT, the microscopic collective parameters of 5DCH are all determined from the CDFT. The moments of inertia are calculated with Inglis-Belyaev formula and the mass parameters from the cranking approximation. V_{coll} is obtained by subtracting the zero-point energy corrections from the total energy that corresponds to the solution of constrained triaxial CDFT.

3 Results and Discussion

To highlight the salient points of the work in these proceedings, we now focus on the data and calculations of a subset of the region, along N=90. The energies of the experimental ground, 0_2^+, gamma and 0_3^+ bands for ^{152}Sm [14, 15], ^{156}Dy [16] and ^{160}Yb [17] are plotted, less a rigid-rotor reference, as a function of spin in Fig. 1.

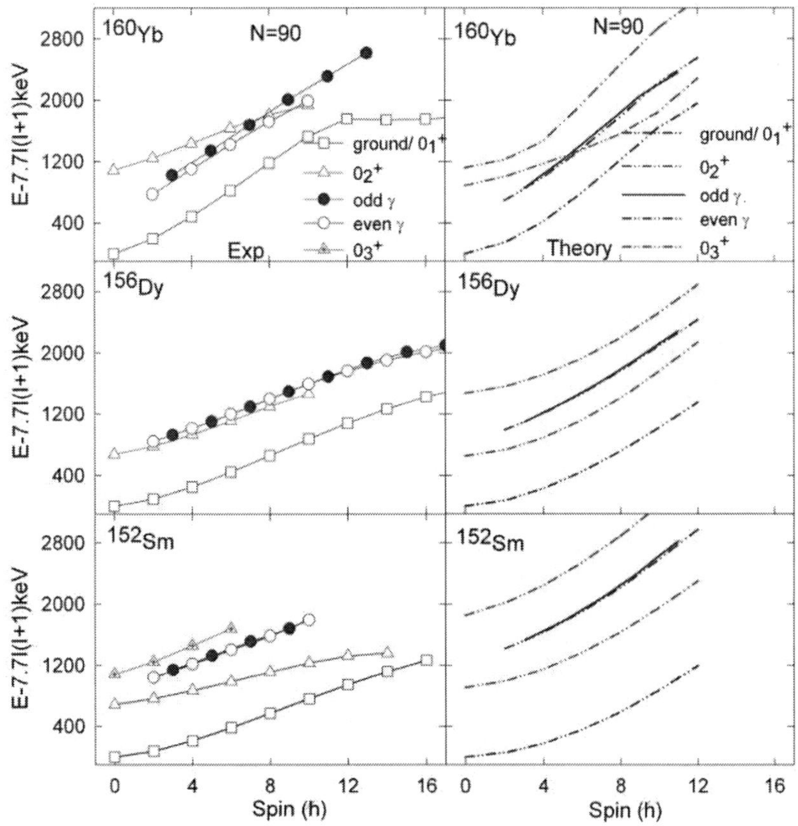

Fig. 1. Experimental bands left, compared with calculated bands, right.

A first observation is that in all cases, the gamma-bands have a very similar slope to that of the ground bands – indicating similar moments-of-inertia for the two bands. The same cannot be said for the 0_2^+ bands, where only in the case of ^{156}Dy does the 0_2^+ band run parallel to the ground band. In ^{160}Yb the 0_2^+ band, has a higher-excitation energy and a larger moment-of-inertia than in ^{156}Dy, which causes it to cross the gamma-band. In ^{152}Sm the 0_2^+ band again has larger moment-of-inertia than the ground band, but no crossing occurs with the gamma-band due to the lowering of the 0_2^+ excitation energy.

To understand the possible role deformation plays in the moments-of-inertia of these bands, we inspect the RMF potential energy surfaces for these nuclei in Fig. 2. The surface of ^{156}Dy shows a single minimum in the β - γ plane that is slightly extended towards higher β and mild triaxiality. This extension develops into a well-deformed triaxial superdeformed minimum in ^{160}Yb at $(\beta,\gamma) \sim (0.45,10°)$.

The corresponding solutions to the 5DCH are shown on the right of Fig. 1, where they are compared to the experimental bands on the left. Note that in all cases, the predicted gamma-bands follow the ground band with a similar moment-of-inertia. Inspection of the odd-even spin staggering in the gamma-bands in both experiment and theory indicates a tendency towards gamma-rigid rotation along *N=90* and *92*, and vibrational character along *N=88*. For the 0_2^+ bands, the theory predicts that sometimes they can be interpreted as collective excitations built on the ground band and in other cases they are the result of a co-existing minimum in the potential energy surface. In ^{160}Yb the calculations predict that the 0_2^+ band should have a very different moment-of-inertia to the ground and gamma-bands.

This results in a crossing of the gamma-band by the 0_2^+ band, in good agreement with the data. Inspection of the probability density distribution of the wavefunction of the 0_2^+ band confirms that it is localized in the minimum near β=0.45. In other words, the β-band of ^{160}Yb should rather be identified as a superdeformed triaxial band.

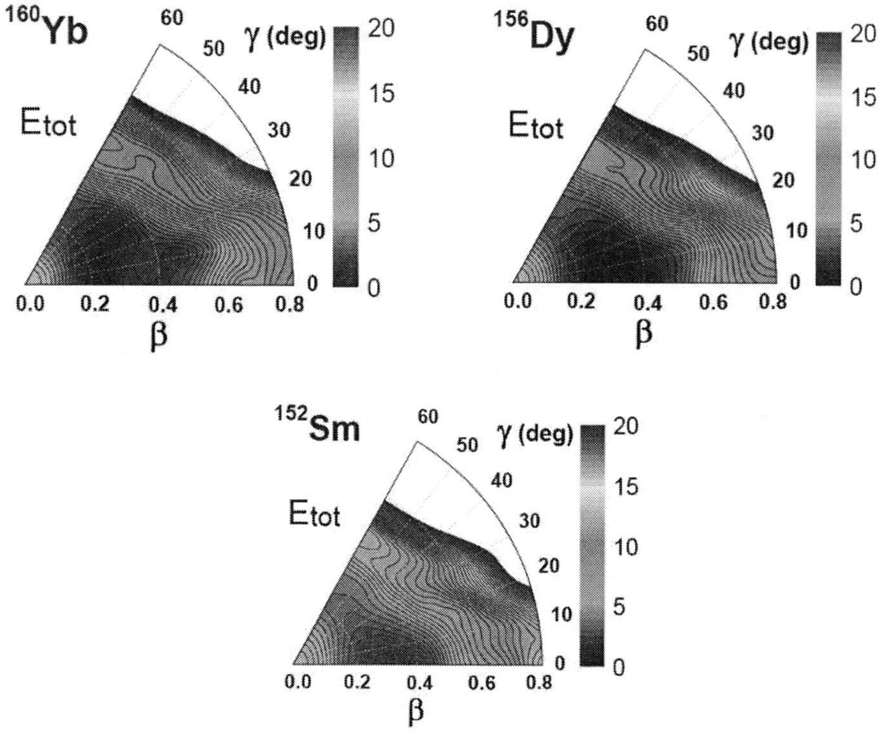

Fig. 2. RMF potential energy surfaces for ^{160}Yb, ^{156}Dy and ^{152}Sm

The crossing of the gamma band by the 0_2^+ band is a general feature observed in the data for the *N=88* to *92* isotopes of Yb and Er [17]. The crossings are well reproduced by the theory in which the so-called β-band is identified with a coexisting triaxial minimum.

In ^{156}Dy the calculated 0_2^+ band is in reasonable agreement with experiment – while the 0_2^+ band lies slightly lower in energy than the experimental one, it has a similar moment-of-inertia to the ground and gamma-bands. This band is better understood as a vibration.

The PES surface for ^{152}Sm has a single minimum on the prolate axis that is soft in the β-direction, reminiscent of the flat X(5) potential[5]. The feature of softness in the PES in the β-direction is also a feature for the *N=90* nucleus ^{154}Gd [17]. Three 0^+ bands are known experimentally in ^{152}Sm, but their theoretical counterparts lie higher in excitation energy. Theoretically, the 0_3^+ band is understood as a two-phonon band; the low excitation energy of the experimental band may point to a role for other degrees of freedom such as pairing[14].

Many other quantities have been calculated using the 5DCH, including E2 transition rates and the staggering in the gamma-bands. These quantities are generally in good agreement with the data. The E0 transition rates are an exception, as they are generally calculated to be stronger than those observed experimentally. A full report of this work will be published elsewhere [17].

We would like to thank the iThemba LABS technical staff for their support in these experiments. This work is supported by a National Research Foundation China-SA grant (No. 92791) and the Natural Science Foundation of China under Grants No. 11461141002, No. 11475140 and No. 11375015.

References

1. A. Bohr and B. R. Mottelson, Nuclear Structure, Vol. II, (1975)
2. P.E. Garrett, J. Phys. G: Nucl. Part. Phys. **27,** R1 (2001)
3. K. Heyde and J. L. Wood, Rev. Mod. Phys. **83**, 1467 (2011).
4. P.E. Garrett et al., Phys. Rev. Lett. **103**, 062501 (2009)
5. J.F. Sharpey-Schafer et al., Eur. Phys. J. A**47**, 5 (2011)
6. R. F. Casten and N.V. Zamfir, Phys. Rev. Lett. **87**, 052503 (2001)
7. T. Nikšić et al.,Phys. Rev. C **79**, 034303 (2009)
8. Z. P. Li et al., Phys. Rev. C **79**, 054301 (2009)
9. L. Próchniak *et al.*, Nucl. Phys. A **648**, 181 (1999)
10. J. Libert, M. Girod, and J.-P. Delaroche, Phys. Rev. C **60**, 054301 (1999)
11. Z. P. Li *et al.*, Phys. Rev. C **81**, 064321 (2010)
12. Z. P. Li *et al.*, Phys. Rev. C **84**, 054304 (2011)
13. Z. P. Li *et al.*, Phys. Lett. B **717**, 470 (2012)
14. W. D. Kulp, *et al.*, Phys. Rev. C **71**, 041303 (R) (2005)
15. M. Martin, Nuclear Data Sheets 114, 1497 (2013)
16. S. Majola, *et al.*, Phys. Rev. C **91**, 034330 (2015)
17. S.N.T Majola *et al.,* to be published

Nuclear shapes: Quest for triaxiality in ^{86}Ge and the shape of ^{98}Zr

V. Werner[1,*], *M.* Lettmann[1], *C.* Lizarazo[1], *W.* Witt[1], *D.* Cline[2], *M.* Carpenter[3], *P.* Doornenbal[4], *A.* Obertelli[1,4,5], *N.* Pietralla[1], *G.* Savard[3], *P.-A.* Söderström[1], *C.-Y.* Wu[6], and *S.* Zhu[3]

[1]Institut für Kernphysik, Technische Universität Darmstadt, 64289 Darmstadt, Germany
[2]Dept. of Physics and Astronomy, University of Rochester, Rochester, NY 14642, USA
[3]Argonne National Laboratory, Argonne, IL 60439, USA
[4]RIKEN Nishina Center, 2-1 Hirosawa, Wako, Saitama 351-0198, Japan
[5]CEA, Centre de Saclay, IRFU/Service de Physique Nucléaire, 91191 Gif-sur-Yvette, France
[6]Lawrence Livermore National Laboratory, Livermore, CA 94550, USA

Abstract. The region of neutron-rich nuclei above the $N = 50$ magic neutron shell closure encompasses a rich variety of nuclear structure, especially shape-evolutionary phenomena. This can be attributed to the complexity of sub-shell closures, their appearance and disappearance in the region, such as the $N = 56$ sub shell or $Z = 40$ for protons. Structural effects reach from a shape phase transition in the Zr isotopes, over shape coexistence between spherical, prolate, and oblate shapes, to possibly rigid triaxial deformation. Recent experiments in this region and their main physics viewpoints are summarized.

1 Introduction

The mass $A \approx 100$ region above $N = 50$ is a region of the nuclear chart which is dominated by the occurrence of multiple major and sub-shell closures. That is, in particular, the $N = 50$ magic neutron shell itself, as well as the $Z = 40$ proton and the $N = 56$ and $N = 58$ neutron sub shells. Among the consequences of these sphericity-driving sub shells is a stabilization of spherical shapes in the ground states of the respective nuclei, and a fast transition toward deformation in the Zr isotopic chain as the sub-shell closures disappear due to changes in the ordering of effective single-particle energies and the early filling of higher-lying orbitals. It is, however, still an open question at what mass number in the Zr isotopic chain the phase transition from a spherical to a deformed ground-state structure occurs. It is known that spherical and deformed structures coexist at low energies in ^{94}Zr [1] and ^{96}Zr [2, 3], but it is not certain at which mass number those structures actually cross toward the ground-state deformed Zr isotopes above $A = 98$. We report on the first measurement of a stringent upper limit for the $B(E2;2^+_1 \rightarrow 0^+_1)$ value of ^{98}Zr, which aids in the assignment of weak collectivity in its ground state.

Furthermore, the coexistence of prolate and oblate deformations has recently been suggested for Se isotopes in the mass $A \approx 90$ region [4]. These structures should cross each other, so a transition of the ground-state deformation from prolate to oblate shape would be

*e-mail: vw@ikp.tu-darmstadt.de

expected. We performed isomer spectroscopy on ^{92,94}Se, which shows a significantly different decay behavior of the respective isomers, which may be a signature for this shape transition. Only two protons below the Se isotopes, namely, in ^{86}Ge, we identify ^{86}Ge as a candidate for displaying rigid ground-state triaxiality [5], a rare phenomenon which has, so far in medium-heavy nuclei, only been observed in its lighter isotope, ^{76}Ge [6].

2 Triaxial ^{86}Ge?

The occurrence of stable triaxiality in nuclei has attracted much attention in nuclear structure research, in terms of triaxial superdeformed bands and their wobbling modes (see, e.g., Ref [7]) or chirality (e.g., Ref. [8]). However, realizations of stabilized triaxiality in the ground state are rare and have heavily been discussed, and the only example in medium-heavy nuclei has recently been identified in ^{76}Ge [6] through the characteristic staggering in the γ band. Namely, in γ-soft nuclei, the 3^+_γ member of the odd-spin γ band is typically located closer to the 4^+_γ than to the 2^+_γ state [9]. In case of rigid triaxiality, the opposite is the case, in addition to the location of the γ band at rather low energies (around or below the 4^+_1 state), which is common for both, rigid and soft triaxial cases.

A recent experimental campaign at RIKEN-RIBF, aimed to locate first excited 2^+ states over a broad range of neutron-rich isotopes (SEASTAR) [10], in many cases quite complicated level schemes have been observed (see, e.g., [4, 11–14]), sometimes allowing to place the lowest members of the γ band. In ^{86}Ge, which was populated from a proton-knockout reaction on ^{87}As in the MINOS active LH_2 target [15], the observation of a low-energy γ-ray at 380 keV in the DALI2 array [16] allowed for the tentative placement of a (3^+_1) state at 1426(16)-keV excitation energy [5], closer to the 2^+_2 state at 1046(14) keV than to the 4^+_2 state at 1911(23) keV (compare Fig. 2 in [5]), in agreement with results from the shell model and from symmetry-conserving configuration mixing Gogny calculations.

3 Prolate-oblate shape transition in ^{92,94}Se

Calculations and recent RIKEN-RIBF data for neutron-rich Se isotopes [4] indicate the possibility for prolate-oblate shape coexistence and a transition from a prolate to an oblate ground-state minimum of the energy potential between ^{90}Se and ^{94}Se. Concurrently with the in-beam SEASTAR experiments we performed isomer spectroscopy using the EURICA array [17]. Isotopes were delivered through the BigRIPS and ZeroDegree separators to the location of EURICA, and energy and time information of delayed γ-ray transitions were recorded. In both, ^{92}Se and ^{94}Se, delayed transitions from an isomer with lifetime in the micro-second range have been observed. From considerations within the Nilsson scheme, an oblate configuration is most likely for those isomers, which would be in line with the observed lifetimes.

A fundamental difference between ^{92}Se and ^{94}Se are the decay behaviors of the respective isomeric states. In ^{92}Se the decay takes place exclusively to the γ band, which is likely due to the lower K hindrance for this decay in comparison to a decay into the $K = 0$ ground state band. In ^{94}Se, however, the isomer decays exclusively into the ground-state band. This could be a signature of oblate deformation in the ground-state band of ^{94}Se, since in that case the decay of an oblate isomer should be less hindered due to a larger overlap of the isomer and the ground-state band wave functions.

4 The critical-point isotope ^{98}Zr

Another case of shape coexistence has been observed in the $^{96-100}$Zr isotopes, as mentioned above. In an experiment at the ATLAS/CARIBU [18] facility at Argonne National Laboratory we employed the combination of the GRETINA [19] and the CHICO2 [20] detectors with a ^{98}Zr beam to perform Coulomb excitation of ^{98}Zr on a ^{196}Pt target at a beam energy of 464 MeV. Unfortunately, the rare-isotope beam was dominated by contaminating ^{98}Mo, which resulted in a γ ray at 1230 keV from the de-excitation of the 3_1^- state of ^{98}Mo. A peak from the decay of the 2_1^+ state of ^{98}Zr was not observed, however, an upper limit of its intensity was obtained from the spectrum. In addition, the beam composition was measured at a decay station behind the experimental setup, yielding a rate of 152(64) pps ^{98}Zr in the beam, as well as rates for other components of the beam. With these values, and with the known excitation cross section of the 2_1^+ state of ^{196}Pt, it was possible to extract a stringent upper limit for the B(E2) excitation strength of the 2_1^+ state of ^{98}Zr. Along with an available lower limit [21] of about 1.8 W.u., and our upper limit of less than 20 W.u. for the B(E2;$2_1^+ \rightarrow 0_1^+$) value of ^{98}Zr, this isotope is not yet on the well-deformed side of the shape-phase transition towards ^{100}Zr with a B(E2;$2_1^+ \rightarrow 0_1^+$) value of about 75 W.u. Hence, ^{98}Zr is the closest to the spherical-deformed phase transtion in the Zr isotopes.

5 Summary

To conclude, we found evidence for three shape-evolutionary aspects in the $A \approx 100$ mass region above $N = 50$. Namely, these are the occurence of a spherical-deformed shape-phase transition in the Zr isotopes, where we identify ^{98}Zr as the transitional isotope, the coexistence of prolate and oblate shapes in the Se isotopes, where we find evidence that ^{94}Se is oblate in its ground-state band, and a first hint for triaxiality in the ground state of ^{86}Ge from the identification of a 3_1^+ state in the γ band. We acknowledge support from the German BMBF Grants 05P15RDFN1, 05P12RDFN8, and 05P15PKFNA, and ERC Grant MINOS-258567.

References

[1] A. Chakraborty *et al.*, Phys. Rev. Lett. **110**, 022504 (2013)
[2] C. Kremer *et al.*, Phys. Rev. Lett. **117**, 172503 (2016)
[3] T. Togashi *et al.*, Phys. Rev. Lett. **117**, 172502 (2016)
[4] S. Chen *et al.*, Phys. Rev. C **96**, 041302(R) (2017)
[5] M. Lettmann *et al.*, Phys. Rev. C **96**, 011301(R) (2017)
[6] Y. Toh *et al.*, Phys. Rev. C **87**, 041304(R) (2013)
[7] D.J. Hartley *et al.*, Phys. Rev. C **80**, 041304(R) (2009)
[8] T. Koike, K. Starosta, and I. Hamamoto, Phys. Rev. Lett. **93**, 172502 (2004)
[9] N.V. Zamfir and R.F. Casten, Phys. Lett. B **260**, 265 (1991)
[10] C. Santamaria *et al.*, Phys. Rev. Lett. **115**, 192501 (2015)
[11] N. Paul *al.*, Phys. Rev. Lett. **118**, 032501 (2017)
[12] L. Olivier *et al.*, Phys. Rev. Lett. **119**, 192501 (2017)
[13] F. Flavigny *et al.*, Phys. Rev. Lett. **118**, 242501 (2017)
[14] C.M. Shand *et al.*, Phys. Lett. B **773**, 492 (2017)
[15] A. Obertelli *et al.*, Eur. Phys. J. A **50**, 8 (2014)
[16] S. Takeuchi *et al.*, Nucl. Instrum. Methods Phys. Res. A **763**, 596 (2014)
[17] P.-A. Söderström *et al.*, Nucl. Instrum. Methods Phys. Res. B **317**, 642 (2013)
[18] G. Savard *et al.*, Nucl. Instrum. Methods Phys. Res. B **376**, 246 (2016)
[19] S. Paschalis *et al.*, Nucl. Instrum. Methods Phys. Res. A **709**, 44 (2013)
[20] C.-Y. Wu *et al.*, Nucl. Instrum. Methods Phys. Res. A **814**, 6 (2016)
[21] S. Ansari *et al.*, Phys. Rev. C **96**, 054323 (2017)

Nuclear shapes studied with low-energy Coulomb excitation

Magda Zielińska[1,*] and *Katarzyna* Hadyńska-Klęk[2]

[1]Irfu, CEA, Université Paris-Saclay, F-91191 Gif-sur-Yvette, France
[2]Department of Physics, University of Surrey, Guildford GU2 7XH, UK

Abstract. Coulomb excitation is one of the rare methods available to obtain information on static electromagnetic moments of short-lived excited nuclear states, including collective non-yrast levels. It is thus an ideal tool to study shape coexistence and shape evolution throughout the nuclear chart. Historically, these experiments were limited to stable isotopes, however the advent of new facilities, providing intense beams of short-lived radioactive species, has opened the possibility to apply this powerful technique to a much wider range of nuclei. Here, we present some recent complex Coulomb-excitation studies and use the example of superdeformed states in ^{42}Ca to demonstrate the sensitivity of the method to second-order effects such as relative signs of electromagnetic matrix elements and quadrupole moments.

1 Introduction

One of the most fundamental properties of the atomic nucleus is its shape, which is governed by the interplay of macroscopic, liquid-drop like properties of the nuclear matter and microscopic shell effects. While closed-shell nuclei always exhibit spherical shapes in their ground states, nucleons residing in open shells tend to deform the mass distribution, with quadrupole deformation being the most important degree of freedom. In some cases, configurations corresponding to different shapes may coexist at similar energies, which results in the wave functions of these states mixing. Experimental observables, such as electromagnetic transition rates between excited states and static quadrupole moments, are closely related to the nuclear shape. The experimental determination of these observables, therefore, represents a stringent test for predictions of the theoretical models.

Coulomb excitation has been shown to be a highly successful method to study nuclear collectivity and deformation. In the scattering of two nuclei, the electromagnetic field that acts between them causes their excitation. At low beam energy (typically 2-5 MeV/A) this process is dominated by the well-known electromagnetic interaction, while contributions from the nuclear force are negligible. This technique has been pioneered in the 1950s and was extensively used during the 1960s and 70s when heavy-ion accelerators and high-resolution germanium detectors became readily available. However, it was limited to stable or very long-lived nuclei. A recent renaissance of this method was related to the advent of ISOL-based RIB facilities, making possible experiments on short-lived radioactive nuclei.

*e-mail: magda.zielinska@cea.fr

Coulomb-excitation cross sections depend in a complex way on signs and magnitudes of electromagnetic matrix elements between the states involved in the excitation process. A given state may be populated directly, or via one or more intermediate states. This leads to appearance of interference terms between possible excitation paths, which can increase or decrease the population of the state in question, depending on the relative signs of the matrix elements involved, see Ref. [1]. Consequently, sophisticated computer codes are needed to extract nuclear structure parameters from the measured cross sections. Among them, the GOSIA code [2] seems to be most widely used as it is the only Coulomb-excitation code able to perform a multi-dimensional χ^2 fit of electromagnetic matrix elements to experimental data. The existing analysis methods have been refined to address common issues present when studying exotic nuclei [1], and currently Coulomb excitation is one of the most important experimental methods used to study nuclear shapes and collectivity, in particular on the neutron-rich side of the nuclear chart. In the present paper, we show some recent examples of complex multi-step Coulomb-excitation studies and illustrate the influence of second-order effects on measured cross sections with the example of superdeformed states in ^{42}Ca.

2 Recent examples of shape coexistence studies using Coulomb excitation of exotic nuclei

The potential of low-energy Coulomb excitation applied to short-lived exotic nuclei was demonstrated in a pioneering experiment on ^{74,76}Kr [3], which was the first to utilise the reorientation effect to measure quadrupole moments with radioactive beams. Opposite signs of the spectroscopic quadrupole moments were found for the states in the ground-state bands and those built on the shape isomers, confirming the shape coexistence scenario. The coexistence of a prolate ground-state band with an excited oblate band built on the low-lying 0^+_2 state for A = 74, 76 and the inversion of oblate and prolate configurations below A = 74 was found to be well described by HFB calculations using the Gogny D1S interaction with configuration mixing (GCM/GOA) [3]. Later, it was shown that allowing for triaxial degrees of freedom is crucial in this mass region to properly reproduce the experimental results [4].

Another region, where important nuclear structure information was obtained using Coulomb excitation of exotic beams, is that of A~100 neutron-rich nuclei. A sudden onset of quadrupole deformation, observed at N = 60 in the neutron-rich Zr and Sr isotopes, triggered an important experimental and theoretical effort to explain the origin of this dramatic shape change. The boundaries of the deformed regions have been firmly established in low-energy Coulomb-excitation experiments at REX-ISOLDE. Regular rotational ground-state bands were observed in for the first time in ^{97,99}Rb [5], and the obtained transition probabilities show that the deformation of these nuclei is essentially the same as that observed inside the well-deformed region. On the other hand, the B($E2$; $2^+_1 \to 0^+_1$) value measured for ^{96}Kr [6, 7] is significantly lower than those for ^{98}Sr and ^{100}Zr, and only slightly higher than that for ^{94}Kr [6, 7]. The recent Coulomb-excitation study of ^{96,98}Sr [8, 9] demonstrated important similarities in terms of transition probabilities and spectroscopic quadrupole moments between the ground-state band in ^{96}Sr and the structure built on the 0^+_2 state in ^{98}Sr. This confirmed the scenario of two distinct configurations associated with different nuclear shapes interchanging at N = 60. However, while in other regions of shape coexistence a strong mixing of underlying structures is the rule, the analysis of the obtained $E2$ matrix elements in ^{98}Sr using the two-state mixing model points to very low mixing between prolate and spherical configurations in the wave functions of the two 0^+ states, in spite of their proximity in energy. This has been related [8] to the abrupt transition at N = 60, in contrast to a smooth evolution of ground-state properties, 2^+_1 energies and B($2^+_1 \to 0^+_1$) transition strengths observed, for example, in $^{182-188}$Hg [10] and ^{74,76}Kr [3].

3 Second-order effects in Coulomb excitation of ^{42}Ca

Recently, the Coulomb-excitation technique has been applied for the first time to study superdeformed states [11, 12]. In the very first experiment using the AGATA array [13], the superdeformed structure in ^{42}Ca was populated following Coulomb excitation of the ^{42}Ca beam on ^{208}Pb and ^{197}Au targets. From the measured γ-ray intensities a complete set of electromagnetic matrix elements between the observed states was obtained, including spectroscopic quadrupole moments and relative signs of transitional $E2$ matrix elements. Those, in turn, were interpreted using the quadrupole sum rules formalism [14] leading to the conclusion that the spherical ground state of ^{42}Ca exhibits large fluctuations into the β-γ plane, while the excited structure has a large quadrupole deformation of $\beta = 0.43(4)$ for 0^+_2, comparable to those measured for other superdeformed bands in this mass region. Additionally, the triaxiality parameter $\langle\cos(3\delta)\rangle$ obtained for the 0^+_2 state provided the first experimental evidence for non-axial character of superdeformed structures around A$\sim$ 40.

The level scheme of ^{42}Ca, taken into account in the Coulomb-excitation analysis, is presented in Fig.1. Transitions observed in the experiment are marked in blue. The analysis yielded two key pieces of information regarding the deformation of the side band, highlighted in red in Fig.1: the $\langle 2^+_2\|E2\|0^+_2\rangle$ matrix element, as well as the spectroscopic quadrupole moment of the 2^+_2 state, are both consistent with a large quadrupole deformation of this band. In the present paper, we analyse the sensitivity of the experimental data to these two observables, as well as to the relative signs of transitional $E2$ matrix elements.

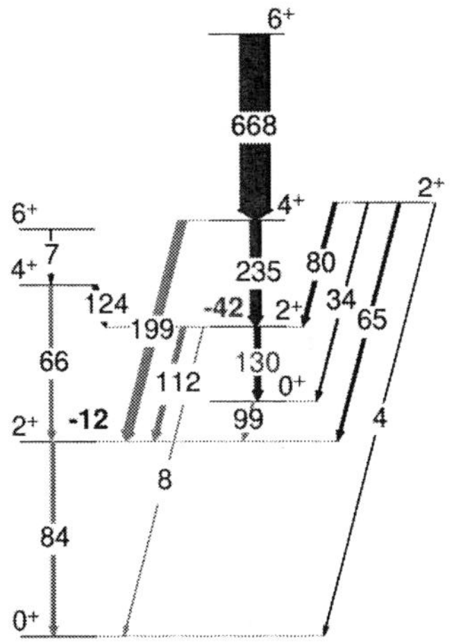

Figure 1. Low-lying excited states in ^{42}Ca, considered in the analysis described in Refs. [11, 12]. Transitions observed in the Coulomb-excitation experiment are marked in blue. The widths and labels of the arrows represent the experimental B($E2$,↓) values in e^2fm^4. Spectroscopic quadrupole moments, expressed in $e$fm^2, are given in bold. Red colour is used to highlight the observables which have been determined for the first time.

The following sign convention has been imposed: the signs of all in-band transitional $E2$ matrix elements, both in the ground state band and in the side band, were assumed to be positive, as well as that of $\langle 0^+_2\|E2\|2^+_1\rangle$. Signs of all other $E2$ matrix elements have been determined relative to those. The signs of two matrix elements, coupling the states of low spin and realtively low excitation energy, strongly influenced the observed excitation cross sections: $\langle 0^+_1\|E2\|2^+_2\rangle$ and $\langle 2^+_1\|E2\|2^+_2\rangle$. Figure 2 presents the influence of the signs of these matrix elements on intensities of γ-ray transitions following Coulomb excitation of ^{42}Ca on ^{208}Pb. The calculations were performed for the magnitudes of all matrix elements obtained in Refs. [11, 12] and all possible combinations of signs of the two matrix elements. It should be noted here that the magnitudes of transitional matrix elements were strongly constrained by known lifetimes of all excited states in ^{42}Ca included in the analysis, measured with precision ranging from 2% (0^+_2) to about 20% (4^+_2, 2^+_3). As shown in Fig. 2, the effect of interference terms on Coulomb-excitation cross section is much larger in the range of scattering angles covered by the particle detector (105-142°), and corresponds to changes of the population of excited states by a factor of two or more. Under these conditions, the sensitivity of the data to the relative signs of matrix elements is undeniable. These signs, obtained for the first time for transitions de-exciting a superdeformed structure, were crucial for the quadrupole sum rules analysis presented in Refs. [11, 12].

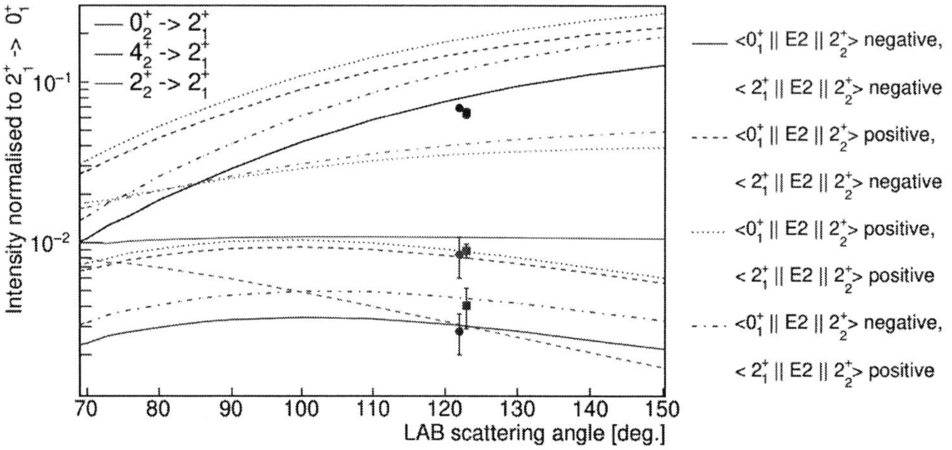

Figure 2. Transition intensities in ^{42}Ca normalised to that of $2_1^+ \to 0_1^+$, following Coulomb excitation of ^{42}Ca on ^{208}Pb in the experimental conditions of the measurement described in Refs. [11, 12]. The solid lines denote the calculations using the final set of matrix elements obtained in the analysis, while for the other lines different signs of the $\langle 0_1^+ \| E2 \| 2_2^+ \rangle$ and $\langle 2_1^+ \| E2 \| 2_2^+ \rangle$ matrix elements were used. The experimental intensity ratios are also shown; those measured for ^{197}Au, denoted by squares, are rescaled to take into account the difference in cross sections, and, for clarity, slightly offset on the x axis.

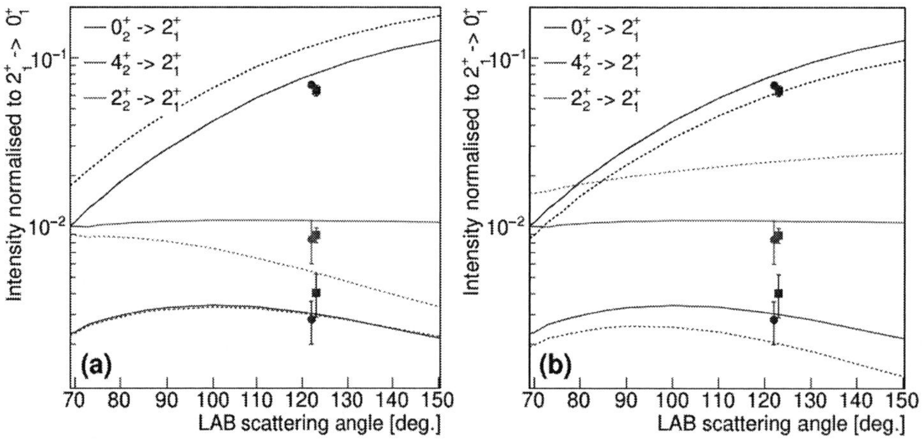

Figure 3. Same as Fig. 2, but the dotted lines correspond to: (a) $\langle 2_2^+ \| E2 \| 0_2^+ \rangle$ equal to zero, (b) quadrupole moment of the 2_2^+ state equal to zero.

Figure 3 illustrates the influence of the $\langle 2_2^+ \| E2 \| 0_2^+ \rangle$ matrix element (a) and the spectroscopic quadrupole moment of the 2_2^+ state (b) on measured γ-ray intensities. Clearly, although the $2_2^+ \to 0_2^+$ transition is too weak to be observed and prior to the present study only an upper limit for the branching ratio has been known, the corresponding matrix element strongly affects excitation cross sections of observed states and hence it could be determined from the intensities of other transitions measured in the Coulomb-excitation experiment. The effects of the $\langle 2_2^+||E2||0_2^+\rangle$ and $\langle 2_2^+||E2||2_2^+\rangle$ matrix elements on the population of both the 0_2^+ and 2_2^+ states are opposite, and consequently an increase of $\langle 2_2^+||E2||0_2^+\rangle$ could be compensated by a larger value of $\langle 2_2^+||E2||2_2^+\rangle$, leading to an equally good reproduction of the $2_2^+ \rightarrow 2_1^+$ and $0_2^+ \rightarrow 2_1^+$ transition intensities. The population of the 4_2^+ state, however, is sensitive only to the latter, thus making it possible to extract both matrix elements independently. The existing correlation between them is reflected in relatively large uncertainties of the two matrix elements ($\langle 2_2^+||E2||0_2^+\rangle = 26\,^{+5}_{-3}$ efm^2, $\langle 2_2^+||E2||2_2^+\rangle$ = -55 $^{+15}_{-15}$ efm^2 [11]). Their values confirm the superdeformed character of the excited structure in ^{42}Ca at low spin. In particular, the measured spectroscopic quadrupole moment of the 2_2^+ state corresponds to β=0.48(16). It should be noted here that improving the precision on the 4_2^+ lifetime would better constrain the magnitude of all matrix elements related to the deexcitation of this state, which would help to evaluate the role of second-order effects more precisely and thus increase the sensitivity to the $\langle 2_2^+||E2||0_2^+\rangle$ and $\langle 2_2^+||E2||2_2^+\rangle$ matrix elements.

4 Summary

The Coulomb-excitation technique can be used to explore many facets of nuclear shape evolution, from the mechanism driving the onset of quadrupole deformation, the development of exotic deformations, such as superdeformed or triaxial shapes, to the coexistence of different shapes in the same nucleus. The sensitivity of Coulomb-excitation data to second-order effects, such as influence of static quadrupole moments or relative signs of matrix elements that are critical for the sum-rule analysis, can be greatly enhanced by precise complementary data, in particular lifetimes and branching ratios. If lifetimes of excited states are unknown, which is often the case for exotic nuclei, integral measurements of Coulomb-excitation cross sections may lead to transition probabilities involving non-yrast states over- or underestimated by as much as a factor of two when a wrong combination of relative signs of matrix elements is assumed. This shows the importance of differential cross-section measurements that are sensitive to relative signs.

References

[1] M. Zielińska *et al.*, Eur. Phys. J. **A52**, 99 (2016)
[2] T. Czosnyka, D. Cline, and C.Y. Wu, Bull. Am. Phys. Soc. **28**, 745 (1983)
[3] E. Clément *et al.*, Phys. Rev. **C75**, 054313 (2007)
[4] M. Girod *et al.*, Phys. Lett. **B676**, 39 (2009)
[5] C. Sotty *et al.*, Phys. Rev. Lett. **115**, 172501 (2015)
[6] M. Albers *et al*, Phys. Rev. Lett. **108**, 062701 (2012)
[7] M. Albers *et al*, Nucl. Phys. **A899**, 1 (2013)
[8] E. Clément *et al.*, Phys. Rev. Lett. **116**, 022701 (2016)
[9] E. Clément *et al.*, Phys. Rev. **C94**, 054326 (2016)
[10] N. Bree *et al.*, Phys. Rev. Lett. **112**, 162701 (2014)
[11] K. Hadyńska-Klęk *et al.*, Phys. Rev. Lett. **117**, 062501 (2016)
[12] K. Hadyńska-Klęk *et al.*, Phys. Rev. **C97**, 024326 (2018)
[13] S. Akkoyun *et al.*, Nucl. Instr. Meth. Phys. Res. **A668**, 26 (2012)
[14] D. Cline, Annu. Rev. Nucl. Part. Sci. **36**, 683 (1986)

Shell model description of heavy nuclei and abnormal collective motions

Chong Qi[1,*]

[1]KTH Royal Institute of Technology, SE-10691 Stockholm

Abstract. In this contribution I present systematic calculations on the spectroscopy and electromagnetic transition properties of intermediate-mass and heavy nuclei around ^{100}Sn and ^{208}Pb. We employed the large-scale configuration interaction shell model approach with realistic interactions. Those nuclei are the longest isotopic chains that can be studied by the nuclear shell model. I will show that the yrast spectra of Te isotopes show a vibrational-like equally spaced pattern but the few known E2 transitions show rotational-like behaviour. These kinds of abnormal collective behaviors cannot be reproduced by standard collective models and provide excellent background to study the competition of single-particle and various collective degrees of freedom. Moreover, the calculated B(E2) values for neutron-deficient and heavier Te isotopes show contrasting different behaviours along the yrast line, which may be related to the enhanced neutron-proton correlation when approaching N=50. The deviations between theory and experiment concerning the energies and E2 transition properties of low-lying 0^+ and 2^+ excited states and isomeric states in those nuclei may provide a constraint on our understanding of nuclear deformation and intruder configuration in that region.

From a nuclear structure physics point of view, one also aims at addressing fundamental nuclear physics questions including: How are complex nuclei built from their basic constituents and how to explain collective phenomena from individual motion [1]? Nuclear theory plays a critical role in explaining those regular emerging phenomena in nuclear many-body systems. In particular, the nuclear shell-model calculations have been shown to be very successful in describing nuclei below ^{100}Sn. The aim of our work is to provide a microscopic shell-model description of the structure and collective behaviours of intermediate-mass and heavy nuclei. In particular, I will show its application of the shell model in nuclei around ^{100}Sn and ^{208}Pb. I will explain the structure and decay studies of those nuclei, regarding both experimental and theoretical opportunities. Although not covered in this contribution, it may be useful to mention that a proper description of the neutron-rich isotones around $N = 82$ and 126 can be important for our understanding of the astrophysical rapid neutron capture process and its possible existence of the third peak.

Nuclear structure models treat the mean field in quite different ways. Within the *ab initio* family, the no-core shell model approach aims at considering the residual correlation between all nuclei in a large space defined by the harmonic oscillator single-particle orbitals. The nuclear shell model, as we call it, is a full configuration interaction approach. It considers the mixing effect of all possible configurations within a given model space. The model space is usually defined by taking a few single-particle orbitals near the Fermi surface. The number of orbitals one can include in above two models is highly restricted due to computation limitation. Extensive studies on algorithm optimization and possible truncation or approximation methods have been carried out. State-of-the-art configuration interaction algorithms are able to diagonalize matrices with dimension up to 2×10^{10}. Other challenges of the no-core and full configuration interaction shell model approaches include for examples the connection of their effective interaction to fundamental forces and the complexity of the their wave function. Despite of these challenges, the nuclear shell model is by far the most accurate and precise theory available on the market.

The residual interaction between valence particles around the Fermi surface is mostly supposed to be of two-body nature. The two-body matrix elements of the effective Hamiltonian can be calculated from realistic nucleon-nucleon potential where one has to consider the effect of its short-range repulsion and the core polarization effects induced by the assumed inert core. However, an optimization of the monopole interaction is necessary in most cases due to the neglect of explicit three-body force and other effects. We have optimized the monopole channel of the effective Hamiltonians for describing the Sn and Pb isotopes, which work well also for most open-shell nuclei. We consider single-particle orbitals $g_{07/2}$, $1d_{5/2,3/2}$, $2s_{1/2}$ and $0h_{11/2}$ (simply denoted as gdsh) for our studies of nuclei above ^{100}Sn [2] and the six orbitals between $N = 82$ and 126, $0h_{9/2}$, $1f_{7/2,5/2}$, $2p_{3/2,1/2}$ and $0i_{13/2}$ (denoted as hfpi) for our studies of neutron-deficient Pb isotopes [3]. But it should be pointed out that the motivation is not to fit data better but to get a better wave function and more reliable predictions on unknowns. We also applied those two effective interactions together for our studies of nuclei involving protons in gdsh and neutrons in hfpi shells.

*e-mail: chongq@kth.se

The neutron single-particle states $d_{5/2}$ and $g_{7/2}$ orbitals in ^{101}Sn are expected to be very close to each other. Our Hamiltonian works well also for Sb, I and Xe isotopes and heavier isotopes [4, 5] isotopes. As examples, in Fig. 1 I plotted the partial calculated level schemes of ^{112,132}Ba (with 6 protons being coupled to 6 neutrons/neutron holes) which are the largest systems we have ever treated in the full/non-truncated shell model space calculation. As can be seen from the figure, the calculations reproduce rather well the experimental spectrum of ^{132}Ba. It is interesting to see that the calculated spectra of ^{112,132}Ba are quite close to each other, even though the former case is expected to be highly influenced by the strong neutron-proton interaction. We are also optimizing the $T = 0$ channel of the effective interaction. In addition, we are extending the model space to include orbitals between the $N = Z = 50$ shell closure as well as neutron orbitals above $N = 82$ in order to study the structure of neutron-rich Cd isotopes.

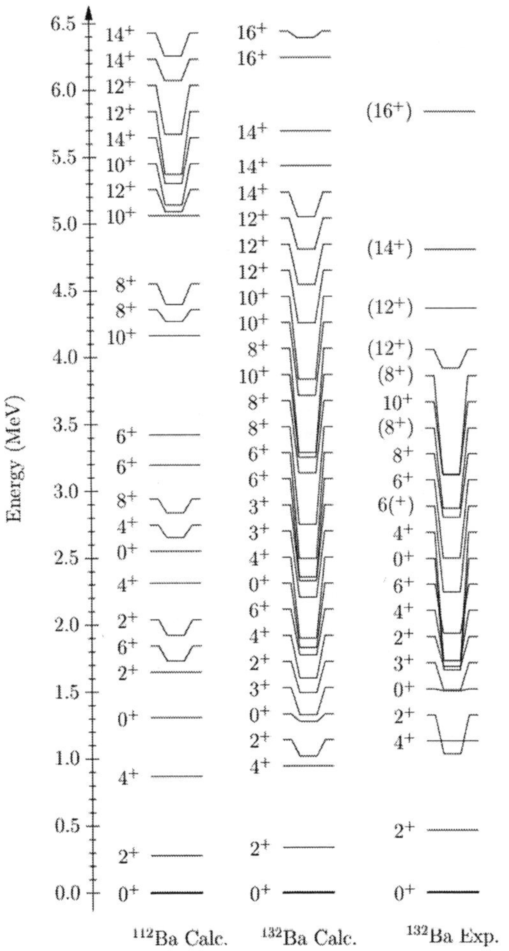

Figure 1. Calculated partial level schemes for ^{112}Ba (6π-6ν) and ^{132}Ba (6p-6h) in comparison with experimental data on ^{132}Ba. We calculated the lowest few states for selected spin values and positive parity.

It was argued that ^{100}Sn may be a soft core in analogy to the soft $N = Z = 28$ core ^{56}Ni. It seems such a possibility can be ruled out based on indirect information from recent measurements in this region [6–10]. The asymmetric electric quadrupole (E2) transition shape in Sn isotopes is suggested to be induced by the Pauli blocking effect [6]. A systematic study on the E2 transition in Te isotopes is done in Ref. [10]. One can state that the limited number of valence protons and neutrons are not expected to induce any significant quadrupole correlation in this region [11–14].

Truncations often have to be applied in order to reduce the size of the shell-model bases. The simplest way of truncation is to restrict the maximal/minimal numbers of particles in different orbitals. This method is applied both to no-core (often being referred to Nmax) and empirical shell model (npnh) calculations. In Ref. [6] we studied the structure and electromagnetic transition properties of light Sn isotopes within the large $gdsh_{11/2}$ model space by restricting the maximal number of four neutrons that can be excited out of the $g_{9/2}$ orbital. However, the convergence can be very slow if there is no clear shell or subshell closure or if single-particle structure are significantly modified by the monopole interaction, as it happens in neutron-rich light nuclei (see, e.g., Ref. [15]). One can evaluate the importance of a given basis vector ψ_i within a partition through a perturbation measure. The challenge here is that the truncated bases may not conserve angular momentum. An angular momentum conserved correlated basis truncation approach is introduced in Ref. [16]. We are implementing this method in the widely distributed shell-model code NuShellX by replacing its projection subroutine with our new correlated basis method. We have also developed an importance truncation based on the total monopole energy. The idea behind is that the Hamiltonian is dominated by the diagonal monopole channel. The monopole interaction can change significantly the (effective) mean field and drive the evolution of the shell structure. We have done truncation calculations for two isotopes ^{200,194}Pb. Convergence can be reached with a small portion (around 10%) of the total M-scheme wave function in both cases. This truncation approach can be easily implemented in most existing shell model codes [17].

We have done a systematic calculation on the structure and electromagnetic transition properties of neutron-deficient Pb isotopes, which may be studied experimentally using the AGATA detector. Our shell-model calculations can reproduce well the excitation energies of the low-lying 0^+ and 2^+ states in isotopes $^{198-206}$Pb [3]. The overall agreement between experiments and calculations are quite satisfactory. For nuclei heavier than ^{196}Pb, the difference between theory and experiment is less than 100 keV. In addition, one can also reproduce well the binding energies as well as the empirical pairing gaps. The empirical pairing gaps can be extracted from the binding energy by using the simple three-point formula, which carry important information on the two-nucleon pair clustering as well as α clustering in the nuclei involved [18, 19]. In nuclear systems the pairing collectivity manifests itself through the coherent contribution of many shell-model configurations which is induced by the strong $T = 1$ pairing interaction. For Pb isotopes, the full shell-model results also agree well with our generalized seniority and nucleon-pair-approximation truncation calculations.

The deviations between theory and experiment concerning the excitation energies and electromagnetic properties of low-lying 0^+ and 2^+ excited states in Pb isotopes indicate that, as suggested by deformed mean field calculations, nuclear deformation as induced intruder configurations from outside the model space should be important for nuclei in that region. Large-scale calculations are being carried out by including by neutron and proton core excited configuration. The excitation energies of the first 2^+ isotopes in Pb isotopes show a rather weak parabolic behaviour. In the lighter Pb isotopes the excitation energy of the second 0^+ state decreases rapidly with decreasing neutron number. It even becomes the first excited state in $^{184-194}$Pb. Within a shell-model context, those low-lying 0^+ states may be interpreted as coexisting deformed states which are induced by proton pair excitations across the $Z = 82$ shell gap. The energy of those core-excited configurations get more favoured in mid-shell Pb isotopes in relation to the stronger neutron-proton correlation in those nuclei (see talks by Garcia-Ramos and Andreyev). The shape-coexisting structure of Cd and Te isotopes is described in Garrett's talk.

In addition to seniority, it is well known that atomic nuclei can exhibit regular and strongly collective motions like vibration and rotation. The rotation and the underlying nuclear deformation in intrinsic frame has been well described by the nuclear shell model. It has its root in the quadrupole-quadrupole neutron-proton correlation. On the other hand, much less is known on the microscopic orgin of the vibration. The low-lying collective excitations of mid-shell Cd and Te isotopes were discussed in terms of quadrupole vibrations [11, 20] in relation to the fact that the even-even isotopes between $N = 56$ and 70 show regular equally-spaced yrast spectra. If that is the case, those isotopes will provide an ideal ground to explore the nature of the elusive nuclear vibration. Another intriguing phenomenon is the nearly constant behaviour of the energies of the 2^+ and 4^+ states in Te and Xe isotopes and their ratios when approaching $N = 50$, in contrast to the decreasing behaviour when approaching $N = 82$ [11]. This was analysed in Ref. [21] based on the quasiparticle random phase approximation approach.

Moreover, the available E2 transition strengths along the yrast line in $^{114,120-124}$Te show an anomalous rotational-like behaviour [22, 23]. It cannot be reproduced by collective models or the interacting boson model. We have done systematic calculations on the E2 transition properties of Te isotopes in Ref. [24]. The calculations reproduce well the excitation energies of the low-lying states as well as the regular and vibrational-like behaviour of the yrast specta of $^{108-130}$Te. In particular, the calculations reproduced reasonably well the nearly constant behaviour of the $B(E2)$ values of ^{114}Te and $^{120-124}$Te along the yrast line. The anomalous constant behaviour is related to the competition between the seniority coupling and the neutron-proton correlations. On the other hand, a squeezed gap between the 6_1^+ and 4_1^+ states is seen in Te isotopes when approaching $N = 82$, resulting in seniority-like spectra. A further experiment is done to study the angular momentum dependence of the E2 transition strengths in ^{112}Te [25]. It shows a similar but more 'vibrational'-like behaviour. Potential-Energy Surface calculations [13] indicate that those nuclei may also be understood from a soft rotor point of view. The yrast spectra of Te isotopes show a vibrational-like equally spaced pattern but the few known E2 transitions show anomalous rotational-like behaviour, which cannot be reproduced by collective models. Moreover, the calculated B(E2) values for neutron-deficient and heavier Te isotopes show contrasting different behaviours along the yrast line, which may be related to the enhanced neutron-proton correlation when approaching $N = 50$.

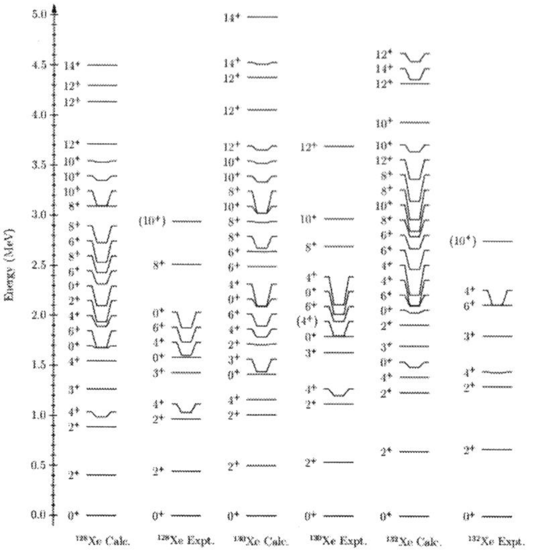

Figure 2. Calculated partial level schemes of $^{128-132}$Xe in comparison with expeirmetal data.

There have been extensive studies on the possible existence of E(5) symmetry in nuclei around $A = 130$. The nucleus ^{134}Ba may be the best E(5) candidate whereas the possibility to see that symmetry in Xe isotopes may have been ruled out in relation to the unexpected observation on the E2 transition properties between nonyrast 2^+ and 0^+ states. We are analysing the calculations on the E2 transition properties of those nuclei. In Fig. 2 we plotted the calculated low-lying states of $^{128-132}$Xe and compared with experimental data.

Further calculations on the E2 transitions of Cd isotopes (with two proton holes instead of particles within $Z = 50$ closure) are underway. The structure properties of Cd isotopes were also reviewed in also Garrett's and Leviatan's talks. The lighter deformed Cd isotopes was studied recently within the shell-model framework in Ref. [26].

To summarize, we have presented selected shell-model results on semi-magic Sn and Pb isotopes and open-shell nuclei including Te and Ba isotopes. I hope that the shell model can be a reliable tool for studying the spectroscopy and (abnormal) collective behaviours of those intermediate mass and heavy nuclei. In the near future, we are in particular interested to have a full understanding the vibrational behaviour of the Cd isotopes as well as the critical-point symmetry behaviours in $A \sim 130$ region.

This work is supported by the Swedish Research Council (VR) under grant Nos. 621-2012-3805, and 621-2013-4323 and the Göran Gustafsson foundation. Computational support provided by the Swedish National Infrastructure for Computing (SNIC) at PDC, KTH, Stockholm is acknowledged. The author also acknowledge the support from KTH through its strategic partnership with Shanghai JiaoTong University.

References

[1] *Nupecc long range plan 2017 "perspectives for nuclear physics"*, http://www.nupecc.org/lrp2016/Documents/lrp2017.pdf
[2] C. Qi, Z.X. Xu, Phys. Rev. C **86**, 044323 (2012)
[3] C. Qi, L.Y. Jia, G.J. Fu, Phys. Rev. C **94**, 014312 (2016)
[4] F. Wang, B.H. Sun, Z. Liu, C. Qi, L.H. Zhu, C. Scholey, S.F. Ashley, L. Bianco, I.J. Cullen, I.G. Darby et al., Phys. Rev. C **96**, 064307 (2017)
[5] F. Wang, B. Sun, Z. Liu, R. Page, C. Qi, C. Scholey, S. Ashley, L. Bianco, I. Cullen, I. Darby et al., Phys. Lett. B **770**, 83 (2017)
[6] T. Back, C. Qi, B. Cederwall, R. Liotta, F.G. Moradi, A. Johnson, R. Wyss, R. Wadsworth, Phys. Rev. C **87**, 031306 (2013)
[7] T. Bäck et al., Phys. Rev. C **84**, 041306 (2011)
[8] C.B. Hinke et al., Nature **486**, 341 (2012)
[9] G. Guastalla et al., Phys. Rev. Lett. **110**, 172501 (2013)
[10] M. Doncel et al., Phys. Rev. C **91**, 061304 (2015)
[11] B. Hadinia et al., Phys. Rev. C **72** (2005)
[12] E. Caurier, F. Nowacki, A. Poves, K. Sieja, Phys. Rev. C **82**, 064304 (2010)
[13] Z.Y. Wu, C. Qi, R. Wyss, H.L. Liu, Phys. Rev. C **92**, 024306 (2015)
[14] E.A. Coello Pérez, T. Papenbrock, Phys. Rev. C **92**, 064309 (2015)
[15] Z.X. Xu, C. Qi, Phys. Lett. B **724**, 247 (2013)
[16] L.F. Jiao, Z.H. Sun, Z.X. Xu, F.R. Xu, C. Qi, Phys. Rev. C **90**, 024306 (2014)
[17] C. Qi, Journal of Physics: Conference Series **742**, 012030 (2016)
[18] A.N. Andreyev et al., Phys. Rev. Lett. **110**, 242502 (2013)
[19] C. Qi, A. Andreyev, M. Huyse, R. Liotta, P.V. Duppen, R. Wyss, Phys. Lett. B **734**, 203 (2014)
[20] Z. Dombrádi et al., Phys. Rev. C **51**, 2394 (1995)
[21] D.S. Delion, R. Wyss, R.J. Liotta, B. Cederwall, A. Johnson, M. Sandzelius, Phys. Rev. C **82**, 024307 (2010)
[22] O. Möller, N. Warr, J. Jolie, A. Dewald, A. Fitzler, A. Linnemann, K.O. Zell, P.E. Garrett, S.W. Yates, Phys. Rev. C **71**, 064324 (2005)
[23] M. Saxena et al., Phys. Rev. C **90**, 024316 (2014)
[24] C. Qi, Phys. Rev. C **94**, 034310 (2016)
[25] M. Doncel, T. Bäck, C. Qi, D.M. Cullen, D. Hodge, B. Cederwall, M.J. Taylor, M. Procter, M. Giles, K. Auranen et al., Phys. Rev. C **96**, 051304 (2017)
[26] T. Schmidt, K.L.G. Heyde, A. Blazhev, J. Jolie, Phys. Rev. C **96**, 014302 (2017)

Systematic shell-model study on spectroscopic properties from light to heavy nuclei

Cenxi Yuan[1,*]

[1]Sino-French Institute of Nuclear Engineering and Technology, Sun Yat-Sen University, Zhuhai, 519082, Guangdong, China

Abstract. A systematic shell-model study is performed to study the spectroscopic properties from light to heavy nuclei, such as binding energies, energy levels, electromagnetic properties, and β decays. The importance of cross-shell excitation is shown in the spectroscopic properties of neutron-rich boron, carbon, nitrogen, and oxygen isotopes. A special case is presented for low-lying structure of ^{14}C. The weakly bound effect of proton $1s_{1/2}$ orbit is necessary for the description of the mirror energy difference in the nuclei around A=20. Some possible isomers are predicted in the nuclei in the southeast region of ^{132}Sn based on a newly suggested Hamiltonian. A preliminary study on the nuclei around ^{208}Pb are given to show the ability of the shell model in the heavy nuclei.

1 Introduction

Many neutron- and proton-rich nuclei are discovered and investigated through the new generation facilities. The nuclear shell model is one of the important methods to study the properties of nuclei from the stability line to the drip line, including the binding energies, energy levels, electromagnetic properties, β decays, and many other properties [1, 2]. But the required calculations are (much) beyond the present computational limitation for many nuclei. Some approximations are suggested to overcome the limitation, such as Monte-Carlo shell model [3], projected shell model [4], and so on. Although it is impossible to investigate all nuclei based on the shell-model calculation in a full model space, it is important to perform systematic shell-model study from light to heavy nuclei. A brief introduction of our recent shell-model works will be presented in the present paper.

2 Shell-model Hamiltonians and results

There are some shell-model Hamiltonians for *psd* model space, such as MK [5], WBT [6], WBP [6], and recently suggested YSOX [7]. YSOX is based on the monopole based universal interaction $V_{\rm MU}$ [8], SFO [9] and SDPF-M [10]. The binding energies, levels, electromagnetic properties, and B(GT) transitions of boron, carbon, nitrogen, and oxygen isotopes are well described by YSOX, especially the neutron drip line of carbon and oxygen isotopes. YSOX reconsidered the off-diagonal cross-shell interaction, which connects the $N\hbar\omega$ and

*e-mail: yuancx@mail.sysu.edu.cn

$(N+2)\hbar\omega$ configurations, which should play an important role in certain states. For example, both the inclusion of $4\hbar\omega$ configuration and the proper strength of the off-diagonal cross-shell interaction are necessary to describe the low-lying levels, electromagnetic and B(GT) transitions of ^{14}C [11]. The recently-observed negative states in ^{17,19}C are better described by YSOX comparing with the previous Hamiltonians [12].

The USD family is one of the most successful Hamiltonians in the *sd* model space, including USD [13], USDA and USDB [14]. But USD family does not include the isospin asymmetry term and can not discribe mirror energy differences (MED) between the mirror partners. In the proton-rich nuclei around A=20, some MED are very large, which comes from the weakly bound proton $1s_{1/2}$ orbit [15, 16]. The USD family can well describe MED in the nuclei around A=20 after modifications on single particle energies and two body matrix elements related to the proton $1s_{1/2}$ orbit [16]. The modified Hamiltonian gives nice description on the recent measured mass, B(GT) strength and branch ratio of ^{22}Si [17].

The nuclei around ^{132}Sn attracts a lot of interests in the recent and coming experimental projects, such as recent works on ^{140}Te [18], ^{140}I [19], and ^{132}Cd [20]. A newly suggested Hamiltonian gives nice descriptions on neutron separation energies, levels, and lifetimes of some isomers in the south-east region of ^{132}Sn [21]. Some possible isomers are suggested in ^{135}Sn, ^{132,133,134}In, ^{131}Cd, ^{130}Ag, and ^{129,130}Pd [21]. Just after the suggestion of the Hamiltonian, the first spectroscopic information in this region was published with the observation of a 618(8) keV 2_1^+ state in ^{132}Cd [20], while the new Hamiltonian gives 708 keV.

The nuclear shell model provides nice descriptions in the nuclei around ^{208}Pb. But taking ^{208}Pb as the core, the number of valence (hole) protons and neutrons can not be very large because of the computational limitation. Some further truncation on the occupancy in each orbit make the calculation on nuclei with more valence (hole) protons and neutrons possible. For example, the spin parity of the ground state of ^{223}Np is observed through α decay and reproduced by the shell-model calculation in a truncated model space [22]. The original computational cost for ^{223}Np is far beyond the present computational limitation. Some preliminary results on the binding energies and levels of lead, thallium, mercury, gold, and platinum were shown during CGS16 based on a Hamiltonian under preparation [23]. The root mean square between observed and shell-model results is only 158 keV for the binding energies of 40 nuclei in these isotopes.

3 Summary

The shell model can calculate the medium and heavy mass nuclei with a core, such as doubly magic nuclei ^{132}Sn and ^{208}Pb. Some nuclei around ^{132}Sn and ^{208}Pb are of highly importance and will be experimentally investigated in the near future. A systematic shell model study should provide many helpful spectroscopic information for the coming experimental study and feedbacks from the nuclear structure to the nuclear force. For example, the shell-model spectroscopic properties for N=82 and 132 isotones are expected to be important for *r*-process.

4 Acknowledgment

The present work has been supported by the National Natural Science Foundation of China (11775316), the Tip-top Scientific and Technical Innovative Youth Talents of Guangdong special support program (2016TQ03N575), the Pearl River S&T Nova Program of Guangzhou (201506010060), the Special Program for Applied Research on Super Computation of the NSFC Guangdong Joint Fund (the second phase) (U1501501), and the Fundamental Research Funds for the Central Universities (17lgzd34).

References

[1] B. A. Brown, Prog. Part. Nucl. Phys. 47, 517 (2001).
[2] E. Caurier, *et al.*, Rev. Mod. Phys. 77, 427 (2005).
[3] T. Otsuka, *et al.*, Prog. Part. Nucl. Phys. 47, 319 (2001).
[4] K. Hara and Y. Sun, Int. J. Mod. Phys. E 04, 637 (1995).
[5] D. J. Millener and D. Kurath, Nucl. Phys. A 255, 315 (1975).
[6] E. K. Warburton and B. A. Brown, Phys Rev. C 46, 923 (1992).
[7] C.X. Yuan, *et al.*, Phys Rev. C 85, 064324 (2012).
[8] T. Otsuka, *et al.*, Phys. Rev. Lett. 104, 012501 (2010).
[9] T. Suzuki, *et al.*, Phys Rev. C 67, 044302 (2003).
[10] Y. Utsuno, *et al.*, Phys Rev. C 60, 054315 (1999).
[11] C.X. Yuan, Chin. Phys. C, 41(10), 104102, (2017)
[12] J.W. Hwang, *et al.*, Phys. Lett. B 769, 503 (2017).
[13] B. A. Brown and B. H. Wildenthal, Annu. Rev. Nucl. Part. Sci. 38, 29 (1988).
[14] B. A. Brown and W. A. Richter, Phys. Rev. C 74, 034315 (2006).
[15] K. Ogawa, *et al.*, Phys. Lett. B 464, 157 (1999).
[16] C.X. Yuan, *et al.*, Phys Rev. C 89, 044327 (2014).
[17] X.X. Xu, *et al.*, Phys. Lett. B 766, 312 (2017); X.X. Xu, *et al.*, arXiv:1610.08291.
[18] B. Moon, *et al.*, Phys Rev. C 95, 044322 (2017).
[19] B. Moon, *et al.*, Phys Rev. C 96, 014325 (2017).
[20] H. Wang, Phys. Rev. C 94, 051301(R) (2016).
[21] C.X. Yuan, *et al.*, Phys. Lett. B 762, 237 (2016).
[22] M.D. Sun, *et al.*, Phys. Lett. B 771, 303 (2017).
[23] C.X. Yuan, *et al.*, in preparation.

EPJ Web of Conferences **178**, 02017 (2018) https://doi.org/10.1051/epjconf/201817802017
CGS16

Vibrational modes of light nuclei

Lorenzo Fortunato[1,2,*]

[1]Dipartimento di Fisica e Astronomia, "G. Galilei", University of Padova, (Italy)
[2]I.N.F.N. -Sez. di Padova, via Marzolo, 8, I-35131, Padova (Italy)

Abstract. A potential energy surface is constructed for various light nuclei, ^{3,4}He with the Argonne V18 potential and then for alpha-conjugate nuclei like ^{12}C,^{16}O, ^{20}Ne with $\alpha - \alpha$ potentials. Despite the nucleons have large momenta and large fluctuations around the equilibrium points, the minimization suggests the underlying geometric structure. We calculate the Hessian matrix determining the normal modes of vibration. This approach might lead to a new way of fitting the alpha-alpha interactions on the spectroscopic data available for alpha-conjugate nuclei, rather than on scattering phase shift.

1 Introduction

Molecular models of nuclei are fundamentally different from models used in molecular physics, where the Born-Oppenheimer approximation is valid and one can think of nuclear motion as a small vibration, happening only close to the minimum of a very deep potential (in molecular energy scales)[1]. Nuclei have large kinetic energy $\langle T \rangle$, comparable to the potential energy $\langle V \rangle$ and the zero point motion inside the Potential Energy Surface (PES, for short) is a large fraction of the well depth, therefore there are large fluctuations around the equilibrium points and we should not expect that the vibrational levels are deeply lying in the potential well as it happens in molecules. At most they can be weakly bound states, close to threshold, or more probably resonances in the continuum. Despite this fact, it is instructive to look at the normal modes, find the "best" internal coordinates to express symmetry-adapted vibrational motio and calculate the energy scale and structure of the vibrational levels.

2 Light nuclei: ^{4}He and ^{3}He

I take Argonne V18 nucleon-nucleon interaction as a well-established potential among all the constituents of the α particle, although in principle all the results are model dependent and the addition of a three body potential still hasn't been investigated in this context. For the moment let's consider only S-wave potentials. It's very interesting to consider also P-waves and other terms, they might lead to conical interesection (see quantum chemistry). With the above interactions, we build up a

*e-mail: fortunat@pd.infn.it

global PES:

$$PES(\vec{r}_1, \vec{r}_2, \vec{r}_3, \vec{r}_4) = \sum_{i<j}^{4} V_{ij}(\mid \vec{r}_i - \vec{r}_j \mid) \qquad \sum^{4} m_i \vec{r}_i = 0 \tag{1}$$

that must be minimized with the constraints imposed by the center of mass. Brute force minimization for the ^{4}He case gives a minimum at -557.531 MeV and a set of equilibrium coordinates. These coordinates are not unique, there are still a number of rotations that give equivalent sets (loci), therefore one can impose certain other conditions, while keeping the minimum at the same energy. The following list of constraints can be imposed on minimization without lifting the minimum:

$$x_1 = -x_2 \quad y_1 = -y_2 \quad z_1 = z_2 \quad x_3 = -x_4 \quad y_3 = -y_4 \quad z_3 = z_4$$

One easily gets the positions of the equilibrium points, around which the nucleons move dynamically with (large) fluctuations.

$$\begin{array}{lll} x_1 = -0.437648 \text{ fm} & y_1 = 0 \text{ fm} & z_1 = -0.31883 \text{ fm} \\ x_3 = 0.0446687 \text{ fm} & y_3 = -0.43476 \text{ fm} & z_3 = 0.318391 \text{ fm} \end{array}$$

They describe a shape that is close to a tetrahedron. There are two differences: the distance between the protons' equilibrium points is larger than the corresponding distance between the neutrons and the differences in spin-spin channels are such that the nn edge is rotated 5.87^o away from being orthogonal to the pp edge. See Fig. 1.

Figure 1. Equilibrium shape of ^{4}He. Protons' equilibrium points in blue.

A clooser look at these values reveals that they satisfy only the geometric requirements of the discrete point-group C_2 that contains only the operations identity (E) and 180° rotation around the symmetry axis (C2). This group has only one even (A) and one odd (B) representations (the multiplication table is trivial).

The diagonalization of the mass-corrected Hessian matrix $\tilde{H}_{lk} = H_{lk}/\sqrt{m_l m_k} = (\partial^2 PES/\partial q_l \partial q_k)/\sqrt{m_l m_k}$ gives the normal modes: it corresponds to pure harmonic vibrations (approximation of the full PES with quadratic functions around the minimum), clearly too drastic for nuclei because the energy levels are almost at threshold. A better approach is of course the full diagonalization of the 4-body problem with kinetic terms + potentials. In any case group theory (as in molecular physics) gives the characters of normal modes of vibrations. They turn out to be very high in energy:

A 689.056 MeV (this is the totally symmetric breathing mode), B 526.534 MeV, A 525.475 MeV, B 392.526 MeV, A 388.906 MeV, A 285.293 MeV, and therefore most problably not observable.

The same minimization with constraints on ^{3}He, gives the minimum at -277.67 MeV with planar geometry, but not quite isosceles! Again the spin-spin interaction is slightly different and the coordinates of the equilibrium points are in this case:

$$\begin{array}{lll} x_1 = \ 0.283102 \text{ fm} & y_1 = \ 0.436916 \text{ fm} & z_1 = 0. \text{ fm} \\ x_2 = \ 0.232502 \text{ fm} & y_2 = -0.436916 \text{ fm} & z_2 = 0. \text{ fm} \\ x_3 = -0.514895 \text{ fm} & y_3 = 0. \text{ fm} & z_3 = 0. \text{ fm} \end{array}$$

The point-group symmetry is just C_s, i.e. the group of identity and reflection upon the horizontal plane and the 3 normal modes are all of A' type with very high energies: 642.033, 1081.92 and 1856.07 MeV (breathing). Once again the eigenenergies are clearly too high for nuclear dynamics (they occur in a regime that is hundreds of MeV above separation threshold), but in principle they should exist anyway.

3 Alpha-conjugate nuclei

In general by incresing the mass of the clusters, their vibrational energies become lower, so maybe these kind of molecular vibrations are observable for heavier nuclei. Bijker and Iachello[2] have clearly demonstrated the succesfull application of the ACM, or algebraic cluster model, to the vibrational-rotational spectrum of $\alpha - \alpha$ conjugate nuclei like ^{12}C and ^{16}O. Therefore it looks promising to try to infer an analytic form for the $\alpha - \alpha$ potential that fits the two lowest vibrations in ^{12}C or the three lowest vibrations of ^{16}O, or simoultaneously both of them and try to compare with other models derived from phase-shifts. A suggested form is similar to the Ali-Bodmer potential, $V_1 e^{-a_1 r^2} + V_2 e^{-a_2 r^2}$ + Coulomb repulsion between charged spheres. Surprisingly, among all possible potentials with two gaussians, ^{12}C is reproduced very well with a potential that is quite shallow and has no nuclear repulsion at short distances. We get the A-bandhead at 7.30 (MeV) and the E-bandhead at 10.40 MeV, to be compared with experimental values: 7.65 and 10.84 MeV.

Instead of looking at the Hessian matrix, one would ideally like to solve the Schroedinger equation for these systems (this has already been done for g.s. and lowest exc.st. in A=3,4 with this potential) but look at the excited states in order to see how low in energies these molecular vibrations occur. Another improvment is to add also P-,D-, etc and see if you have other minima (conical intersections). From the knowledge of equilibrium configurations, one can start thinking on how to include rotations (moments of inertia) and what types of bands they generate. One can use the symmetry-adapted wavefunctions as the best basis to optimize the calculation of the whole vibrational spectrum. It is very promising to try to apply the same molecular machinery to complex cluster structure in light nuclei.

References

[1] L. Fortunato, Journal of Physics: Conf. Series **876** 012008 (2017)
[2] R. Bijker, F. Iachello , Phys. Rev. Lett. **112**, 152501 (2014)

Isomers and shell evolution in neutron-rich nuclei below the doubly magic nucleus ^{132}Sn

Hiroshi Watanabe[1,2,*]

[1]IRCNPC, School of Physics and Nuclear Energy Engineering, Beihang University, Beijing 100191 China
[2]RIKEN Nishina Center, 2-1 Hirosawa, Wako, Saitama 351-0198, Japan

Abstract. The level structures of the very neutron-rich nuclei ^{128}Pd$_{82}$ and ^{126}Pd$_{80}$ have been investigated for the first time. A new isomer with a half-life of 5.8(8) μs in ^{128}Pd is proposed to have a spin and parity of 8^+ and is associated with a maximally aligned configuration arising from the $g_{9/2}$ proton subshell with seniority $v = 2$. The level sequence below the 8^+ isomer is similar to that in the $N = 82$ isotone ^{130}Cd, but the electric quadrupole transition that depopulates the 8^+ isomer is more hindered in ^{128}Pd than in ^{130}Cd, as expected in the seniority scheme for a semi-magic, spherical nucleus. For ^{126}Pd, three new isomers with $J^\pi = (5^-), (7^-)$, and (10^+) have been identified with half-lives of 0.33(4) μs, 0.44(3) μs, and 23.0(8) ms, respectively. The smaller energy difference between the 10^+ and 7^- isomers in ^{126}Pd than in the heavier $N = 80$ isotones can be interpreted as being ascribed to the monopole shift of the $h_{11/2}$ neutron orbit. The nature of the $N = 82$ shell closure scrutinized with these characteristic isomers is discussed.

1 Introduction

The concept of magicity is of supreme importance for many-body fermionic systems in a confined space. The stability of atomic nuclei, which consist of a number of sub-atomic particles called protons and neutrons (nucleons), is much influenced by a shell structure and its resulting magic numbers; nuclei with specific numbers of nucleons (2, 8, 20, 28, 50, 82 both for protons and for neutrons, and also 126 for neutrons) near the β-stability line necessitate relatively high energies to remove one or two nucleons compared to the neighboring isotopes. For the last few decades, however, the study of exotic nuclei using radioactive isotope (RI) beams revealed that the aforementioned magic numbers are not necessarily universal and are subjected to a change in some regions of light-mass nuclei with highly unbalanced ratios of protons and neutrons [1, 2]. Such a paradigm shift is one of the frontier issues in nuclear physics, and whether it will also take place at the heavier traditional magic numbers is an open question worth investigating.

The present work focuses on neutron-rich nuclei below the doubly magic nucleus ^{132}Sn. It is still under debate that the $N = 82$ *spherical* shell gap is quenched as approaching the neutron drip line. From an astrophysical point of view, it has been argued that the property of the $N = 82$ shell closure far from the valley of β-stability affects the Solar System abundance pattern particularly around the prominent $A \approx 130$ peak in the *r*-process nucleosynthesis [3], and hence, a proper understanding of the underlying nuclear structure of the *r*-process isotopes is highly required. The characteristic isomers that involve high-j orbits, such as $\pi g_{9/2}$ and $\nu h_{11/2}$, in their configurations can serve as a sensitive probe for the evolution of shell structures in this exotic region. We have performed decay spectroscopy experiments at the RI Beam Factory (RIBF) facility [4], which has the capability of providing the world's strongest RI beams, as part of the EURICA campaign using high-intensity ^{238}U beams. New results obtained include the seniority isomer with $J^\pi = (8^+)$ in ^{128}Pd [5], the long-lived (10^+) isomeric state in ^{126}Pd [6], and a high-spin β-decaying isomer in ^{127}Ag [7]. In this report, the nature of the $N = 82$ shell closure will be discussed in terms of the seniority isomerism in the $N = 82$ isotones, as well as the effect of the monopole interaction between the $g_{9/2}$ proton and $h_{11/2}$ neutron subshells.

2 Experimental procedures

Neutron-rich isotopes below the doubly magic nucleus ^{132}Sn have been produced using in-flight fission of a ^{238}U^{86+} beam accelerated up to 345 MeV/u by a series of accelerators at RIBF facility [4], cooperated by RIKEN Nishina Center and CNS, University of Tokyo. The primary beam with the intensity ranging from 7 to 12 pnA impinged on a 3-mm thick beryllium target. The nuclei of interest were separated and transported through the BigRIPS separator and the following ZeroDegree spectrometer [9], operated with wedge-shaped aluminum degraders with thicknesses of 2.9 and 2.5 mm at the first and second dispersive focal planes, respectively, for purification of the secondary beams. Identification of particles with the atomic number (Z) and the mass-to-charge ratio (A/Q) was achieved on the basis of the ΔE-TOF-$B\rho$ method, in which the energy loss (ΔE), time of flight (TOF), and mag-

*e-mail: hiroshi@ribf.riken.jp

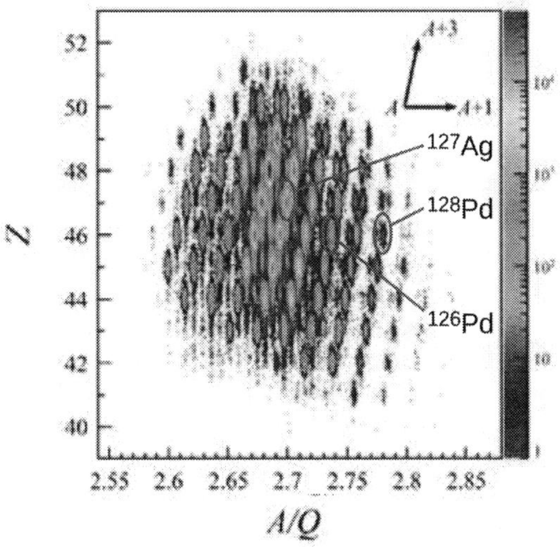

Figure 1. Particle-identification spectrum obtained in the present work. $^{127}Ag^{47+}$ and $^{126,128}Pd^{46+}$ ions are indicated with circles. The figure is taken from Ref. [8].

netic rigidity ($B\rho$) were measured using the focal-plane detectors on the beam line [10]. A particle-identification spectrum obtained in the present experiment is displayed in Fig. 1.

The identified fragments were implanted into a highly segmented active stopper, named WAS3ABi [11], which consisted of eight double-sided silicon-strip detectors (DSSSD) stacked compactly. Each DSSSD had a thickness of 1 mm with an active area segmented into sixty and forty strips (1-mm pitch) on each side in the horizontal and vertical dimensions, respectively. The DSSSDs also served as detectors for electrons following β-decay and internal conversion (IC) processes. Gamma rays were detected by the EURICA array [12] that consisted of twelve Cluster-type detectors, each of which contained seven HPGe crystals packed closely. The γ-ray measurements were carried out with a time condition up to 100 μs relative to the trigger signal generated either from a plastic scintillation counter placed at the end of the beam line or from WAS3ABi.

The beam, electron, and γ-ray events were time-stamped and recorded by independent data-acquisition systems. For the analysis of beam-γ (delayed) coincidence, the γ-ray data sets were combined with those of the beam particles on an event-by-event basis using information on the time stamp. Isomeric states with (sub)microsecond lifetimes were identified with appropriate time gates. All data sets containing beam, electron, and γ-ray events were used for β-γ and IC-γ coincidence analyses, in which the implantation of an identified particle was associated with the subsequent electron events that were detected in the same DSSSD pixel. Decay half-lives ($T_{1/2}$) in the millisecond range were extracted from the time distributions of γ-ray gated electron events with respect to the fragment implantation.

3 Results and discussion

3.1 $^{128}Pd_{82}$

Figure 2(a) shows a γ-ray energy spectrum measured in delayed coincidence with ^{128}Pd ions. Four γ rays at energies of 75, 260, 504, and 1311 keV have been unambiguously observed. These γ rays are found to be in mutual coincidence and exhibit consistent time behavior. Therefore, we conclude that they proceed through a single cascade originating from one isomeric state. A least-squares fit of the summed gated time spectra of the isomeric-decay transitions yields $T_{1/2} = 5.8(8)$ μs. The total internal conversion coefficient (α_T) for the 75-keV transition derived from a comparison with the 1311-keV γ-ray intensity is 2.6(17), which is consistent with the theoretical value of $\alpha_T = 3.88$ for an $E2$ multipolarity. Based on these arguments, the level scheme of ^{128}Pd is proposed as displayed in Fig. 3, where the spin and parity of the 5.8-μs isomeric state at 2151 keV is assigned as $J^\pi = 8^+$. A transition strength of $B(E2; 8^+ \rightarrow 6^+) = 0.22(3)$ W.u. can be obtained from the measured half-life of the 2151-keV isomeric state.

Assuming that the excited states in ^{128}Pd are the same multiplet members (with seniority $\upsilon = 2$, as will be discussed later), the half-life of the $J^\pi = 6^+$ state is estimated to be 22(3) ns from that of the 8^+ isomer. Since this value is significantly shorter than the flight time of fragments from the production target to the implantation position (~ 650 ns), there would be little ^{128}Pd ions remaining at this state during the flight. Hence, the observed intensities of the four cascade γ rays should depend only on the population of the 8^+ isomeric state.

The excitation energies of the $J^\pi = 2^+ - 8^+$ states in ^{128}Pd are comparable to those in ^{130}Cd [13]. The constancy of level energies is characteristic of the seniority scheme, where seniority υ counts the number of nucleons that are not in pairs coupled to spin zero. In case of a n-particle (or n-hole) system in a single-j shell, the level energies with identical J^π and υ are independent of n. Thus, the excited states in ^{128}Pd can be interpreted in terms of the $\upsilon = 2$ configuration of the $\pi g_{9/2}$ subshell, which locates just below the Z = 50 shell gap.

A direct implication of the $N = 82$ shell closure can be corroborated from the behavior of the $E2$ transition strength for the $8^+ \rightarrow 6^+$ isomer decay in ^{128}Pd. The $E2$ matrix elements between $\upsilon = 2$ states are close to zero near the middle of the valence shell in semi-magic nuclei, giving rise to seniority isomerism. For the $N = 82$ isotones, it turns out that the $B(E2; 8^+ \rightarrow 6^+)$ value is much smaller in ^{128}Pd than in ^{130}Cd, as expected in the exact seniority classification, namely, $B(E2; \upsilon = 2, n = 4) = \frac{1}{9}B(E2; \upsilon = 2, n = 2)$. This observation indicates that both the $J^\pi = 8^+$ and 6^+ states in the $N = 82$ isotones have good seniority $\upsilon = 2$ in the well isolated $\pi g_{9/2}$ subshell, in consequence of the robust shell closure.

3.2 $^{126}Pd_{80}$

The level scheme of ^{126}Pd constructed in the current work is also shown in Fig. 3. The 693-keV γ ray, which is

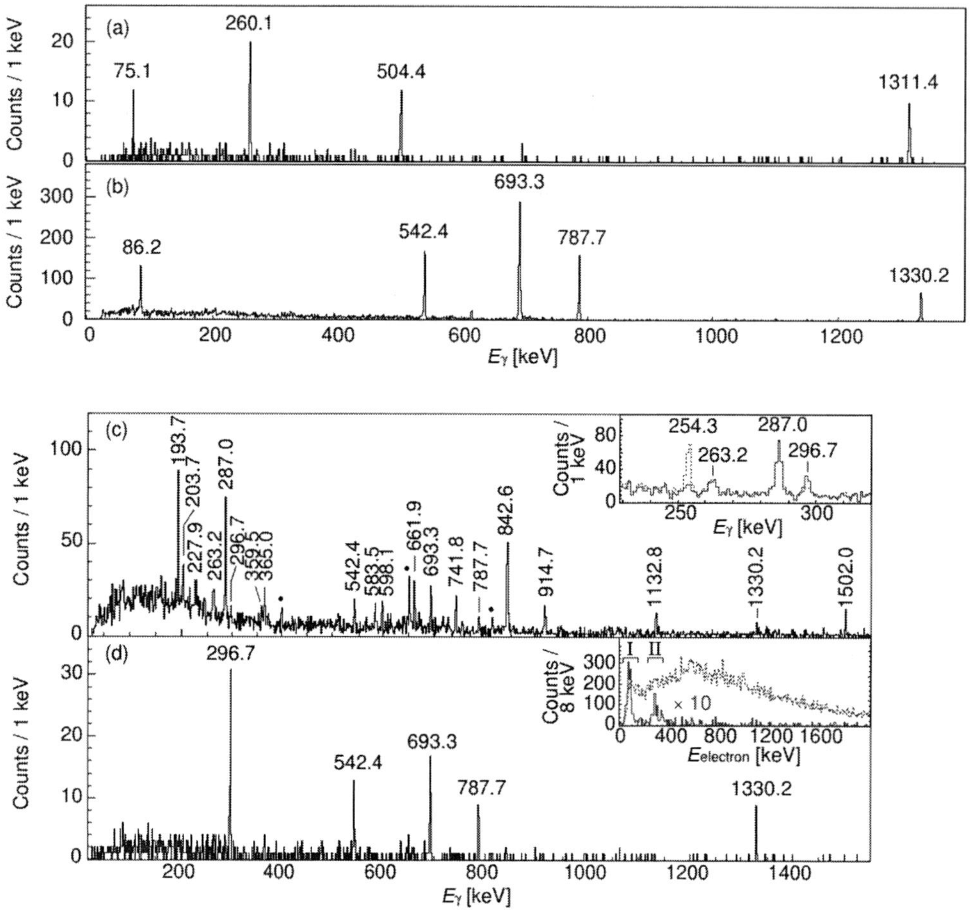

Figure 2. Gamma-ray spectra measured (a) in coincidence with ^{128}Pd ions within a time range of 0.15–25 μs, (b) in coincidence with ^{126}Pd ions within 0.15–5 μs, and (c) within 50 ms after the ^{126}Pd implantation with a gate on an electron-γ time difference of $-0.5 \le \Delta t_{e\gamma} \le 0.5$ μs. Contaminants from the granddaughter ^{126}Cd are marked with filled circles. The inset magnifies the energy region from 230 to 310 keV measured with $-4.0 \le \Delta t_{e\gamma} \le 0.5$ μs (red) and $-0.5 \le \Delta t_{e\gamma} \le 50$ μs (blue). (d): γ rays measured with $-4.0 \le \Delta t_{e\gamma} \le 0.5$ μs, and additionally, with a gate on the electron peak marked with "I" in the inset, where the spectrum depicted with the red line (multiplied by a factor of ten) is obtained with a sum of gates on the γ rays below the (5$^-$) isomer in ^{126}Pd, while the blue line represents an electron spectrum without γ-ray gates.

the most intense peak of those observed in delayed coincidence with ^{126}Pd ions [Fig. 2(b)], is assigned as the $2^+ \rightarrow 0^+$ transition. The coincidence spectrum with a gate on the 542-keV γ ray clearly exhibited γ rays at energies of 693, 788, and 86 keV, but not at 1330 keV, indicating that two parallel cascades stem from a common level at 2023 keV. The 86-keV transition is placed just above the 2023-keV level, because this γ ray was observed to precede all the four transitions.

A half-life of 0.44(3) μs is derived from the time distribution of the 86-keV γ ray relative to the beam implantation, while the time difference between the 86-keV γ ray and the subsequent transitions provides $T_{1/2}$ = 0.33(4) μs. Thus, it is found that two isomeric states lie at 2023 and 2110 keV. The assignment of spins and parities for the two isomeric states is based on the γ-ray feeding pattern and transition probabilities. Given the $J^\pi = 5^-$ and 7^- assignments for the 2023- and 2110-keV levels, respectively, the strengths of the isomeric decay transitions are $B(E1; 5^- \rightarrow 4^+) = 2.9(4) \times 10^{-9}$ W.u. (542 keV), $B(E3; 5^- \rightarrow 2^+) = 0.24(3)$ W.u. (1330 keV), and $B(E2; 7^- \rightarrow 5^-) = 2.13(14)$ W.u. (86 keV). These strengths are comparable with the values of the corresponding transitions in the $N = 80$ isotone ^{128}Cd [14].

The γ rays below these isomers, except for the 86-keV line, have been also observed in coincidence with electrons that were associated with the prior implantation of ^{126}Pd, as demonstrated in Fig. 2(c). This observation implies the existence of a long-lived, higher-spin isomer which decays via the cascades that include electromagnetic transitions with relatively large total conversion coefficients. With gates on these γ rays, a prominent peak can be found

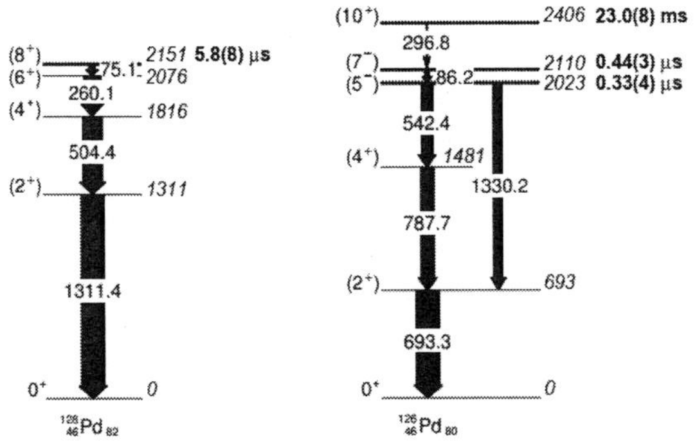

Figure 3. Partial level schemes of ^{128}Pd (left) and ^{126}Pd (right) established in the present work [5, 6].

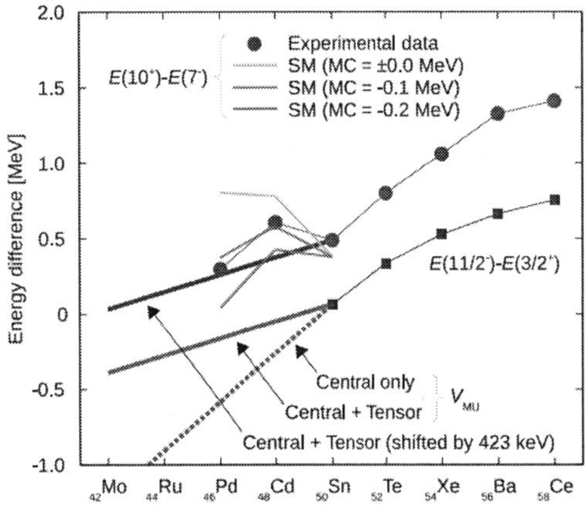

Figure 4. Energy differences between the 10$^+$ and 7$^-$ states, $\Delta E^{10^+}_{7^-}$, in N = 80 (filled circles) and between the 11/2$^-$ and 3/2$^+$ states in N = 81 (filled squares). For Z ≤ 50, the differences in energy between the $1h_{11/2}$ and $2d_{3/2}$ neutron orbits calculated by $V_{\rm MU}$ interaction [15] and $\Delta E^{10^+}_{7^-}$ by shell-model (SM) calculations with monopole corrections (MC) are also depicted.

in an electron spectrum [marked with "I" in the inset of Fig. 2(c)]; this corresponds to the conversion electrons for the 86-keV, $E2$ transition (α_T = 2.374). In Fig. 2(d), a γ-ray at 297 keV is clearly visible in addition to the γ rays below the (5$^-$) isomer by gating on the 86-keV IC peak. The appearance of the 297-keV peak is emphasized by taking a γ-ray time condition earlier than electron events, as is evident from the inset of Fig. 2(c), suggesting that this new γ ray precedes the highly converted 86-keV transition. Thus, the long-lived isomer can be identified at an excitation energy of 2406 keV. A peak marked with "II" in the inset of Fig. 2(d) is expected to arise from the conversion electrons for the 297-keV transition, being most likely of an $E3$ character (α_T = 0.1197). In addition to the internal-decay branch, β decay from the long-lived isomer was observed to populate excited states at high spins in ^{126}Ag. The isomeric half-life was determined to be 23.0(8) ms by taking a weighted average of those derived from the time distributions of the internal- and external-decay branches.

The main surprise in the present work is the small energy difference between the (10$^+$) and (7$^-$) isomers, $\Delta E^{10^+}_{7^-}$, in ^{126}Pd, compared to the analogous levels in the heavier N = 80 isotones, as demonstrated in Fig. 4. Since these two levels consist predominantly of maximally aligned two neutron-hole configurations, $(\nu 1h^{-2}_{11/2})_{10^+}$ and $(\nu 1h^{-1}_{11/2}2d^{-1}_{3/2})_{7^-}$, their level energies depend on the single-particle energies (SPEs) of the $\nu 1h_{11/2}$ and $\nu 2d_{3/2}$ neutron orbits, as well as the strength of interactions between them. It can be seen in Fig. 4 that the energy of the 11/2$^-$ level relative to the 3/2$^+$ ground state in the neighboring N = 81 isotones decreases when approaching Z = 50. Note that these states are of one neutron-hole nature, and that the smooth reduction in energy can be essentially interpreted in terms of a short-range proton-neutron interaction as follows: The $\pi 1g_{7/2}$ proton orbit lies just above the Z = 50 shell closure. The monopole interaction between $\pi 1g_{7/2}$ and $\nu 1h_{11/2}$, a spin-flip pair with $\Delta\ell$ = 1 and Δn = 0, is stronger than the $\pi 1g_{7/2}$-$\nu 2d_{3/2}$ pair, due to the larger overlap between the radial wave functions of the two orbits. Therefore, when emptying the $\pi 1g_{7/2}$ subshell, the $\nu 1h_{11/2}$ orbit is relatively less bound than the $\nu 2d_{3/2}$ one. The moderation in the slope at higher Z may be ascribed to the partial occupation of the $\pi 2d_{5/2}$ orbit, which exerts a weaker (stronger) monopole interaction on the $\nu 1h_{11/2}$ ($\nu 2d_{3/2}$) neutrons than $\pi 1g_{7/2}$. As is evident from the comparison in Fig. 4, a similar trend is observed for $\Delta E^{10^+}_{7^-}$. This finding supports that these isomers can serve as sensitive probes for the evolution of the constituent neutron shell orbits.

Below Z = 50, protons in the $\pi 1g_{9/2}$ subshell play a major role in changing the neutron SPEs, because this orbit is near the Fermi surface. The neutron SPEs are estimated by the monopole-based universal interaction, $V_{\rm MU}$, which consists of the Gaussian central force and the tensor force based on the π+ρ meson exchanges, using the parameters fixed in Ref. [15]. The $V_{\rm MU}$ interaction has been applied to other regions of exotic nuclei and successfully reproduced various types of the shell evolution [15]. Figure 4 shows the evolutions of the relative 11/2$^-$ energy starting from the experimental value in ^{131}Sn, predicted by the $V_{\rm MU}$ calculation with A = 130. In the calculation, the $\pi 1g_{9/2}$ subshell is considered to be well isolated between Z = 50 and 40. In Fig. 4, only with the central force the 11/2$^-$ energy rapidly decreases to negative values as the proton number decreases, indicating that the $\nu 1h_{11/2}$ orbit moves away from the $\nu 2d_{3/2}$ one towards the N = 82 gap. If there is the tensor force effect, however, the slope becomes less steep. The observed reduction in $\Delta E^{10^+}_{7^-}$ from ^{130}Sn to ^{126}Pd is consistent with the expectation of $V_{\rm MU}$ including both the central and tensor forces, given the same trend as the 11/2$^-$ relative energies. Based on these findings, it can be concluded that the tensor force slows down the upward drift of the $\nu 1h_{11/2}$ SPE when protons are removed from the $\pi 1g_{9/2}$ orbit below ^{132}Sn.

One can notice a kink in the systematics of $\Delta E_{7^-}^{10^+}$ at ^{128}Cd in Fig. 4. While the overall trend can be accounted for by V_{MU}, other multipole components have to be introduced in a shell-model framework to describe the details of the level evolution. To confirm the effect of configuration mixing, a shell-model (SM) study based on the j-j coupling scheme was performed. With the doubly magic nucleus ^{132}Sn as a closed core, the model space adopted includes the $2d_{3/2}$, $1h_{11/2}$, $3s_{1/2}$, $2d_{5/2}$, and $1g_{7/2}$ orbits for neutron holes, and the $1g_{9/2}$, $2p_{1/2}$, $2p_{3/2}$, $1f_{5/2}$ orbits for proton holes. The SPEs were taken from the experimentally known levels in the single-hole nuclei, ^{131}Sn and ^{131}In. The surface-delta interaction was employed for the proton-proton and neutron-neutron interactions, which were adjusted to reproduce the observed level energies of ^{130}Cd and ^{130}Sn, respectively. The proton-neutron two-body matrix elements (TBME) were derived from the empirical parameterization of interaction potentials [16]. For the $\pi 1g_{9/2}$-$\nu 1h_{11/2}$ multiplet, the monopole correction (MC) was introduced by adding a constant to the diagonal TBME. The same idea of MC was adopted in large-scale shell-model calculations for neutron-rich nuclei in the south-west quadrant of ^{132}Sn [17].

The results of SM calculations with different values of the monopole correction for $Z \leq 50$ are displayed in Fig. 4. The kink manifests in the calculated $\Delta E_{7^-}^{10^+}$, as well as in the experimental result. Moreover, it can be seen that the variation in $\Delta E_{7^-}^{10^+}$ is very sensitive to the value of the monopole correction. According to the present SM calculation, the wave functions of the 10^+ and 7^- states in each $N = 80$ isotone are scarcely affected by the monopole corrections. These arguments indicate that the lowering of $\Delta E_{7^-}^{10^+}$ in ^{126}Pd is ascribed to the monopole effect between the $\pi 1g_{9/2}$ and $\nu 1h_{11/2}$ orbitals rather than the configuration mixing of any kind.

4 Summary

This paper reports on the experimental results of characteristic isomers in ^{126,128}Pd studied as part of the EURICA experimental campaign in 2012 at RIBF. The data analysis and shell-model calculations for ^{127}Ag are still ongoing.

References

[1] O. Sorlin, M.G. Porquet, Prog. Part. Nucl. Phys. **61**, 602 (2008)

[2] T. Nakamura, H. Sakurai, H. Watanabe, Prog. Part. Nucl. Phys. **97**, 53 (2017)

[3] K. Kratz, B. Pfeiffer, O. Arndt, Eur. Phys. J. A **25**, 633 (2005)

[4] Y. Yano, Nucl. Instrum. Methods Phys. Res. B **261**, 1009 (2007)

[5] H. Watanabe, G. Lorusso, S. Nishimura, Z.Y. Xu, T. Sumikama, P.A. Söderström, P. Doornenbal, F. Browne, G. Gey, H.S. Jung et al., Phys. Rev. Lett. **111**, 152501 (2013)

[6] H. Watanabe, G. Lorusso, S. Nishimura, T. Otsuka, K. Ogawa, Z.Y. Xu, T. Sumikama, P.A. Söderström, P. Doornenbal, Z. Li et al., Phys. Rev. Lett. **113**, 042502 (2014)

[7] H. Watanabe, (to be submitted)

[8] Y. Shimizu, T. Kubo, N. Fukuda, N. Inabe, D. Kameda, H. Sato, H. Suzuki, H. Takeda, K. Yoshida, G. Lorusso et al., Journal of the Physical Society of Japan **87**, 014203 (2018)

[9] T. Kubo, D. Kameda, H. Suzuki, N. Fukuda, H. Takeda, Y. Yanagisawa, M. Ohtake, K. Kusaka, K. Yoshida, N. Inabe et al., Progress of Theoretical and Experimental Physics **2012**, 03C003 (2012)

[10] N. Fukuda, T. Kubo, T. Ohnishi, N. Inabe, H. Takeda, D. Kameda, H. Suzuki, Nucl. Instrum. Methods B **317**, 323 (2013)

[11] S. Nishimura, Progress of Theoretical and Experimental Physics **2012**, 03C006 (2012)

[12] P.A. Söderström, S. Nishimura, P. Doornenbal, G. Lorusso, T. Sumikama, H. Watanabe, Z. Xu, H. Baba, F. Browne, S. Go et al., Nucl. Instrm. Methods B **317**, 649 (2013)

[13] A. Jungclaus, L. Cáceres, M. Górska, M. Pfützner, S. Pietri, E. Werner-Malento, H. Grawe, K. Langanke, G. Martínez-Pinedo, F. Nowacki et al., Phys. Rev. Lett. **99**, 132501 (2007)

[14] L. Cáceres, M. Górska, A. Jungclaus, M. Pfützner, H. Grawe, F. Nowacki, K. Sieja, S. Pietri, D. Rudolph, Z. Podolyák et al., Phys. Rev. C **79**, 011301 (2009)

[15] T. Otsuka, T. Suzuki, M. Honma, Y. Utsuno, N. Tsunoda, K. Tsukiyama, M. Hjorth-Jensen, Phys. Rev. Lett. **104**, 012501 (2010)

[16] J.P. Schiffer, W.W. True, Rev. Mod. Phys. **48**, 191 (1976)

[17] H.K. Wang, K. Kaneko, Y. Sun, Phys. Rev. C **89**, 064311 (2014)

Nuclear structure study for the neutron-rich nuclei beyond ^{132}Sn: In-beam gamma-ray spectroscopy of ^{136}Sn and ^{132}Cd

He Wang[1,*], *Nori* Aoi[2], *Satoshi* Takeuchi[3], *Masafumi* Matsushita[4], *Pieter* Doornenbal[1], *Tohru* Motobayashi[1], *David* Steppenbeck[1], *Kenichiro* Yoneda[1], *Hidetada* Baba[1], *Zsolt* Dombrádi[5], *Kota* Kobayashi[6], *Yosuke* Kondo[3], *Jenny* Lee[7], *Hong-Na* Liu[8,1], *Ryogo* Minakata[3], *Daiki* Nishimura[9], *Hideaki* Otsu[1], *Hiroyoshi* Sakurai[1], *Dora* Sohler[5], *Ye-Lei* Sun[8], *Zheng-Yang* Tian[8], *Ryuki* Tanaka[3], *Zsolt* Vajta[5], *Zai-Hong* Yang[1], *Tetsuya* Yamamoto[2], *Yan-Lin* Ye[8], and *Rin* Yokoyama[4]

[1]RIKEN Nishina Center, 2-1 Hirosawa, Wako, Saitama 351-0198, Japan
[2]Research Center for Nuclear Physics, Osaka University, Ibaraki, Osaka 567-0047, Japan
[3]Department of Physics, Tokyo Institute of Technology, 2-12-1 Ookayama, Meguro, Tokyo 152-8551, Japan
[4]Center for Nuclear Study, University of Tokyo,RIKEN campus, Wako, Saitama 351-0198, Japan
[5]Institute for Nuclear Research, H-4001 Debrecen, P.O. Box 51, Hungary
[6]Department of Physics, Rikkyo University, 3-34-1 Nishi-Ikebukuro, Toshima, Tokyo 172-8501, Japan
[7]Department of Physics, University of Hong Kong, Pokfulam Road, Hong Kong
[8]School of Physics and State Key Laboratory of Nuclear Physics and Technology, Peking University, Beijing 100871, China
[9]Department of Physics, Tokyo University of Science, 2641 Yamazaki, Noda, Chiba 278-8510, Japan
[10]Department of Physics, University of Tokyo, 7-3-1 Hongo, Bunkyo, Tokyo 113-0033, Japan

Abstract. The neutron-rich nuclei ^{136}Sn and ^{132}Cd have been studied in the purpose of nuclear structure for the nuclei beyond the doubly-magic nucleus ^{132}Sn. The $2^+_1 \rightarrow 0^+_{gs}$ transitions were identified for these two nuclei using in-beam γ-ray spectroscopy in coincidence with one- and two-proton removal reactions, respectively, at the RIKEN Radioactive Isotope Beam Factory. The 2^+_1 state in ^{136}Sn is found to be similar to that for ^{134}Sn indicating the seniority scheme may also hold for the heavy tin isotopes beyond $N = 82$. For ^{132}Cd, the 2^+_1 state provides the first spectroscopic information in the even-even nuclei locating in the region "southeast" of ^{132}Sn and the result is discussed in terms of proton-neutron configuration mixing. In both these two nuclei, it was found that the valence neutrons play an essential role in their low-lying excitations.

1 Introduction

Nuclear structure study for the very neutron-rich isotopes is one of major topics in nuclear physics. In particular, nuclei with a few valence particle and/or holes outside ^{132}Sn ($Z = 50, N = 82$) have attract much attention as ^{132}Sn is a doubly-magic nucleus while locating far away from the line of stability. It thus provides an opportunity to explore possible exotic nuclear structure beyond doubly-magic nucleus towards the neutron-drip line. In addition, elements around tin have long isotopic chains ranging from two neutron magic numbers $N = 50$ and $N = 82$ and it thus enables us to investigate systematically the structures as a function of neutron number in a wide range. However, the experimental studies are limited for the isotopes beyond $N = 82$ because of the difficulties in accessing this exotic region of the nuclear chart.

Spectroscopic information, especially the first 2^+ states (2^+_1) in the even-even nuclei reveal interesting features in this region. Beyond $N = 82$, the energy of the first 2^+ excited state [$E_x(2^+_1)$] was found to be much lower in ^{134}Sn ($Z = 50, N = 84$) [1] than those for the lighter Sn isotopes below $N = 82$ where $E_x(2^+_1)$ keeps almost constant. At the "northeast" of ^{132}Sn, a neutron dominance nature in the 2^+_1 excitation was reported for ^{136}Te ($Z = 52, N = 84$) from the decrease of both $E_x(2^+_1)$ and the reduced transition probabilities [$B(E2)$] across $N = 82$ [2]. These studies suggest that valence neutrons outside $N = 82$ may play an important role in the 2^+_1 excitations in the neutron-rich nuclei beyond ^{132}Sn. It is therefore intriguing to extend the spectroscopic studies in more neutron-rich side in order to investigate more on the role of neutron in the low-lying excitations.

In the present work, we report on the 2^+_1 states in two neutron-rich nuclei: ^{136}Sn ($Z = 50, N = 86$) following the tin isotopic chain and ^{132}Cd ($Z = 48, N = 84$) along the $N = 84$ isotones.

2 Experimental Setup

The experiment was performed at the Radioactive Isotope Beam Factory (RIBF) operated by the RIKEN Nishina Center and the Center for Nuclear Study, University of Tokyo. Secondary beams were produced via in-flight fission reactions of ^{238}U on a tungsten located at the objective

*e-mail: wanghe@ribf.riken.jp

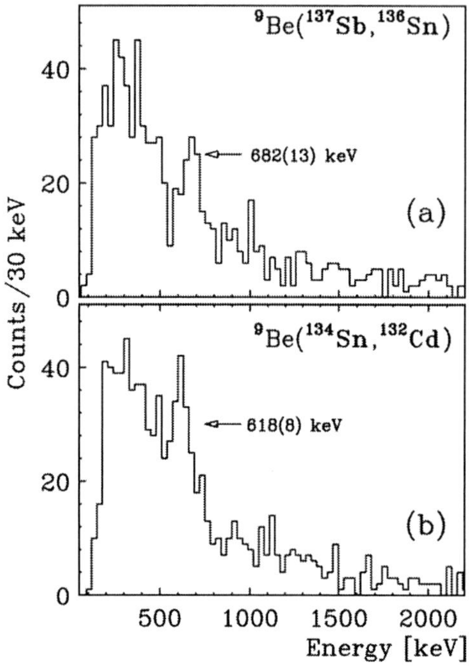

Figure 1. Doppler-shift corrected γ-ray energy spectra in coincidence with (a) ^{9}Be(^{137}Sb, ^{136}Sn) and (b) ^{9}Be(^{134}Sn, ^{132}Cd) reaction channels.

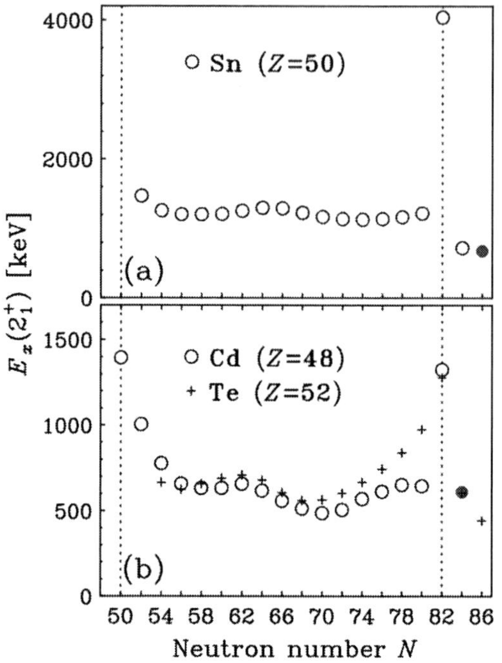

Figure 2. $E_x(2_1^+)$ systematics for (a) Sn and (b) Cd isotopic chain. In Panel (b), the $E_x(2_1^+)$ values for Te are displayed for comparison. The results obtained in the present work are highlighted by red color in both panels.

point of the BigRIPS separator [3]. The fission products including nuclei around ^{137}Sb and ^{134}Sn were selected and purified in the first stage of BigRIPS by inserting a wedge-shaped aluminum degrader with thicknesses of 0.8 g/cm^2. Further purification for the secondary beams was made in the second stage of BigRIPS using a 0.4-g/cm^2-thick degrader. The particles were identified event-by-event by using the TOF-$B\rho$-ΔE method, in which time of flight (TOF), magnetic rigidity ($B\rho$) and energy loss (ΔE) were measured to deduce the atomic number (Z) and mass-to-charge ratio (A/Q) [4, 5].

The secondary reactions were induced by a ^{9}Be target with a thickness of 1.1 g/cm^2. Reaction residues were collected and analyzed by the ZeroDegree spectrometer [3]. The particle identification was made using again the TOF-$B\rho$-ΔE method in a similar way to BigRIPS. In addition, the total kinetic energy was measured to deduce the mass number (A) for the identification of different charge states for these heavy ions. Detailed information on the experimental setup and particle identifications can be found in Refs. [6, 7].

Gamma rays emitted from the decay of the excited states of the reaction products were detected by the DALI2 array [8] surrounding the secondary target. DALI2 consisted of 186 NaI(Tl) scintillation detectors and covered the polar angle from 14 to 148 degrees with respect to the beam direction. The energy resolution and full energy peak efficiency were 9% (FWHM) and 22%, respectively, for the 0.662 MeV γ-ray measured using a ^{137}Cs source.

3 Results and Discussions

The γ-ray energy spectra in coincidence with the respective one- and two-proton removal reactions to ^{136}Sn and ^{132}Cd are displayed in Fig. 1. Prominent transitions observed at 682(13) and 618(8) keV in Fig. 1(a) and (b), respectively, correspond to the $2_1^+ \rightarrow 0_{gs}^+$ transitions for ^{136}Sn and ^{132}Cd.

The experimental $E_x(2_1^+)$ values as a function of neutron number are shown for the even-even Sn and Cd isotopes in Fig. 2(a) and (b), respectively. As displayed in Fig. 2(a), the $E_x(2_1^+)$ value for ^{136}Sn is close to that for ^{134}Sn. In another word, $E_x(2_1^+)$ keeps constant in the Sn isotopes beyond $N = 82$. This behavior is similar to that for the lighter Sn isotopes below $N = 82$, where $E_x(2_1^+)$ is almost constant reflecting the consequence of the seniority $\nu = 2$ configuration [9]. The similar $E_x(2_1^+)$ values in ^{136}Sn and ^{134}Sn suggest that the seniority scheme holds also for the tin isotopes beyond $N = 82$. The similar $E_x(2_1^+)$ values as well as the constant 4_1^+ and 6_1^+ states are also found by identifying the 6^+ seniority isomers in ^{136,138}Sn [10], supporting further the dominant seniority $\nu = 2$ scheme.

For ^{136}Sn, a mixed seniority is suggested in the 4_1^+ state from the $B(E2; 6^+ \rightarrow 4^+)$ value [10]. To reproduce the $B(E2; 6^+ \rightarrow 4^+)$ systematics in ^{134,136,138}Sn, shell-model calculations with an empirically redued neutron-pairing strength was applied and such shell-model calculations suggest that the 4_1^+ state in ^{136}Sn has mixed seniority $\nu = 2$

and $\nu = 4$ components. The reduction of $\nu = 2$ component in the 4_1^+ is supported by another shell-model calculations [11]. While the 4_1^+ state in ^{136}Sn has mixed seniority, the 2_1^+ and 6_1^+ states are predominant seniority $\nu = 2$ as demonstrated by the shell-model calculations [11].

Although $E_x(2_1^+)$ keeps constant below and above $N = 82$ along the Sn isotopic chain, the $E_x(2_1^+)$ systematics show an asymmetric pattern around ^{132}Sn. The $E_x(2_1^+)$ values in ^{134,136}Sn are about 500 keV lower than those for the lighter ones below $N = 82$. Such reduction in $E_x(2_1^+)$ beyond $N = 82$ might be due to a quenching of neutron pairing in the $N > 82$ region, because the 2_1^+ excitation is mainly determined by the strength of neutron pairing under the seniority scheme. Indeed, the neutron gap was suggested to be reduced from the mass measurement for the Sn isotopes above $N = 82$ [12]. The reduction of neutron pairing is also supported by Ref. [10] because a reduced neutron-pairing strength was used to reproduce the $B(E2; 6^+ \rightarrow 4^+)$ systematics as mentioned above.

The 2_1^+ state in ^{132}Cd provides a further insight on the role of neutrons in the low-lying excitation in more neutron-rich region. In order to investigate the nature of the 2_1^+ state in ^{132}Cd more quantitatively, a simple seniority-two empirical model was employed [2, 13]. With respect to ^{132}Sn, ^{132}Cd has a two-proton-hole (π^{-2}) and two-neutron-particle (ν^2) configuration. As ^{132}Sn exhibits a robust doubly-magic nature [14], the wave function of the 2_1^+ state in ^{132}Cd can be therefore described by taking the 2_1^+ states of ^{130}Cd (π^{-2}) and ^{134}Sn (ν^2) as the basis states. In this model, an interaction of $|V|$ = 275 keV is required to generate the 618-keV state in ^{132}Cd from the mixing of the 1325- and 725-keV levels in ^{130}Cd and ^{134}Sn, resulting in the relationship

$$|2_1^+; {}^{132}\mathrm{Cd}\rangle = 0.93|\nu^2\rangle \pm 0.36|\pi^{-2}\rangle.$$

The respective neutron- and proton-excitation contributions were determined to be 87% and 13%, indicating a neutron dominance nature of the 2_1^+ state in ^{132}Cd.

Neutron excitation is also known in the $N = 84$ isotone ^{136}Te [2] from the asymmetric $B(E2)$ pattern in the Te isotopes around $N = 82$. By using the same seniority-two model, the neutron contribution and the interaction of $|V|$ were deduced to be 85% and 282 keV, respectively, for ^{136}Te by taking the experimental 2_1^+ states of ^{134}Te and ^{134}Sn as the basis states [2]. In addition, it is worth to note that the $E_x(2_1^+)$ values in Cd and Te agree with each other beyond $N = 82$ as displayed in Fig. 2(b). The similarities of the $E_x(2_1^+)$ value, the interaction and the neutron contributions between ^{132}Cd and ^{136}Te suggest that the neutrons play a similar role in the 2_1^+ excitation in these two isotopes.

The neutron-dominance nature might be attributed to the reduction of neutron pairing strength as suggested by the QRPA calculations [15, 16]. As shown by these calculations, the neutron-pairing gap becomes smaller across $N = 82$ while the proton one remains almost the same. The reduction in the neutron-pairing gap lead to an enhancement in the neutron amplitudes in the wave function of the 2_1^+ state, resulting in a predominant neutron contribution to the 2_1^+ excitation. Both the 2_1^+ states in ^{136}Sn and ^{132}Cd suggest a quenching of neutron-pairing strength above the $N = 82$ gap. The pairing weakening might be related to the nature of nuclear structure around $N = 82$. One possible reason for the neutron-pairing decrease might be a lower level density above $N = 82$ [12, 15].

4 Summary

In summary, the 2_1^+ states in ^{136}Sn and ^{132}Cd were studied using the in-beam γ-ray spectroscopy in coincidence with one- and two-proton removal reactions, respectively. The measured $E_x(2_1^+)$ value of ^{136}Sn is found to be similar as that for ^{134}Sn, indicating the seniority scheme holds beyond $N = 82$ for the tin isotopes. Neutron-excitation dominance of the 2_1^+ state in ^{132}Cd is suggested from the proton-neutron configuration mixing and, is supported by a seniority-two empirical model. Both the asymmetric $E_x(2_1^+)$ pattern around $N = 82$ in Sn and the neutron-dominance nature in ^{132}Cd suggest a reduction of neutron-pairing strength above $N = 82$. Such reduction might be due to a lower level density above $N = 82$. Further studies on the neutron-pairing gap are encouraged to explore more on the neutron dominance in the low-lying excitations.

5 Acknowledgment

We thank the accelerator staff of RIKEN Nishina Center for providing the U primary beam and the BigRIPS team for tuning the secondary beam.

References

[1] A. Korgul et al., Eur. Phys. J. A **7**, 167 (2000).
[2] D. C. Radford et al., Phys. Rev. Lett. **88**, 222501 (2002).
[3] T. Kubo et al., Prog. Theor. Exp. Phys. **2012**, 03C003 (2012).
[4] T. Ohnishi et al., J. Phys. Soc. Jpn. **79**, 073201 (2010).
[5] N. Fukuda et al., Nucl. Instrum. Methods Phys. Res. B **317**, 323 (2013).
[6] H. Wang et al., Prog. Theor. Exp. Phys. **2014**, 023D02 (2014).
[7] H. Wang et al., Phys. Rev. C **94**, 051301 (2016).
[8] S. Takeuchi et al., Nucl. Instrum. Methods Phys. Res. A **763**, 596 (2014).
[9] R. F. Casten and B. M. Sherrill, Prog. Part. Nucl. Phys. **45**, S171 (2000).
[10] G. S. Simpson et al., Phys. Rev. Lett. **113**, 132502 (2014).
[11] M. P. Kartamyshev et al., Phys. Rev. C **76**, 024313 (2007).
[12] J. Hakala et al., Phys. Rev. Lett. **109**, 032501 (2012).
[13] N. Imai et al., Phys. Rev. Lett. **92**, 062501 (2004).
[14] J.M. Allmond et al., Phys. Rev. Lett. **112**, 172701 (2014).
[15] J. Terasaki et al., Phys. Rev. C **66**, 054313 (2002).
[16] J. Terasaki, J. Phys.: Conf. Ser. **533**, 012059 (2014).

New nuclear structure data beyond ^{132}Sn

Radomira Lozeva[1,2,*]

[1]CSNSM, CNRS/IN2P3, Université Paris-Saclay, 91405 Orsay Campus, France
[2]IPHC, CNRS/IN2P3, Université de Strasbourg, 67037 Strasbourg, Cedex 2, France

Abstract. Exotic nuclei beyond the ^{132}Sn double shell-closure are influenced by both the Sn superfluity and the evolving collectivity only few nucleons away. Toward even more neutron-rich nuclei, especially at intermediate mass number, the interplay between single-particle and collective particle-hole excitations competes. In some cases with the extreme addition of neutrons also other effects as the formation of neutron skin, stabilization as sub-shell gaps or orbital crossings may be expected. The knowledge of nuclear ingredients is especially interesting beyond ^{132}Sn and little is known on how the excitation modes develop with the addition of both protons and neutrons and for example systematic prompt and decay studies can be such very sensitive probe. Recently, we have approached this region of nuclei in several experimental measurements following ^{238}U projectile fission on ^{9}Be and n-induced fission on ^{241}Pu and ^{235}U. Consistent data analysis allows to access various spins and excitation energies and provide new input to theory. Examples from these studies on several nuclei in the A~140 region were presented during the conference together with the possible interpretation of the new data. Here, we will illustrate one example on ^{136}I using two complementary data sets.

1 Introduction and Motivation

Nuclei beyond ^{132}Sn present very particular properties. On one hand they are exotic, being far from stability, on the other - simple, as they are located close to the doubly-magic spherical nucleus ^{132}Sn. Thus, generally single-valence particle excitations of few neutrons as in ^{135}Sb, ^{136}Sb [1,2] up to few protons away as in ^{135}I, ^{136}I [3,4] present rather simple descriptions. An onset of collectivity is expected for these nuclei as soon as both protons and neutrons are added onto the core system. Currently available experimental data compared to theory shows that e.g. ^{138}Te [5] is at the border to develop some deformation [6]. Experimental excitation energies, compared to theory report on quadrupole and possible octupole collectivity above Xe, toward the heavy Ba region [7,6] despite expectations in ^{139}Te [8,9].

To date, only a few shell-model descriptions are available in this region, where ^{132}Sn is treated as a core and few-valence particle excitations in the lower-lying $\nu f_{7/2}$, $\nu p_{3/2}$ and $\pi g_{7/2}$, $\pi d_{5/2}$ reasonably describe the beginning [8,10] of these shells. In addition, some polarisation of the core has to be assumed. More monopole corrections are needed e.g. at the end of the $\nu f_{7/2}$ shell, around N=90, to describe our recent experimental data on possible crossing between $\pi g_{7/2}$ and $\pi d_{5/2}$ orbitals [11]. Even if some of these descriptions do reasonably well close to ^{132}Sn, they show discrepancies e.g. in the position of the generally isolated $\pi g_{7/2}$ and $\pi d_{5/2}$ orbitals. In particular, this relates to the calculated depressed position of $\pi d_{5/2}$ for ^{136}I, ^{138}I [8,12] believed to be responsible for their compressed low-energy spectra, resulting in the existence of long-lived isomers [13]. On the neutron side, differences exist in the predictions for N=90 sub-shell gap between the $\nu f_{7/2}$ and $\nu p_{3/2}$ orbitals [14], an assumption inconsistent with [15]. Having worked in this region in several works to date, we provide indispensable ingredients to understand many of the open questions. For this, it is important to further investigate specifically the species close to N=82, to compare them with and explore the entire proton-neutron particle valence space beyond.

2 Measuremens and Data

Several experimental investigations were recently performed using various methods, as on the very neutron-rich nuclei - in isomer and β-decay spectroscopy at EURICA, the less neutron-rich in prompt γ-ray spectroscopy at EXILL and in purely β-decay measurements at LOHENGRIN. Details of these studies were discussed elsewhere [2,5,11,16,17]. The strategy in these complementary measurements is to find a common description of the entire region on orbital evolution and collectivity, while searching for single-particle and collective excitations and their persistence with the addition of particles. Here, we will give one example on the observation of high-energy low spin states for Z=53 nuclei, that we traced towards the end of the $\nu f_{7/2}$ shell at $^{140}I_{87}$ [18] and at its beginning at $^{136}I_{83}$ [17]. In the ^{140}I

* Corresponding author: radomira.lozeva@csnsm.in2p3.fr

case, the states were populated in pure β-decay of Te projectiles, while for ^{136}I, they were populated in the prompt data of fission fragments, which accessed also β-decay states in the daughter nucleus, as well as in another β-decay experiment of Te→I nuclei.

Particular attention for these iodine nuclei shall be drawn to the excitations taking place at high energy and low-spin for the (1$^+$) states, which are changing parity with respect to the negative parity g.s. They are interesting because they represent changing configurations, different from those in Sb and Te isotopes. Appearing between $\nu h_{9/2} \rightarrow \pi h_{11/2}$ orbitals, which are open at Z=53 for the iodine nuclei, below Z=52 these allowed-type transitions are completely blocked [19].

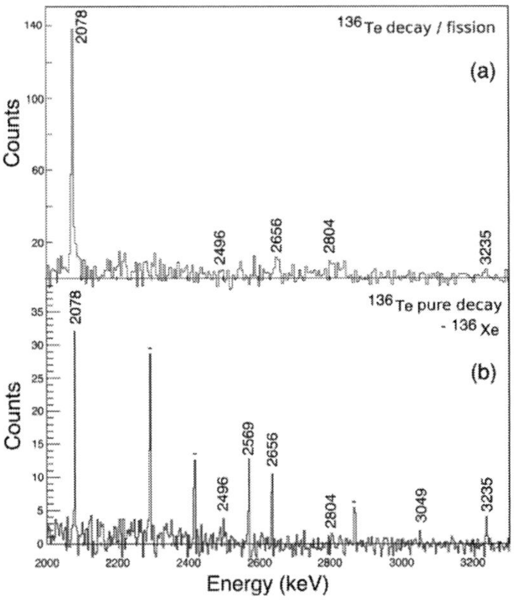

Fig. 1. Decay γ-ray transitions from 1$^+$ levels in ^{136}I observed in β-decay of products of prompt fission data (a) and in pure ^{136}Te β-decay (b).

This leads to possible structures of type $[(\nu h_{9/2})(\pi h_{11/2})(\pi g_{9/2})^2_{0+}]_{1+}$ and $[(\nu f_{7/2})(\pi h_{11/2})(\pi g_{7/2})^2_{2+}]_{1+}$ [20]. For ^{136}I this was suggested to be due possible configuration mixing of the $(\nu h_{9/2})^2$ and $(\nu f_{7/2})(\nu h_{9/2})$ in the g.s. of ^{136}Te. Three such candidate states were than proposed in the β-decay level scheme at relatively high excitation energy of 2656.4, 3137.1 and 3235.2 keV. They were identified by evaluators as 1$^+$ states [13], although no firm justification has been experimentally presented except for the first excited 1$^+$ state (at 2656.4 keV) [20]. The previously observed γ-ray transitions depopulating these levels in ^{136}I represent the main decay path of the mother nucleus to the 1$^-$ g.s. through either direct high-energy transitions or through the population of intermediate negative-parity levels. It is interesting to note here, that two of these 1$^+$ candidate states at 2656.4 and 3235.2 keV have rather strong decay branches, while the third candidate placed in between (at 3137.1 keV) has a largely suppressed intensity of its decay transitions. It may thus be possible that this state has a different assignment e.g. leaving only two 1$^+$ states present in ^{136}I. This is supported also by the observed I_r and *logft* values.

We populated these states in prompt data following the n-induced fission of ^{241}Pu, where non-yrast β-decaying states could be accessed. These transitions are detected in a coincidence gate with the lowest transitions (87+135 keV) in ^{136}I, de-exciting its first and second excited states. The resulting spectrum is presented in Fig.1(a). Independently, in a purely β-decay experiment, we have identified the ^{136}Te nucleus and collected data, corresponding to its decay to ^{136}I with appropriate $T_{1/2}$. The resulting relevant spectrum is presented in Fig.1.(b), where all previously known transitions in ^{136}I are shown. In addition, we have identified transitions from the grand-daughter ^{136}Xe, present in the time gate.

3 Discussion

Except the strong decay from the first 1$^+$ state by the 2077.9 keV γ-ray, it is interesting to underline the detection of a 2656.0 keV transition, expected [13], but unobserved as a direct decay from this first 1$^+$ state to the 1$^-$ g.s. state. It appears in the gated spectrum in Fig.1(a), therefore it is a transition depopulating another state at e.g. 3235.2 keV, proposed in [20]. The 3235.2 keV transition, proposed as a similar direct branch decay to the g.s. by [20] is also present in this particular gated spectrum. The relative intensities between the transitions within the same data set may be compared. For this, one can neglect the different type of population between them for these levels, which will be discussed elsewhere together other new spectroscopy data on ^{136}I [21].

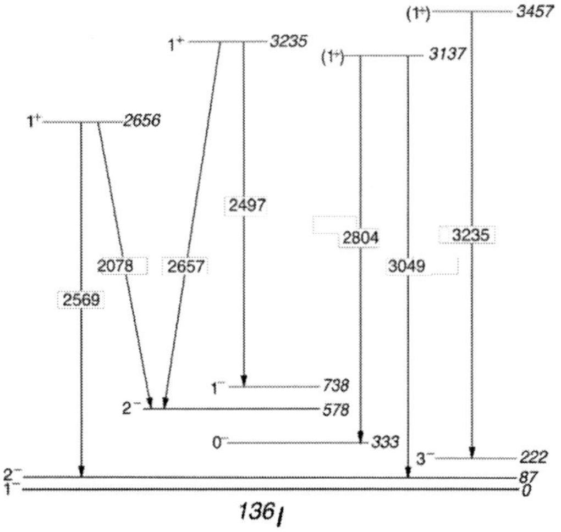

Fig. 2. Partial level scheme of ^{136}I with the relevant 1$^+$ states.

Related to the current discussion, it is interesting to point out, that one may possibly observe the 2656.0 keV transition with the same energy decaying out of the third

1^+ level, while the 3235.2 keV transition may represent depopulation of another fourth 1^+ level in ^{136}I. The intensities of the transitions de-exciting the second 1^+ level (at 3137.1 keV) of 2804.0 keV and a new fourth 1^+ level (at 3457.3 keV) are thus comparable. Partial level scheme of ^{136}I [13,20] with decay branches from all possible 1^+ levels is presented in Fig. 2.

Furthermore, the observation of three 1^+ candidates in ^{140}I [18], support these assignments. According to the observed I_β and *logft* values indeed the first and third (1^+) levels are populated stronger than the second (1^+) candidate state. This points at the a different origin of the second 1^+ state, similarly to ^{136}I. The population of allowed transitions is observed by the strong feeding of first and third 1^+ states in both isotopes, which are suppressed in energy with the increase of N, consistent with [18].

4 Future plans

The newly obtained data, related to new spectroscopy in ^{136}I, as well as other I nuclei in the region are subject of dedicated articles.

References

1. A. Korgul *et al*, Phys. Rev. C **91**, 027303 (2015)
2. R. Lozeva *et al*, Phys. Rev. C **92**, 024304 (2015)
3. P. Spagoletti *et al*, Phys. Rev. C **95**, 021302 (2017)
4. W. Urban *et al*, Eur. Phys. Journ. A **27**, 257 (2006)
5. P. Lee *at al*, Phys. Rev. C **92**, 044320 (2015)
6. H. Naidja *et al*, Phys. Rev. C **96**, 034312 (2017)
7. D. Bianco *et al*, Phys. Rev. C **88**, 024303 (2013)
8. W. Urban *et al*, Phys. Rev. C **65**, 024307 (2002)
9. R. Kshetri *et al*, Phys. Rev. C **74**, 034314 (2006)
10. L. Corraggio *et al*, Phys. Rev. C **80**, 021305(R) (2009)
11. R. Lozeva *et al*, Phys. Rev. C **93**, 014316 (2016)
12. T. Rzaca-Urban *et al*, Phys. Rev. C **75**, 054319 (2007)
13. www.nndc.bnl.gov
14. S. Sarkar *et al*, Phys. Rev. C **78**, 024308 (2008)
15. H. Naidja *et al*, Journ. Phys. **580**, 012030 (2015)
16. M. Jentschel *et al*, Journ. Instr. **12**, P11003 (2017)
17. R. Lozeva *et al*, Eur. Phys. Journ., in press (2017)
18. B. Moon *et al*, Phys. Rev. C **95**, 044322 (2017)
19. I. Borzov *et al*, Phys. Rev. C **67**, 025802 (2003)
20. F. Schussler *et al*, Z. Phys. A **283**, 43 (1977)
21. R. Lozeva *et al*, Phys. Rev. C (to be published)

Triaxial deformations in neutron-rich nuclei with Z = 41 – 46, A ~ 100 – 116 based on prompt fission γ spectroscopy

Y.X. Luo[1,2], J.H. Hamilton[1], J.O. Rasmussen[2,3], A.V. Ramayya[1], S. Frauendorf[4,5], E. Wang[1], J.K. Hwang[1], S.J. Zhu[1,6], Y.X. Liu[7], F.R. Xu[8], Y. Sun[9,10], S.H. Liu[1,11], Yu. Oganessian[12], W. C. Ma[13]

[1]*Physics Department, Vanderbilt University, Nashville, TN 37235, USA*
[2]*Lawrence Berkeley National Laboratory, Berkeley, CA 94720, USA*
[3]*Department of Chemistry, U.C. Berkeley, Berkeley, CA 94720, USA*
[4]*Department of Physics, University of Notre Dame, Notre Dame IN 46556, USA*
[5]*Institut für Strahlenphysik, FZD-Rossendorf, Postfach, D-01314 Dresden, Germany*
[6]*Department of Physics, Tsinghua University, Beijing, 100084, China*
[7]*School of Science, Huzhou Teachers College, Huzhou, 31300, China*
[8]*School of Physics, Peking University, Beijing 100871, China*
[9]*Department of Physics and Astronomy, Shanghai JiaoTong Univ., Shanghai, China*
[10]*Institute of Modern Physics, Chinese Academy of Sciences, Lanzhou, China*
[11]*Chemistry Department, University of Kentucky, Lexington, KY 40505, USA*
[12]*Flerov Laboratory for Nuclear Reactions, JINR, Dubna, Russia*
[13]*Physics Department, Mississippi State University, MS 39762, USA*

Abstract. The paper reviews the systematic studies of triaxial deformations, new mode excitations and shape evolutions with regard to triaxial deformation in the neutron-rich nuclei with Z = 41-46, A~100-116.

1 Introduction

The neutron-rich nuclei in this region are intermediate between the strongly deformed Sr (Z=38) – Y (Z=39) – Zr (Z=40) nuclei and the spherical doubly magic ^{132}Sn [1-4]. The Fermi levels relative to the high-j subshell, $\nu h_{11/2}$, $\pi g_{9/2}$ locate at bottom, middle, to upper half of the shells, favoring triaxial prolate, through large triaxial deformations to triaxial oblate, and oblate shapes providing good opportunities for studying shape transition and new excitations with regard to triaxial deformations. The studies are based on the fission γ spectroscopy using spontaneous fission of ^{252}Cf at Gammasphere [2,4].

2 Systematic studies of triaxial deformations of the nuclei in the region [4, and the references therein]

Great successes have been achieved in intensive systematic work in identifications of unknown neutron-rich nuclei, studies of triaxial deformations, shape evolutions with regard to triaxial deformations in the nuclear region. Maximum triaxiality, rigid rotors, $\gamma \sim -30^0$ were identified in Ru (Z=44) isotopes [4, 5]. Less pronounced triaxiality was found in Pd isotopes [6]. Stable $\gamma \sim -41^0$ was deduced in ^{114}Pd, a minimum energy gain 0.32 MeV, in contrast to 0.67 MeV in Ru isotopes. In Ag (Z=47) isotopes, one sees softness towards triaxiality and very rich structure. In Cd (Z=48) isotopes, quasi-particle couplings, vibrations, onset of collectivity, quasi-rotations, soft triaxiality $\gamma \sim -10^0$ and so on were proposed [4]. In the region below Ru ($Z \leq 44$) [4], Tc (Z=43) with large triaxiality, $\gamma \sim -22^0 - -26^0$, and Mo (Z=42) with large triaxiality, rigid rotors, $\gamma \sim 20^0$ were deduced. In Nb (Z=41) isotopes coexisting triaxiality and transitional behavior from $\gamma \sim 2^0$ to 15^0 were proposed. However, axially-symmetric shapes with large $\varepsilon_2 \sim 0.40$ were deduced in Y (Z=39) and Zr (Z=40) for A < 104, and onset of triaxiality was confirmed in ^{104,106}Zr.

3 New excitations and shape evolutions found in the triaxial nuclei in the region

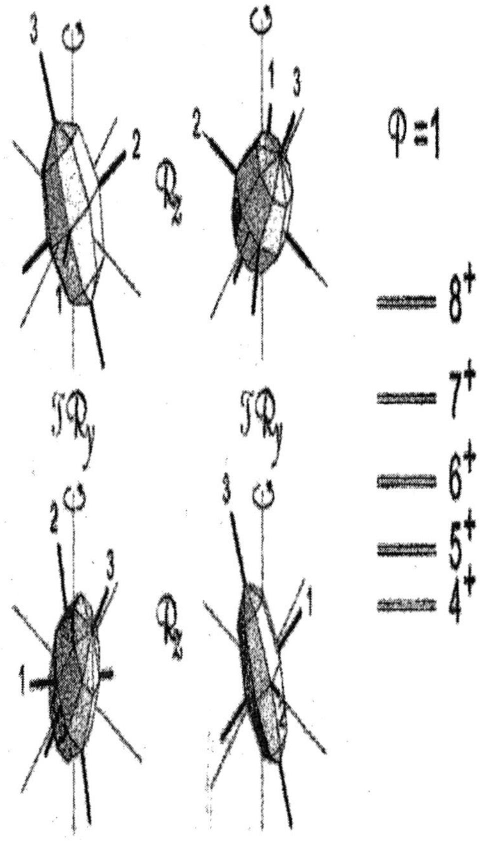

Fig. 1 Chiral breaking in triaxial nuclei.

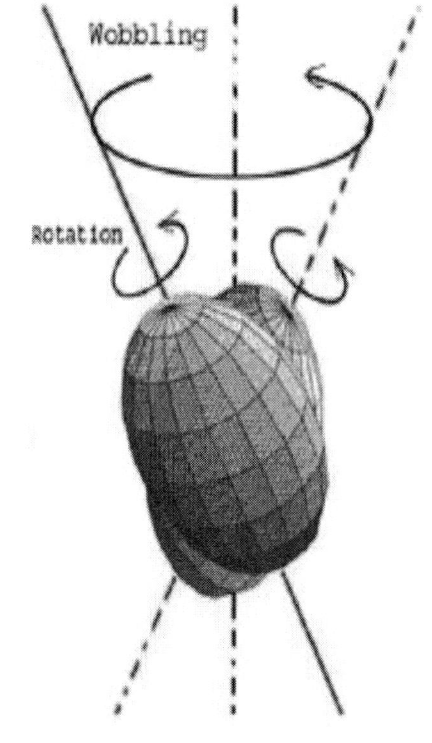

Fig. 2 Wobbling motion in triaxial nuclei.

In this neutron-rich region chiral symmetry breaking has been identified in ^{104,106,108}Mo, ^{100}Tc, ^{110,112}Ru, $^{103\text{-}106}$Rh and ^{104,105}Ag, with disturbed chirality, by soft- and less-pronounced triaxiality, proposed in ^{108}Ru and ^{112,114,116}Pd, respectively [4-6]. The fingerprints of chirality, that is energy degeneracy, similar electro-magnetic properties of the partner levels, and small and nearly constant signature splitting of the doublet bands have been studied in chiral nuclei. While one has seen the evolution of chiral symmetry breaking with maximum triaxiality in ^{110,112}Ru to disturbed chirality by γ - softness in ^{108}Ru (from N = 66, 68, to N=64), the evolution of chiral symmetry breaking along the N=66, 68 isotonic chain from Z=44 to Z=46 has also been seen in chiral ^{110,112}Ru with maximum triaxiality to disturbed chirality in Pd isotopes with less-pronounced triaxial deformations. The doublet bands in ^{104}Mo recently identified (Fig. 3) show the best energy degeneracy, that is the smallest and most stable energy differences of the partner levels (Fig. 4), and near zero level staggering in this nuclear region (The latter not shown here in the paper because of the page constraint) [7].

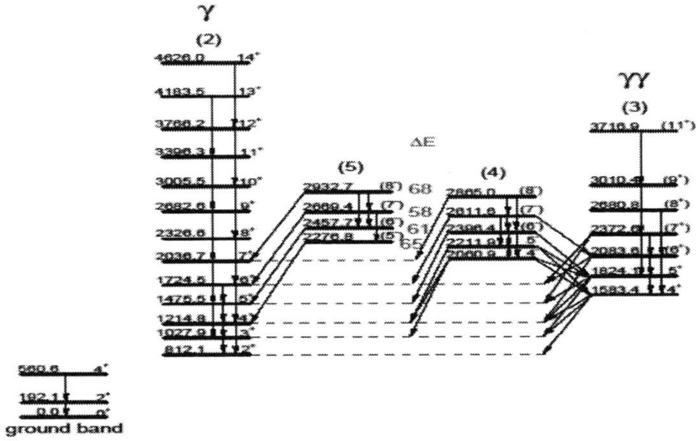

Fig. 3 Chiral symmetry breaking was recently identified in ^{104}Mo. Fingerprints for chirality of a nucleus were confirmed in the doublet bands 4 and 5 [7].

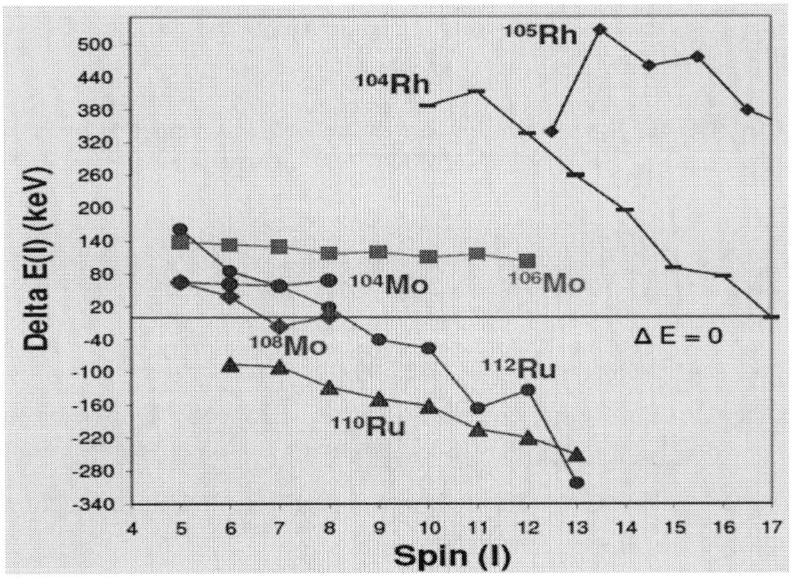

Fig. 4 Energy degeneracy of the partner levels in the doublet bands of the chiral nuclei identified in this nuclear region. The doublet bands in ^{104}Mo show the smallest and most stable energy differences of the partner levels among all the chiral nuclei.

The wobbling motion predicted in triaxial nuclei constitute a revolving motion of J about an axis of a triaxial nucleus. Wobbling motion manifest itself as a fingerprint that the excitations of the $\alpha = 0$ wobbling (even-spin members of the γ band) are above those of the $\alpha = 1$ wobbling (odd-spin members of the γ band) in an even-even nucleus. The N=68 isotones ^{112}Ru and ^{114}Pd were identified as the first and second even-even wobbler at moderate spins in the region [4].

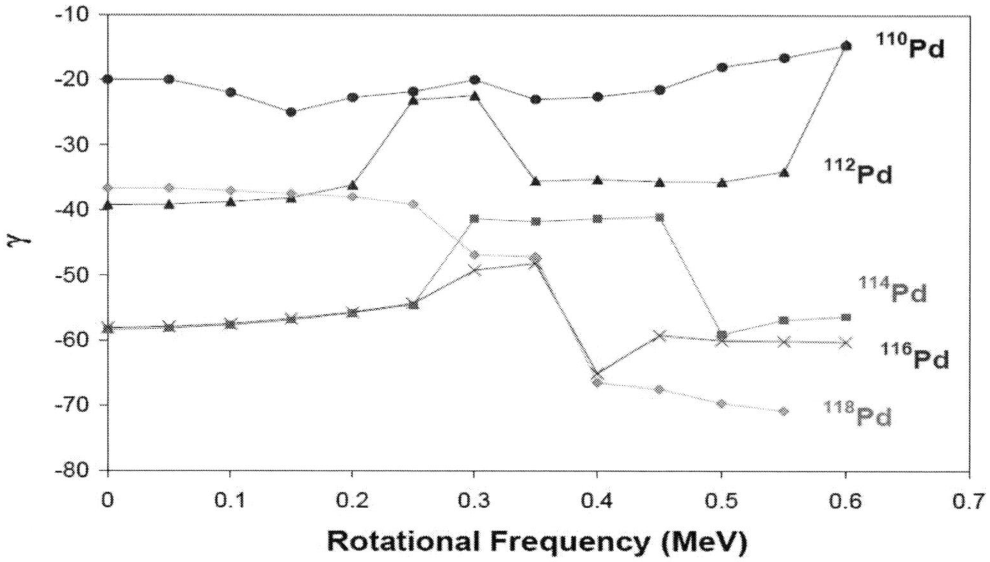

Fig. 5 γ values corresponding to the minima of TRS calculations, indicating the triaxial deformations changing with the rotational frequency and neutron number in Pd [6].

An overall shape evolution from triaxial prolate via triaxial oblate to oblate was identified in even-even $^{110-118}$Pd (see Fig. 5), showing a more complete and complex shape transition than predicted long time ago [6].

Acknowledgment The work on the present review was supported by the US DOE Grants and Contract No. DE-FG-05-88ER40407, DF-FG02-95ER40934 and DE-FG02-95EER4093.

References

1. J. SKALSKI *et al.* Nucl. Phys. A**617**, 282(1997)
2. J.H. Hamilton, Prog in Particle and Nucl. Phys. **15**, 107(1985); Treatise on Heavy Ion Science, ed. D. ALLAN BROMLEY. Plenum, NY, **8**, 363(1989)
3. Y.X. Luo *et al.* J Phys. G**31**, 1303(2005) ; S.J. Zhu *et al.* Inter. Jour. of Modern Phys. E**18**(8), 1717(2009)
4. Y.X. Luo *et al.* Chinese Nucl. Phys. Rev. **32**, 1, 1(2015); Nucl. Phys. Rev. **27**, 3, 229(2010)
5. Y.X. Luo *et al.* Phys. Lett. B**670**, 307(2009)
6. Y.X. Luo *et al.* Nucl. Phys. A**919**, 67(2013)
7. B. Musangu *et al.* Proceedings of the ICFN 6, p255, Novenber 6-12, 2014, Sanibal, USA

Probing the $E2$ properties of the scissors mode with real photons

Tobias Beck[1,*], *Norbert* Pietralla[1], *Jacob* Beller[1], *Vera* Derya[2], *Bastian* Löher[1], *Deniz* Savran[3], *Werner* Tornow[4], *Volker* Werner[1], and *Andreas* Zilges[2]

[1]Institut für Kernphysik, TU Darmstadt, Schlossgartenstr. 9, D-64289 Darmstadt, Germany
[2]Institut für Kernphysik, Universität zu Köln, Zülpicher Str. 77, D-50937 Köln, Germany
[3]GSI Helmholtzzentrum für Schwerionenforschung, Planckstr. 1, D-64291 Darmstadt, Germany
[4]Department of Physics, Duke University and Triangle Universities Nuclear Laboratory, Durham, North Carolina 27708-0308, USA

Abstract. The $E2/M1$ multipole mixing ratio $\delta_{1\to2}$ of the $1^+_{sc} \to 2^+_1$ γ-ray transition of ^{156}Gd and ^{164}Dy has been measured using the linearly polarized photon beams of the HIγS facility. The employed method of photon-scattering experiments in combination with polarized, quasi-monochromatic beams and a dedicated detector setup is highly sensitive to the electric quadrupole-decay properties of the scissors mode.

1 Introduction

The study of the properties of the nuclear scissors mode [1–3] provides an essential insight into the nature of the restoring forces between the proton and neutron subsystems. Hence, the isovector low-lying $J^\pi_K = 1^+_1$ scissors mode of deformed nuclei has been studied extensively in the past with main focus on strong $M1$ transitions to the ground-state band [4, 5, and Refs. therein]. Despite the quadrupole-collective origin of the nuclear scissors mode, the $E2$ properties of the scissors mode were unknown until recently [6]. Likewise, information on the predicted existence of the scissors mode's rotational band with the $J = 1$ state as band head is limited [7, 8].
First information on the $E2$ decay transition strength of the scissors mode was extracted [6] from a high-statistics photon-scattering experiment on ^{156}Gd using quasi-monochromatic photon beams provided by the High Intensity γ-ray Source (HIγS) [9] at Duke University, Durham, NC, USA. The data allowed for measuring a finite value of the $E2/M1$ multipole mixing ratio and, thus, first measurement of an $E2$ transition between mixed-symmetric and a fully-symmetric states in axially deformed nuclei. The obtained results indicate that highest-precision photon-scattering experiments [4, 10] with linearly polarized photons [11] are highly sensitive to the electric quadrupole-decay properties of the scissors mode. It is the purpose of this manuscript to present first results obtained from a subsequent experiment on the well-deformed nucleus ^{164}Dy at HIγS.

2 Experimental details and results

The investigation of the $E2$ properties of the scissors mode is based on the well-known properties of the electromagnetic decay of the 1^+_{sc} scissors mode state to the 2^+_1 state. The linear polarization of the impinging photon beam causes an anisotropic azimuthal distribution of the scattered photons which is detected using the γ^3 setup [12]. It includes four high-purity Germanium (HPGe) detectors forming a setup which is especially sensitive to the $E2/M1$ mixing ratio

$$\delta_{1\to2} = \frac{\sqrt{3}}{10}\frac{E_\gamma}{\hbar c}\frac{\langle 2^+_1 \,\|\, \hat{T}(E2) \,\|\, 1^+_{sc}\rangle}{\langle 2^+_1 \,\|\, \hat{T}(M1) \,\|\, 1^+_{sc}\rangle} \quad (1)$$

of the $1^+_{sc} \to 2^+_1$ transition [13, 14]. Here, the multipole mixing ratio is given in the phase convention of Krane *et al.* [15]. The quantities $\hat{T}(E2)$ and $\hat{T}(M1)$ denote the electric quadrupole and magnetic dipole transition operators, respectively.
The ratio N_i/N_j of γ-ray intensities observed in detectors i and j is sensitive to the multipole mixing ratio $\delta_{1\to2}$ of the $1^+_{sc} \to 2^+_1$ transition. Subsequently, the multipole mixing ratio is obtained by comparison of the observed ratio to the respective ratio of angular distributions $W(\vartheta_i, \varphi_i, \delta_{1\to2})/W(\vartheta_j, \varphi_j, \delta_{1\to2})$ for the $0^+_1 \to 1^+_{sc} \to 2^+_1$ sequence. This method produces two equitable solutions; one close to zero, the other corresponding to dominant $E2$ character. Hence, further constraints based on the comparison of experimental decay intensity ratios to expectations from the Alaga rule [16] are needed to identify [6] the consistent solution for the multipole mixing ratio.

2.1 ^{156}Gd

The results of the pioneering experiment on ^{156}Gd are presented in detail elsewhere [6]. In the following, the robustness of the method towards systematic uncertainties shall be addressed.
Due to the compact geometry of the detector setup the angular distribution functions have to be integrated over the solid angles of the individual detectors considering the

*e-mail: tbeck@ikp.tu-darmstadt.de

mean free path of 3 MeV photons in the detector material. The exact positioning of the target sample relative to the impinging γ-ray beam and the detector setup was achieved using a cross-line laser system. Nevertheless, an uncertainty contribution associated with a displacement of the target parallel to the beam axis remains and is estimated to be smaller than ±5 mm. Naturally, this results in altered solid angles of the respective detectors. However, for small values of the multipole mixing ratio $\delta_{1\to2}$ the resulting systematic uncertainty is of comparable size to the statistical uncertainty. Figure 1 shows the result for the multipole mixing ratio of the decay of the strongest fragment of the scissors mode of ^{156}Gd indicating the robustness of the method towards a displacement of the target sample along the beam axis.

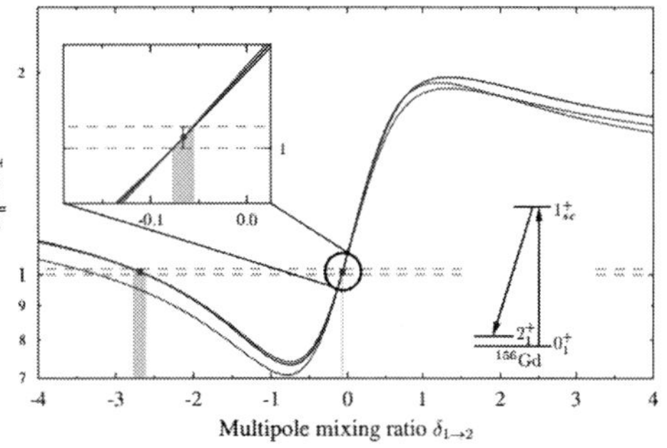

Figure 1. Comparison of the measured intensity ratio to the ratio of integrated angular distributions $W(135°, 0°, \delta_{1\to2})/W(135°, 90°, \delta_{1\to2})$ for the $0^+_1 \to 1^+_{sc} \to 2^+_1$ sequence indicated by the blue line. The two possible solutions for the multipole mixing ratio are marked in red. The brown and green lines mark the ratio of angular distributions integrated over the altered detector solid angles caused by a displacement of the target sample by ±5 mm in the direction of the γ-ray beam.

2.2 ^{164}Dy

Following the first successful application of the presented method, its advancement to cases with different detector setups and lower statistics is currently pursued. While the experiment on ^{156}Gd featured a polarimeter-like setup of four HPGe detectors at backward angles, an advanced geometry had been implemented for the measurement of the $E2$ decay of the scissors mode of ^{162,164}Dy. Two detectors were mounted at a polar angle of $\vartheta = 90°$ with respect to the incoming beam and at azimuthal angles φ of 0° and 90° with respect to the horizontal polarization plane. In addition, two HPGe detectors were placed at $(\vartheta, \varphi) = (135°, 225°)$ and $(135°, 315°)$, resulting in three unique detector combinations defined by the symmetry properties of the angular distribution functions.

Figure 2 shows the resulting ratios of integrated angular distributions $W(\vartheta_i, \varphi_i, \delta_{1\to2})/W(\vartheta_j, \varphi_j, \delta_{1\to2})$ for two detectors i and j as a function of the $E2/M1$ multipole mixing ratio. In the following, first results of the two detectors in polarimeter geometry, i.e. at a polar angle of $\vartheta = 90°$ and azimuthal angles φ of 0° and 90° corresponding to the

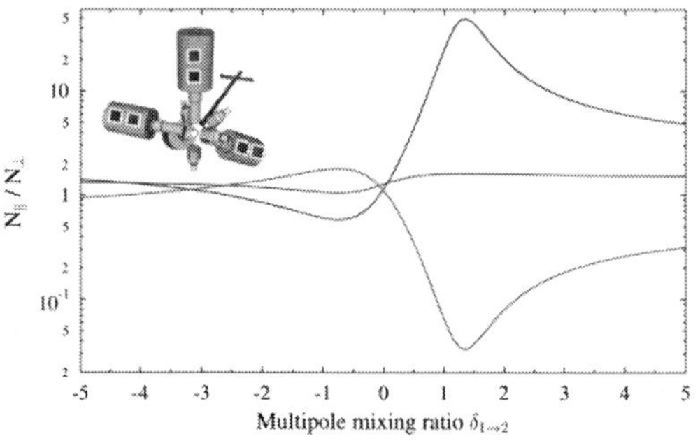

Figure 2. Evolution of the ratio of integrated angular distributions $W(\vartheta_i, \varphi_i, \delta_{1\to2})/W(\vartheta_j, \varphi_j, \delta_{1\to2})$ with the multipole mixing ratio of the $1^+_{sc} \to 2^+_1$ transition for different detector combinations color coded in the schematic setup (upper left corner). Due to the symmetry properties of the angular distribution functions three different detector combinations have to be considered.

blue curve of Figure 2, are presented.

The $(\vec{\gamma}, \gamma')$ spectra of ^{164}Dy measured in the polarization plane (red) and perpendicular to it (blue) are shown in Figure 3. The energy profile of the impinging γ-ray beam with a width of about 3.5% of the centroid energy 3.180 MeV is indicated by a dashed Gaussian. The decay of three scissors mode states to the ground state are prominently observed in the polarization plane corresponding to the red spectrum. Their mixed $E2/M1$ transition to the 2^+_1 state is found 0.073 MeV below, respectively.

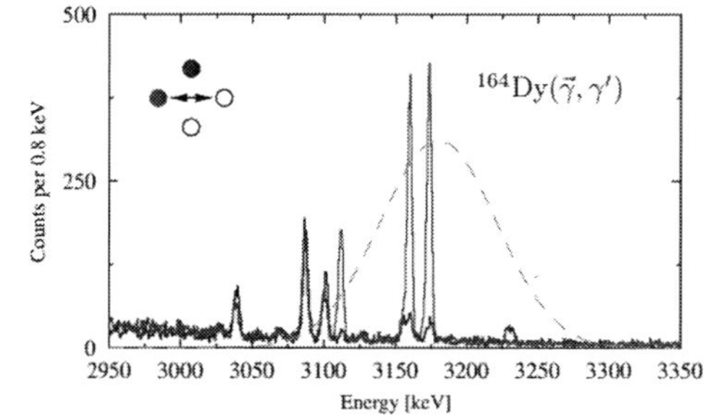

Figure 3. Gamma-ray spectra of the ^{164}Dy$(\vec{\gamma}, \gamma')$ reaction. Detectors were placed at a polar angle of $\vartheta = 90°$ and in the horizontal polarization plane (red) of the incident γ-ray beam and perpendicular to it (blue). The spectra are dominated by the decays of three scissors mode states to the ground state and to the 2^+_1 state. The energy profile of the $\vec{\gamma}$-ray beam is indicated by the dashed Gaussian curve.

For the 1^+_{sc} state located at 3.173 MeV a finite value of the multipole mixing ratio $\delta_{1\to2}$, which significantly differs from zero, is obtained from the $1^+_{sc} \to 2^+_1$ decay transition at 3.100 MeV. From the squared multipole mixing ratio $\delta^2_{1\to2} = \Gamma_{1^+_{sc}\to2^+_1,E2}/\Gamma_{1^+_{sc}\to2^+_1,M1}$ and the partial decay width $\Gamma_{1^+_{sc}\to2^+_1} = \Gamma_{1^+_{sc}\to2^+_1,M1} + \Gamma_{1^+_{sc}\to2^+_1,E2}$ [17] a preliminary transition strength $B(E2; 1^+_{sc} \to 2^+_1)$ well below 1 W.u. is determined. Naturally, the integration of the results of the remaining detector combinations (cf. Figure 2) will slightly alter this result.

3 Summary and Outlook

The results of ^{156}Gd and ^{164}Dy indicate that properties of the scissors mode's electric quadrupole decay can be obtained from highest-precision photon-scattering experiments with linearly polarized photons. However, further experiments are essential to shed light on the distribution of $E2$ strength compared to $M1$ strength in a single nucleus as well as its evolution with the number of valence neutrons.

4 Acknowledgments

The authors thank the HIγS accelerator crew for providing perfect experimental conditions. This work was supported by the Deutsche Forschungsgemeinschaft under Grants No. SFB 634, No. SFB 1245, and No. ZI 510/7-1, partly by the BMBF under Grant No. 05P15RDEN9, by the Alliance Program of the Helmholtz Association (HA216/EMMI) and by the U.S. DOE Grant No. DE-FG02-97ER41033. We further acknowledge U.S. DOE for the supply of isotopically enriched materials through ORNL.

References

[1] N. Lo Iudice, F. Palumbo, Phys. Rev. Lett. **41**, 1532 (1978)

[2] F. Iachello, Nucl. Phys. A **358**, 89 (1981)

[3] D. Bohle, A. Richter, W. Steffen, A.E.L. Dieperink, N. Lo Iudice, F. Palumbo, O. Scholten, Phys. Lett. B **137**, 27 (1984)

[4] U. Kneissl, H. Pitz, A. Zilges, Prog. Part. Nucl. Phys. **37**, 349 (1996)

[5] K. Heyde, P. von Neumann-Cosel, A. Richter, Rev. Mod. Phys. **82**, 2365 (2010)

[6] T. Beck, J. Beller, N. Pietralla, M. Bhike, J. Birkhan, V. Derya, U. Gayer, A. Hennig, J. Isaak, B. Löher et al., Phys. Rev. Lett. **118**, 212502 (2017)

[7] Y. Sun, C.L. Wu, K. Bhatt, M. Guidry, D.H. Feng, Phys. Rev. Lett. **80**, 672 (1998)

[8] Y. Sun, C.L. Wu, K. Bhatt, M. Guidry, Nucl. Phys. A **703**, 130 (2002)

[9] H.R. Weller, M.W. Ahmed, H. Gao, W. Tornow, Y.K. Wu, M. Gai, R. Miskimen, Prog. Part. Nucl. Phys. **62**, 257 (2009)

[10] U. Kneissl, N. Pietralla, A. Zilges, J. Phys. G: Nucl. Part. Phys. **32**, R217 (2006)

[11] N. Pietralla, H. Weller, V. Litvinenko, M. Ahmed, A. Tonchev, Nucl. Instr. Meth. Phys. Res. A **483**, 556 (2002)

[12] B. Löher, V. Derya, T. Aumann, J. Beller, N. Cooper, M. Duchêne, J. Endres, E. Fiori, J. Isaak, J. Kelley et al., Nucl. Instr. Meth. Phys. Res. A **723**, 136 (2013)

[13] N. Pietralla, M.W. Ahmed, C. Fransen, V.N. Litvinenko, A.P. Tonchev, H.R. Weller, AIP Conf. Proc. **656**, 365 (2003)

[14] G. Rusev, A.P. Tonchev, R. Schwengner, C. Sun, W. Tornow, Y.K. Wu, Phys. Rev. C **79**, 047601 (2009)

[15] K. Krane, R. Steffen, R. Wheeler, At. Data Nucl. Data Tables **11**, 351 (1973)

[16] G. Alaga, K. Alder, A. Bohr, B.R. Mottelson, Dan. Mat. Fys. Medd. **29** (1955)

[17] V. Werner, Ph.D. thesis, Universität zu Köln (2004)

β-γ and isomeric decay spectroscopy of ^{168}Dy

G.X. Zhang[1], *H. Watanabe*[1,2], *F.G. Kondev*[4], *G.J. Lane*[3], *P.H. Regan*[5,6], *P.-A. Söderström*[2], *P.M. Walker*[5], *H. Kanaoka*[7], *Z. Korkulu*[8], *P.S Lee*[9], *J.J Liu*[10], *S. Nishimura*[2], *J. Wu*[2,11], *A. Yagi*[7], *D.S. Ahn*[2], *T. Alharbi*[12], *H. Baba*[2], *F. Browne*[13], *A.M. Bruce*[13], *R.J. Carroll*[5], *K.Y. Chae*[14], *Zs. Dombradi*[8], *P. Doornenbal*[2], *A. Estrade*[15], *N. Fukuda*[2], *C. Griffin*[15], *E. Ideguchi*[16], *N. Inabe*[2], *T. Isobe*[2], *S. Kanaya*[7], *I. Kojouharov*[17], *T. Kubo*[2], *S. Kubono*[2], *N. Kurz*[17], *I. Kuti*[8], *S. Lalkovski*[5], *C.S. Lee*[9], *E.J. Lee*[14], *G. Lorusso*[2,5,18], *G. Lotay*[5], *C.-B Moon*[19], *I. Nishizuka*[20], *C.R. Nita*[21], *A. Odahara*[7], *Z. Patel*[5], *V.H. Phong*[2,21], *Zs. Podolyák*[5], *O.J. Roberts*[22], *H. Sakurai*[2], *H. Schaffner*[17], *C.M. Shand*[5], *Y. Shimizu*[2], *T. Sumikama*[20], *H. Suzuki*[2], *H. Takeda*[2], *S. Terashima*[1], *Zs. Vajta*[8], *J.J. Valiente-Dóbon*[24], and *Z.Y. Xu*[10]

[1]*IRCNPC, School of Physics and Nuclear Energy Engineering, Beihang University, Beijing 100191, China*
[2]*RIKEN Nishina Center, 2-1 Hirosawa, Wako, Saitama 351-0198, Japan*
[3]*Department of Nuclear Physics, R.S.P.E., Australian National University, Canberra, ACT 0200, Australia*
[4]*Nuclear Engineering Division, Argonne National Laboratory, Argonne, Illinois 60439, USA*
[5]*Department of Physics, University of Surrey, Guildford GU2 7XH, United Kingdom*
[6]*National Physical Laboratory, Teddington, Middlesex, TW11 0LW, United Kingdom*
[7]*Department of Physics, Osaka University, Machikaneyama-machi 1-1, Osaka 560-0043 Toyonaka, Japan*
[8]*MTA Atomki, P. O. Box 51, Debrecen, H-4001, Hungary*
[9]*Department of Physics, Chung-Ang University, Seoul 156-756, Republic of Korea*
[10]*Department of Physics, the University of Hong Kong, Pokfulam Road, Hong Kong*
[11]*Department of Physics, Peking University, Beijing 100871, China*
[12]*Department of Physics, College of Science in Zulfi, Almajmaah University, P.O. Box 1712, 11932, Saudi Arabia*
[13]*School of Computing Engineering and Mathematics, University of Brighton, Brighton, BN2 4GJ, United Kingdom*
[14]*Department of Physics, Sungkyunkwan University, Suwon 440-746, Republic of Korea*
[15]*School of Physics and Astronomy, University of Edinburgh, Edinburgh EH9 3JZ, United Kingdom*
[16]*Research Center for Nuclear Physics (RCNP), Osaka University, Ibaraki, Osaka 567-0047, Japan*
[17]*GSI Helmholtzzentrum für Schwerionenforschung GmbH, 64291 Darmstadt, Germany*
[18]*National Physical Laboratory, Teddington, Middlesex, TW11 0LW, United Kingdom*
[19]*Department of Display Engineering, Hoseo University, Chung-Nam 336-795, Republic of Korea*
[20]*Department of Physics, Tohoku University, Aoba, Sendai, Miyagi 980-8578, Japan*
[21]*School of Computing Engineering and Mathematics, University of Brighton, Brighton, BN2 4GJ, United Kingdom*
[22]*VNU Hanoi University of Science, 334 Nguyen Trai, Thanh Xuan, Hanoi, Vietnam*
[23]*School of Physics, University College Dublin, Belfield, Dublin 4, Ireland*
[24]*Istituto Nazionale di Fisica Nucleare, Laboratori Nazionali di Legnaro, 35020 Legnaro, Italy*

Abstract. This contribution will report on the experimental work on the level structure of ^{168}Dy. The experimental data have been taken as part of the EURICA decay spectroscopy campaign at RIBF, RIKEN in November 2014. In the experiment, a ^{238}U primary beam is accelerated up to 345 MeV/u with an average intensity of 12 pnA. The nuclei of interest are produced by in-flight fission of ^{238}U impinging on Be target with a thickness of 5 mm. The excited states of ^{168}Dy have been populated through the decay from a newly identified isomeric state and via the β decay from ^{168}Tb. In this contribution, scientific motivations, experimental procedure and some preliminary results for this study are presented.

1 Introduction

Atomic nuclei consisting of a number of protons and neutrons are driven towards non-spherical equilibrium shapes when moving away from shell closures. Being located in the close vicinity of the double midshell at Z = 66, N = 104, ^{168}Dy (Z = 66, N = 102) is excepted to have a large quadrupole deformation for the ground state. The excited states in such a well-deformed nucleus are characterized by collective rotation and surface oscillations, including quadrupole (β and γ) and even higher-order vibrations, as well as by other quasiparticle excitations. The intrinsic state with a multi-quasiparticle configuration is likely to be a metastable state (isomer), when the projection of the total nuclear spin on the symmetry axis, denoted by K, is largely different from that of the lower-lying levels to which the isomer decays. The presence of these collective and intrinsic excitations within a narrow range of energy results in an interplay among them to a greater or lesser extent, giving rise to a rich variety of structural aspects in deformed nuclei. Therefore, spectroscopic studies of low-energy excitations in this doubly mid-shell region will provide a good testing ground for various collective model calculations.

This article will report on the spectroscopic results of $^{168}_{66}$Dy. For N = 102 isotones, $K^\pi = 4^-$ isomers had been identified in $^{170}_{68}$Er (E_x = 1269 keV, $T_{1/2}$ = 43 ns) [1], $^{172}_{70}$Yb (1641 keV, 0.5 ns) [2], and $K^\pi = 6^-$ isomers in $^{164}_{62}$Sm (1486 keV, 0.6 μs), $^{166}_{64}$Gd (1601 keV, 0.95 μs) [3], $^{170}_{68}$Er (1591 keV, 4.0 ns) [1], $^{172}_{70}$Yb (1550 keV, 3.6 μs) [4], $^{174}_{72}$Hf (1714 keV, 0.45 ns) [5]. The K^π = 6^- and 4^- isomers are interpreted as the neutron two-quasiparticle configurations $\nu^2 7/2^+[633] \otimes 5/2^-[512]$ and $\nu^2 7/2^+[633] \otimes 1/2^-[521]$, respectively. Yet isomeric states have not been identified in ^{168}Dy.

The octupole vibrational excitations have been identified in the doubly midshell region at relatively low energy, for example, $^{164}_{66}$Dy (976.9 keV, $K^\pi = 2^-$) [6], $^{166}_{66}$Dy (1029.9 keV, $K^\pi = 2^-$) [6], $^{170}_{66}$Dy (861 keV, $K^\pi = 2^-$) [7] and $^{172}_{70}$Yb (1154.9 keV, $K^\pi = 1^-$) [6]. Thus the aforementioned K isomers with negative parity are probably mixed with the rotational band-members built on such low-lying octupole vibrations. On the other hand, the $K^\pi = 3^+$ states have been identified in $^{170}_{68}$Er (1217.5 keV) [1] and $^{172}_{70}$Yb (1172.4 keV) [4]. In $^{174}_{72}$Hf, the $K^\pi = 3^+$ state has the possibility that the hexadecapole vibrational excitation may also have a contribution, which has already been studied before [8].

2 Experimental Procedure

In order to study the level structure of the neutron-rich rare-earth nuclei around double midshell, an experiment was performed in November 2014 at the RI-Beam Factory (RIBF) at RIKEN [9]. The $^{238}U^{86+}$ primary beam was accelerated up to 345 MeV/u by a sequential acceleration system consisting of a linac injector (RILAC) and four ring cyclotrons (RRC-fRC-IRC-SRC). The secondary beams were produced by in-flight fission of the U beam incident on a Be target with a thickness of 5 mm. The nuclei of interest were separated and identified through the BigRIPS spectrometer [10]. In this experiment, the secondary beams were transported with two different settings of the slits on the beam line; one is optimized for $^{170}Dy^{66+}$, and the other for $^{172}Dy^{66+}$. The respective particle identification spectra are shown in Fig. 1. About 1.2×10^4 ions and 1.0×10^4 ions were collected for $^{168}Tb^{65+}$ and the sum of $^{168}Dy^{65+,66+}$, respectively.

The identified particles were implanted into the WAS3ABi active stopper [11] which was comprised of two layers of double-sided silicon-strip detectors (DSSSD). Each DSSSD had a thickness of 1 mm with an active area of 60×40 mm^2 with 2400 segmentations of 1 mm^2 pitch. The WAS3ABi was placed at the end of the beam line for the measurement of heavy-ion implantation and electrons following β-decay and internal conversion processes.

Gamma rays emitted from the implanted RIs were detected by the EUROBALL-RIKEN Cluster Array (EURICA) [12] that had a full energy-peak efficiency of about 10 % for 1 MeV γ ray. The EURICA array consisted of 84 HPGe crystals arranged in nearly 4π geometry in the form of twelve 7-element CLUSTER-type detectors. The γ rays emitted from the implanted RIs were detected within a coincidence time window of 100 μs with respect to the trigger signals generated either when the heavy ions passed through a plastic scintillator mounted ~1 m upstream of the WAS3ABi stopper or when the DSSSDs were fired by the decay electrons.

The beam, electron, and γ-ray events were time-stamped and recorded by independent data-acquisition systems. Isomeric states with (sub)microsecond lifetimes were identified by delayed coincidence between γ-ray and beam signals on an event-by-event basis. Meanwhile, all the data sets containing beam, electron and γ-ray events were needed for electron-γ coincidence analyses, in which the ion implantation of an identified particle was associated with the subsequent electron events that were detected in the same or neighboring DSSSD pixels where the beam particle was implanted.

3 Preliminary results

Figure 2 shows the β-delayed γ-ray spectrum measured within 20 s after the implantation of ^{168}Tb. Before the present work, the excited states of ^{168}Dy had been studied by the β decay of ^{168}Tb [13] and multi-nucleon transfer reactions with a ^{82}Se beam incident on an ^{170}Er target [14]. The γ rays at energies of 75 and 173 keV, which were previously assigned as $2^+_1 \to 0^+_1$ and $4^+_1 \to 2^+_1$, respectively, have been confirmed in the present work. In addition to the transitions reported previously, several new γ rays are clearly visible at 216, 322, 915 and 1131 keV in the present work.

Figure 3 shows the γ-ray spectrum measured within 350 ns ~ 3 μs after the implantation of ^{168}Dy. New γ rays at 236 and 348 keV are clearly visible. Note that in Fig. 3, the γ-ray at 405 keV is a contaminant from an isomeric state in ^{174}Ho, which could not be separated from the hydrogen-like component of the ^{168}Dy ions in terms of Z on the particle identification plot due to the poor resolution of the ion chamber. Based on the present work, we

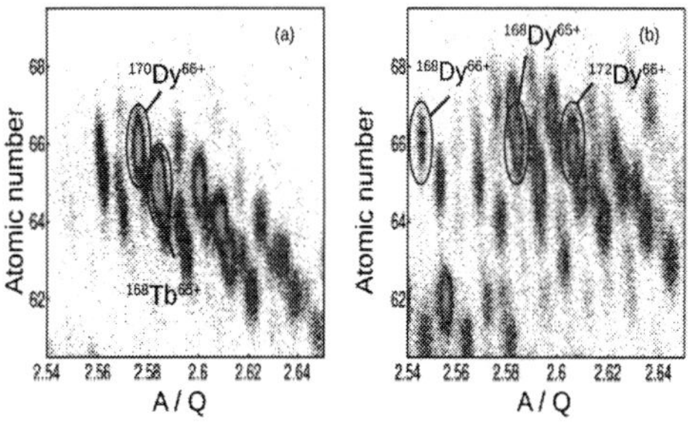

Figure 1. Particle identification spectra obtained with two different setting optimized for (a) $^{170}Dy^{66+}$ and (b) $^{172}Dy^{66+}$ ions.

have established a new level scheme of ^{168}Dy, the details of which will be presented elsewhere along with the discussion about the spin-parity assignment, transition hinderances, multi-quasiparticle configurations, and so on.

4 Summary

Decay spectroscopy experiment of ^{168}Dy has been carried out as part of the EURICA experimental campaign at RIBF, RIKEN. Following the production of neutron-rich isotopes by in-flight fission of ^{238}U, the nuclei relevant to the present work were separated and identified through the BigRIPS spectrometer. The nuclei of interest were implanted into the WAS3ABi array, which also served as a detector for β rays and internal conversion electrons accompanying the decay of the implanted radioactive isotopes. The heavy-ion implantation was associated with the subsequent decay electrons based on the position correlation in WAS3ABi. Gamma rays following β decay were measured by the EURICA array in coincidence with electrons detected by WAS3ABi. Meanwhile, isomeric states with half-lives ranging from several tens of nanoseconds to (sub)microseconds could be unambiguously identified by taking delayed coincidence between γ-ray and identified particles on an event-by-event basis. New results obtained in the present work include a number of γ rays following the β decay from ^{168}Tb and the isomeric decay in ^{168}Dy. The detailed analysis including evaluation of γ-ray intensities, feeding patterns, coincidence relationship, and transition strengths will allow us to establish a level scheme of ^{168}Dy, which will be presented elsewhere.

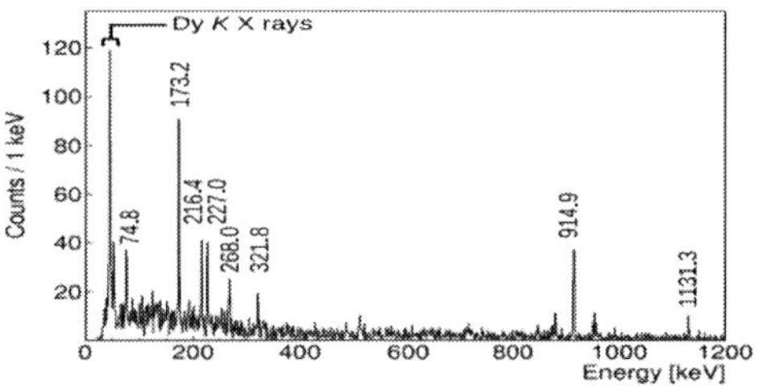

Figure 2. γ-ray energy spectrum measured in coincidence with electrons within 20 s after the implantation of ^{168}Tb.

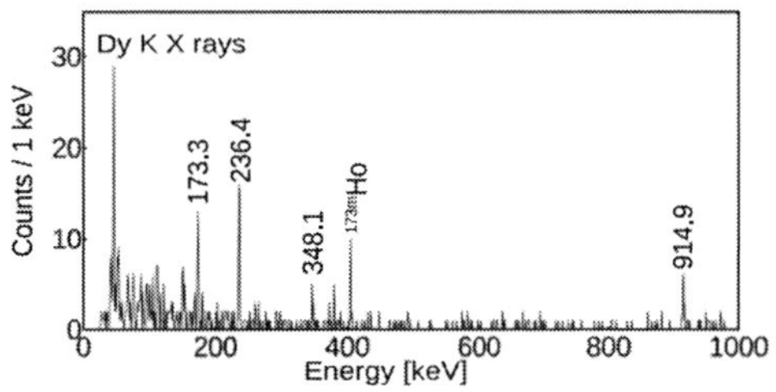

Figure 3. γ-ray energy spectrum measured in EURICA after the implantation of ^{168}Dy ions within a time range of 0.35 – 3 μs.

References

[1] G.D. Dracoulis, G.J. Lane, F.G. Kondev, H. Watanabe, D. Seweryniak, S. Zhu, M.P. Carpenter, C.J. Chiara, R.V.F. Janssens, T. Lauritsen et al., Phys. Rev. C **81**, 054313 (2010)

[2] L. Kostov, W. Andrejtscheff, H. Rotter, H. Prade, F. Stary, Nucl. Phys. A **406**, 541 (1983)

[3] Z. Patel, P.A. Söderström, Z. Podolyák, P.H. Regan, P.M. Walker, H. Watanabe, E. Ideguchi, G.S. Simpson, H.L. Liu, S. Nishimura et al., Phys. Rev. Lett. **113**, 262502 (2014)

[4] R. Nordhagen, R. Diamond, F. Stephens, Nucl. Phys. A **138**, 231 (1969)

[5] W. Andrejtscheff, L.K. Kostov, L.G. Kostova, P. Petkov, J. Doring, L. Kaubler, H. Rotter, E. Will, Tech. Rep. ZfK-621,P40 (1987)

[6] http://www.nndc.bnl.gov/ensdf/

[7] Z. Patel, Z. Podolyák, P. Walker, P. Regan, P.A. Söderström, H. Watanabe, E. Ideguchi, G. Simpson, S. Nishimura, F. Browne et al., Phys. Lett. B **753**, 182 (2016)

[8] N. Gjorup, P. Walker, G. Sletten, M. Bentley, B. Fabricius, J. Sharpey-Schafer, Nuclear Physics A **582**, 369 (1995)

[9] Y. Yano, Nucl. Instrum. Methods B **261**, 1009 (2007)

[10] N. Fukuda, T. Kubo, T. Ohnishi, N. Inabe, H. Takeda, D. Kameda, H. Suzuki, Nucl. Instrum. Methods B **317**, 323 (2013)

[11] S. Nishimura, Prog. Theor. Exp. Phys. p. 03C006 (2012)

[12] P.A. Söderström, S. Nishimura, P. Doornenbal, G. Lorusso, T. Sumikama, H. Watanabe, Z. Xu, H. Baba, F. Browne, S. Go et al., Nuclear Instruments and Methods in Physics Research Section B: Beam Interactions with Materials and Atoms **317**, 649 (2013)

[13] M. Asai, S. Ichikawa, K. Tsukada, M. Sakama, M. Shibata, Y. Kojima, A. Osa, I. Nishinaka, Y. Nagame, K. Kawade et al., Phys. Rev. C **59**, 3060 (1999)

[14] P.A. Söderström, J. Nyberg, P.H. Regan, A. Algora, G. de Angelis, S.F. Ashley, S. Aydin, D. Bazzacco, R.J. Casperson, W.N. Catford et al., Phys. Rev. C **81**, 034310 (2010)

High-K Isomers in Light Superheavy Nuclei by PNC-CSM method

Xiao-Tao He[1,*]

[1]College of Material Science and Technology, Nanjing University of Aeronautics and Astronautics, Nanjing 210016, China

Abstract. The high-K isomeric states in light superheavy nuclei around A = 250 mass region are investigated by the Cranked Shell Model (CSM) with pairing treated by a Particle-Number Conserving (PNC) method. With including the higher-order deformation ε_6, both of the high−K multi-particle state energies and the rotational bands in ^{254}No and N = 150 isotone are reproduced well. The isomeric state energies and the microscopic mechanism of kinematic moment of inertia variations versus rotational frequency are discussed. The irregularity of the two-neutron $K^\pi = 8^-$ state band at $\hbar\omega \approx 0.17$ in ^{252}No is caused by the configuration mixing with the two-proton $K^\pi = 8^-$ band. .

1 Introduction

In recent decades, intensive studies of the in-beam and decay spectroscopic have been performed on the light well-deformed superheavy nuclei around A = 250 mass region. Increasing experimental spectroscopic data have been reported [1, 2]. Moreover, A = 250 region is one of the typical mass region for presence of the high−K isomers. Numerous isomers (mainly two-particle states) are identified now (see Table 1 in Ref.[3]). Beside the two-particle states, four-particle states are also observed, like the 2.928(3) MeV 184(2)μs isomer in ^{254}No [4] and 247(73)μs isomer in ^{254}Rf [5]. These experimental data provide valuable informations to explore the structure of the light superheavy nuclei and to test the nuclear theoretical models. Moreover, through the investigations of single-particle structure of the light superheavy nuclei, one can predict more precisely the next spherical magic number beyond ^{208}Pb. This is crucial for producing the superheavy elements. In the present work, the high−K multi-particle state bands in ^{254}No and the $K^\pi = 8^-$ isomeric bands observed systematically in N = 150 isotone are investigated by Cranked Shell Model with pairing correlation treated by Particle-Number-Conserving method.

2 Theoretical method

The cranked shell model hamiltonian of an axially symmetric nucleus in the rotating frame is $H_{\rm CSM} = H_0 + H_{\rm P} = \sum_n (h_{\rm Nil} - \omega j_x)_n + H_{\rm P}(0) + H_{\rm P}(2)$, where $h_{\rm Nil}$ is the Nilsson Hamiltonian, $-\omega j_x$ is the Coriolis force with the cranking frequency ω about the x axis (perpendicular to the nuclear symmetry z axis). $H_{\rm P}$ is the pairing including monopole and quadrupole pairing

*e-mail: hext@nuaa.edu.cn

Table 1. Multi-particle states predicted by the PNC-CSM calculations.

nuclei	K^π	Configuration	E_x(MeV)	E_x^{exp}(MeV)
^{250}Fm	8^-	$\nu9/2^-[734]\otimes\nu7/2^+[624]$	1.213	1.199
^{252}No	8^-	$\nu9/2^-[734]\otimes\nu7/2^+[624]$	1.139	1.254
^{254}No	3^+	$\pi7/2^-[514]\otimes\pi1/2^-[521]$	1.154	0.988
^{254}No	8^-	$\pi9/2^+[624]\otimes\pi7/2^-[514]$	1.272	1.297
^{254}No	10^+	$\nu9/2^-[734]\otimes\nu11/2^-[725]$	2.526	2.013
^{254}No	16^+	$\{\nu9/2^-[734]\otimes\nu7/2^+[613]$ $\pi9/2^+[624]\otimes\pi7/2^-[514]\}$	3.213	2.928

correlations. $H_{\rm CSM}$ is diagonalized directly in a truncated Cranked Many-Particle Configuration (CMPC) (Fock) space. Therefore, particle number is conserved and Pauli blocking effects are taken into account exactly.

The Nilsson parameters (κ,μ) which are optimized to reproduce the experimental level schemes for heavy nuclei around $A = 250$ mass region in Refs. [6, 7] are used. The axially symmetric deformation parameters $\varepsilon_{2,4,6}$ are taken from Refs. [7, 8]. The CMPC space is constructed in the proton $N = 4, 5, 6$ and neutron $N = 6, 7$ shells. The dimensions of the CMPC space are about 1000 and the corresponding effective monopole and quadrupole pairing strengths are $G_0 = 0.25$ MeV and $G_2 = 0.02$ MeV for both of proton and neutron for all the studied nuclei.

3 High$-K$ multi-particle states

Multi-particle state configurations and energies in ^{254}No and in the $N = 150$ isotone calculated by the PNC-CSM are listed in Table 1. A better reproduction of the experimental state energies will be obtained by including of ε_6 deformation. The $K^\pi = 8^-$ isomer state is observed systematically in this mass region. Unlike the $K^\pi = 8^-$ isomer in the $N = 150$ isotones, whose configuration is accepted as two-neutron $\nu9/2^-[734]\otimes\nu7/2^+[624]$ state in the literature, the configuration of the observed $K^\pi = 8^-$ isomer state in ^{254}No is in dispute up to now. In the PNC-CSM calculation, the lowest 8^- state in ^{254}No has the two-proton $\pi9/2^+[624]\otimes\pi7/2^-[514]$ configuration with energy 1.272 MeV, which reproduces very well the experimental value 1.297(2) MeV.

The experimental kinematic moment of inertia $J^{(1)}$ among the ground state band (GSB) and the $K^\pi = 8^-$ state band in ^{254}No and in $N = 150$ isotone (^{252}No and ^{250}Fm) are compared with the PNC-CSM calculations in Figure 1. The theoretical results reproduce the experimental data well. The authors in Ref. [9] claimed that there is an irregularity of the experimental dynamical moments of inertia $J^{(2)}$ at $\hbar\omega \approx 0.17$ MeV (spin $I \sim 17\hbar$) of the $K^\pi = 8^-$ band in ^{252}No (see Figure 6 in Ref.[9]), and they suggested that this irregularity is due to the band crossing with a $K^\pi = 2^-$ octuple band. The PNC-CSM calculations show that the behaviors of $J^{(1)}$ for two-neutron $K^\pi = 2^-$ (or 7^-), $\nu9/2^-[734]\otimes\nu5/2^+[622]$ state at $\hbar\omega \geq 0.15$ MeV is similar with that for $K^\pi = 8^-$ state. But its energy is too high. This might imply that the irregularity in MoI is not from the crossing with the two-neutron $K^\pi = 2^-$ (or 7^-) state. The $J^{(1)}$ for the two-proton $K^\pi = 8^-, \pi9/2^+[624]\otimes\pi7/2^-[514]$ state shows a slight upbending at $\hbar\omega \approx 0.17$ MeV, and the gradual increase afterwards can reproduce the experimental variations well. Thus, if the assumption that the irregularity of MoI at $\hbar\omega \geq 0.17$ MeV coming from band-crossing is correct, it might result from the crossing with the two-proton $\pi9/2^+[624]\otimes\pi7/2^-[514]$ configuration, which itself is mixed with $\pi9/2^+[624]\otimes\pi1/2^-[521]$ configuration.

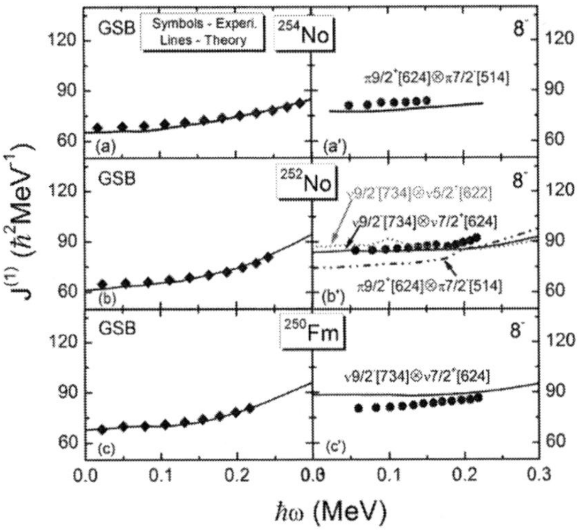

Figure 1. (Color online) Kinematic moment of inertia $J^{(1)}$ versus rotational frequency $\hbar\omega$ of the ground state bands [(a),(b),(c)] and the $K^{\pi} = 8^{-}$ state bands [(a'),(b'),(c')] in ^{254}No and in $N = 150$ isotone (^{252}No and ^{250}Fm). Theoretical calculations are denoted by lines and experiment data are denoted by solid symbols.

4 Summary

In summary, the Cranked Shell Model with pairing correlations treated by a Particle-Number Conserving method is used to study the multi-particle states observed in ^{254}No and in the $N =$ 150 isotone. Calculations including high-order deformation ε_6 lead to a better reproduction of the experimental data. The irregularity of the two-neutron $K^{\pi} = 8^{-}$ state band in ^{252}No is analysed together with other low-lying two-particle state bands. The abnormal behavior at $\hbar\omega \approx 0.17$ of the MoI is caused by the band cross with two-proton $K^{\pi} = 8^{-}$ band.

Acknowledgments

This work was supported by the National Natural Science Foundation of China under Grant No. 11775112 and 11275098.

References

[1] R.D. Herzberg, P.T. Greenlees, Prog. Part. Nucl. Phys. **61**, 674 (2008)
[2] R.D. Herzberg, J. Phys. G: Nucl. Part. Phys. **30**, R123 (2004)
[3] F. Kondev, G. Dracoulis, T. Kibédi, At. Data Nucl. Data Tables **103–104**, 50 (2015)
[4] R. Clark, K. Gregorich, J. Berryman, M. Ali, J. Allmond, C. Beausang, M. Cromaz, M. Deleplanque, I. Dragojević, J. Dvorak et al., Phys. Lett. B **690**, 19 (2010)
[5] H.M. David, J. Chen, D. Seweryniak, F.G. Kondev, J.M. Gates, K.E. Gregorich, I. Ahmad, M. Albers, M. Alcorta, B.B. Back et al., Phys. Rev. Lett. **115**, 1 (2015)
[6] Z.H. Zhang, J.Y. Zeng, E.G. Zhao, S.G. Zhou, Phys. Rev. C **83**, 1 (2011)
[7] Z.H. Zhang, X.T. He, J.Y. Zeng, E.G. Zhao, S.G. Zhou, Phys. Rev. C **85**, 1 (2012)
[8] P. Möller, J. Nix, At. Data and Nucl. Data Tables **59**, 185 (1995)
[9] B. Sulignano, C. Theisen, J.P. Delaroche, M. Girod, J. Ljungvall, D. Ackermann, S. Antalic, O. Dorvaux, A. Drouart, B. Gall et al., Phys. Rev. C **86**, 1 (2012)

Investigation of excited 0^+ states in ^{160}Er populated via the (p, t) two-neutron transfer reaction

C. Burbadge[1*], *P.E.* Garrett[1], *G.C.* Ball[2], *V.* Bildstein[1], *A.* Diaz Varela[1], *M.R.* Dunlop[1], *R.* Dunlop[1], *T.* Faestermann[3], *R.* Hertenberger[4], *D.S.* Jamieson[1], *D.* Kisliuk[1,5], *K.G.* Leach[1,6], *J.* Loranger[1], *A.D.* MacLean[1], *A.J.* Radich[1], *E.T.* Rand[1], *C.E.* Svensson[1], *S.* Triambak[7,8], and *H.-F.* Wirth[4]

[1]Department of Physics, University of Guelph, Guelph, Ontario N1G 2W1, Canada
[2]TRIUMF, 4004 Wesbrook Mall, Vancouver, British Columbia V6T 2A3, Canada
[3]Technische Universität München, D-85748 Garching, Germany
[4]Fakultät für Physik, Ludwig-Maximilians-Universität München, D-85748 Garching, Germany
[5]Department of Physics, University of Toronto, Toronto, Ontario M5S 1A7, Canada
[6]Department of Physics, Colorado School of Mines, Golden, Colorado 80401, USA
[7]Department of Physics and Astronomy, University of the Western Cape, P/B X17, Bellville ZA-7535, South Africa
[8]iThemba Laboratory for Accelerator Based Sciences, P.O. Box 722, Somerset West 7129, South Africa

Abstract. Many efforts have been made in nuclear structure physics to interpret the nature of low-lying excited 0^+ states in well-deformed rare-earth nuclei. However, one of the difficulties in resolving the nature of these states is that there is a paucity of data. In this work, excited 0^+ states in the $N = 92$ nucleus ^{160}Er were studied via the ^{162}Er$(p, t)^{160}$Er two-neutron transfer reaction, which is ideal for probing $0^+ \rightarrow 0^+$ transitions, at the Maier-Leibnitz-Laboratorium in Garching, Germany. Reaction products were momentum-analyzed with a Quadrupole-3-Dipole magnetic spectrograph. The 0^+_2 state was observed to be strongly populated with 18% of the ground state strength.

1 Introduction

The nature of excited 0^+ states in well-deformed nuclei remains an open question in nuclear structure physics. Traditionally, deformed nuclei are suggested to have β-vibrational and γ-vibrational bands which result from collective low-lying surface vibrational excitations [1]. Recent data in nuclei near $N = 90$, such as Gd [2–4], Sm [5–8] and Dy [9–11], have put into question whether the nature of low-lying excited 0^+ states should be re-examined [12, 13].

Two-neutron transfer reactions are excellent probes to study excited 0^+ states as they are sensitive to pairing correlations in nuclei [14–17]. This is demonstrated by the strongly-populated 0^+_2 states which emerged in both (p, t) and (t, p) reactions in the $N = 90$ region [7]. In particular, the cross sections of the first excited $L = 0$ excitations were comparable to that of their ground states in $N = 88 - 90$ nuclei, indicating a rapid onset of deformation [7, 13].

Evidence of collectivity in the $N = 90$ region is also demonstrated by the simultaneous increase of the $B(E2; 0^+_1 \rightarrow 2^+_1)$ value and the $E_{4^+_1}/E_{2^+_1}$ energy ratio, plotted in Figure 1, indicating that a rapid transition between the vibrational and rotational limits occurs near $N = 90$.

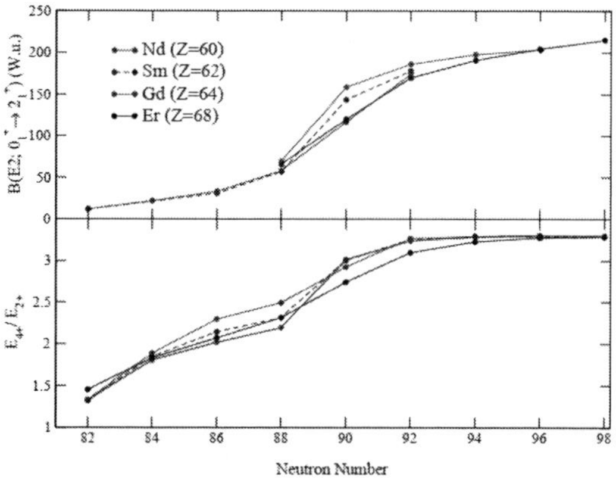

Figure 1. $B(E2; 0^+_1 \rightarrow 2^+_1)$ value and $E_{4^+_1}/E_{2^+_1}$ ratio systematics in the $N = 90$ region plotted as a function of neutron number. The dramatic increase in both quantities suggests that these isotopes lie in a transitional region of rapid shape change. The Er isotopic chain also possesses similar characteristics.

2 Experimental Details

Excited states in ^{160}Er have been studied via the ^{162}Er(p, t) reaction at the Maier-Leibnitz-Laboratorium (MLL) in Garching, Germany. Proton beams up to 2 μA were accelerated to 22 MeV or 24 MeV using a 14 MV tandem

*e-mail: cburbadg@uoguelph.ca

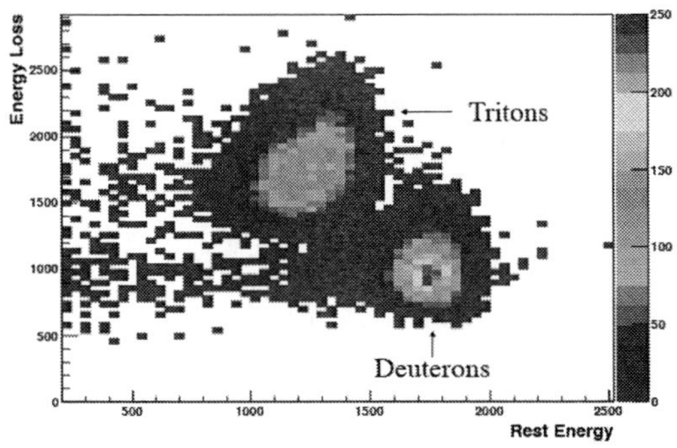

Figure 2. ΔE- E histogram for the $^{162}Er(p,t)$ reaction at 30°. A gate is placed on the outgoing tritons (left), to eliminate coincidence tagging of a deuteron contaminant (right).

Van de Graaff accelerator. The proton beam impinged on a highly-enriched ^{162}Er target. Reaction products were momentum-analyzed with a Quadrupole-3-Dipole (Q3D) magnetic spectrograph. In the focal plane of the Q3D, two proportional counters produce two energy-loss signals, ΔE and ΔE_1. A thick plastic scintillator located behind the proportional counters stop the particles to determine their energy, E. An example of a ΔE- E histogram used to identify and gate on reaction ejectiles is shown in Figure 2.

An elastic scattering angular distribution was collected from 15° to 115° to determine the target thickness and select an appropriate global optical model potential (OMP) [18–22] to be used in the Distorted Wave Born Approximation (DWBA) calculation. The target thickness was determined to be 61(3) μg/cm^2 by normalizing the cross section at 15° to the Becchetti and Greenlees OMP [18], which best reproduced the distribution minima. In this work, the DWBA calculations were performed using FRESCO, a coupled-channel reactions code [26]. The isotopic purity of 99% is remarkable given the 0.14(1)% natural abundance [24] of ^{162}Er.

3 Results and Conclusions

Evaluated data for levels in ^{160}Er [23] were used to calibrate the triton spectrum, plotted in Figure 3. Members up to $J^\pi = 4^+$ were assigned by comparison of angular distributions to DWBA calculations. It is worth noting that some of the members of the higher-lying $K^\pi = 0^+$ and $K^\pi = 2^+$ bands are speculative, motivated by the similarity of the ^{160}Er band structures to those of ^{162}Er and ^{152}Sm, but are in agreement with those observed in unevaluated works [10, 25].

To report the strength of the excited 0^+ states relative to the ground state, the relative cross section strength, S, is defined by

$$S = \frac{\frac{d\sigma}{d\Omega}^{LAB}_{0^+_{ex}}}{\frac{d\sigma}{d\Omega}^{DWBA}_{0^+_{ex}}} \cdot \left(\frac{\frac{d\sigma}{d\Omega}^{LAB}_{0^+_{gs}}}{\frac{d\sigma}{d\Omega}^{DWBA}_{0^+_{gs}}} \right)^{-1}$$

Table 1. Energy and relative strengths of excited 0^+ states, normalized at 5° to the DWBA calculation.

E_{exp} (keV)	$S(0^+_{ex}/0^+_1)$(%)
0	100
894	18(1)
1279	1.0(1)
1528	1.0(1)
1864	1.4(5)
1930	0.10(3)
2032	1.7(1)
2129	0.7(1)

where the differential cross sections are stated in the centre-of-mass frame. Normalizing both the excited and ground states to the DWBA calculation applies a Q-value correction to account for the dependence of the reaction cross section on excitation energy. Figure 4 shows the agreement between the 0^+_1 and 0^+_2 state cross sections and their DWBA calculations.

The relative cross section strength of excited 0^+ states in this work are listed in Table 1. The low-lying $K^\pi = 0^+$ band head is strongly populated with 18% of the ground state strength, while higher excited 0^+ states have a relative strength of less than 2%. The strong population of the 0^+_2 in (p,t) reactions in the $N = 92$ region, reminiscent of the strong (p,t) strength in the $N = 90$ region [7], may suggest that the same mechanism is responsible for the strength of 0^+ states in both $N = 90$ and $N = 92$ nuclei.

References

[1] A. Bohr, B.R. Mottelson, *Nuclear Structure Volume II: Nuclear Deformations* (W.A. Benjamin, Inc. Advanced Book Program Reading, 1975)
[2] J.F. Sharpey-Schafer et al., Eur. Phys. J. A **47**, 5 (2011)
[3] J.F. Sharpey-Schafer et al., Eur. Phys. J. A **47**, 6 (2011)
[4] W.D. Kulp et al., Phys. Rev. Lett. **91**, 102501 (2003)
[5] P.E. Garrett et al., Phys. Rev. Lett. **103**, 062501 (2009)
[6] W.D. Kulp et al., Phys. Rev. C **77**, 061301(R) (2008)
[7] W.D. Kulp et al., Phys. Rev. C **71**, 041303(R) (2005)
[8] W.D. Kulp et al., Phys. Rev. C **76**, 034319 (2007)
[9] S.N.T. Majola et al., Phys. Rev. C **91**, 034330 (2015)
[10] N. Blasi et al., Phys. Rev. C **90**, 044317 (2014)
[11] M.A. Caprio et al., Phys. Rev. C **66**, 054310 (2002)
[12] P.E. Garrett, J. Phys. G **27**, R1 (2001)
[13] K. Heyde, J.L. Wood, Rev. Mod. Phys. **83**, 1467 (2011)
[14] C.Y. Wu et al., Annu. Rev. Nucl. Part. Sci. **40**, 285 (1990)
[15] D. Bès, R. Broglia, Nucl. Phys. **80**, 289 (1966)
[16] S. Yoshida, Nucl. Phys. **33**, 685 (1962)
[17] A.A. Shihab-Eldin et al., Int. J. of Mod. Phys. E **04**, 411 (1995)

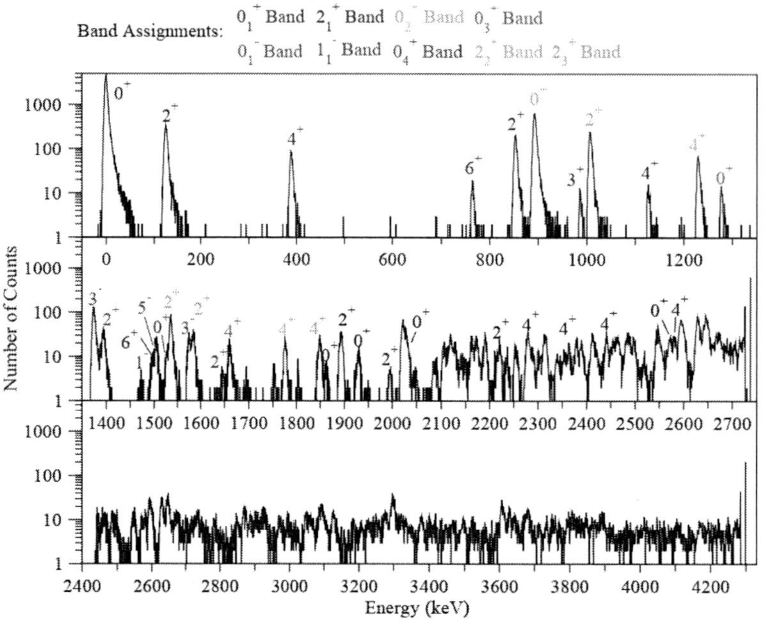

Figure 3. Energy-calibrated spectrum from the ^{162}Er(p,t) reaction at a beam energy of 24 MeV and Q3D angle of 30°. The angular momentum of some of the more prominent peaks are labelled.

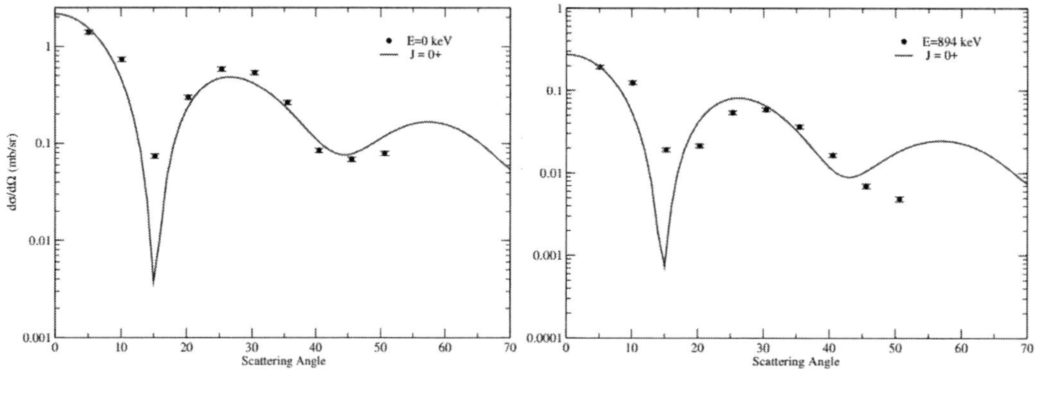

Figure 4. Comparison of the angular distributions of the 0_1^+ state (left) and 0_2^+ state (right) to their DWBA calculations at a beam energy of 24 MeV, demonstrating the level of certainty in assigning J=0^+ states due to the unique shape of its angular distribution.

[18] F.D. Becchetti, G.W. Greenlees, Phys. Rev. **182**, 1190 (1969)
[19] R.L. Varner et al., Phys. Rep. **201**, 57 (1991)
[20] B. Morillon, P. Romain, Phys. Rev. C **76**, 044601 (2007)
[21] R.L. Walter, P.P. Guss, Rad. Effects **95**, 73 (1986)
[22] A.J. Koning, J.P. Delaroche, Nucl. Phys. A **713**, 231 (2003)
[23] C.W. Reich, Nuclear Data Sheets **105**, 557 (2005)
[24] P. Holliger, C. Devillers, Earth Planet. Sci. Lett. **52**, 76 (1981)
[25] K. Dusling et al., Phys. Rev. C **73**, 014317 (2006)
[26] I.J. Thompson, *Fresco coupled reaction channels calculation*, http://www.fresco.org.uk/, version 2.9

New low-spin states of ^{122}Xe observed via high-statistics β-decay of ^{122}Cs

B. Jigmeddorj[1,a], P. E. Garrett[1], C. Andreoiu[2], G. C. Ball[3], T. Bruhn[3,2], D. S. Cross[2], A. B. Garnsworthy[3], B. Hadinia[1,b], M. Moukaddam[3,c], J. Park[3,4,d], J. L. Pore[2,e], A. J. Radich[1], M. M. Rajabali[3,f], E. T. Rand[1,g], U. Rizwan[2], C. E. Svensson[1], P. Voss[2,h], Z. M. Wang[2,3], J. L. Wood[5], and S. W. Yates[6]

[1]*University of Guelph, Guelph, ON, Canada*
[2]*Simon Fraser University, Burnaby, BC, Canada*
[3]*TRIUMF, Vancouver, BC, Canada*
[4]*University of British Columbia, Vancouver, BC, Canada*
[5]*Georgia Institute of Technology, Atlanta, GA, USA*
[6]*University of Kentucky, Lexington, KY, USA*

Abstract. Excited states of ^{122}Xe were studied via the β^+/EC decay of ^{122}Cs with the 8π γ-ray spectrometer at the TRIUMF-ISAC facility. Compton-suppressed HPGe detectors were used for measurements of γ-ray intensities, $\gamma\gamma$ coincidences, and γ-γ angular correlations. Two sets of data were collected to optimize the decays of the ground (21.2 s) and isomeric (3.7 min) states of ^{122}Cs. The data collected have enabled the observation of about 505 new transitions and about 250 new levels, including 51 new low-spin states. Spin assignments have been made for 58 low-spin states based on the deduced β-decay feeding and γ-γ angular correlation analyses.

1 Introduction

It is known that nuclei in the $Z > 50$, $N < 82$ region exhibit a very smooth evolution of collectivity. This region is ideal for studying the development of collectivity but the collective properties of many nuclei in this region are not fully characterized because of a general lack of available spectroscopic data for low-spin states. This work is part of a program to systematically study collectivity in the Xe isotopes. Proton sub-shell gaps were suggested in $^{124-132}$Xe isotopes [1] based on very strong population of excited 0_3^+ states in (^{3}He,n) reactions. Recent work on ^{124}Xe [2] has refined the $B(E2)$ values and established that nearly identical quadrupole collectivity exists for the 0_3^+ band and the ground state band, and suggested that the 0_3^+ band may represent a significant fragment of the proton pairing vibration.

Our first result on ^{122}Xe [3] was the establishment of the 2^+ band members of the 0_2^+ and 0_3^+ bands and observation of important, but previously unobserved, $2_3^+ \to 0_2^+$ and $2_3^+ \to 0_3^+$ transitions. The spin of the 0_3^+ state in ^{122}Xe was suggested based on energy systematics in Ref. [2]. Our second report on ^{122}Xe [4, 5] confirmed the spins of the 0_3^+ state and of the previously suggested 2^+ members of the 0_2^+ and 0_3^+ bands using angular correlation analysis. The present work is focused on newly observed low-spin states of ^{122}Xe.

2 Experimental Detail

The high-statistics ^{122}Cs β^+/EC decay experiment was performed at the TRIUMF-ISAC facility located in Vancouver, BC, Canada. The ^{122}Cs beam was produced through spallation reactions induced by a 65-μA, 500-MeV proton beam bombarding a thick ^{nat}Ta foil target. The atoms were ionized with a Re surface-ion source and a magnetic mass separator selected the ions of interest with $A = 122$. The high-intensity beam of 1.1×10^7 ions/s of ^{122}Cs in the 1^+ ground state with a 21.2 s half life, and 2.1×10^6 ions/s of ^{122}Cs in the 8^- isomeric state with a 3.7 minute half life, was delivered to the centre of the 8π γ-ray spectrometer [6–8] and implanted into a FeO-coated mylar tape. The 8π spectrometer consisted of 20 High-Purity Ge (HPGe) detectors surrounded by bismuth-germanate (BGO) Compton-suppression shields.

Two sets of data were collected in repeated cycles for short and long durations to favour the decays of the two β-decaying states in ^{122}Cs in order to have the best sensitivity of potential excited states in ^{122}Xe. Analyses of the γ-γ matrices and fitting of the spectra were performed with the `Radware` package [9]. An example of the γ-ray coincidence data is shown in Fig. 1.

[a]e-mail: bjigmedd@uoguelph.ca
[b]Present Address: McMaster University, Hamilton, ON, Canada
[c]Present Address: University of Surrey, Guildford, UK
[d]Present Address: Lund University, Lund, Sweden
[e]Present Address: Lawrence Berkeley National Laboratory, USA
[f]Present Address: Tennessee Technological University, TN, USA
[g]Present Address: Chalk River Laboratories, ON, Canada
[h]Present Address: Albion College, Michigan, USA

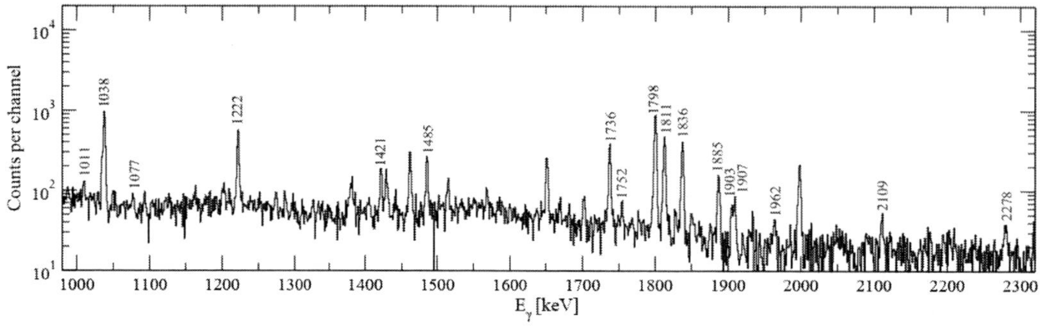

Figure 1. Portion of the γ-ray energy spectrum in coincidence with the 843-keV γ-ray $2^+ \rightarrow 0^+$ transition in ^{122}Xe. The data were taken with a short tape cycle to optimise the observation of the ground state decay of ^{122}Cs.

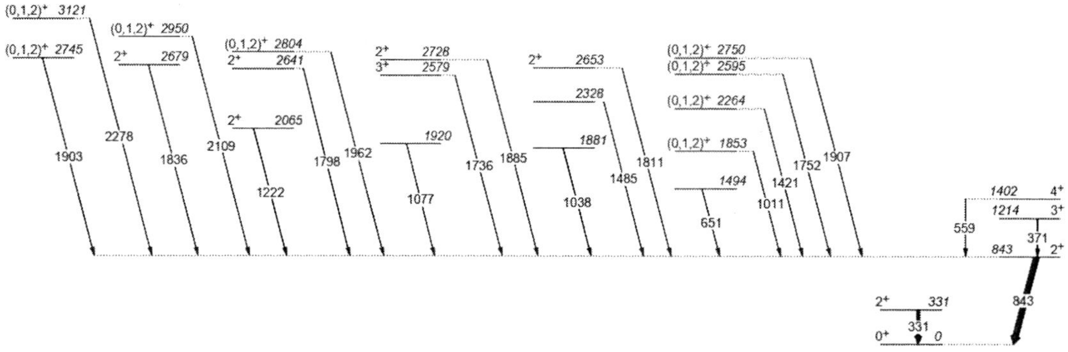

Figure 2. Partial level scheme of ^{122}Xe displaying those transitions that are in coincidence with the 843-keV $2^+ \rightarrow 0^+$ γ-ray transition. Blue colour indicates newly-observed transitions and newly-assigned labels of the states and spins.

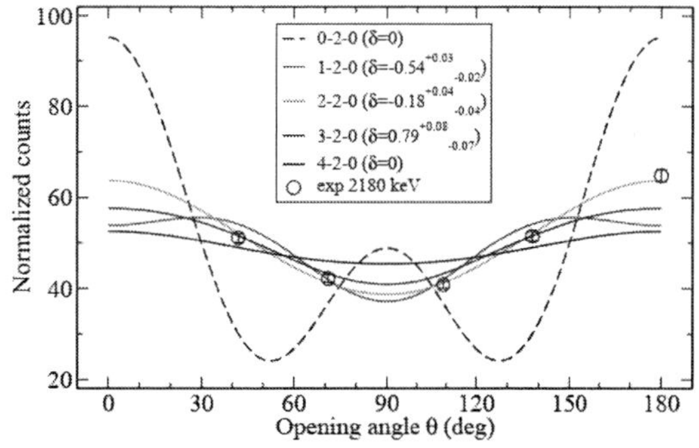

Figure 3. Angular correlation of 1798- and 843-keV γ rays.

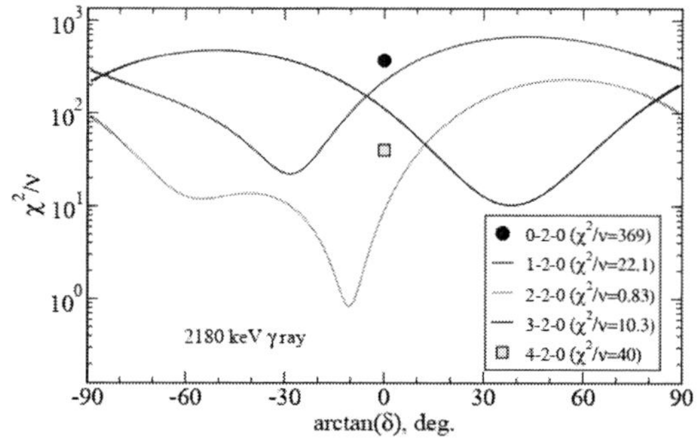

Figure 4. χ^2/ν plot of the angular correlation of 1798- and 843-keV γ rays.

3 Results and Discussions

Spin assignments have been made using data for 58 γ-ray cascades from low-spin levels using an angular correlation analysis and based on β-decay feeding. Two γ-ray coincidence gates, placed on the 331- and 843-keV γ-ray transitions, were analyzed for transitions from low-spin levels. The partial level scheme for levels decaying to the 843-keV 2^+ level is shown in Fig. 2, constructed in part using the data presented in Fig. 1. Levels are labelled with their energies in keV and their spin and parity, J^π. The transition arrows are labelled with their energies in keV and with arrow widths proportional to the intensity. Newly observed transitions and newly assigned labels of the states and spins are indicated in blue.

Spin assignments were made by γ-γ angular correlation analysis, details of which were given in Ref. [5]. As

Table 1. Newly-observed transitions with the spin assignments of the initial levels and deduced mixing ratios with 1σ uncertainty from the analyses of γ-γ angular correlations. All cascades proceed through the 331-keV $2_1^+ \rightarrow 0_1^+$ γ ray.

E_i (lit.)	E_i (new)	J_i^π (lit.)	J_i^π (new)	E_γ	δ (new)
1214.3	1213.5	$(3)^+$	$1^+, 3^+$	882.4	
1495.0	1493.8	2^+	2^+	1163.0	$5.1^{+1.0}_{-0.74}$
	1853.0		$(0,1,2)^+$	1522.1	
1882.3	1881.0		2^+	1550.0	$-1.18^{+0.12}_{-0.11}$
	1919.7		$(0,1,2)^+$	1588.9	
	2046.6		2^+	1715.1	$-0.45^{+0.07}_{-0.09}$
2065.5	2064.6	2^+	2^+	1733.8	0.10(3)
2264.4	2263.8	$0^+, 1, 2$	2^+	1932.4	$-0.84^{+0.23}_{-0.16}$
2343.1	2342.4	2^+	$3^+, 1^+$	2010.0	
	2406.4		$2^+, 4^+$	2074.9	
	2511.4		2^+	2180.0	0.18(4)
	2596.6		2^+	2265.2	0.32(5)
	2619.7		$(0,1,2)^+$	2288.4	
	2642.8		2^+	2311.6	$1.86^{+0.14}_{-0.15}$
	2654.9		2^+	2323.8	$22^{+\infty}_{-9}$
	2680.0		2^+	2349.2	$0.45^{+0.08}_{-0.09}$
	2705.2		$(0,1,2)^+$	2374.2	
	2730.3		2^+	2398.8	$0.05^{+0.03}_{-0.02}$
	2752.5		$(0,1,2)^+$	2421.1	
	2806.9		$2^+, 3^+, 4^+$	2475.8	
	2842.7		$(0,1,2)^+$	2511.2	
	2858.8		$(0,1,2)^+$	2527.6	
	2898.0		$(0,1,2)^+$	2566.4	
	2935.5		2^+	2604.4	0.15(4)
	2979.2		$(0,1,2)^+$	2647.7	
	3002.0		$(0,1,2)^+$	2670.6	
	3014.6		$3^+, 1^+$	2683.5	
	3047.4		$(0,1,2)^+$	2716.7	
	3070.8		$(0,1,2)^+$	2939.1	
	3085.2		$(0,1,2)^+$	2754.1	
	3097.5		$(0,1,2)^+$	2766.4	
	3111.4		$(2)^+$	2780.4	-1.17(12)
	3139.5		$(0,1,2)^+$	2808.3	
	3180.5		$(2)^+$	2848.8	$-0.52^{+0.10}_{-0.25}$
	3205.2		$(0,1,2)^+$	2874.5	
	3243.3		$(0,1,2)^+$	2912.3	
	3256.7		$(0,1,2)^+$	2925.7	
	3271.0		$(0,1,2)^+$	2939.1	
	3275.2		$(0,1,2)^+$	2944.7	
	3306.3		$(0,1,2)^+$	2975.4	
	3438.5		$(0,1,2)^+$	3107.4	
	3450.0		$(0,1,2)^+$	3118.8	
	3550.8		$(0,1,2)^+$	3220.3	
	3664.2		$(0,1,2)^+$	3333.7	
	4136.4		$(0,1,2)^+$	3805.0	
	4140.9		$(0,1,2)^+$	3810.6	

an example, the plots for the angular correlation of the 1798- and 843-keV γ-ray cascade, and minimization of the χ^2/ν as a function of the $E2/M1$ mixing ratio, are given in Fig. 3 and Fig. 4, respectively. The sign convention of the mixing ratio follows Ref. [10].

Despite the high statistics obtained in the experiment, spin assignments by angular correlation analysis is limited to the intense γ-γ cascades. For levels that were weakly populated, spin assignments were made based on the direct β-decay branch to the level of interest. The investigated γ-ray transitions are tabulated in Table 1 and Table 2 for γ-ray cascades involving the 331- and 843-keV γ rays, respectively.

Table 2. Newly-observed transitions with the spin assignments of the initial levels and mixing ratios with 1σ uncertainty deduced from the γ-γ correlation analysis. All cascades proceed through the 843-keV $2_2^+ \rightarrow 0_1^+$ γ ray.

E_i (lit.)	E_i (new)	J_i^π (lit.)	J_i^π (new)	E_γ	δ (new)
1495.0	1493.8	2^+		651.4	
	1853.0		$(0,1,2)^+$	1010.7	
1882.3	1881.0			1038.2	
	1919.7		$(0,1,2)^+$	1076.8	
2065.5	2064.6	2^+	2^+	1222.1	$-3.4^{+0.32}_{-0.39}$
2264.4	2263.8	$0^+, 1, 2$	$(0,1,2)^+$	1421.2	
	2327.7			1485.0	
	2578.8		3^+	1735.7	$-5.2^{+0.49}_{-0.60}$
	2595.5		$(0,1,2)^+$	1752.2	
2642.3	2641.4	1,2	2^+	1798.5	$-0.58^{+0.08}_{-0.09}$
	2653.3		2^+	1810.8	$35^{+\infty}_{-18}$
	2678.6		2^+	1835.7	$-0.12^{+0.04}_{-0.03}$
	2727.8		2^+	1885.2	-1.38(11)
	2745.2		$(0,1,2)^+$	1902.7	
	2750.3		$(0,1,2)^+$	1906.7	
	2804.1		$(0,1,2)^+$	1961.8	
	2949.8		$(0,1,2)^+$	2109.0	
	3121.3		$(0,1,2)^+$	2278.3	

Levels that are fed from the 1^+ ground state of ^{122}Cs and appear to have no discrete γ-ray feeding from higher-lying levels are assigned spins of $(0,1,2)^+$ based on the assumption of allowed β decays.

4 Conclusions

The level structure of ^{122}Xe was studied via the β^+/EC decay of ^{122}Cs. The data were analyzed using $\gamma\gamma$ coincidence and angular correlation methods. To date, about 505 new transitions and about 250 new levels have been observed. Among them, spin assignments have been made for 58 low-spin states. For weakly populated levels that are directly fed from the β-decaying state, suggested spin assignments were made assuming allowed transitions.

References

[1] W. P. Alford *et. al.*, Nucl. Phys. A **323**, 339 (1979)
[2] A.J. Radich *et. al.*, Phys. Rev. C **91**, 044320 (2015)
[3] B. Jigmeddorj *et. al.*, Web of Conf. 107, 03014 (2016)
[4] P. E. Garrett *et. al.*, A Phys. Pol. B. Vol. 48. 523 (2017)
[5] B. Jigmeddorj *et. al.*, Physics Procedia 90 (2017)
[6] P. E. Garrett *et al.*, Nucl. Inst. Meth. B **261**, 1084 (2007)
[7] A.B. Garnsworthy and P.E. Garrett, Hyper. Int. **225**, 121 (2014)
[8] P. E. Garrett *et. al.*, J. Phys. Conf. Ser. **639**, 012006 (2015)
[9] D.C. Radford, RadWare software repository maintained at http://radware.phy.ornl.gov
[10] H. J. Rose and D. M. Brink. Rev. Mod. Phys., 39, (1967).

Studying Nuclear Structure at the extremes with S^3

Julien Piot for the S^3 Collaboration[1,*]

[1]Grand Accélérateur National d'Ions Lourds, Boulevard Becquerel, 14000 Caen, France

Abstract. The in-depth study of the regions of Superheavy elements and the proton drip line around ^{100}Sn are two major challenges of today's Nuclear Physics. Performing detailed spectroscopic studies on these nuclei requires a significant improvement of our detection capabilities. The Super-Separator-Spectrometer S^3 is part of the SPIRAL2 facility at GANIL. Its aim is to use the high stable beam currents provided by the new LINAC to reach rare isotopes by fusion-evaporation.

1 Introduction

The SPIRAL2 project [1] will be a breakthrough for the study of nuclei far from stability. This large scale international instrument belongs to the new generation of nuclear experimental facilities and is presently under construction at GANIL. Its injector will be constituted of a latest generation A-PHENIX ECR ion source coupled to a superconducting linear accelerator (LINAG). This device will be able to provide, in addition to its proton and deuteron beam, stable beams with intensities up to 100 pμA for ions with $A < 60$. The Phase 1 of the project is now under tests. Proton beams have been successfully delivered to the exit of the Radio Frequency Quadrupole and deuteron beams are now being tested. The cryogenic accelerating cavities have been cooled down and are being proven for temperature stability.

The LINAG will provide beams to two experimental halls in the first phase. The first is dedicated to Neutrons For Science (NFS) and the second, the Super Separator Spectrometer (S^3) [2], to synthesis and study of the heaviest elements and N=Z nuclei around ^{100}Sn.

2 The Super Separator Spectrometer

S^3 will use the high intensity stable beams of the LINAG to produce exotic nuclei through fusion-evaporation reactions and select them. Thus, S^3 is designed to offer the possibility to access the present spectroscopic no man's land such as the vicinity of the Superheavy Elements and areas beyond the drip lines, especially around ^{100}Sn.

2.1 Physics Cases

The design of S^3 was guided by its general purpose : studying rare isotopes and phenomena in nuclei produced by fusion-evaporation reactions. Two regions of interest can be reached through these reactions : Transactinide and Superheavy nuclei and N=Z isotopes near ^{100}Sn. The study of Transactinide and Superheavy nuclei has been pursued for many decades now and yet our understanding of this region remains limited. The objectives of S^3 are to provide production yields increased by a factor of 10 to 100 compared to the current capabilities. Under these conditions, decay spectroscopy of nuclei with production cross-sections of the order of the picobarn become possible. Therefore the investigation of nuclear structure closer to the so-called Island of Stability becomes possible and with it the availability of models with an increased predictive power.

The region of ^{100}Sn is equally challenging if not more. In addition to the extremely low cross-sections, the multiplicity of the evaporation channels produced a background that clouds the identification of the nuclei of interest. Access to the nuclear structure information in this region is however very much of interest regarding the double magicity of ^{100}Sn. Isomersim, shape and exotic decay modes all command an in-depth study.

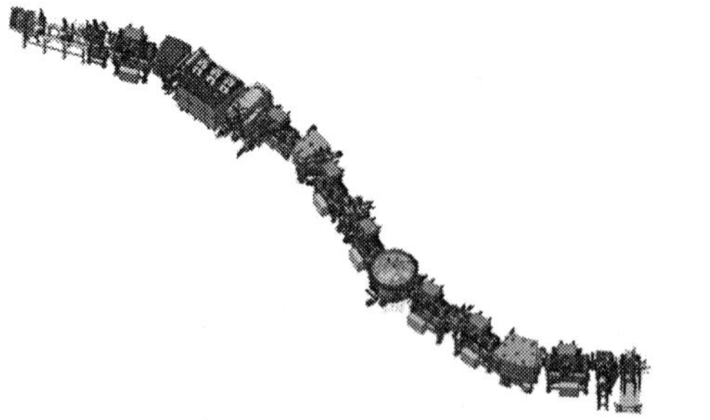

Figure 1. CAD layout of S^3 with the decay spectroscopy setup SIRIUS.

*e-mail: piot@ganil.fr

2.2 Instrumentation

In order to achieve these objectives, the present limitations due to low production cross-sections far from stability need to be overcome. The design of S^3 was focused on maximizing the transmission of fusion-evaporation products to the final focal plane. It is constituted of a momentum achromat followed by a mass spectrometer with a mass resolution approaching 1/400 and a beam rejection power of 10^{13}. For this reason triplets of cryogenic multipoles are used for the optics to enhance the opening gap and provide corrections up to the octupolar level. Seven of these modules make the optical lattice of S^3. They are powered by custom made power supplies designed for quench protection. In addition to these elements, the selection is made by three room-temperature magnetic dipoles and an electric dipole.

In order to handle the high power deposition generated by the beams driven by the LINAG, the targets are mounted on a 70 cm wheel to maintain the temperature of the target material below the melting point. This large wheel will be used for stable targets such as Lead or Bismuth and Uranium. The control of the thickness and integrity of the target throughout the experiment will be done by three complementary diagnostics : an electron-gun, an alpha source and a thermal camera. The use of actinide targets is obviously foreseen to reach the heaviest elements. The scarcity of these material and their radioactivity impose the design of a smaller wheel with a dedicated handling procedure to ensure the confinement of the radioactive material. A prototype of this target station has already been tested.

S^3 has been designed to accommodate different working modes depending on the experimental requirements. The first mode to be commissioned and used is the Convergent Mode that focuses on the efficiency by maximizing the transmission. This mode uses the Mass spectrometer to focus all five transmitted charge states into a 5 cm spot at the final focal plane. Relaxing the ion optics in the second part of the spectrometer provides an enhanced transmission up to 68 % for the reaction $^{58}Ni+^{40}Ca \rightarrow^{94} Ag+p3n$.

The second mode is the High-Resolution Mode where the emphasis is put on the Mass separation at the final focal plane. The Mass resolution of $\Delta M/M = 1/460$ will allow to separate 3 or 5 charge states and three mass-units for identification by the position on the final focal plane. The simulated transmission is 50% for the reaction $^{58}Ni +^{46} Ti \rightarrow^{100} Sn + 4n$. This Mode will be used to identify or select by mass the ions in addition to decay-tagging or laser ionization.

The final mode, called High Transmission Mode is a compromise between the two previously cited. It aims at maximizing the transmission while maintaining a mass resolution of $\Delta M/M = 260$ for low velocity recoiling nuclei from asymetric reactions such as $^{22}Ne +^{238} U \rightarrow^{255} No + 5n$. In this case where the recoiling velocity is of the order of 10 MeV, the transmission of S^3 will be between 15 and 20 %.

The full power of S^3 will be used at focal plane where it will be possible to select precisely incoming ions with a gas-cell for laser spectroscopy or to implant the selected recoiling nuclei at focal-plane for nuclear structure studies.

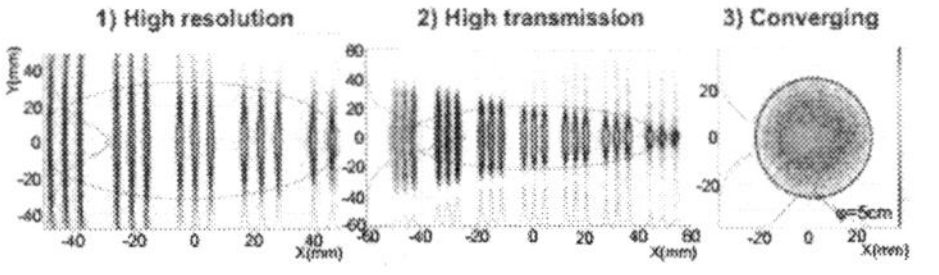

Figure 2. Simulated emittance at the final focal plane for the three operating modes of S^3.

3 Detection setups

From the start of the operation, S^3 will be equipped with two complementary detection setups. The first one will be dedicated to decay spectroscopy with alpha particles, gamma-rays, conversion electrons and spontaneous fission. It is called SIRIUS for System for the Identification of Recoiling Ions Using S^3. The second one will be dedicated to the selection and study of ground-state properties through laser spectroscopy. It is called S^3 Low Energy Branch (S^3-LEB).

3.1 Decay spectroscopy with SIRIUS

SIRIUS is designed to perform the most efficient recoil-decay tagging for nuclei separated by S^3. Its configration is based on the time-proven achitecture of detection setups such as GABRIELA [3], GREAT [4] or the focal plane of SHIP [5]. Nuclei are implanted in the 128x128 channels Double Sided Silicon Strip Detector (DSSD) with a thin entrance window (inferior to 50nm). Their decay is recorded at the same pixel as the implantation site, allowing a time correlation with little possibility of contamination. Four custom made "Strippy-pad" silicon dectectors [8] are placed upstream in a tunnel configuration to recover escaping alpha particles, conversion electrons and spontaneous fission products. The time correlation between the implantation event and its decay recorded by all silicon detectors provide the recoil-decay tagging identification and the spectroscopy. An array of five EXOGAM clover detectors with BGO Shields [6] complete the system for gamma-ray detection. Upstream a Secondary electron detector (SeD) is used to provide time and position information [7]. In conjunction with the DSSD, it provides the time of flight of the recoiling ions and their tracking thanks to its position resolution.

The acquisition of SIRIUS is based on the NUMEXO digitizers adapted to our needs. The preamplifiers of the DSSD are using ASIC chips accommodating a Floating Point Charge Sensitive Amplifier (FPCSA) to accommodate high energy implantation events followed rapidly by low energy decay. The tunnel detectors use a preamplifier compensated by a feedback loop to accommodate the two gains. The preamplifier uses a high gain by default and will be sent successive step functions in the case of a saturation to restore the baseline in the dynamic range of the amplifier.

The chambers hosting the detector have been designed to shorten as much as possible the distance between the exit of S^3 and the DSSD and to enhance the efficiency and energy resolution of the gamma-ray detection (see fig.3). The preamplifiers are housed outside the chamber on a dedicated flange as close as possible to their respective detectors to reduce the noise contribution of the cabling.

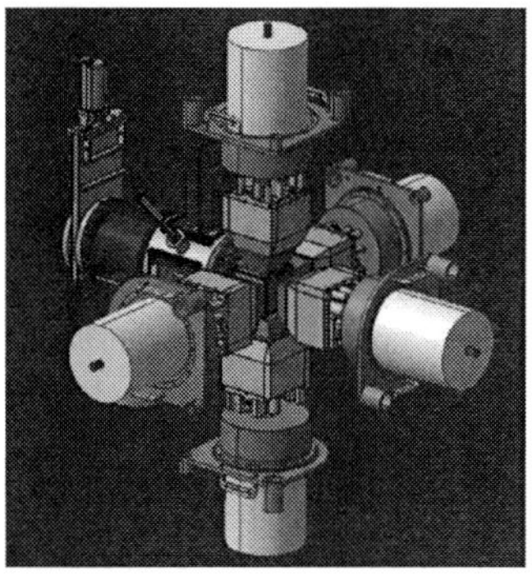

Figure 3. Layout of SIRIUS.

3.2 Laser spectroscopy with S3-LEB

$S^3 - LEB$ is the versatile combination of multiple but complementary detectors (see fig.4) [9, 10]. Its aim is to measure the ground state properties (mean square charge radii differences, magnetic dipole and quadrupole moments) of exotic nuclei using laser ionization spectroscopy. Thanks to the selectivity and efficiency of in-gas jet laser ionization spectroscopy in combination with S^3, studies of transactinide nuclei, refractory elements and medium-mass N=Z nuclei is reachable.

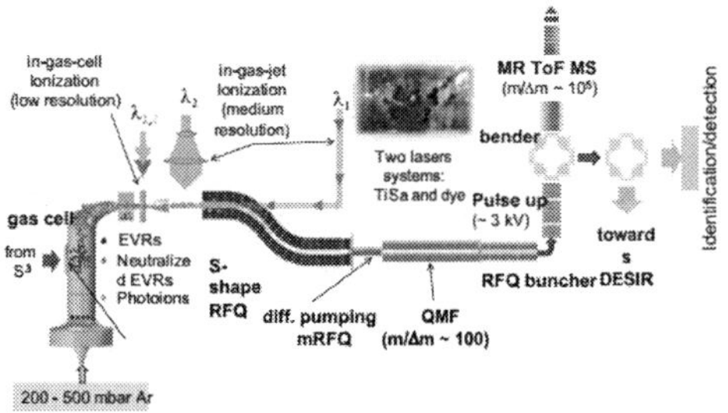

Figure 4. Schematic drawing of the Low Energy Branch of S^3.

The first part is the laser ionization spectroscopy setup installed at the final focal plane of S^3 to selectively ionize the atoms of interest. It consists in a gas-cell separated from the spectrometer by a thin foil, a laser system for the ionization and radio frequency based ion optical elements to prepare the ion bunches for further studies. The nuclei are thermalised in the gas cell and evacuated with the gas flow to a de Laval nozzle designed to accelerate the gas jet up to mach 7. The size and shape of the gas jet is optimized to maximize the interaction of the nuclei with the lasers. Titanium-Sapphire or Dye Lasers can be used depending on the frequency range of the ionization scheme of the ion of interest. The ionized atoms are then extracted to a S-shaped RFQ followed by a Quadrupole Mass Filter (QMF) and a buncher. The selected ion bunches can then be sent for ion counting to a silicon detector, to a Multi reflection Time of Flight Mass Spectrometer or to another detection facility such as the DESIR hall.

4 Conclusions

The Super Separator Spectrometer is designed to make the best use of the high intensity stable beams available from the SPIRAL2 injector. It is designed for the selection and identification of rare ions produced by fusion-evaporation. Two complementary detection systems are constructed to study the nuclei selected by S^3 : SIRIUS to investigate nuclear structure using decay-tagging and S^3-LEB to measure ground state properties using laser spectroscopy. These two complementary devices will allow to investigate rare nuclei regions such as Superheavy Elements or neutron-deficient nuclei close to ^{100}Sn.

References

[1] *Spiral2 White Book Ganil (2006).*

[2] F. Dechery *et al. Eur. Phys. J. A* **31**, 56 (2015).

[3] K. Hauschild *et al., Nucl. Instrum. Methods Phys. Res. A* **560**, 388-394 (2006).

[4] R. D. Page *et al., Nucl. Instrum. Methods Phys. Res. B* **204**, 634-637 (2003).

[5] S. Hofmann *et al., Rev. Mod. Phys.* 72, 733-767 (2000).

[6] J. Simpson *et al., APH N.S., Heavy Ion Physics* **11**, 159-188 (2000).

[7] A. Drouart *et al., Nucl. Instrum. Methods Phys. Res. A* **579**, 1090-1095 (2007);
J. Pancin *et al., J. Inst.* **4**, P12012 (2009).

[8] P. Brionnet *et al., to be submitted to Nucl. Inst. Meth. Phys. Res. A.*

[9] R. Ferrer *et al., Nucl. Instrum. Methods Phys. Res. B* **317**, 570-581 (2013).

[10] Yu. Kudryavtsev *et al., Nucl. Instrum. Methods Phys. Res. B* **297**, 7 (2013).

Probing the structure of the stable Xe isotopes with inelastic neutron scattering

Erin E. Peters[1,*], *Timothy J.* Ross[1], *Benjamin P.* Crider[2], and *Steven W.* Yates[1,**]

[1]Departments of Chemistry and Physics & Astronomy, University of Kentucky, Lexington, Kentucky 40506, USA

[2]Department of Physics & Astronomy, Mississippi State University, Mississippi State, Mississippi, 39762, USA

Abstract. The stable isotopes of xenon, which have attracted interest for a number of reasons, span a transitional region that evolves from γ-soft structures for the lighter mass isotopes to nearly spherical ^{136}Xe with a closed neutron shell. The nature of this transition, which is gradual, is not well understood. To provide detailed spectroscopic information on the Xe isotopes, we have studied ^{130,132,134,136}Xe at the University of Kentucky Accelerator Laboratory using inelastic neutron scattering and γ-ray detection. These measurements yielded γ-ray angular distributions, branching ratios, multipole mixing ratios, and level lifetimes (from the Doppler-shift attenuation method), which allowed the determination of reduced transition probabilities and provided insight into the structure of these nuclei.

1 Introduction

The stable isotopes of xenon span a region of nuclear structure that is transitional in nature. Based on the ratio of the energies of the 4_1^+ and 2_1^+ states in these nuclei, it could be inferred that the lighter mass isotopes are γ-soft rotors, and ^{134}Xe is vibrational in character, as shown in Fig. 1. Indeed, the former interpretation was shown in the work by Radich et al. [1]. ^{130,132}Xe lie in a transitional region between the γ-soft and vibrational structures and can lend insight into how this transition occurs. One potential description is that a critical point exists and is represented by the E(5) symmetry. Finally, ^{136}Xe has a closed shell of 82 neutrons and should be well-described in terms of the shell model. Thus, there is much to be learned by investigating these Xe isotopes. Moreover, information about these nuclei is lacking because they are difficult to study, as the required targets for many scattering and reaction methods are gases requiring high pressures, cryogenics, or measurements in inverse kinematics.

In addition to being interesting from a structural point of view, these nuclei play prominent roles in the ongoing searches for neutrinoless double-beta decay ($0\nu\beta\beta$). ^{136}Xe is a candidate in the search for $0\nu\beta\beta$, ^{134}Xe is a contaminant in these experiments providing additional backgrounds [2], and ^{130}Xe is the daughter nucleus in the double-β decay of ^{130}Te. Comprehensive structural information provides crucial tests of the nuclear structure models used in calculating the nuclear matrix element for $0\nu\beta\beta$ and the neutrino mass, if this exotic decay process is observed.

*e-mail: fe.peters@uky.edu
**e-mail: yates@uky.edu

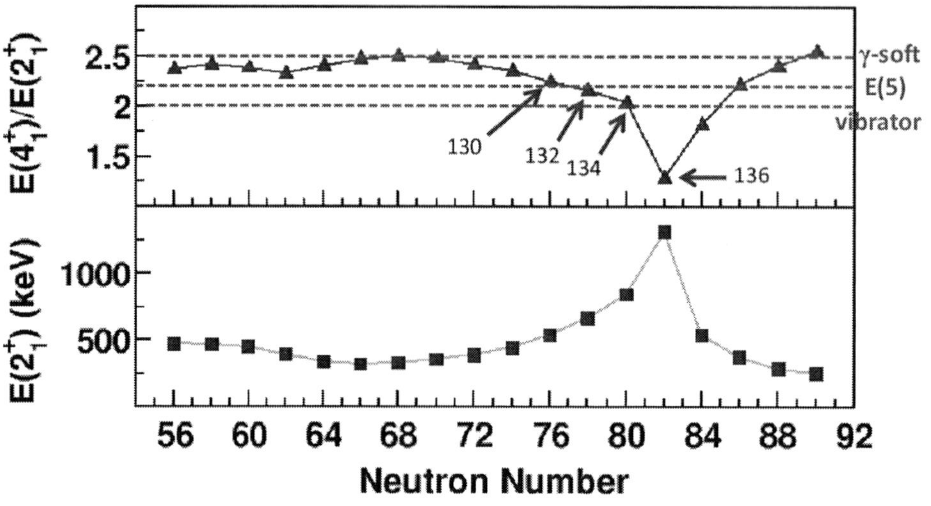

Figure 1. Systematics of the Xe isotopes showing the limits for γ-soft, E(5), and vibrational structures.

2 Experiments

With seemingly so much knowledge to gain, we chose to perform inelastic neutron scattering (INS) measurements at the University of Kentucky Accelerator Laboratory (UKAL) on ^{130,132,134,136}Xe. For each of these isotopes, highly enriched gases were converted to approximately 10 grams of solid XeF_2 and γ-ray spectroscopic measurements were performed following INS with nearly monoenergetic neutrons produced by the ^{3}H(p,n)^{3}He reaction. Both angular distribution and excitation function measurements were performed, yielding γ-ray thresholds, which aided in the placement of γ rays from the level of origin, spins and parities, multipole mixing ratios, and level lifetimes via the Doppler-shift attenuation method (DSAM). This extracted information was used to obtain the reduced transition probabilities for many transitions.

3 Results and discussion

Through comparison with the theoretical predictions for the reduced transition probabilites for the E(5) symmetry, it was determined that neither ^{130}Xe, nor ^{132}Xe is a good representation of this critical-point symmetry. Both exhibit transitions with large E2 strengths that differ from the allowed transitions of the model. The complete details of these studies can be found in Ref. [3]. It was further determined that ^{134}Xe is not a good representation of a vibrational nucleus due to the lack of E2 strength when compared with the harmonic vibrator model [4]. All three of these isotopes are transitional and clear descriptions of their structures are still lacking.

As for ^{136}Xe, this closed-neutron-shell nucleus can be well understood within the shell model. With four protons in excess of the Z=50 closed shell, ^{136}Xe should exhibit excitations with $\pi(g_{7/2})^2$, $\pi(g_{7/2})^4$, $\pi(g_{7/2}d_{5/2})$, and $\pi(d_{5/2})^2$ configurations. Candidates for such states have been identified and are shown in Fig. 2. This work is still in progress and the complete results will be relayed in a future publication.

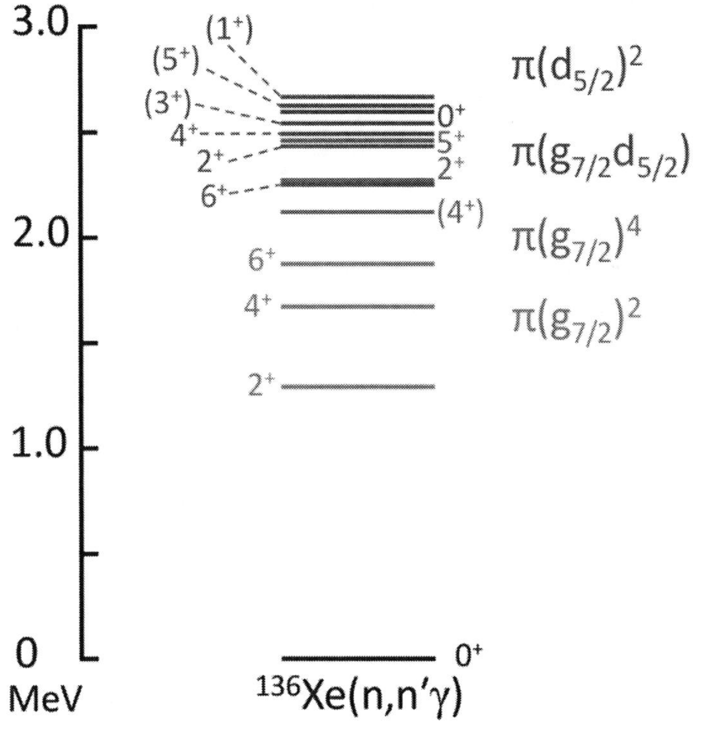

Figure 2. Excited states in ^{136}Xe with preliminary configuration assignments.

4 Summary

In summary, the structures of ^{130,132,134,136}Xe have been studied with inelastic neutron scattering at UKAL. Adequate model descriptions for the lighter three transitional isotopes are still lacking, but the structure of ^{136}Xe can be understood in terms of a shell model description.

We wish to thank H.E. Baber for his continuing contributions to UKAL. This material is based upon work supported by the U. S. National Science Foundation under Grant No. PHY-1606890.

References

[1] A.J. Radich, P.E. Garrett, J.M. Allmond, C. Andreoiu, G.C. Ball, L. Bianco, V. Bildstein, S. Chagnon-Lessard, D.S. Cross, G.A. Demand et al., Phys. Rev. C **91**, 044320 (2015).

[2] E.E. Peters, T.J. Ross, S.H. Liu, M.T. McEllistrem, S.W. Yates, Phys. Rev. C **95**, 014325 (2017).

[3] E.E. Peters, T.J. Ross, S.F. Ashley, A. Chakraborty, B.P. Crider, M.D. Hennek, S.H. Liu, M.T. McEllistrem, S. Mukhopadhyay, F.M. Prados-Estévez et al., Phys. Rev. C **94**, 024313 (2016).

[4] E.E. Peters, A. Chakraborty, B.P. Crider, S.F. Ashley, E. Elhami, S.F. Hicks, A. Kumar, M.T. McEllistrem, S. Mukhopadhyay, J.N. Orce et al., Phys. Rev. C **96**, 014313 (2017).

Strong coupling in ^{65,67}Mn

Xiaoyu Liu[1,2,*], *Zhong* Liu[1], and *Bing* Ding[1]

[1]Key Laboratory of High Precision Nuclear Spectroscopy, Institute of Modern Physics, Chinese Academy of Sciences, Lanzhou 730000, People's Republic of China
[2]University of Chinese Academy of Sciences, Beijing 100049, China

Abstract. Excited states in ^{63,65,67}Mn were studied via in-beam γ-ray spectroscopy following knockout reactions from ^{68}Fe. Similar level schemes, consisting of the $11/2^-$, $9/2^-$, $7/2^-$ and $5/2^-_{g.s.}$ level sequence, connected by $I \to I-1$ transitions, were established. Their level structures show features consistent with strongly-coupled rotational bands with $K = 5/2$.

1 Introduction

At the $N = 40$ harmonic oscillator shell closure, ^{68}Ni has the appearance of being doubly magic based on the high 2^+_1 energy and a low transition probability. However, well deformed low-lying states with considerable particle-hole excitation strength across the $N = 40$ gap have been observed and/or predicted in the nuclei around ^{68}Ni, described as a new "island of inversion" [1]. Lying in between Cr and Fe, Mn isotopes are expected to be well deformed around $N = 40$, so it is very intriguing and challenging to reveal the coupling mode in these odd-mass Mn isotopes.

2 Experiment and Results

The experiment was performed at the Radioactive Isotope Beam Factory (RIBF). A 15 pnA ^{238}U primary beam, accelerated to 345 MeV/u, was impinged on a 3-mm-thick ^{9}Be primary target for the production of secondary radioactive isotope beam at the entrance of the BigRIPS separator. The isotopes of interest were selected and separated from the secondary beams, and then impinged on a 102(1)-mm-thick liquid hydrogen (LH_2) secondary target. The secondary reaction residues were produced via proton-induced knockout reactions. The particle identification before and after the LH_2 target was performed in the BigRIPS and ZeroDegree spectrometer. The Time-Projection Chamber (TPC) of the MINOS device surrounding the LH_2 target was used to reconstruct the vertex of the knockout reactions. The γ rays emitted in-flight were measured by the DALI2 γ-ray spectrometer surrouding the MINOS device.

In the present work, excited states of ^{63,65,67}Mn were populated by the fragmentation of ^{68}Fe via the ^{68}Fe$(p, 2pxn)^{67-x}$Mn reaction channels, with $x = 4$, 2 and 0, respectively. The Doppler-corrected γ-ray spectra are displayed in figure 1. The γ-transition energies, relative intensities after efficiency calibration, and the tentative spin-parity assignments of the initial and final states are summarized in table 1. The spin-parity assignments were made following the systematics of low-lying levels in the lighter odd-mass Mn isotopes.

*e-mail: liuxiaoyu@impcas.ac.cn

https://doi.org/10.1051/epjconf/201817802029

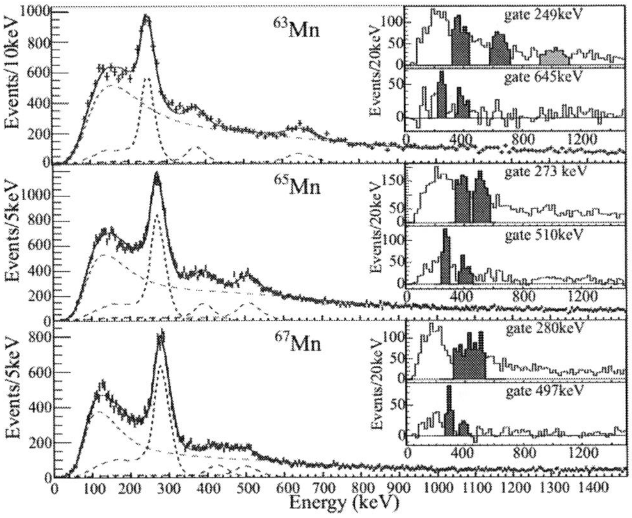

Figure 1. (Color online) Doppler-corrected γ-ray spectra of ^{63,65,67}Mn detected by DALI2, and the background subtracted γ-γ coincidence spectra labeled with the gating transitions are displayed in the insets.

Table 1. The γ-transition energies (E_γ), relative intensities after efficiency calibration (I_γ) observed in ^{63,65,67}Mn, and the tentative spin-parity assignments of the initial and final states.

	^{63}Mn		^{65}Mn		^{67}Mn	
$J_i^\pi \rightarrow J_f^\pi$	E_γ (keV)	I_γ	E_γ (keV)	I_γ	E_γ (keV)	I_γ
$(7/2^-) \rightarrow 5/2^-_{g.s.}$	249(5)	100(4)	273(5)	100(2)	280(5)	100(2)
$(9/2^-) \rightarrow (7/2^-)$	645(6)	20(2)	510(6)	22(1)	497(7)	13(1)
$(11/2^-) \rightarrow (9/2^-)$	376(7)	20(2)	394(6)	16(1)	427(8)	11(1)

3 Summary

In summary, we studied the spectroscopy of ^{63,65,67}Mn. Very similar level schemes with $11/2^- \rightarrow 9/2^- \rightarrow 7/2^- \rightarrow 5/2^-_{g.s.}$ transition cascades were proposed. With the excitation energies following the $I(I+1)$ law and the decay pattern dominated by $\Delta I = 1$ transitions, the first cases of strongly-coupled rotational bands in the $N = 40$ "island of inversion" have been identified in ^{65,67}Mn.

4 Acknowledgement

We express our gratitude to the RIKEN Nishina Center accelerator staff the BigRIPS team. The development of MINOS and the core MINOS team were supported by the European Research Council through the ERC Grant No. MINOS-258567. The work at Institute of Modern Physics was supported by the National Natural Science Foundation of China (Grant Nos. 11635003, 11675225, 11405224, U1632144 and 11435014), the Hundred Talented Project of the Chinese Academy of Sciences and the National Key Basic Research Development Program of China under Grant Nos. 2013CB834403 and 2013CB834404.

References

[1] J. Ljungvall *et al.*, Phys. Rev. C **81**, 061301(R) (2010).

Triaxiality in the odd-A nuclei $^{109-117}$I studied through a microscopic rotation-particle coupling

Swati Modi[1]*

[1]Department of Physics, Indian Institute of Technology, Roorkee 247667, India

Abstract. A systematic study of ground state spectrum with the triaxial deformation γ for odd-A Iodine isotopes $^{109-117}$I is carried out with the nonadiabatic quasiparticle approach. The rotation-particle coupling is accomplished microscopically such that the matrix elements of a particle-plus-rotor system are written in terms of the rotor energies. The $5/2^+$ state is confirmed as ground state for odd-A $^{111-117}$I and also coming out as lowest in energy for ^{109}I.

1 Introduction

Iodine with $Z = 53$ lies in the transitional region between near-spherical Sn ($Z = 50$) and deformed Ce ($Z = 58$) nuclei. The transitional nuclei exhibit many interesting phenomena like shape coexistence, triaxiality, rapid variation in shape with changing proton and neutron number, band crossing, etc. Ground state structure of odd-A Iodine isotopes ($^{109-117}$I) from the proton drip line to the region near the β-stability line are analysed in this work with the microscopic nonadiabatic quasiparticle approach. Study of Iodine chain is important to see the effect of a successive increment of neutron numbers on the shape of the isotopes. ^{109}I is a ground state proton emitter and well studied in Refs. [1, 2]. The deformed shell model predicts that the odd proton lies in the $d_{5/2}$ and $g_{7/2}$ orbitals near the Fermi level [3]. These states are coupled to the states of the corresponding even-even core to study the rotational bands.

2 Formalism

In the microscopic rotation particle-coupling, the wave function of an odd-A nucleus with the particle at position $\vec{r}$, and orientation ω of the rotor is given by

$$\Psi_{IM}(\vec{r},\omega) = \sum_{ljR\tau} \frac{\phi^I_{ljR\tau}(r)}{r} |ljR\tau, IM\rangle, \qquad (1)$$

where (I, M, K) are quantum numbers for the particle-plus-rotor system, (l, j, Ω) are related to the particle and (R, K_R, M_R) are rotor quantum numbers. $\phi^I_{ljR\tau}(r)$ and $|ljR\tau, IM\rangle$ are the radial and angular parts of the wave function, respectively. The total Hamiltonian is written as

$$H = H_{\rm avg} + H_{\rm pair} + H_{\rm rot}\ , \qquad (2)$$

where $H_{\rm avg}$ corresponds to the intrinsic energy of the odd particle. The pairing interaction is given by $H_{\rm pair}$, and $H_{\rm rot}$ is the rotor Hamiltonian.

The matrix element of $H_{\rm rot}$ can be written in K representation by utilizing the experimental rotor energies E_{TRi} and the calculated rotor wave functions $c^{Ri}_{K_R}$ as [2]

$$\begin{aligned}&\langle lj\Omega_p, K', IM|H_{\rm rot}|lj\Omega_p K, IM\rangle \\ &= \sum_{R,K_R,K'_R} A^{IK'}_{j\Omega'_p,RK'_R} \sum_i c^{Ri}_{K'_R} E_{TRi} c^{Ri}_{K_R} A^{IK}_{j\Omega_p,RK_R} \\ &= W^{K'K}_{j\Omega_{p'}\Omega_p}\ . \end{aligned} \qquad (3)$$

Here, the amplitude, $A^{IK}_{j\Omega_p,RK_R}$ is used to transforms the wave function from R to K representation and vice versa. The matrix element of the total Hamiltonian H can be written as

$$\begin{aligned}&\langle q'K', IM|H|qK, IM\rangle = \epsilon_q\, \delta_{K'K}\delta_{q'q} + \\ &\sum_{lj\Omega_{p'}\Omega_p} W^{K'K}_{j\Omega_{p'}\Omega_p} \times \int dr\, f_{uv}\, \phi^{IK'*}_{lj\Omega_{p'}}(r)\, \phi^{IK}_{lj\Omega_p}(r)\ , \end{aligned} \qquad (4)$$

where q defines particle state and ϵ_q is the quasiparticle energy obtained through the single-particle energy [4]. The quantity f_{uv} is used to transform the matrix element from single-particle states to quasiparticle states.

3 Results and discussions

We analyze the systematics of the energy levels of even N iodine isotopes ^{109}I to ^{117}I with different parameter sets for the Woods-Saxon potential. The corresponding rotor (^{108}Te to ^{116}Te) states [5] are taken as input in the nonadiabatic quasiparticle approach. The positive parity spectrum of Iodine isotopes is presented in Fig. 1 as a function of γ. In Fig. 1(a-e), the Esbensen-Davids parameter set [6] is used to calculate the energy levels of the considered iodine isotopes. The value of deformation parameters β_2 and β_4 are taken from Ref. [7]. Our calculations with the Esbensen-Davids parameters suggest that the state with $I^\pi = 5/2^+$ is the ground state of ^{109}I to ^{117}I, in consistence with Ref. [1] where the experimental results are presented for ^{111}I to ^{117}I along with a prediction for ^{109}I. The

*e-mail: swatimodi2@gmail.com

https://doi.org/10.1051/epjconf/201817802030

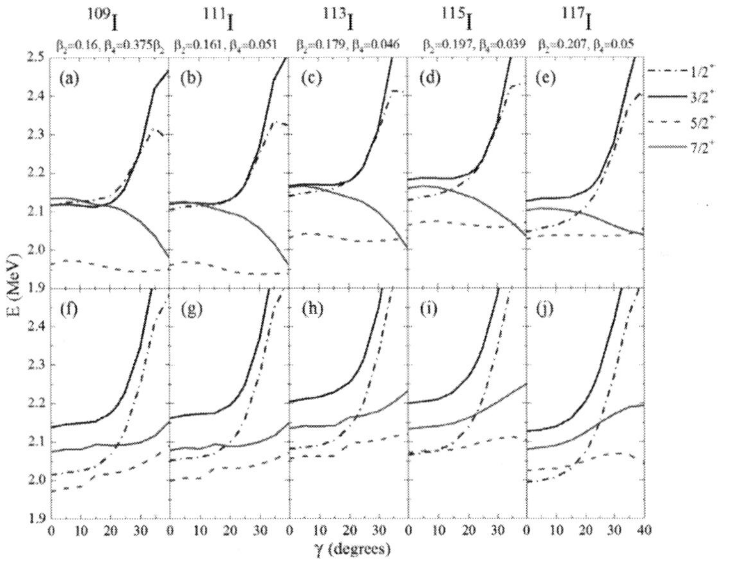

Figure 1. The yrast states of ^{109}I to ^{117}I as a function of γ. In the upper panel (a-e) the results are shown calculated with the Esbensen-Davids [6] parameter set and in the lower panel (f-j) the Chepurnov [8] parameter set is utilised. Rotor (^{108,110,112,114,116}Te) energies are taken from Ref. [5].

other lower angular momentum states ($1/2^+$ and $3/2^+$) are nearly degenerate with $7/2^+$ at lower γ. This degeneracy is lifted with an increase in the neutron number. In Ref. [1], the energy gap between $I^\pi = 5/2^+$ and $I^\pi = 7/2^+$ is reported to be increasing with neutron number which exhibits decrement in triaxiality with our results. From Fig. 1(a), it can be seen that the calculated results for ^{109}I following the same trend of the results for other isotopes.

To check the impact of the parameters of the mean field potential on the above inferences, we repeat the calculations with the Chepurnov parameter set [8] for which the results are presented in Fig. 1(f-j). The major difference between above two parameter sets is the strength of the spin-orbit potential. With the Chepurnov parameter set also, the energy of $I^\pi = 5/2^+$ turns out to be the lowest for ^{109,111,113}I but not for ^{115,117}I at low γ ($< 15°$).

We infer that the results from Esbensen-Davids parameter set are consistent with the data and thus this parameter set is more reliable than the Chepurnov parameter set. The state $5/2^+$ is lowest in energy for all these isotopes but since ^{109}I is the ground state proton emitter, the analysis of decay width proves $3/2^+$ state as its proton emitting state and the ground state [2].

4 Summary

A systematic study of a chain of nuclei is important to understand the change in nuclear interaction with the one species of nucleons. Iodine isotopes $^{109-117}$I with even-N are analysed with the triaxial degree of freedom. Despite the triaxiality decreasing with increasing N, the ground state calculated with the Esbensen-Davids parameter set for ^{111}I to ^{117}I shows a good agreement with the experimental data compared to the Chepurnov parameter set. For the proton emitter ^{109}I, the decay analysis suggests $3/2^+$ as the ground state.

Acknowledgements

S.M. would like to thankfully acknowledge the significant contributions of Dr. P. Arumugam, Dr. M. Patial, Prof. E. Maglione and Prof. L. Ferreira. This work is supported by the Council of Scientific and Industrial Research, Government of India, via project no. 03(1338)/15/EMR-II.

References

[1] M. Petri et al., Phys. Rev. C **76**, 054301 (2007)
[2] S. Modi, M. Patial, P. Arumugam, E. Maglione, L.S. Ferreira, Phys. Rev. C **95**, 054323 (2017); Phys. Rev. C **95**, 024326 (2017)
[3] E. Maglione, L.S. Ferreira, R.J. Liotta, Phys. Rev. C **59**, R589 (1999)
[4] S. Modi, M. Patial, P. Arumugam, L.S. Ferreira, E. Maglione, Phys. Rev. C **96**, 064308 (2017)
[5] *Brookhaven national nuclear data center*, `http://www.nndc.bnl.gov/chart/`
[6] H. Esbensen, C.N. Davids, Phys. Rev. C **63**, 014315 (2000)
[7] P. Möller, J.R. Nix, W.D. Myers, W.J. Swiatecki, At. Data Nucl. Data Tables **59**, 185 (1995)
[8] V.A. Chepurnov, Yadern. Fiz. **7** (1967)

Attempt to connect the nuclear charge radii with the experimental α decay data for superheavy nuclei

Yibin Qian[1,*]

[1]Department of Applied Physics, Nanjing University of Science and Technology, Nanjing 210094, China

Abstract. Significant progresses have been made so far for the synthesis of the heaviest elements, while the knowledge of them appears to be quite limited even when it comes to basic properties, e.g., their size. On the other side, the observation of α decay chains is the main tool to identify the newly produced elements. In this report, we propose to make use of the available experimental α decay data to extract the nuclear charge radii of superheavy nuclei. Within the density dependent cluster model, the nucleon density distribution of the target nucleus is determined by exactly reproducing the measured α decay half-life of its parent, finally leading to the nuclear radii. Encouraged by the agreement between theory and experiment for heavy nuclei, we extend the study to the region of superheavy nuclei as well.

1 Background and motivation

With the advent of improved facilities and technologies, a great deal of experimental efforts have been devoted to synthesize the superheavy nuclei (SHN), leading to the makeup of nuclidic chart up to $Z = 118$ [1, 2]. The α decay chains are not only taken as the main source to identify the newly synthesized SHN, but also provide insight knowledge about their properties. As well known, there are available methods to measure the nuclear charge radius such as particles scattering on the target nuclei, Kα X rays, measurements of transitions energies in muonic atoms and optical isotopes shifts [3]. Unfortunately, the detection on the SHN radii is still difficult using these methods due to short lifetimes and tiny cross sections of heaviest nuclei. It is therefore interesting to employ an alternative approach using the available experimental α decay data to extract more information on structural properties, e.g., the nuclear charge radii, for superheavy nuclei. In fact, α decay has been taken as a tool to recognize the nuclear radii at an early stage of nuclear physics, despite the rough assumptions. Owing to the advanced experimental facilities and theoretical approaches, improved experimental data have been accumulated for the α decay process [1, 2, 4] and considerable progresses have also been made for the α decay calculations [5–13]. With these in mind, we propose to establish a bridge between the nuclear radii and the experimental α decay data. The main logic of the present model will be presented with initial results in the following section.

*e-mail: qyibin@njust.edu.cn

2 Brief introduction of theoretical approach and results

Given that the α decaying nucleus is pictured as an α particle interacting with an axially symmetric deformed core nucleus, the total α-core potential is composed of the nuclear and Coulomb terms,

$$V(r,\theta) = \lambda V_N(r,\theta) + V_C(r,\theta), \qquad (1)$$

where θ is the orientation angle of the emitted α particle with respect to the symmetric axis of the core, and λ is the renormalization factor for the depth of nuclear potential. Based on the double folding integral, the nuclear and Coulomb potentials can be constructed as [14]

$$V_{NorC}(\mathbf{r},\theta) = \iint d\mathbf{r}_1\, d\mathbf{r}_2\, \rho_1(\mathbf{r}_1)\rho_2(\mathbf{r}_2)\, \upsilon(\mathbf{s} = |\mathbf{r}_2 + \mathbf{r} - \mathbf{r}_1|), \qquad (2)$$

where $\upsilon(\mathbf{s})$ denotes the M3Y-Reid type nucleon-nucleon interaction and the standard proton-proton Coulomb interaction for nuclear and Coulomb potentials, respectively. Besides the widely-used Gaussian form ρ_1 for the α particle, the mass and charge density distributions of the daughter nucleus are supposed to behave in the deformed two parameter Fermi (2pF) form. Once the total α-core interaction potential is confirmed, the final decay width can be obtained via the modified two-potential approach plus the averaging procedure [7, 15]. Undoubtedly, the density distribution $\rho_2(r_2)$ is an important input for this whole calculation leading to the final α decay half-lives. In turn, one can get the fact that the parameters of ρ_2 can be determined by exactly reproducing the experimental α decay half-lives. Subsequently, the required root-mean-square charge radius can be achieved via the corresponding charge density distribution. In this way, the rms nuclear charge radii

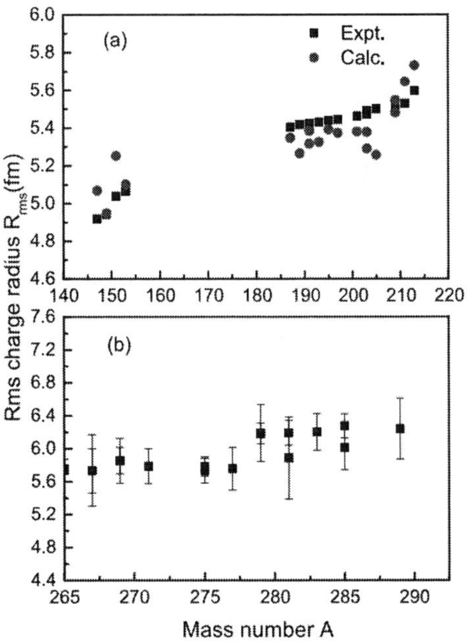

Figure 1. (a) Comparison of our extracted rms nuclear charge radii with the available experimental values for odd-A nuclei with $Z = 65 - 87$. (b) Predicted rms charge radii of odd-A isotopes with $Z = 102 - 115$, along with error bars coming from the uncertainties of the experimental α decay data.

is linked to the experimental α decay data, and more details can be found in Ref. [16]. Initially, we have paid attention to the odd-A nuclei with available measured decay data and related radii values. The upper panel (a) of fig. 1 represents the detailed comparison of our extracted nuclear charge radii with measured ones, which indicates that the satisfactory agreement between theory and experiment has been reached. Encouraged by this, we extend the study to heaviest odd-A nuclei, as shown in the lower panel (b) of fig. 1. It is shown that the present approach may provide an alternative way to pursue the fundamental feature, namely nuclear radii, for those unmeasured nuclei.

3 Summary

In summary, the rms nuclear charge radii of target nuclei have been extracted from the experimental α decay data of their parents via the density-dependent cluster model. During this process, the density distribution of daughter nucleus is determined by exactly reproducing the measured half-life of α decaying nucleus, resulting in the focused nuclear charge radii. It is found that our presently deduced nuclear radii are in good agreement with available experimental data for odd-A nuclei with $Z = 65 - 87$. The predictions on nuclear charge radii are then made for odd mass superheavy nuclei, which may be expected to give a possible way to recognize this fundamental property of SHN.

This work is supported by the National Natural Science Foundation of China (Grant No. 11605089), by the Natural Science Youth Fund of Jiangsu Province (Grant No. BK20150762), and by the Fundamental Research Funds for the Central Universities (Grant No. 30916011339).

References

[1] Yu Ts Oganessian, V K Utyonkov, Rep. Prog. Phys. **78**, 036301 (2015).
[2] S. Hofmann, J. Phys. G: Nucl. Part. Phys. **42**, 114001 (2015).
[3] I. Angeli, K. P. Marinova, At. Data Nucl. Data Tables **99**, 69 (2013).
[4] G. Audi, F. G. Kondev, Meng Wang, W. J. Huang, S. Naimi, Chin. Phys. C **41**, 030001 (2017).
[5] B. Buck, A. C. Merchant, and S. M. Perez, Phys. Rev. Lett. **65**, 2975 (1990).
[6] G. Royer, J. Phys. G: Nucl. Part. Phys. **26**, 1149 (2000).
[7] V. Yu. Denisov and H. Ikezoe, Phys. Rev. C **72**, 064613 (2005).
[8] Chang Xu and Zhongzhou Ren, Phys. Rev. C **68**, 034319 (2003).
[9] P. Mohr, Phys. Rev. C **73**, 031301(R) (2006).
[10] D. N. Poenaru, I. H. Plonski, and W. Greiner, Phys. Rev. C **74**, 014312 (2006).
[11] D. S. Delion, S. Peltonen, and J. Suhonen, Phys. Rev. C **73**, 014315 (2006).
[12] C. Qi, F. R. Xu, R. J. Liotta, and R. Wyss, Phys. Rev. Lett. **103**, 072501 (2009).
[13] M. Ismail, W. M. Seif, A. Y. Ellithi, and A. Abdurrahman, Phys. Rev. C **92**, 014311 (2015).
[14] A. M. Kobos, B. A. Brown, P. E. Hodgson, G. R. Satchler, A. Budzanowski, Nucl. Phys. A **384**, 65 (1982).
[15] Yibin Qian, Zhongzhou Ren, Phys. Rev. C **84**, 064307 (2011).
[16] Dongdong Ni, Zhongzhou Ren, Tiekuang Dong, and Yibin Qian, Phys. Rev. C **87**, 024310 (2013)

New developments in reaction theory: preparing for the FRIB era

F.M. Nunes[1,2,⋆], *P.C.* Capel[3,4], *Ch.* Elster[5], *L.* Hlophe[1], *Jin* Lei[1], *Weichuan* Li[1], *A.E.* Lovell[1,2], *G.* Potel[1], *J.* Rotureau[1], and *T.* Poxon-Pearson[1,2]

[1]*National Superconducting Cyclotron Laboratory, Michigan State University, East Lansing, MI 48824*
[2]*Department of Physics and Astronomy, Michigan State University, East Lansing, MI 48824-1321*
[3]*Physique Nucléaire et Physique Quantique, Université Libre de Bruxelles (ULB), B-1050 Brussels*
[4]*Institut für Kernphysik, Technische Universität Darmstadt, 64289 Darmstadt, Germany*
[5]*Institute of Nuclear and Particle Physics, and Department of Physics and Astronomy, Ohio University, Athens, OH 45701, USA*

Abstract. This is a brief report on the progress made towards an exact theory for (d,p) on heavy nuclei, which is important to determine neutron capture rates for r-process nuclei. We first discuss the role of core excitation in the framework of Faddeev equations. Following that, we provide the status of the Faddeev theory being developed in the Coulomb basis with separable interactions. We then present some recent developments on nonlocal nucleon optical potentials. Finally, the progress on the theory transfer to the continuum is summarized.

1 Introduction

Since the early eighties, reaction experiments with radioactive beams have shed light on so many intriguing phenomena, from nuclear halos [1] to a strong dependence of the expected shell occupancy with the asymmetry of the system [2]. This is an exciting time for low energy nuclear physics because there are new facilities coming online worldwide, like the Facility for Rare Isotope Beams (FRIB), which will increase substantially the beam intensity and enable studies in unexplored regions of the nuclear chart, particularly toward neutron-rich heavy nuclei. Reaction probes are versatile and provide detailed information on many different nuclear properties ranging from shell structure, skins and pairing to the extraction of capture rates for the r-process and rp-process, as well as many applications such as stockpile stewardship. For example, neutron capture reactions on unstable nuclei cannot be measured directly. An attractive indirect method involves extracting this information from transfer (d,p) reactions. The essential element in the analysis of these experiments is reaction theory: reaction theory is crucial to disentangle the desired structure or astrophysical information from experiments.

In this work, we provide an overview of a few recent developments that are taking place in the field of nuclear reactions, with focus on nucleon induced and deuteron induced reactions. In Section 2 we discuss recent results including core excitation in transfer (d,p) reactions [3]. Next, in Section 3, we present recent results for solving the Faddeev equations for (d,p) using separable interactions. We also discuss two different approaches to obtain the nucleon optical potentials (Section 4) and mention

⋆e-mail: nunes@nscl.msu.edu

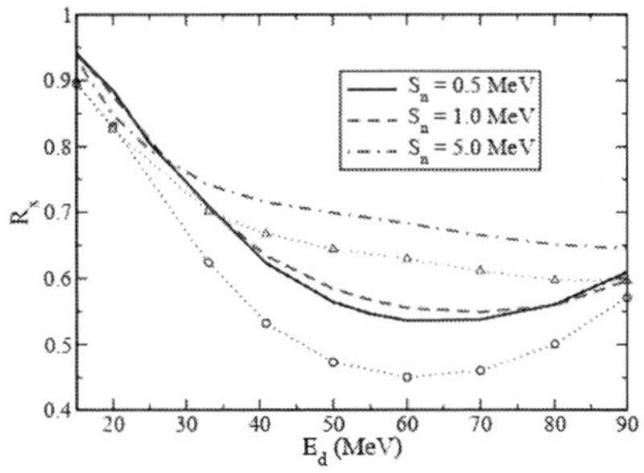

Figure 1. Spectroscopic factor ratio R_x as a function of the beam energy extracted from ^{10}Be(d,p)^{11}Be for various separation energies of the final nucleus. The lines with symbols are the results obtained when the excitation energy of ^{10}Be is strongly reduced. Figure from [3].

a recent review on inclusive (d,p) populating continuum states (Section 5). Finally, the conclusions are presented in Section 6.

2 Core excitation in (d,p) reactions

Deuteron induced reactions on targets are most often cast as a three-body problem. Over the last decade, much work has gone into establishing the methods used to solve the three-body scattering problem. Examples are the benchmarks of the adiabatic wave approximation method and the continuum discretized coupled-channel method against the exact Faddeev approach for transfer reactions [4, 5]. In those studies, the system was treated as $n + p + A$, and no excitation was allowed for A. When is this approximation adequate?

Core-excitation effects have been studied in detail in the context of the breakup of loosely bound two-body-like projectiles (e.g. [6]). Results from these breakup studies demonstrate that, when including core excitation dynamically in the reaction, observables are significantly modified. Until recently, no equivalent work using non-perturbative approaches has been done for the transfer channel.

A couple of years ago, the full Faddeev formalism was extended to include core excitation [7]. This method was used to explore the role of core excitation in (d,p) reactions, and understand the origin of the dynamical effects [3]. Based on the particle-rotor model of n+^{10}Be, a number of ^{11}Be-like models were generated with a range of separation energies, while maintaining a significant core-excited component. We used the n+^{10}Be system as a toy model, included the first two states of ^{10}Be and coupled the additional neutron to a total spin $J = \frac{1}{2}$ just as in ^{11}Be(g.s.), but then adjusted the interaction to produce a number of models with different separation energies ranging $S_n = 0.1 - 5.0$ MeV. We then applied the method of [7], which includes dynamical core excitation in the reaction mechanism to all orders, to the ^{10}Be(d,p)^{11}Be like reactions, and studied the excitation effects for beam energies from the range $E_d = 15 - 90$ MeV [3]. We compared the resulting angular distributions and the spectroscopic factor that would be extracted from the cross sections, when including dynamical core excitation in the reaction model (S^{fadd}), to that of the original structure model S^{th}. We introduced the ratio $R_x = \frac{S^{fadd}}{S^{th}}$, whereby a deviation from $R_x = 1$ provides a measure of the importance of the dynamical excitation effects in the reaction. Fig.1 shows the behavior of R_x as a function of beam energy for ^{11}Be-like final states with several binding energies. Our results in [3] show a strong beam-energy dependence of the extracted spectroscopic factor. In fact, for intermediate beam energies

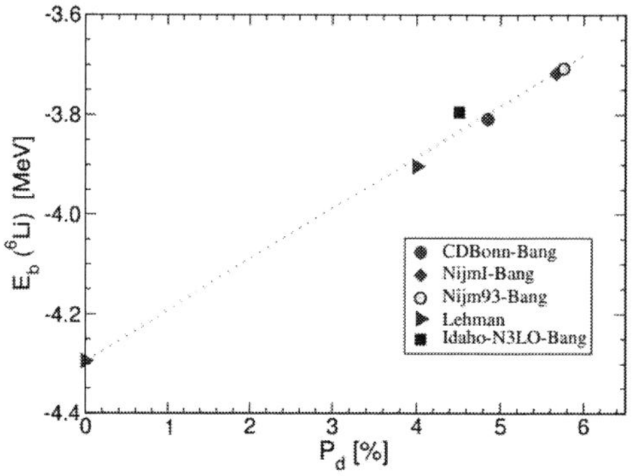

Figure 2. Three-body binding energy of ^{6}Li as a function of the deuteron d-state probability for a variety of NN interactions.

dynamical effects can be so large that the spectroscopic factor extracted by taking the ratio of data and theory is reduced to half the value of the spectroscopic factor in the original structure model. This strong dependence is most important for loosely bound systems. Given that the predicted effect is so large, it should be possible, through dedicated experiments, to test it experimentally.

3 Toward a Faddeev theory for (d,p) on heavy nuclei

The Faddeev method of [7] cannot be applied to intermediate and heavy mass targets due to the complications that arise with screening the Coulomb interaction. Mukhamedzhanov et al. [8] rewrote the Faddeev equations in the Coulomb basis using separable interactions. Separable forms for the optical potentials were developed ([9] and references therein). Recently, the three-body problem with separable interactions was implemented and applied to ^{6}Li [10]. The main conclusion in that work is that separable interactions can reproduce the exact solutions to very high accuracy. The next step in this work is to generalize the equation for positive energies, and consider elastic, breakup and rearrangement channels in detail. This work in ongoing.

As a sideline, we also explored the dependence on the bound state properties on the choice of the NN interaction. In Fig.2 the symbols show the results obtained for the three-body binding energy of ^{6}Li for a variety of NN interactions commonly used in our field, as a function of their prediction for the probability of the deuteron d-state. Note that only the NN interaction is varied, the ^{4}He-N system is kept fixed with the parameters described in Ref.[10]. The dashed line serves to guide the eye and suggests that there is a linear correlation between the three-body binding energy and the deuteron d-state probability. It appears that the three-body binding energy is only sensitive to long-range properties of the NN interaction. Further investigation is needed to determine universality in the system.

4 Nucleon optical potentials and the effects of nonlocality

An essential input to nuclear reaction theories for (d,p) is the nucleon optical potential. Although in the past, nucleon optical potentials have been taken as local interactions, we understand they should be non-local and that non-locality has an important effect on reaction observables (e.g. [11, 12]). Recently we derived this potential from ab-initio coupled cluster theory. The method starts with one-particle Green's functions, and then inverts the Dyson equation to obtain the self energy, identified with the optical potential [13]. The microscopic optical potentials obtained lacked absorption but the

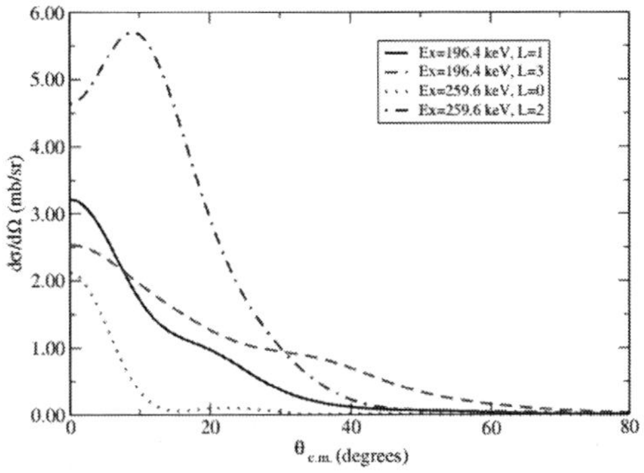

Figure 3. Angular distributions for $^{30}P(d,n)^{31}S$ at 30 MeV/u populating a couple of final states: dependence on the angular momentum transferred.

real part exhibited robust properties, including strong non-locality with a shape that deviates from Gaussian and varies with the position inside the nucleus. In addition, the potential generated in [13] is strongly energy dependent.

While an ab-initio approach cannot currently be directly applied to the interpretation of data, it can provide guidance to phenomenological approaches. There are two nonlocal global optical potentials available and both are energy *independent*. The ab-initio results call for both, nonlocality and energy dependence in the optical potential. One obvious shortcoming of the Perey and Buck potential [14] is that it was only fit to neutrons on Pb at low energies and does not account well for elastic scattering at $E > 20$ MeV. As a preliminary test of whether elastic scattering data calls for an explicit energy dependence in the non-local optical potential, we have started to investigate the problem using the standard χ^2 minimization techniques [15]. This study shows that a new nonlocal phenomenological optical potential with a linear energy dependence can provide an overall χ^2 that is significantly smaller than the one obtained when using the energy independent Perey and Buck potential [14], for data in the energy range $E = 10-50$ MeV. While promising, this problem needs to be revisited using a wider range of data and a more comprehensive parameter space.

5 Transfer reactions to the continuum

Several applications to astrophysics require transfer to the continuum. An example is the analysis of $^{30}P(d,n)^{31}S$ data to extract the proton capture on ^{30}P and inform abundance predictions for novae ejecta [16]. Because those measurements contain only integrated transfer cross sections, it was difficult to match the states predicted by shell model with those measured by GRETINA. In order to improve this analysis, it is desirable to measure the neutron angular distributions.

Transfer to the continuum is also important on the neutron dripline. One can use (d,p) reactions to populate the same continuum compound states as the (n,γ) reaction. This indirect approach requires an understanding of how probable it is to populate these compound states, as opposed to breakup. A recent review of the efforts in this area [17] discusses three independent implementations of the reaction formalism, all relying on a two step approach: first the deuteron breaks up in the field of the target, and then the neutron either gets captured (non-elastic) or flies away (elastic breakup). In [17] the three methods are benchmarked and applied to (d,p) reactions on Ca isotopes. As one moves toward the neutron dripline, the elastic breakup component increases but even for well bound systems this component is never negligible, and needs to be appropriately subtracted from the measured (d,p) cross section for a meaningful connection to (n,γ).

6 Concluding remarks

In these proceedings, we discuss a number of recent developments in the field of reaction theory, with particular emphasis on nucleon and deuteron induced reactions. Recent Faddeev results for (d,p) on a light target show that core excitation is important for single-nucleon transfer, and the magnitude of the effect is strongly dependent on the beam energy. This theoretical prediction calls for dedicated experiments. We also report on progress made toward a full Faddeev approach to (d,p) reactions on heavy nuclei using separable interactions and the Coulomb basis. The results for the ^{6}Li bound state demonstrate that separable representations of realistic two-body potentials can reproduce the exact solution for the three-body bound state problem with high precision. Work on the extension to positive energy is ongoing.

In the past we found that nonlocality is important in (d,p) reactions, affecting not only the magnitude of angular distributions but also the shape. We discuss recent results in extracting an ab-initio optical potential from coupled cluster theory as well as a phenomenological approach which includes both nonlocality and energy dependence in the optical potential. Finally, a few important developments concerning transfer to the continuum are also presented in connection to extracting neutron and proton capture rates on exotic nuclei. Theories for predicting cross sections have recently been benchmarked and several improvements have been identified. [1]

References

[1] K. Riisager, Rev. Mod. Phys. **66**, 1105 (1994)
[2] J.A. Tostevin, A. Gade, Phys. Rev. C **90**, 057602 (2014)
[3] A. Deltuva, A. Ross, E. Norvaišas, F.M. Nunes, Phys. Rev. C **94**, 044613 (2016)
[4] F.M. Nunes, A. Deltuva, Phys. Rev. C **84**, 034607 (2011)
[5] N.J. Upadhyay, A. Deltuva, F.M. Nunes, Phys. Rev. C **85**, 054621 (2012)
[6] N.C. Summers, F.M. Nunes, Phys. Rev. C **77**, 049901 (2008)
[7] A. Deltuva, Phys. Rev. C **92**, 064613 (2015)
[8] A. Mukhamedzhanov, V. Eremenko, A. Sattarov, Phys.Rev. **C86**, 034001 (2012)
[9] L. Hlophe, C. Elster, Phys. Rev. C **95**, 054617 (2017)
[10] L. Hlophe, J. Lei, C. Elster, A. Nogga, F. Nunes, Phys. Rev. C, submitted (arXiv:1710.02602) (2017)
[11] L.J. Titus, F.M. Nunes, G. Potel, Phys. Rev. C **93**, 014604 (2016)
[12] A. Ross, L.J. Titus, F.M. Nunes, Phys. Rev. C **94**, 014607 (2016)
[13] J. Rotureau, P. Danielewicz, G. Hagen, F.M. Nunes, T. Papenbrock, Phys. Rev. C **95**, 024315 (2017)
[14] F. Perey, B. Buck, Nuclear Physics **32**, 353 (1962)
[15] A. Lovell, P. Bacq, P. Capel, F. Nunes, L. Titus, Phys. Rev. C, in press (2017)
[16] A. Kankainen, P. Woods, H. Schatz, T. Poxon-Pearson, D. Doherty, V. Bader, T. Baugher, D. Bazin, B. Brown, J. Browne et al., Physics Letters B **769**, 549 (2017)
[17] G. Potel, G. Perdikakis, B.V. Carlson, M.C. Atkinson, W.H. Dickhoff, J.E. Escher, M.S. Hussein, J. Lei, W. Li, A.O. Macchiavelli et al., The European Physical Journal A **53**, 178 (2017)

[1]This work was supported by National Science Foundation under Grants No. NSF-PHY-1403906, NSF-PHY-1520929 and NSF-PHY-1520972, the Stewardship Science Graduate Fellowship program under Grant No. DE-NA0002135, and the Department of Energy under Contract No. DE-FG52-08NA28552. P. C. acknowledges the support of the Deutsche Forschungsgesellschaft (DFG) with the Collaborative Research Center 1245. Ch.E. acknowledges the U. S. Department of Energy under contract No. DE-FG02-93ER40756 with Ohio University.

One-nucleon pickup reactions and compound-nuclear decays

J. E. Escher[1,*], J. T. Burke[1], R. J. Casperson[1], R. O. Hughes[1], and N. D. Scielzo[1]

[1]Lawrence Livermore National Laboratory, Livermore, CA 94550, USA

Abstract. One-nucleon transfer reactions, long used as a tool to study the structure of nuclei, are potentially valuable for determining reaction cross sections indirectly. This is significant, as many reactions of interest to astrophysics and other applications involve short-lived isotopes and cannot be measured directly. We describe a procedure for obtaining constraints for calculations of neutron capture cross sections using observables from experiments with transfer reactions. As a first step toward demonstrating the method, we outline the theory developments used to properly describe the production of the compound nucleus $^{88}Y^*$ via the one-nucleon pickup reaction $^{89}Y(p,d)^{88}Y^*$ and test the description with data from a recent experiment. We indicate how this development can be used to extract the unknown $^{87}Y(n,\gamma)$ cross section from $^{89}Y(p,d\gamma)$ data. The example illustrates a more generally applicable method for determining unknown cross sections via a combination of theory and transfer (or inelastic scattering) experiments.

1 Introduction

One-nucleon pickup and stripping reactions have been used for decades in nuclear-structure studies. Most studies have focused on the low-energy ($\lesssim$2-3 MeV) properties of nuclei where single-particle properties are most clearly distinguished. Understanding excitations at higher energies, created by removing a nucleon from a deeply-bound orbital or by populating a resonance, is more challenging [1]. With the recent interest in using transfer reactions to indirectly infer cross sections for reactions involving very short-lived nuclei [2–6], it becomes relevant to revisit the issue.

The motivation for carrying out such transfer experiments stems from a need to determine unknown cross sections for nuclear astrophysics and other applications. Neutron capture cross sections, for instance, are crucial for understanding the production of the elements from iron to uranium that we observe today. It is well known that nucleosynthesis of heavy elements beyond ^{56}Fe takes place primarily by neutron capture on lighter seed nuclei in the s (slow neutron capture) and r (rapid neutron capture) processes, with other processes contributing to the abundances of some specific isotopes. Near stability, neutron captures on s-process branch points, unstable nuclei with a lifetime long enough to allow the s process to proceed by either neutron capture or β decay, are of particular interest. The competition between capture and decay paths depends on environmental variables, such as neutron densities, temperatures, and pressure, as well as on nuclear properties. Measured s-process abundances, therefore, yield valuable insights into the detailed conditions of the astrophysical s process. Further away from stability, one finds isotopes for which the capture cross sections affect the isotopic abundance patterns predicted by r-process models. While the r process is known to take place in a high-temperature, high-flux environment, neither the exact path along the nuclear chart nor the astrophysical site where it takes place is known. Capture cross sections significantly affect the isotopic abundance patterns predicted by astrophysical models and need to be known to draw conclusions about stellar evolution and nucleosynthesis [7–9].

In this contribution, we will describe how transfer reactions can be employed to indirectly determine cross sections for reactions involving unstable (or highly-radioactive) isotopes. We will focus on understanding the $^{89}Y(p,d)$ pickup reaction and indicate how this knowledge can be used to determine the cross section for neutron capture on the short-lived ^{87}Y isotope. In the next section, we will briefly review the formalism for compound neutron capture and show how a transfer reaction measurement, such as the (p,d) pickup reaction, can provide constraints for calculations of capture cross sections. In Section 3, we will outline the theory developments needed to describe compound-nucleus formation in the (p,d) reaction. We will show results from a recent experiment that serve as a test of the theory. Concluding remarks are offered in Section 4.

2 Cross section calculations and constraints from indirect measurements

Cross sections for many short-lived isotopes can be calculated, but suffer from large uncertainties due to poorly-constrained inputs. Much effort has been devoted to develop models to calculate nuclear level densities and transmission coefficients for particles, photons, and fission, and

*e-mail: escher1@llnl.gov

to formulate parameter recommendations for compound-nuclear (CN) reactions [10]. For low-energy neutron capture, nuclear level densities and γ-ray strength functions play a crucial role and need to be constrained. Auxiliary experiments provide important constraints for these models and their parameters. Most notably, neutron resonance measurements provide average level spacings and thus constraints for the level densities near the neutron separation energy [11]. Average radiative widths provide information on the product of the level density and the γ-ray strength function. However, for reactions involving unstable targets, the requisite neutron resonance spacings and average radiative capture widths are not available, and modeling the decay of the CN becomes the primary source of uncertainty in calculations of capture cross sections.

The CN capture reaction $n + A \rightarrow B^* \rightarrow \gamma + B$ is described in the statistical Hauser-Feshbach formalism [12, 13], which accounts for conservation of angular momentum and parity:

$$\sigma_{n+A,\chi}(E_n) = \sum_{J,\pi} \sigma^{CN}_{n+A}(E_{ex}, J, \pi)\ G^{CN}_{\chi}(E_{ex}, J, \pi)\ . \quad (1)$$

Here $\chi = \gamma + B$ denotes the exit channel, and the excitation energy E_{ex} of B^* is related to the neutron energy via $E_n = \frac{A+1}{A}(E_{ex} - S_n)$, where S_n is the neutron separation energy of the nucleus B. Width fluctuation corrections have been omitted here, but have to be included in the calculations. The CN formation cross section, $\sigma^{CN}_{n+A} = \sigma(n + A \rightarrow B^*)$, can be calculated using existing optical potentials. The theoretical decay probabilities G^{CN}_{χ} for the different decay channels χ are often quite uncertain, since they contain transmission coefficients, $T^J_{\chi ls}$, for all competing exit channels, the associated level densities, $\rho_I(U)$, in the various residual nuclei, as well as information on discrete levels:

$$G^{CN}_{\chi}(E_{ex}, J, \pi) = \quad (2)$$
$$\frac{\sum_{\ell s I} \int T^J_{\chi \ell s} \rho_I(U) dE_{\chi}}{\sum'_{\chi' \ell' s'} T^J_{\chi' \ell' s'} + \sum_{\chi' \ell' s' I'} \int T^J_{\chi' \ell' s'}(E_{\chi'}) \rho_{I'}(U') dE_{\chi'}}.$$

The quantities ℓ and ℓ' are the relative orbital angular momenta in the exit channels and $\rho_I(U)$ is the density of levels of spin I at energy U in the residual nucleus. The denominator includes contributions from decays to discrete levels and to regions described by a level density. All sums over quantum numbers must respect parity conservation, although this is not explicitly expressed here.

In the absence of constraints on the level densities and γ-ray strength functions, the decay probabilities $G^{CN}_{\chi}(E_{ex}, J, \pi)$ become the primary source of uncertainty in calculations of capture cross sections. It is the purpose of indirect methods to provide constraints that reduce the uncertainties in the calculated cross sections.

In an approach that has become known as the 'surrogate method' [2], the compound nucleus B^* is produced by an inelastic scattering or transfer reaction $d + D \rightarrow b + B^*$, and the desired decay channel is observed in coincidence with the outgoing particle b at angle θ_b. The idea is schematically illustrated in Figure 1 for a one-nucleon pickup reaction, p + $^{A+2}$Z → d + $^{A+1}Z^*$.

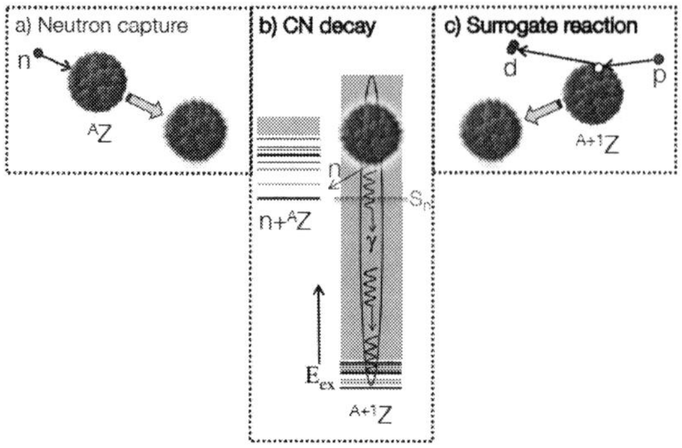

Figure 1. Surrogate approach, applied to a neutron-capture reaction. Neutron and target nucleus ^{A}Z fuse to form the CN $^{A+1}Z^*$ (panel a), which subsequently decays via a γ-cascade (panel b). Insufficient knowledge of nuclear properties results in a poor description of possible decay channels. In a surrogate experiment, the CN is produced by an alternative reaction, here a $^{A+2}$Z(p,d)$^{A+1}Z^*$ reaction (panel c). Observation of the subsequent decay of $^{A+1}Z^*$ provides crucial constraints for the decay models.

The probability for forming $B^* = {}^{A+1}Z^*$ in the surrogate reaction (with specific values for E_{ex}, J, π) is $F^{CN}_{\delta}(E_{ex}, J, \pi, \theta_b)$, where δ refers to the surrogate reaction $d + D \rightarrow b + B^*$. The quantity

$$P_{\delta\chi}(E_{ex}, \theta_b) = \sum_{J,\pi} F^{CN}_{\delta}(E_{ex}, J, \pi, \theta_b)\ G^{CN}_{\chi}(E_{ex}, J, \pi)\ , \quad (3)$$

which gives the probability that the CN B^* was formed with energy E_{ex} and decayed into channel χ, can be obtained experimentally by detecting a discrete γ-ray transition characteristic of the residual nucleus (or some other suitable observable).

The distribution $F^{CN}_{\delta}(E_{ex}, J, \pi, \theta_b)$, which may be very different from the CN spin-parity populations following the absorption of a neutron in the desired reaction, has to be determined theoretically, so that the branching ratios $G^{CN}_{\chi}(E_{ex}, J, \pi)$ can be extracted from the measurements. In practice, the decay of the CN is modeled using a Hauser-Feshbach-type decay model and the $G^{CN}_{\chi}(E_{ex}, J, \pi)$ are obtained by adjusting parameters in the model to reproduce the measured probabilities $P_{\delta\chi}(E_{ex}, \theta_b)$. Subsequently, the sought-after cross section can be obtained by combining the calculated cross section $\sigma^{CN}_{\alpha}(E_{ex}, J, \pi)$ for the formation of $B^* = {}^{A+1}Z^*$ (from n+^{A}Z) with the extracted decay probabilities $G^{CN}_{\chi}(E_{ex}, J, \pi)$ for this state, see Eq. 1.

3 Surrogate (p,d) reactions: theory and experimental tests

Generating accurate predictions of the surrogate spin-parity (J^{π}) distribution requires a model for the reaction mechanisms that are involved in the formation of the compound nucleus. Here we consider the example of ^{89}Y(p,d)^{88}Y* which produces the CN ^{88}Y*. This CN is relevant to the ^{87}Y(n,γ) reaction, to which we wish to apply the surrogate method. ^{87}Y has a half-life of 79.8 hours and current cross section calculations have to rely on regional systematics.

In the example considered here, the surrogate reaction produces ^{88}Y* by removing neutrons from inner shells of the ^{89}Y nucleus: deep hole states are involved in the production of ^{88}Y* near S_n. Their location and fragmentation as a function of E_{ex} was obtained using the dispersive optical model approach of Mahaux and Sartor [14]. At the high excitation energies involved, one-step (p,d) pickup processes have to be complemented by contributions from two-step processes such as (p,p')(p',d) and (p,d')(d',d), in which the initial ^{89}Y or the final ^{88}Y are inelastically excited. The calculations were carried out using the coupled-channels code Fresco [15]. The proton-nucleus optical potential by Koning and Delaroche [16] was used to describe the entrance channel, and the deuteron-nucleus potential by Daehnick *et al.* [17] was employed for the exit channel. Due to the large number of states in the energy band populated, the different contributions can be assumed to add incoherently. Angle-integrated (p,d) cross sections can be calculated and compared to measured (p,d) cross sections.

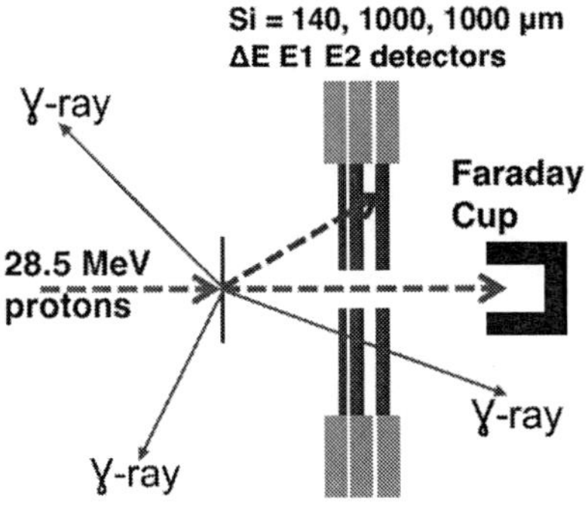

Figure 2. Experimental setup. A 28.5-MeV beam was incident on an ^{89}Y target. Outgoing deuterons were identified and their energy measured in a silicon telescope consisting of one thin and two thick detectors at angles covering 30 – 60°. Coincident γ rays were detected in five HPGe clover detectors surrounding the target chamber.

The requisite experiment was carried out at the Texas A&M University Cyclotron Institute, where the K150 Cyclotron was employed to provide a 28.5-MeV proton beam. The setup is shown schematically in Figure 2. Particle-γ coincidence data was collected using the STAR-LiTeR array, a combination of a silicon telescope array and five HPGe clover detectors [18]. The outgoing deuteron was detected at angles between 30° and 60° with 1°-2° angular resolution. From a reconstruction of the deuteron energies it was determined that the CN ^{88}Y was populated at excitation energies up to about 12 MeV. Contributions from carbon and oxygen contaminants were subtracted after measuring their characteristics using a carbon target within this particular setup. To provide information on the decay of the CN, the experiment determined coincidences between the outgoing deuteron and γ-rays for transitions between low-lying states (E_n < 2 MeV) in either ^{88}Y or ^{87}Y, as a function of the excitation energy. More details about the experiment are given in Refs. [18–20].

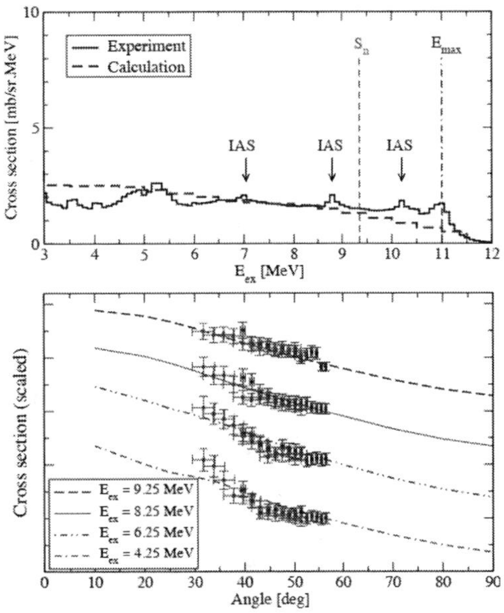

Figure 3. Comparison of the predicted ^{89}Y(p,d)^{88}Y* cross section to measured results. Top panel: The angle-integrated (p,d) cross section is shown as a function of excitation energy in ^{88}Y*. The calculation takes into account both one-step (p,d) and two-step (p,p')(p',d) and (p,d')(d',d) mechanisms. The cross section is integrated over the 30°-60° angular range covered by the experiment. The experiment also populated known isobaric analog states ('IAS'); these are not explicitly included in the reaction calculation. E_{max} indicates the experimental upper limit and S_n = 9.35 MeV is the neutron separation energy in ^{88}Y. Bottom panel: Angular differential cross sections for the (p,d) reaction are shown for four excitation energies in the CN ^{88}Y. The energies were selected to give a representative sample, without including the effects of the IAS. Data from two different detector segments are shown as blue and green circles with error bars. The cross sections were scaled for ease of comparison.

Preliminary results are shown in Figure 3, where we compare theoretical predictions to data from the (p,d) experiment. The dispersive optical model, which gives energy-averaged quantities, is not expected to reproduce the details of the rich nuclear structure present at low excitation energies (E_{ex} ≲2-3 MeV). Hence we consider the results at higher energies (E_{ex}≈3-10 MeV). At those energies, two-step contributions are found to dominate the calculated (p,d) cross section, with one-step mechanisms contributing only about 20-25%. The figure shows the sum of the two contributions as a dashed curve. Also, the experiment was seen to populate isobaric analog states, indicated with 'IAS' in the figure. The current reaction formalism does not predict cross sections for populating these states, but their effect on the spin distribution populated in the compound nucleus can be accounted for, since their spin and parity quantum numbers are known. Overall, the combination of the dispersive optical model with the two-step approach is found to give results that are in good agreement with the measurements, as can be seen in the top panel of Figure 3.

The characteristic angular shapes of one-nucleon transfers to low-lying discrete states are regularly used to extract useful information on the angular-momentum transfer in the reaction and, hence, on the possible spin and parity values of the final states. Here we focus on transfers to finite energy bins at higher excitation energies and see that the cross sections show little structure. Measurements for E_{ex} = 4.25 MeV, 6.25 MeV, 8.25 MeV, and 9.25 MeV (which are free of IAS contributions) show that the cross section is slightly decreasing, but relatively flat, in the 30°-60° range covered in the experiment, see bottom panel of Figure 3. The calculations reproduce this behavior well. Note that both the calculated and measured cross sections were scaled in order to allow for a clear angular comparison within the figure. The agreement of the absolute cross sections can be assessed by the comparison shown in the top panel.

The comparison of the predicted cross section to the data shows good agreement. This is an important step towards understanding the formation of a compound nucleus in a transfer reaction. For surrogate applications, the weights $F_{\delta}^{CN}(E_{ex}, J, \pi, \theta_b)$ of the final spins populated in the reaction have to be determined. This can be accomplished by calculating the relative contributions of the individual cross sections to the total (p,d) cross section.

The spin-parity distributions $F_{\delta}^{CN}(E_{ex}, J, \pi, \theta_b)$ obtained in this manner can then be combined with a model of the CN decay to reproduce the measured coincidence probabilities $P_{\delta\chi}(E_{ex}, \theta_b)$, as indicated in Equation 3. By varying the parameters in the decay models entering the expression for $G_{\chi}^{CN}(E_{ex}, J, \pi)$ and comparing the calculated $P_{\delta\chi}(E_{ex}, \theta_b)$ with the data, we can obtain constraints on these models and hence on the $G_{\chi}^{CN}(E_{ex}, J, \pi)$. This information is then used to calculate the desired neutron capture cross section for the short-lived ^{87}Y nucleus according to Equation 1. The results will be reported elsewhere [21].

4 Conclusions

One-nucleon transfer reactions are a valuable tool, not only for studying the structure of nuclei, but also for indirectly determining reaction cross sections. Many reactions of interest to astrophysics and other applications involve short-lived isotopes and cannot be measured directly. We discussed a procedure for obtaining constraints for calculations of neutron capture cross sections using observables from experiments with transfer reactions. Our specific example focused on the one-nucleon pickup reaction ^{89}Y(p,d)^{88}Y* which can be used to determine the neutron capture cross section for the short-lived ^{87}Y nucleus. We outlined the theory developments needed to properly describe the CN production via the transfer reaction and showed comparisons with experimental data.

This is a first, but critical step toward extracting the desired cross section from indirect (p,dγ) data. The method outlined here properly takes into account the spins and parities populated in the surrogate reaction. Previous attempts to obtain (n,γ) cross sections from surrogate measurements invoked the approximation that either the decay probabilities $G_{\chi}^{CN}(E_{ex}, J, \pi)$ are independent of spin and parity or that the surrogate and desired reactions populate the CN in a similar manner. A number of theoretical [22–24] and experimental investigations [25–28] have demonstrated that these assumptions are not valid, with the region near ^{90}Zr being particularly sensitive to spin effects [18, 19, 22, 23, 25].

The method outlined here for one-nucleon pickup reactions can also be employed with stripping reactions, such as (d,p), or with inelastic scattering. Several groups have revisited the formalism for inclusive (d,p) reactions [29–33], with the goal of describing transfers to both low-lying bound and highly-excited states above particle emission thresholds. All approaches have to be carefully tested and extended to apply to not only spherical but also deformed nuclei [34–36].

The longer-term goal is to develop a method that can be used for a wider range of reactions and a variety of projectile-target combinations. This makes it possible to reach a large number of isotopes, in particular when the technique can be combined with radioactive beams in inverse-kinematics experiments [37–39].

Acknowledgments

This contribution is dedicated to the memory of our friend and collaborator Cornelius Beausang, whose enthusiasm, optimism, and collegiality has been and will remain an inspiration. This work was performed under the auspices of the U.S. Department of Energy (DOE) by Lawrence Livermore National Laboratory under contract DE-AC52-07NA27344.

References

[1] S. Gales, C. Stoyanov, A.I. Vdovin, Phys. Rep. **166**, 125 (1988)
[2] J.E. Escher, J.T. Burke, F.S. Dietrich, N.D. Scielzo, I.J. Thompson, W. Younes, Rev. Mod. Phys. **84**, 353 (2012)
[3] R. Raut, A.P. Tonchev, G. Rusev, W. Tornow, C. Iliadis, M. Lugaro, J. Buntain, S. Goriely, J.H. Kelley, R. Schwengner et al., Phys. Rev. Lett. **111**, 112501 (2013)
[4] A. Spyrou, S.N. Liddick, A.C. Larsen, M. Guttormsen, K. Cooper, A.C. Dombos, D.J. Morrissey, F. Naqvi, G. Perdikakis, S.J. Quinn et al., Phys. Rev. Lett. **113**, 232502 (2014)
[5] S.N. Liddick, A. Spyrou, B.P. Crider, F. Naqvi, A.C. Larsen, M. Guttormsen, M. Mumpower, R. Surman, G. Perdikakis, D.L. Bleuel et al., Phys. Rev. Lett. **116**, 242502 (2016)

[6] Escher, J. E., Tonchev, A. P., Burke, J. T., Bedrossian, P., Casperson, R. J., Cooper, N., Hughes, R. O., Humby, P., Ilieva, R. S., Ota, S. et al., EPJ Web of Conferences **122**, 12001 (2016)
[7] A. Arcones, D.W. Bardayan, T.C. Beers, L.A. Bernstein, J.C. Blackmon, B. Messer, B.A. Brown, E.F. Brown, C.R. Brune, A.E. Champagne et al., Progress in Particle and Nuclear Physics **94**, 1 (2017)
[8] Nuclear Science Advisory Committee, *Reaching for the Horizon: The 2015 Long Range Plan for Nuclear Science* (2015)
[9] M. Mumpower, R. Surman, G. McLaughlin, A. Aprahamian, Progress in Particle and Nuclear Physics **86**, 86 (2015)
[10] R. Capote, M. Herman, P. Oblozinský, P. Young, S. Goriely, T. Belgya, A. Ignatyuk, A. Koning, S. Hilaire, V. Plujko et al., Nuclear Data Sheets **110**, 3107 (2009)
[11] S.F. Mughabghab, *Atlas of Neutron Resonances, Resonance Parameters and Thermal Cross Sections Z=1-100*, 5th edn. (Elsevier, Amsterdam, 2006)
[12] W. Hauser, H. Feshbach, Phys. Rev. **87**, 366 (1952)
[13] P. Fröbrich, R. Lipperheide, *Theory of Nuclear Reactions* (Clarendon Press, Oxford, 1996)
[14] C. Mahaux, R. Sartor, in *Advances in nuclear physics* (Springer US, 1991), Vol. 20, pp. 1–223
[15] I.J. Thompson, Computer Physics Reports **7**, 167 (1988)
[16] A.J. Koning, J.P. Delaroche, Nucl. Phys. **A713**, 231 (2003)
[17] W.W. Daehnick, J.D. Childs, Z. Vrcelj, Phys. Rev. C **21**, 2253 (1980)
[18] S. Ota, J.T. Burke, R.J. Casperson, J.E. Escher, R.O. Hughes, J.J. Ressler, N.D. Scielzo, I.J. Thompson, R.A.E. Austin, B. Abromeit et al., Phys. Rev. C **92**, 054603 (2015)
[19] S. Ota, J. Burke, R. Casperson, J. Escher, R. Hughes, J. Ressler, N. Scielzo, I. Thompson, R. Austin, E. McCleskey et al., EPJ Web of Conferences **93**, 02001 (2015)
[20] R.O. Hughes, J.T. Burke, R.J. Casperson, J.E. Escher, S. Ota, J.J. Ressler, N.D. Scielzo, R.A.E. Austin, B. Abromeit, N.J. Foley et al., Phys. Rev. C **93**, 024315 (2016)
[21] J.E. Escher, J.T. Burke, R.O. Hughes et al., *Constraining neutron capture cross sections for unstable nuclei with surrogate reaction data and theory*, to be published (2018)
[22] C. Forssén, F. Dietrich, J. Escher, R. Hoffman, K. Kelley, Phys. Rev. C **75**, 055807 (2007)
[23] J.E. Escher, F.S. Dietrich, Phys. Rev. C **81**, 024612 (2010)
[24] S. Chiba, O. Iwamoto, Phys. Rev. C **81**, 044604 (2010)
[25] N.D. Scielzo, J.E. Escher, J.M. Allmond, M.S. Basunia, C.W. Beausang, L.A. Bernstein, D.L. Bleuel, J.T. Burke, R.M. Clark, F.S. Dietrich et al., Phys. Rev. C **81**, 034608 (2010)
[26] R. Hatarik, L.A. Bernstein, J.A. Cizewski, D.L. Bleuel, J.T. Burke, J.E. Escher, J. Gibelin, B.L. Goldblum, A.M. Hatarik, S.R. Lesher et al., Phys. Rev. C **81**, 011602 (2010)
[27] G. Boutoux, B. Jurado, V. Méot, O. Roig, L. Mathieu, M. Aïche, G. Barreau, N. Capellan, I. Companis, S. Czajkowski et al., Physics Letters B **712**, 319 (2012)
[28] Q. Ducasse, B. Jurado, M. Aïche, P. Marini, L. Mathieu, A. Görgen, M. Guttormsen, A.C. Larsen, T. Tornyi, J.N. Wilson et al., Phys. Rev. C **94**, 024614 (2016)
[29] G. Potel, F.M. Nunes, I.J. Thompson, Phys. Rev. C **92**, 034611 (2015)
[30] J. Lei, A.M. Moro, Phys. Rev. C **92**, 044616 (2015)
[31] J. Lei, A.M. Moro, Phys. Rev. C **92**, 061602 (2015)
[32] B.V. Carlson, R. Capote, M. Sin, Few-Body Systems **57**, 307 (2016)
[33] G. Potel, G. Perdikakis, B.V. Carlson, M.C. Atkinson, W. Dickhoff, J.E. Escher, M.S. Hussein, J. Lei, W. Li, A.O. Macchiavelli et al., Eur. Phys. J. A **53**, 178 (2017)
[34] B.L. Andersen, B.B. Back, J.M. Bang, Nucl. Phys. A **147**, 33 (1970)
[35] B.B. Back, O. Hansen, H.C. Britt, J.D. Garrett, Phys. Rev. C **9**, 1924 (1974)
[36] V. Gueorguiev, J. Escher, Bulg. J. Phys. **44**, 489 (2017)
[37] J. Cizewski, R. Hatarik, K. Jones, S. Pain, J. Thomas, M. Johnson, D. Bardayan, J. Blackmon, M. Smith, R. Kozub, Nuclear Instruments and Methods in Physics Research Section B: Beam Interactions with Materials and Atoms **261**, 938 (2007)
[38] A. Ratkiewicz, J. Cizewski, S. Pain, A. Adekola, J. Burke, R. Casperson, N. Fotiades, M. McCleskey, S. Burcher, C. Shand et al., EPJ Web of Conferences **93**, 02012 (2015)
[39] J.A. Cizewski, A. Ratkiewicz, J.E. Escher, A. Lepailleur, S.D. Pain, G. Potel, EPJ Web Conf. **165**, 01013 (2017)

Neutron capture on short-lived nuclei via the surrogate (d,pγ) reaction

Jolie A. Cizewski[1*,*] and *Andrew* Ratkiewicz[1,2]

[1]Department of Physics and Astronomy, Rutgers University, New Brunswick, NJ 08901 USA
[2]Nuclear & Chemical Sciences Division, Lawrence Livermore National Laboratory, CA, 94550 USA

Abstract. Rapid *r*-process nucleosynthesis is responsible for the creation of about half of the elements heavier than iron. Neutron capture on short-lived nuclei in cold processes or during freeze out from hot processes can have a significant impact on the final observed *r*-process abundances. We are validating the (d,pγ) reaction as a surrogate for neutron capture with measurements on ^{95}Mo targets and a focus on discrete transitions. The experimental results have been analyzed within the Hauser-Feshbach approach with non-elastic breakup of the deuteron providing a neutron to be captured. Preliminary results support the (d,pγ) reaction as a valid surrogate for neutron capture. We are poised to measure the (d,pγ) reaction in inverse kinematics with unstable beams following the development of the experimental techniques.

1 Introduction

Neutron capture is responsible for the synthesis of almost all of the elements heavier than iron. While the slow, *s* process proceeds close to stability, the rapid, *r* process occurs on nuclei far from stability. Studies by Surman and her coworkers [1,2] have demonstrated how unknown (n,γ) rates on nuclei near the *r*-process path, and in particular near closed neutron shells, could have significant impact on predicting abundances with *r*-process network calculations. Constraining (n,γ) rates could also serve to inform our knowledge of the site of the *r* process. This is especially important for synthesis of A≈80 elements [2] where both neutrino-driven winds from core-collapse supernovae and neutron-star mergers could be sites for nucleosynthesis.

Neutron capture reactions have been studied for decades. The importance of these studies has been highlighted in the long series of (Neutron) Capture Gamma-Ray Spectroscopy symposia over the past 38 years. More recently, the techniques [e.g., 3] have been developed to measure (n,γ) reactions on rare isotopes with relatively long half lives (>100 days) where the need to suppress the decay radiation is critical. However, all of the *r*-process waiting point isotopes have very short half lives, and for many are not even known

* Corresponding author : cizewski@physics.rutgers.edu

to exist. The *r*-process path is characterized by isotopes where (n,γ) and (γ,n) reactions are in equilibrium and the neutron separation energies are typically ≈3.5 MeV. Most *r*-process scenario environments are hot, with temperatures of $T>10^9$ K and high neutron densities, $>10^{20}$ n/cm^3. However, in cold scenarios, and especially in freeze out from hot scenarios, the environments are no longer characterized by (n,γ) ↔ (γ,n) equilibrium. These are conditions where (n,γ) on daughters of the *r*-process waiting points become important. Unknown (n,γ) rates on specific nuclei, especially near neutron shell closures, have been shown by Surman and colleagues to significantly impact the final observed *r*-process abundances [1,2]. Measurements of (n,γ) reactions on scores of unstable nuclei near the r-process path are required to constrain models of *r*-process nucleosynthesis, including freeze out from hot processes.

The neutron is unstable – therefore it will probably always be impossible to measure (n,γ) reactions directly on key nuclei that live for << 100 days. The only way to deduce (n,γ) cross sections will be via a validated surrogate for these reactions and with accelerated radioactive ion beams. The present manuscript outlines the status of the validation of a very promising surrogate for neutron capture: the (d,pγ) reaction.

2 (d,pγ) as (n,γ) surrogate

Neutron capture can be modelled by using an optical model to calculate the entrance n+A system that forms the compound nucleus at a given excitation energy and spin-parity and a Hauser-Feshbach (HF) approach to model the decay. The (n,γ) cross section can then be written as [4]:

$$\sigma_{n\gamma}(E_n) = \sum_{J,\pi} \sigma_n^{CN}(E_{ex}, J, \pi) G_\gamma^{CN}(E_{ex}, J, \pi) \qquad (1)$$

The formation of the n+A compound nucleus (CN), σ_n^{CN}, and decay of the compound nucleus, $G_\gamma^{CN}(E_{ex}, J, \pi)$, need to be determined. While the optical model used to calculate σ_n^{CN} is robust, the level density and gamma-ray strength function models needed to calculate the HF decay are on less solid footing, especially away from stability and as the nucleus becomes less bound. A surrogate reaction is chosen to form the same compound nucleus, with character as close as possible to the desired n+A system. The reaction framework must compensate for the differences between the desired and surrogate reactions. The measured gamma-ray decay of the surrogate CN then provides the information needed to calculate the level density (LD) and gamma-ray strength function (GSF) that reproduce the surrogate data and provide input to calculate G_γ^{CN} in equation 1.

Gregory Potel and coworkers [5] have recently demonstrated that the (d,p) reaction is an excellent candidate for forming a compound nucleus that is a surrogate for the n+A system. Because the deuteron is weakly bound, it can readily break up via both non-elastic and elastic processes. The non-elastic breakup provides a neutron that can be captured by the "target" nucleus with a specific orbital angular momentum transferred. The Potel framework predicts as a function of excitation energy the balance between non-elastic and elastic breakup, as well as the distribution of transferred angular momentum. The (d,p) reaction at ≈7-10 MeV/u (14 to 20 MeV deuteron energy) brings in relatively little angular momentum, typically no more than about ℓ=3, reducing the angular momentum mismatch between the surrogate and desired (n,γ) reaction which is dominated by low ℓ (e.g., ℓ=0,1) capture.

Jutta Escher and coworkers [4] have been developing the protocol for using surrogate discrete gamma-ray spectroscopy as a function of excitation energy to deduce the level density and γ-ray strength function parameters needed to extract (n,γ) cross sections. In the surrogate approach, the LD and GSF formulations needed to calculate the (n,γ) cross section are deduced from the measured gamma-decay probabilities as a function of excitation energy and angular momentum [4]:

$$P_{dp\gamma}(E_x) = \sum_{J,\pi} F_{dp}^{CN}(E_{ex}, J, \pi) G_{\gamma}^{CN}(E_{ex}, J, \pi) \qquad (2)$$

$$P_{dp\gamma_i}(E_x) = {N_{dp\gamma_i}} \Big/ {\varepsilon_\gamma N_p} \qquad (3)$$

The left-hand side of eq. 2 is the measured surrogate gamma-ray decay probability deduced from eq. 3 for each transition, the ratio of measured (d,pγ) coincidence to singles proton data as a function of excitation energy, corrected for appropriate efficiencies in the system. Several discrete transitions, from states with different initial angular momenta are used as input. The Potel model [5] is used to calculate the formation of the surrogate compound nucleus, *F*, through non-elastic deuteron breakup; the angular momentum decomposition as a function of excitation energy is taken into account. In the Escher framework, a Bayesian approach is used to deduce the level density and strength function parameters that best fit the data and hence the gamma-decay *G* of the surrogate compound nucleus. This deduced *G* as a function of excitation energy and angular momentum is then used as input into equation 1 to calculate the surrogate (n,γ) cross section, with an optical model used to provide the formation of the (n,γ) CN.

3 Status of (d,pγ) surrogate validation: the ^{95}Mo(d,pγ) reaction

Our collaboration measured the (d,pγ) reaction at the Texas A&M cyclotron with stable ^{95}Mo targets and an array of annular silicon strip detectors and clover Compton-suppressed HPGe detectors, the STARLiTeR system [6]. ^{95}Mo was chosen because the final nucleus is even-even (and therefore a relatively simple gamma-ray spectrum at low excitation energies) and the ^{95}Mo(n,γ) cross section had been measured to ≈200 keV [7,8]. A schematic of the ^{96}Mo level scheme and preliminary spectra from both ^{95}Mo (d,pγ) at TAMU and (n,γ) at LANSCE reaction measurements are presented in Figure 1. In both the (d,pγ) and (n,γ) studies, transitions depopulating low-lying levels (up to 4^+) are dominant with relatively little population of the 6^+ yrast state, and no observation of the 8^+ to 6^+ yrast transition in the (n,γ) reaction. Note that for the $5/2^+$ ^{95}Mo ground state, ℓ=0 capture populates 2^+,3^+ compound nuclear states.

Figure 2 displays preliminary gamma-ray decay probabilities for six states in ^{96}Mo and two states in ^{95}Mo, the (n,n'γ) surrogate that becomes important above the neutron separation energy. The decay probabilities of these ^{96}Mo states are being used by Escher to deduce (n,γ) cross sections, combined with Potel predictions of the non-elastic breakup, and optical model predictions of the n+A compound nucleus formation. Preliminary results reproduce the previously measured and evaluated cross sections [7,8]. The excitation function of the transitions in ^{95}Mo have a similar shape to (n,n'γ) excitation functions measured with E>175 keV neutrons at LANSCE with the GEANIE array [12].

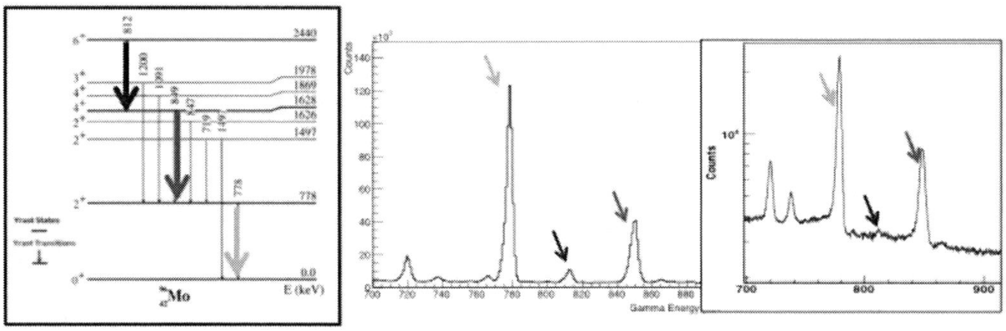

Figure 1. (Left) Schematic of the level scheme of ^{96}Mo. (Center) Preliminary ^{95}Mo(d,pγ) spectrum measured with STARLiTeR. (Right) Preliminary ^{95}Mo(n,γ) spectrum for the 3^+ resonance at 45 eV [9,10] measured on FP12 at LANSCE.

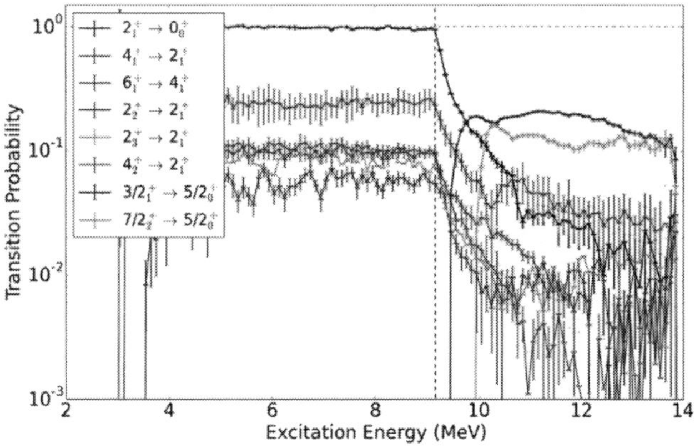

Figure 2. Preliminary gamma-decay probabilities for six states in ^{96}Mo and two states in ^{95}Mo measured with the STARLiTeR system at TAMU. Adopted from [11].

4 Developing techniques to measure (d,pγ) reactions with RIBs

We are part of the Gammasphere ORRUBA: Dual Detectors for Experimental Structure Studies (GODDESS) collaboration [13] that has developed the techniques to couple the Oak Ridge Rutgers University Barrel Array (ORRUBA) of position-sensitive silicon strip detectors [14] to the Gammasphere array of up to 110 Compton-suppressed HPGe detectors at the ATLAS accelerator at Argonne National Laboratory. The commissioning experiment exploited ^{134}Xe beams with ≈1 mg/cm^2 CD_2 targets to measure excitations in ^{135}Xe including candidates for 2f7/2 and 3p3/2 excitations above the N=82 shell gap. Preliminary results have been reported [13] and show gamma-ray events in coincidence with protons well above the neutron separation energy, the surrogate (n,n'γ) excitation region. ORRUBA can also be coupled to GRETINA [15] and a new LaBr3(Ce)-based Hybrid Array of Gamma-Ray Detectors (HAGRiD) currently being developed under the leadership of the University of Tennessee [16].

5 Summary and conclusions

In summary, we are demonstrating that the (d,pγ) reaction is a valid surrogate for the (n,γ) reaction important in understanding observed abundances in *r*-process nucleosynthesis. The Potel theoretical framework has shown how non-elastic breakup of the deuteron provides a neutron to be captured in the (d,p) reaction and predicts the decomposition of the transferred angular momentum as a function of excitation energy. The framework developed by Escher [4] is used to take extracted gamma-ray decay probabilities for several discrete transitions as a function of excitation energy to model level density and gamma-ray strength functions needed as input to predict gamma decay of the n+A compound nucleus calculated with an optical model. The techniques to measure (d,pγ) reactions with RIBs have been developed. First experiments with N≈82 nuclei near the *r*-process path of nucleosynthesis are being planned.

We would like to thank our STARLiTeR collaborators for help in realizing the TAMU results and Jutta Escher, Gregory Potel, Aderemi Adekola, Sean Burcher and Nicholas Fotiades for sharing with us preliminary results. This work is supported in part by the U.S. Department of Energy National Nuclear Security Administration under the Stewardship Science Academic Alliances program DE-FG52-09NA29467 and DE-NA0000979 (Rutgers); DE-AC52-07NA27344 and LDRD 16-ERD-022 (LLNL) and the National Science Foundation.

References

1. M. Mumpower *et al.*, Prog. Nucl. Part. Phys. **86**, 86 (2016)
2. R. Surman *et al.*, AIP Advances **4**, 041008 (2014)
3. R. Reifarth *et al.*, Nucl. Instrum. Meth. Phys. Res. B **241**, 176 (2005).
4. J. E. Escher *et al.*, Rev. Mod. Phys. **84**, 353 (2012) ; J. E. Escher *et al.*, EPJ Web of Conferences **122**, 12001 (2016)
5. G. Potel, F. M. Nunes, and I. J. Thompson, Phys. Rev. C **92**, 034611 (2015)
6. S. Lesher *et al.*, Nucl. Inst. & Meth. in Phys. Res. A **621**, 286 (2010) and R. J. Casperson *et al.*, Phys. Rev. C **90**, 034601 (2014)
7. A. De L. Musgrove *et al.*, Nucl. Phys. A **270**, 108 (1976)
8. ENDF and ENSDF databases http://www.nndc.bnl.gov
9. J.A. Cizewski et al., BAPS.2014.HAW.KB.8 (2014)
10. A.S. Adekola, private communication
11. A. Ratkiewicz *et al.*, to be published (2018).
12. S. Burcher and N. Fotiades, private communication
13. S.D. Pain, *et al.*, Physics Procedia **90**, 455 (2017)
14. S.D. Pain et al., Nucl. Instrum. and Meth. B **261**, 1122 (2007)
15. P. Fallon, A. Gade and I.Y. Lee, Annu. Rev. Nucl. Part. Sci. **66**, 321 (2016)
16. K. Smith, et al., Nucl. Instrum. Meth. Phys. Res. B **414**, 190 (2018).

Measurement of the radiative capture cross section of the *s*-process *branching points* ^{204}Tl and ^{171}Tm at the n_TOF facility (CERN)

A. Casanovas[1], C. Domingo-Pardo[2], C. Guerrero[3], J. Lerendegui-Marco[3], F. Calviño[1], A. Tarifeño-Saldivia[1], R. Dressler[4], S. Heinitz[4], N. Kivel[4], J. M. Quesada[3] D. Schumann[4], O. Aberle[5], V. Alcayne[6], J. Andrzejewski[7], L. Audouin[8], V. Bécares[6], M. Bacak[5,9,10], M. Barbagallo[5,11], F. Bečvář[12], G. Bellia[13,14], E. Berthoumieux[10], J. Billowes[15], D. Bosnar[16], A. Brown[17], M. Busso[11,18,19], M. Caamaño[20], L. Caballero-Ontanaya[2], M. Calviani[5], D. Cano-Ott[6], F. Cerutti[5], Y. H. Chen[8], E. Chiaveri[5,15,3], N. Colonna[11], G. Cortés[1], M. A. Cortés-Giraldo[3], L. Cosentino[13], S. Cristallo[11,18,21], L. A. Damone[11,22], M. Diakaki[23,5], M. Dietz[24], E. Dupont[10], I. Durán[20], Z. Eleme[25], B. Fernández-Domínguez[20], A. Ferrari[5], P. Ferreira[26], P. Finocchiaro[13], V. Furman[27], K. Göbel[28], A. Gawlik[7], S. Gilardoni[5], T. Glodariu[29], I. F. Gonçalves[26], E. González-Romero[6], F. Gunsing[10], J. Heyse[30], D. G. Jenkins[17], F. Käppeler[31], Y. Kadi[5], T. Katabuchi[32], A. Kimura[33], M. Kokkoris[23], Y. Kopatch[27], M. Krtička[12], D. Kurtulgil[28], I. Ladarescu[2], C. Lederer-Woods[24], S. Lo, Meo[34,35], S. J. Lonsdale[24], D. Macina[5], T. Martínez[6], A. Masi[5], C. Massimi[35,36], P. Mastinu[37], M. Mastromarco[5], F. Matteucci[38,39], E. A. Maugeri[4], A. Mazzone[11,40], E. Mendoza[6], A. Mengoni[34], V. Michalopoulou[23], P. M. Milazzo[38], F. Mingrone[5], A. Musumarra[13,14], A. Negret[29], R. Nolte[41], F. Ogállar[42], A. Oprea[29], N. Patronis[25], A. Pavlik[43], J. Perkowski[7], L. Persanti[11,18,21], I. Porras[42], J. Praena[42], D. Radeck[41], D. Ramos[8], T. Rauscher[44,24], R. Reifarth[28], D. Rochman[4], M. Sabaté-Gilarte[5,3], A. Saxena[45], P. Schillebeeckx[30], S. Simone[13], A. G. Smith[15], N. V. Sosnin[15], A. Stamatopoulos[23], G. Tagliente[11], J. L. Tain[2], T. Talip[4], L. Tassan-Got[5,8], A. Tsinganis[5], J. Ulrich[4], S. Valenta[12], G. Vannini[35,36], V. Variale[11], P. Vaz[26], A. Ventura[35], V. Vlachoudis[5], R. Vlastou[23], A. Wallner[46], P. J. Woods[24], T. Wright[15], P. Žugec[16], and U Köster[47]

[1]Universitat Politècnica de Catalunya, Spain
[2]Instituto de Física Corpuscular, CSIC-Universidad de Valencia, Spain
[3]Universidad de Sevilla, Spain
[4]Paul Scherrer Institut (PSI), Villingen, Switzerland
[5]European Organization for Nuclear Research (CERN), Switzerland
[6]Centro de Investigaciones Energéticas Medioambientales y Tecnológicas (CIEMAT), Spain
[7]University of Lodz, Poland
[8]Institut de Physique Nucléaire, CNRS-IN2P3, Univ. Paris-Sud, Université Paris-Saclay, F-91406 Orsay Cedex, France
[9]Technische Universität Wien, Austria
[10]CEA Irfu, Université Paris-Saclay, F-91191 Gif-sur-Yvette, France
[11]Istituto Nazionale di Fisica Nucleare, Sezione di Bari, Italy
[12]Charles University, Prague, Czech Republic
[13]INFN Laboratori Nazionali del Sud, Catania, Italy
[14]Dipartimento di Fisica e Astronomia, Università di Catania, Italy
[15]University of Manchester, United Kingdom
[16]Department of Physics, Faculty of Science, University of Zagreb, Zagreb, Croatia
[17]University of York, United Kingdom
[18]Istituto Nazionale di Fisica Nucleare, Sezione di Perugia, Italy
[19]Dipartimento di Fisica e Geologia, Università di Perugia, Italy

[20]University of Santiago de Compostela, Spain
[21]Istituto Nazionale di Astrofisica - Osservatorio Astronomico di Teramo, Italy
[22]Dipartimento di Fisica, Università degli Studi di Bari, Italy
[23]National Technical University of Athens, Greece
[24]School of Physics and Astronomy, University of Edinburgh, United Kingdom
[25]University of Ioannina, Greece
[26]Instituto Superior Técnico, Lisbon, Portugal
[27]Joint Institute for Nuclear Research (JINR), Dubna, Russia
[28]Goethe University Frankfurt, Germany
[29]Horia Hulubei National Institute of Physics and Nuclear Engineering, Romania
[30]European Commission, Joint Research Centre, Geel, Retieseweg 111, B-2440 Geel, Belgium
[31]Karlsruhe Institute of Technology, Campus North, IKP, 76021 Karlsruhe, Germany
[32]Tokyo Institute of Technology, Japan
[33]Japan Atomic Energy Agency (JAEA), Tokai-mura, Japan
[34]Agenzia nazionale per le nuove tecnologie (ENEA), Bologna, Italy
[35]Istituto Nazionale di Fisica Nucleare, Sezione di Bologna, Italy
[36]Dipartimento di Fisica e Astronomia, Università di Bologna, Italy
[37]Istituto Nazionale di Fisica Nucleare, Sezione di Legnaro, Italy
[38]Istituto Nazionale di Fisica Nucleare, Sezione di Trieste, Italy
[39]Dipartimento di Astronomia, Università di Trieste, Italy
[40]Consiglio Nazionale delle Ricerche, Bari, Italy
[41]Physikalisch-Technische Bundesanstalt (PTB), Bundesallee 100, 38116 Braunschweig, Germany
[42]University of Granada, Spain
[43]University of Vienna, Faculty of Physics, Vienna, Austria
[44]Department of Physics, University of Basel, Switzerland
[45]Bhabha Atomic Research Centre (BARC), India
[46]Australian National University, Canberra, Australia
[47]Institute Laue Langevin (ILL), Grenoble, France

Abstract. The neutron capture cross section of some unstable nuclei is especially relevant for s-process nucleosynthesis studies. This magnitude is crucial to determine the local abundance pattern, which can yield valuable information of the *s*-process stellar environment. In this work we describe the neutron capture (n,γ) measurement on two of these nuclei of interest, ^{204}Tl and ^{171}Tm, from target production to the final measurement, performed successfully at the n_TOF facility at CERN in 2014 and 2015. Preliminary results on the ongoing experimental data analysis will also be shown. These results include the first ever experimental observation of capture resonances for these two nuclei.

Introduction and motivations

The nucleosynthesis of elements heavier than iron in the Universe is mainly produced by a series of neutron capture reactions and beta-decays in the so-called *slow* (s) and *rapid* (r) processes. The main features and basic characteristics of both processes were already well sketched in the seminal papers by Burbidge et al. [1] and Cameron [2]. An up to date review of the *s*-process, including the main stellar sites where it takes place, can be found in Käppeler et al. [3].The *s*-process mechanism operates during core He-burning and shell C-burning in massive stars of $M>8M_{sun}$ (also called *weak s*-process), as well as in H-burning and He-burning layers of Thermally-Pulsing low-mass stars (1-3 M_{sun}) of the Asymptotic Giant Branch (TP-AGBs), in what is known as the *main s*-process. While the weak *s*-process is the main contributor to the abundances of the elements from Fe to Sr, the main *s*-process is the dominant source for elements of A>90.

Any reliable model of the *s*-process nucleosynthesis aims to reproduce faithfully the solar elemental abundances, and for this requires a precise knowledge of the neutron capture cross sections and, in some cases also the beta decay rates, of ideally all nuclei involved. Among all these nuclei, of particular importance are some nuclides which are radioactive, with half-lives from years to Gy. This means that during the *s*-process its decay process competes with the neutron capture; since they effectively split the *s*-process flow, these nuclides are known as *branching points*. The determination of the neutron capture cross section of these nuclei is crucial to determine the elemental abundances around the branching point, and also enables us to give some constraints on several variables of the *s*-process stellar environment, such as temperature, neutron density or pressure. Their radioactive nature, which is what makes them relevant, is also what makes the measurement of the capture cross section especially challenging. In this work we describe the first ever capture measurement on two of these branching points, ^{171}Tm and ^{204}Tl, including the target preparation, the experimental setup, as well as preliminary results for both nuclei. Preliminary results for $^{203}Tl(n,\gamma)$, necessary for the ^{204}Tl data analysis procedure, are also shown.

2. Target preparation

The ^{204}Tl and ^{171}Tm samples were produced by neutron irradiation, at the high neutron flux of the nuclear reactor at ILL. Samples of 263 mg of $^{203}Tl_2O_3$, enriched to 99.5%, and 238 mg of $^{170}Er_2O_3$, enriched to 98.1% were used as primary materials. These samples had been pressed into cylindrical pellets at PSI, and enclosed in quartz ampoules to make them suitable for irradiation at ILL. While the Er irradiated seed could be chemically purified and then electroplated to produce a 22 mm diameter target containing 3.47 mg of ^{171}Tm, the ^{204}Tl had to be left untouched inside the ampoule due to the very high beta activity of the sample, 200 GBq. By the time of the experiment the sample contained 9 mg of ^{204}Tl (4% relative Tl concentration), plus 4 mg (2%) of its daughter ^{204}Pb. An impurity of 370kBq of ^{60}Co was also present in the ^{204}Tl sample. Additionally, since the precise shape and spatial distribution of the $^{203\text{-}204}Tl$ sample inside the ampoule was uncertain after irradiation, a gamma scanning procedure of the sample was performed prior to the capture measurement [4].

3. Capture measurement

Both ^{171}Tm and ^{204}Tl capture measurements were carried out in the Experimental Area 1 (EAR1) of the n_TOF facility at CERN [5], in the fall of 2014 and summer of 2015, respectively. At n_TOF a pulsed neutron beam is produced by spallation of a 20 GeV/c pulsed proton beam from the PS impinging on a massive lead target. The energy of the neutrons is determined with a high resolution via the time-of-flight technique. A set of 4 C_6D_6 liquid scintillation detectors were used to detect the prompt capture γ-rays. These detectors are optimized for a very low sensitivity to the neutrons scattered by the sample. In the case of the ^{204}Tl, the high beta activity, together with the high decay energy of these electrons (763 keV), produced an additional and very intense gamma-ray background due to bremsstrahlung in the quartz ampoule. This made it necessary to shield the detectors with a 2 mm thick lead foil, which also helped to overcome a few other experimental effects, like possible shifts in detector gain observed during the ^{171}Tm campaign (up to 15%) or in the $^{241}Am(n,\gamma)$ measurement [6].

The counting rate TOF (or neutron-energy equivalent) spectra were transformed, after background subtraction, into capture yield by applying the so-called Pulse Height Weighting Technique (PHWT) [7]. The gold saturated resonance method [8] was employed for absolute yield normalization. For the analysis of the ^{204}Tl measurement the special sample details, such as the quartz encapsulation and the uncertain geometry of the activated material involved also the additional measurement of a $^{203}Tl_2O_3$ pure sample, nearly identical to the one irradiated.

4. Preliminary results

The capture yield analysis in the Resolved Resonance Region (RRR) is analysed and parameterized using the Bayesian R-matrix analysis code SAMMY [9]. In the case of ^{171}Tm, it has been possible to analyse the RRR up to 700 eV, which allowed to extract individual resonance parameters for 28 *s*-wave resonances for the first time ever [10].

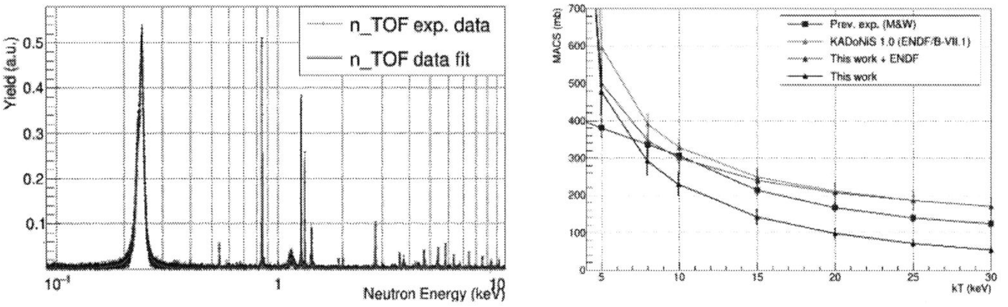

Figure 1. *Left:* Capture Yield analysis for the ^{203}Tl(n,γ) reaction. *Right*: In black, the MACS obtained employing the new resonance data from 30 eV up to 26 keV. This is compared to the MACS calculated with the ENDF evaluation data (green), and the MACS obtained in the experimental measurement by Macklin and Winters (blue).

For the analysis of the $^{203\text{-}204}$Tl sample (n,γ) yield, the first step was to analyze the ^{203}Tl pure sample yield with the highest possible accuracy, in order to reliably account for it in the later analysis of the ^{204}Tl(n,γ). In ^{203}Tl(n,γ) more than 70 resonances, from 37 eV up to 26 keV, have been identified and analysed with SAMMY. This included the first experimental measurement for resonances at a neutron energy lower than 3 keV (fig. 1). These new

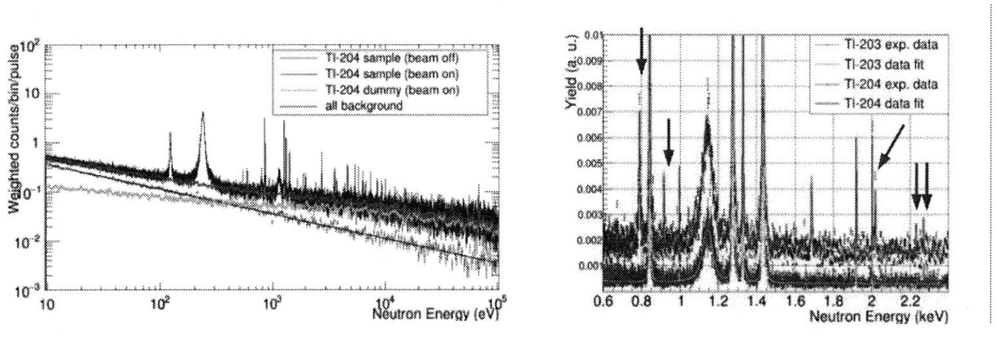

Figure 2. *Left:* Counting rate vs neutron energy spectra for the $^{203\text{-}204}$Tl(n,γ) measurement, showing the different background contributions. *Right*: Comparison of the ^{203}Tl pure sample and $^{203\text{-}204}$Tl sample capture yields, with arrows indicating the ^{204}Tl(n,γ) resonances identified up to now.

resonances have a significant impact in the Maxwellian Averaged Cross Section (MACS) at 5-8 keV when compared to the commonly accepted MACS in KADoNiS v0.3 [11], which is based on the previous measurement [12]. Finally, the analysis of the ^{204}Tl sample is ongoing; at the present time, six resonances of ^{204}Tl in the energy range from 122 eV up to 2.2 keV have been identified for the first time (fig. 2, right). A few additional candidates are under evaluation.

The authors acknowledge financial support by the Spanish FPA2014-52823-C2-2-P project, by the EC Marie Curie Action "NeutAndalus" (FP7-PEOPLE-2012-CIG- 334315), by the ARGOS scholarship of the Spanish Nuclear Safety Council (CSN) and the Universitat Politècnica de Catalunya, and by the University of Sevilla via the VI PPIT-US program.

References

[1] E. Burbidge, G. Burbidge, W. Fowler, F. Hoyle, Rev. Mod. Phys. **29**, 547 (1957)

[2] A. Cameron, A.E.C.L. Chalk River, Canada, Technical Report No. CRL–41 (1957)

[3] F. Käppeler, R. Gallino, S. Bisterzo, and W. Aoki, Rev. Mod. Phys. **83**, 157 (2011)

[4] A. Tarifeño-Saldivia, *Characterization of the spatial distribution of high radioactive targets for capture cross section measurements of s-process branching nuclei at CERN n_TOF*, CHANDA workshop, (XI-2015)

[5] C. Guerrero, et al., Eur. Phys. J. A **49**, 27 (2013)

[6] K. Fraval, F. Gunsing et al., Phys. Rev. C **89**, 044609 (2014)

[7] U. Abbondanno, et al., Nucl. Instrum. Meth. A **521**, 454–467 (2004)

[8] R. Macklin, J. Halperin and R. Winters, Nucl. Instrum. Meth. A **164**, 213-214 (1979)

[9] N. M. Larson, *Updated Users' Guide for SAMMY: Multilevel R-Matrix Fits to Neutron Data Using Bayes' Equations*, ORNL/TM-9179/R8 ENDF-364/R2 (2008)

[10] J. Lerendegui-Marco et al., *Proceedings of the 6th Workshop on Nuclear Fission and Spectroscopy of Neutron Rich Nuclei* (2017, to be published)

[11]I. Dillmann, R. Plag,F.Käppeler, T.Rauscher, *Proceeding of the workshop "EFNUDAT Fast Neutrons - scientific workshop on neutron measurements, theory & applications"* (2009)

[12] R. Macklin and R. Winters, Astrophys. J. **208**, 812 (1976)

Neutron capture cross section of ^{185}Re leading to ground and isomer states of ^{186}Re in the keV-neutron energy region

T. Katabuchi[1,*], K. Takebe[1], S. Umezawa[1], R. Fujioka[1], T. Saito[1], and M. Igashira[1]

[1]Research Laboratory for Nuclear Reactors, Tokyo Institute of Technology, 2-12-1 Ookayama, Meguro-ku, Tokyo, 152-8550, Japan

Abstract. The neutron capture cross section of ^{185}Re was measured in the astrophysically important energy region. Measurements were made using a neutron beam from a ^{7}Li$(p,n)^7$Be neutron source with energies ranging from 3 to 90 keV. Two different experimental techniques, time-of-flight (TOF) and activation methods, were employed. In the TOF experiments, the total neutron capture cross section of ^{185}Re was determined by the pulse-height weighting technique. In the activation method, the partial capture cross section leading to the ground state of ^{186}Re was measured by detecting decay γ-rays from neutron activated samples. The present cross section values were compared with evaluated cross section data and previous measurements. The difference between the TOF and activation results was smaller than experimental uncertainties. This suggests that the production cross section of isomer states of ^{186}Re is very small.

1 Introduction

A Re/Os nucleo-cosmochronometer has been proposed to date a rapid neutron-capture process (r-process) in nucleosynthesis [1]. Rhenium-187 is primarily considered a pure r-process nuclide and has a long half-life of 43.5 Gyr. Thus, the ^{187}Re/^{187}Os abundance ratio changing with the ^{187}Re half-life can be a good chronometer for the r-process. However, slow neutron-capture process (s-process) through an isomer state of ^{186}Re ($T_{1/2}$ = 0.2 Myr) may contaminate the ^{187}Re/^{187}Os abundance ratio [2]. It is necessary to evaluate a contribution from ^{186m}Re created from the ^{185}Re$(n,\gamma)^{186m}$Re reaction.

In this work, we experimentally determined the total capture cross section of ^{185}Re by the time-of-flight (TOF) method, and the partial capture cross section leading to the ground state of ^{186}Re by the activation method. The production cross section of the isomer state of ^{186}Re is discussed from a difference between the two different technique measurements.

2 Experiments and Data Analysis

Experiments were performed at the Laboratory for Advanced Nuclear Energy at the Tokyo Institute of Technology. Incident neutrons were generated through the ^{7}Li$(p,n)^7$Be reaction by a pulsed proton beam from a Pelletron accelerator bombarding a lithium target. The incident neutron energy distributed from a few keV to 90 keV, determined by the TOF method.

For the TOF experiments to determine the total capture cross section of ^{185}Re, an isotopically enriched ^{185}Re sample (enrichment: 97.7%) in a physical form of metal powder was used. The sample weight was 0.2967 g. The sample powder was encased in a graphite container with an inner diameter of 10 mm. A gold sample (10 mm diam. and 2.15 g) was used as standard for cross section measurements. The flight length from the neutron source to the sample was 12 cm. Capture γ-rays from the sample were detected with an anti-Compton NaI(Tl) spectrometer in the TOF experiments. The details of the experimental method and the data analysis procedure can be found elsewhere[3]. The (n,γ) cross sections were obtained from the pulse height spectra by the pulse-height weighting technique[4]. The absolute cross sections were determined from the ratio of ^{185}Re to ^{197}Au yields and the JENDL-4.0 evaluated ^{197}Au(n,γ)^{198}Au cross section. Corrections for neutron self shielding, multiple scattering and impurities in the sample were made. The neutron capture cross section was derived in four energy regions: 15-25, 25-35, 35-55, and 55-90 keV.

In the activation experiments, a natural rhenium sample (^{185}Re: 37.4%, ^{187}Re: 62.6%) was used. Rhenium metal powder was pressed to form a pellet with a diameter of 10 mm. The weight of the sample was 0.2848 g. The rhenium sample was sandwiched between two gold samples with the same diameter and placed at a distance of 2 cm from the ^{7}Li$(p,n)^7$Be neutron source for irradiation. The gold samples were standard for determining the neutron capture cross section of ^{185}Re. After irradiation, γ-rays from the activated rhenium and gold samples were measured with a HP Ge detector that was well shielded from natural background γ-rays. Decay γ-rays of 136 keV (^{186}Re) and 412 keV (^{198}Au) were clearly observed but transitions from isomer states of ^{186}Re did not appear in the spectra due to much longer half-life than its ground state. The partial neutron capture cross section leading to

*e-mail: buchi@lane.iir.titech.ac.jp

the ground state of ^{186}Re were determined from measured activities.

3 Results

The results are plotted in Fig. 1. The present experimental results were compared with evaluated cross section [5] and previous experimental data [6]-[11]. For activation experimental results, the averaged cross section over the incident neutron energy distributing from 3 keV to 90 keV is plotted at 42 keV, the average incident neutron energy. To compare the TOF data with the activation average cross section, the averaged cross section was calculated from the present TOF data. The averaged cross section calculated from the TOF results was 1.22 ± 0.05 b while the activation experiments results in 1.29 ± 0.11 b. The difference between the two values is smaller than experimental uncertainties. This suggests that the isomer production cross section is smaller than the uncertainties.

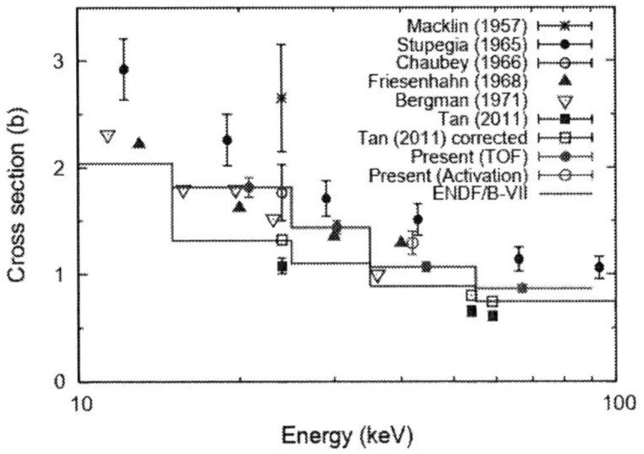

Figure 1. Neutron capture cross section of ^{185}Re.

4 Summary

We measured the neutron capture cross section of ^{185}Re in the keV energy region. To discuss the production cross section of astrophysically important ^{186}Re isomers, two different experimental techniques, time-of-flight and activation methods, were employed. In the TOF experiment, the total capture cross section of ^{185}Re was measured. On the other hand, using the activation method, the partial neutron capture cross section leading the ground state of ^{186}Re was determined. The difference between the two experimental cross sections was smaller than the experimental uncertainties. We conclude that the production cross section of ^{186}Re isomers is very small. In future work, we can reduce the experimental uncertainties, for example, uncertainty of the detection efficiency of the HP Ge detector in the activation measurement that can be improved with better standard γ-ray sources. Then, we can discuss more on astrophysical impact of our experimental results.

References

[1] D. D. Clayton, Nature, **224**, 56 (1969).
[2] T. Hayakawa, Astrophys. J., **628**, 533 (2005).
[3] S. Mizuno et al., J. Nucl. Sci. Technol., **36** 493 (1999).
[4] R. L. Macklin et al., Nucl Instrum Meth. Phys. Res., **91** 565 (1971).
[5] ENDF/B-VII.1 data file for ^{185}Re (MAT = 7525), evaluated by I. J. Thompson and N. C. Summers, (2011).
[6] R. L. Macklin et al., Phys. Rev., **107** 504 (1957).
[7] D. C. Stupegia et al., J. Nucl. Energ. Parts A/B., **19** 767 (1965).
[8] A. K. Chaubey and M. L. Sehgal, Phys. Rev., **22** 191 (1968).
[9] S. J. Friesenhahn, J. Nucl. Energ., **22** 191 (1968).
[10] A. A. Bergman et al., Neutron Phys. Conf., **1** 144 (1971).
[11] V. H. Tan et al., J. Korean Phys. Soc., **59** 1757 (2011).

Study of photon strength functions via $(\vec{\gamma}, \gamma', \gamma'')$ reactions at the γ^3-setup

Johann Isaak[1,*], *Deniz* Savran[2], *Tobias* Beck[3], *Udo* Gayer[3], *Krishichayan*[4], *Bastian* Löher[3], *Norbert* Pietralla[3], *Marcus* Scheck[5], *Werner* Tornow[4], *Volker* Werner[3], and *Andreas* Zilges[6]

[1]Research Center for Nuclear Physics, Osaka University, Japan
[2]GSI Helmholtzzentrum für Schwerionenforschung, Darmstadt, Germany
[3]Institut für Kernphysik, TU Darmstadt, Germany
[4]Department of Physics, Duke University and Triangle Universities Nuclear Laboratory
[5]School of Engineering, University of the West of Scotland
[6]Institut für Kernphysik, Universität zu Köln

Abstract. One of the basic ingredients for the modelling of the nucleosynthesis of heavy elements are so-called photon strength functions and the assumption of the Brink-Axel hypothesis. This hypothesis has been studied for many years by numerous experiments using different and complementary reactions. The present manuscript aims to introduce a model-independent approach to study photon strength functions via γ-γ coincidence spectroscopy of photo-excited states in ^{128}Te. The experimental results provide evidence that the photon strength function extracted from photoabsorption cross sections is not in an overall agreement with the one determined from direct transitions to low-lying excited states.

1 Introduction

The photon strength function (PSF) serves as an essential input for nuclear astrophysical model calculations. It plays an important role in capture and photo-disintegration reactions as well as in astrophysical scenarios describing the nucleosynthesis. In the past, different experimental methods and approaches have been used to study the PSF (see, e.g., Refs. [1–5] and references therein). However, many of these methods are model dependent either in the reaction mechanism itself or in the data analysis. In this contribution, we present a model-independent approach, exemplarily for ^{128}Te, to extract the PSF in real-photon scattering experiments using quasi-monochromatic photon beams provided by the High Intensity γ-ray Source (HIγS) [6] at Duke University, Durham, NC, USA.

2 Methods

In the following, two independent methods are introduced to determine the PSF for the excitation as well as for the decay channel in a single experiment exploiting the monochromatic character of the photon beam provided by the HIγS facility.

*e-mail: jisaak@rcnp.osaka-u.ac.jp

The photoabsorption cross section σ_γ ($= \sigma_{\gamma\gamma} + \sigma_{\gamma\gamma'}$) is linked to the PSF built on the ground state by $f(E_\gamma) \propto \sigma_\gamma / E_\gamma$ assuming predominantly dipole transitions. The procedure to determine σ_γ as a function of the excitation energy at HIγS was discussed in detail in previous works, such as [5, 7–9]. The idea is sketched in Fig. 1.a). After photoexcitation of the nucleus in a given enery range defined by the energy and the width of the quasi-monochromatic photon beam, σ_γ is reconstructed by measuring all ground-state transitions ($\sigma_{\gamma\gamma}$) and all events cascading via intermediate levels ($\sigma_{\gamma\gamma'}$), where $\sigma_{\gamma\gamma'}$ can be approximated by the intensity observed for the $2^+_1 \rightarrow 0^+_1$ transition [7].

The second approach is illustrated in Fig. 1.b) and c) and was firstly introduced in Ref. [2] using proton-γ-γ correlations in ^{94}Mo$(d, p)^{95}$Mo reactions. Due to the novel γ-γ coicidence setup γ^3 [10] it is feasible for the first time to apply this method in $(\vec{\gamma}, \gamma'\gamma'')$ reactions, here shown for the example of ^{128}Te. Primary transitions from excited states at E_{x_1} to low-lying excited levels yield information on the PSF at the corresponding transition energies: $f(E_\gamma = E_{x_1} - E_{2^+_1}) \propto I_{x_1 \rightarrow 2^+_1}$, with $I_{x_1 \rightarrow 2^+_1}$ being the associated transition intensity. The observation of several direct transitions to low-lying states per excitation energy and beam-energy setting, respectively, allows to reconstruct the PSF over a broad γ-ray energy range, which is schematically shown in Fig. 1.c) for a hypothetical PSF. These two outlined approaches allow to independently study the PSF in the excitation and the decay channel, respectively.

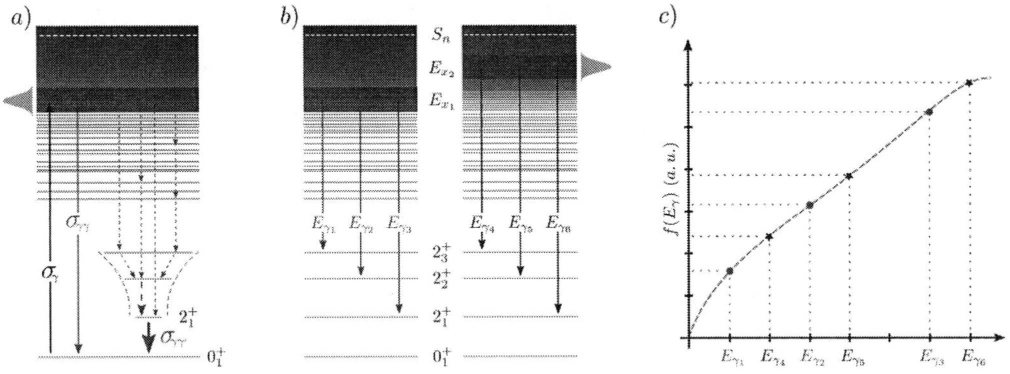

Figure 1. For details see text.

3 Experiment & Results

For the present work a metallic and highly-enriched (99.8 %) ^{128}Te target was used to perform photon-scattering experiments with quasi-monochromatic γ-ray beams with energies between 3 MeV and 9 MeV in steps of about 250 keV. The spectral distribution of the beam is usually about FWHM $\approx$ 3 % of the beam energy.

Typical γ-ray spectra for the measurement of primary transitions to low-lying excited states after photo-excitation via an 8 MeV γ-ray beam are shown in Fig. 2.a). The blue spectrum is obtained from a gate on the energy of the $2^+_1 \rightarrow 0^+_1$ transition (E_γ = 743 keV). The peak at ~ 7.26 MeV corresponds to the full-energy events of the direct population of the 2^+_1 level from excited states at 8 MeV. Primary transitions to other levels, such as 2^+_3 (green spectrum) and 2^+_4 (red spectrum) are determined in a similar fashion. The individual transition intensities can be converted into values of the PSF at the corresponding γ-ray energy.

All measured transition intensities for beam energies from 5.8 MeV to 8.5 MeV are shown in Fig. 2.b) as a function of the γ-ray energy (black filled squares). For the measurements with beam energies above 6.4 MeV decays into up to the 2_8^+ state are observed. Due to the steps of ~ 250 keV between two measurements it is possible to obtain data points at the same or similar γ-ray energy from different beam-energy settings. The fluctuations of the data points exhibit a factor of about 2-3, which is larger than expected from Porter-Thomas fluctuations of around 5 % to 15 %. This is one indication that the average decay properties of photo-excited states in ^{128}Te below the neutron separation threshold (S_n = 8.78 MeV) cannot be described by a single excitation-energy independent PSF.

Nevertheless, the current data set is used to compute a moving average weighted by a Gaussian distribution with FWHM = 300 keV (grey shaded band). That averaged PSF is compared to the PSF extracted from photoabsorption cross sections (blue filled squares) shown in Fig. 2.c). A deviation of both PSFs as a function of the γ-ray energy is observed. This observation additionally indicates that the PSF built on the ground state (photo-excitation) differs from the PSF built on excited states (photo-deexcitation) for the present case of ^{128}Te, which is in contradiction to the Brink-Axel hypothesis [11, 12]. However, additional systematic studies applying the outlined approach and comparison to data from complementary experiments are crucial before general conclusions on the Brink-Axel hypothesis can be drawn.

This work was supported by EMMI (HA216/EMMI), the DFG under Grant Nos. SFB 634 and ZI 510/7-1, and the U.S. DoE under Grant No. DE-FG02-97ER41033.

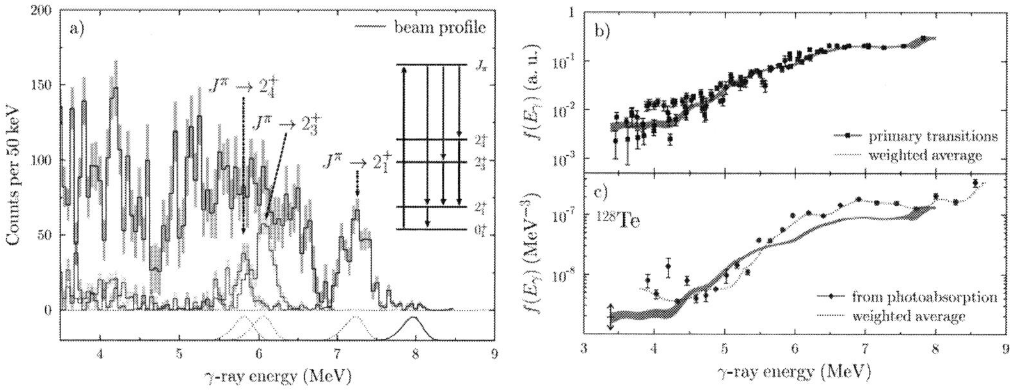

Figure 2. For details see text.

References

[1] J. Kopecky and M. Uhl, Phys. Rev. C **41**, 1941 (1990).
[2] M. Wiedeking et al., Phys. Rev. Lett. **108**, 162503 (2012).
[3] M. Guttormsen et al., Phys. Rev. Lett. **116**, 012502 (2016).
[4] R. Schwengner et al., Phys. Rev. C **78**, 064314 (2008).
[5] J. Isaak et al., Phys. Lett. B **727**, 361 (2013).
[6] H. R. Weller et al., Prog. Part. Nucl. Phys. **62**, 257 (2009).
[7] A. P. Tonchev et al., Phys. Rev. Lett. **104**, 072501 (2010).
[8] C. Romig et al., Phys. Rev. C **88**, 044331 (2013).
[9] B. Löher et al., Phys. Lett. B **756**, 72 (2016).
[10] B. Löher et al., Nucl. Instr. and Meth. A **723**, 136 (2013).
[11] D. M. Brink, Ph.D. thesis, Oxford University (1955).
[12] P. Axel, Phys. Rev. **126**, 671 (1962).

Cross-section measurements of the ^{94}Mo(γ,n) and ^{90}Zr(γ,n) reactions using real photons at the HIγS facility

Adriana Banu[1,*], *Jack Silano*[2,3], *Hugon Karwowski*[2,3], *Evan Meekins*[1], *Megha Bhike*[2,4], *Werner Tornow*[2,4], and *Mathew McCleskey*[5]

[1]Department of Physics and Astronomy, James Madison University, Harrisonburg, Virginia 22807, USA
[2]Triangle Universities Nuclear Laboratory, Durham, North Carolina 27708, USA
[3]University of North Carolina at Chapel Hill, North Carolina 27516, USA
[4]Department of Physics, Duke University, Durham, North Carolina 27708, USA
[5]University of Maryland Medical Center, Baltimore, Maryland 21201, USA

Abstract. The photodisintegration reaction cross-sections for ^{94}Mo(γ,n) and ^{90}Zr(γ,n) have been experimentally investigated with quasi-monochromatic photon beams at the High Intensity γ-Ray Source (HIγS) facility, Triangle University Nuclear Laboratory (TUNL). The measurements were focused primarily on studying the energy dependence of the photoneutron cross sections, which is the most direct way of testing statistical models, and were performed close to the respective neutron thresholds and above up to ~ 20 MeV. Neutrons from the (γ,n) reactions were detected using a 4π assembly of ^{3}He proportional counters developed at Los Alamos National Laboratory and presently available at TUNL. While the ^{94}Mo(γ,n) cross section measurement aims to contribute to a broader investigation for understanding the γ-process (the mechanism responsible for the nucleosynthesis of the so-called *p*-nuclei), the information from the ^{90}Zr(γ,n) data is relevant to constrain QRPA calculations of γ-ray strength functions in this mass region. In this contribution, we will present our preliminary results of the total (γ,n) excitation functions for the two photoneutron reactions on ^{94}Mo and ^{90}Zr.

1 Introduction

How nuclear reactions that occur in stars and stellar explosions have been forging the elements out of hydrogen and helium leftover from the Big Bang is a longstanding, still timely research topic in nuclear astrophysics [1]. Although there is fairly complete understanding of the production of elements up to iron by nuclear fusion reactions in stars, important details concerning the production of the elements beyond iron remain puzzling.

Current understanding is that the nucleosynthesis beyond iron proceeds mainly *via* neutron capture reactions and subsequent β^- decays in the *s*-process and *r*-process [2]. But some 35 *proton-rich stable* isotopes, between ^{74}Se and ^{196}Hg cannot be synthesized by neutron-capture processes since they are located on the neutron-deficient side of the valley of β-stability. These proton-rich nuclides are generally referred to as *p*-nuclei [3]. The mechanism responsible for their synthesis is termed the *p*-process. The astrophysical details of the *p*-process are still under discussion. So far it has been impossible to reproduce the solar abundances of all *p*-isotopes using a single process. Several different sites and (independently operating) processes seem to be required. However, the largest fraction of *p*-isotopes is created by sequences of photodisintegrations and β^+ decays. Because of the dominance of photodisintegrations, this mechanism of the *p*-process is referred to as the γ-process [4]. It is generally accepted that the γ-process occurs mainly in explosive O/Ne burning during supernova Type II explosions at temperatures in the range of $T \approx$ 2-3 GK, but supernovae Type Ia and Ib/c have also been considered [5]. Calculations based on the γ-process concept can produce the bulk of the *p*-nuclei within a factor of ≈ 3 [6]. However, the most abundant *p*-isotopes, ^{92,94}Mo and ^{96,98}Ru, as well as the whole region A < 124 are underproduced, making their nucleosynthesis one of the great outstanding mysteries in nuclear astrophysics. It is not yet clear whether the observed underproductions are due to deficiencies in the astrophysical models or in the nuclear physics input, *i.e.* the reaction rates used in the model.

Contrary to the *s*- or *r*-process, the concepts of steady flows or reaction rate equilibria cannot be applied to the *p*-process, which operates far from equilibrium. As a result, an extended network of some 20000 reactions linking about 2000 nuclei in the A ≤ 210 mass region must be computed in detail. It is impossible to measure all these reaction rates in the laboratory. Hence, it becomes obvious that the vast majority of reaction rates

* Corresponding author: banula@jmu.edu

must be determined theoretically. Usually the missing reaction rates are calculated within the framework of Hauser-Feshbach (HF) statistical model [5] with typical uncertainties of about 20-30% [6]. The model requires input based on nuclear structure physics, optical model potentials, and nuclear level densities to calculate transmission coefficients (average widths) which, in turn, determine the reaction cross sections, and thus, the reaction rates. The uncertainties involved in any HF cross section calculation are not related to the model of formation and de-excitation of the compound nucleus itself, but rather to the evaluation of the nuclear quantities necessary for the calculation of the transmission coefficients. The photon transmission coefficient is particularly relevant in case of photonuclear reactions and is calculated assuming the dominance of dipole *E1* transitions. The transmission coefficient for γ emission with multipolarity *L* is related to the (downward) strength function *f* as follows:

$$T_\gamma^L = 2\pi E_\gamma^{2L+1} f(E_\gamma) \quad (1)$$

Much effort has been and it is still devoted to measuring and understanding the electric dipole strength function that exhibits a pronounce peak at the Giant Dipole Resonance (GDR) energy. There are many approaches to derive *f*, each leading to an energy dependence of the E1 transmission given by a Lorentzian, around the GDR energy. Experimental photoabsorptian data confirm the simple semi-classical prediction of a Lorentzian shape at energies around the resonance energy but this description is less satisfactory at lower energies, and especially near the reaction threshold [5]. Therefore, it is of substantial interest to develop models of the microscopic type which are expected to provide a reasonable reliability and predictive power for the *E1*-strength function. Attempts in this direction, such as QRPA calculations [7,8], have been applied successfully to several photoneutron cross section measurements carried out recently with quasi-monochromatic laser-Compton scattering γ rays [9-12].

Despite the endemic problem of reproducing the solar abundances of ^{92,94}Mo and ^{96,98}Ru, as well as the whole region $A < 124$, most recently a study performed by Travaglio *et al.* [13] in supernova Type Ia using both deflagration and detonation models achieved a remarkable breakthrough. For the first time, a stellar source has been shown to produce the solar abundances of both light and heavy *p*-nuclei almost at the same level as ^{56}Fe, including the so much debated isotopes ^{92}Mo and ^{96,98}Ru. And yet the isotope ^{94}Mo is obtained with a much lower abundance with respect to all the other light *p*-nuclei. Another remarkable finding of [13] points out that the γ-process can make important contributions to the production of the neutron magic ^{90}Zr, previously known as genuine *s*-process nuclide.

In light of the intriguing findings of [13], we were motivated to investigate ^{94}Mo and ^{90}Zr, and in this contribution, we will present our preliminary results of the total (γ,n) excitation functions for the two photoneutron reactions.

2 Experiment

The measurements of the ^{94}Mo(γ,n)^{93}Mo were performed at the γ-ray energy range from 9 MeV to ~20 MeV, in steps of 50-100 keV. In the case of ^{94}Mo(γ,n)^{93}Mo reaction, the unstable product ^{93}Mo besides having a long half-life of $T_{1/2} = 3500$ y decays by electron capture without γ-ray emission. The only way to study ^{94}Mo(γ,n)^{93}Mo experimentally is by direct neutron counting. Neutrons from the (γ,n) reaction were detected using an assembly of 4π ^{3}He proportional counters [14], developed at Los Alamos National Laboratory and presently available at HIγS. The neutron array is readout by two TTL electronic output signals corresponding to the detected events in the inner and outer rings of the array that are recorded on a scaler in the local VM-USB based data acquisition system. Thus, neutron multiplicity of each neutron counter was not possible. The ^{94}Mo target was placed in the center of the ^{3}He counter array. A deuterated benzene target located downstream from the ^{94}Mo target and two liquid scintillator neutron detectors, each at 90° on either side of the target, were used to determine the γ-ray beam intensity event by event via the d(γ,n)p reaction.

The integrated photon flux was determined to high precision based on well-known total cross sections [15,16] and angular distributions [17] of the neutrons produced by deuteron photodisintegration. This procedure was cross-calibrated at a few γ-ray energies using the activation of a ^{197}Au target. The advantages of using ^{197}Au(γ,n) as a monitor reaction are threefold: (i) the cross section has smooth energy dependence even near the threshold region due to the high-level density of the resultant ^{196}Au reaction product; (ii) this reaction has one of the highest photoneutron cross sections; (iii) the half-life and decay energy of the reaction product nucleus are very convenient for off-line measurements. After the irradiation a low-background 60% HPGe detector, combined with a Canberra Multiport II multichannel analyzer, was used to count off-line the activity of the gold foil monitor. At each photon beam energy, the beam distribution and intensity was measured separately, at low flux, with a 123% HPGe detector positioned on/off along the beam axis. For this purpose precisely machined copper attenuators were remotely inserted into the beam.

Typical photon beam intensities were ~10^7 – 10^8 γ/s while a typical beam energy spread was ~4-5% (FWHM). The target thickness and enrichment were ~600 mg/cm^2, 98.97% for ^{94}Mo and ~1g/cm^2, 97.7% for ^{90}Zr.

3 Preliminary Results

In Figure 1 we present preliminary results for the measured excitation function in the case of the ^{94}Mo(γ,n) reaction.

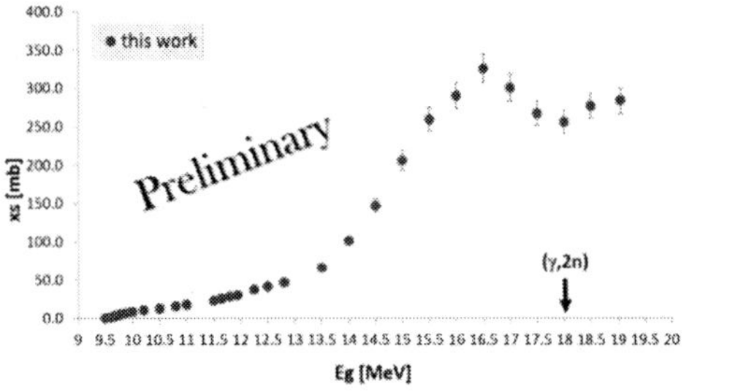

Fig. 1. Excitation function measured for the ^{94}Mo(γ,n) reaction.

In Figure 2 we present a comparison of our work with a recent measurement published by H. Utsunomiya *et al.* [10].

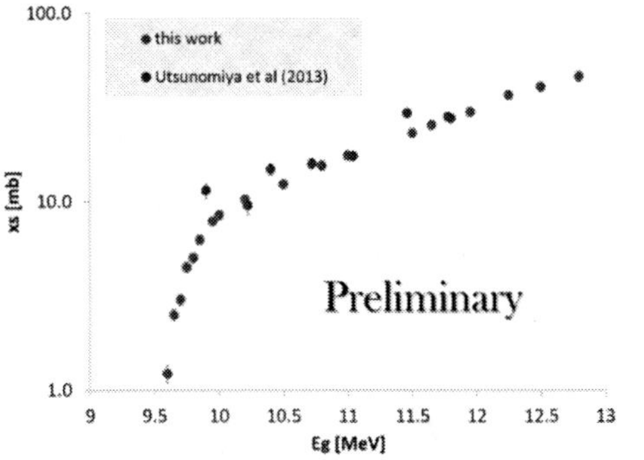

Fig. 2. Excitation function measured for the ^{94}Mo(γ,n) reaction (this work) illustrated in comparison with a recent work [10]; data is presented in logarithmic scale.

In Figure 3 we present preliminary results for the measured excitation function in the case of the ^{90}Zr(γ,n), shown in comparison with past work.

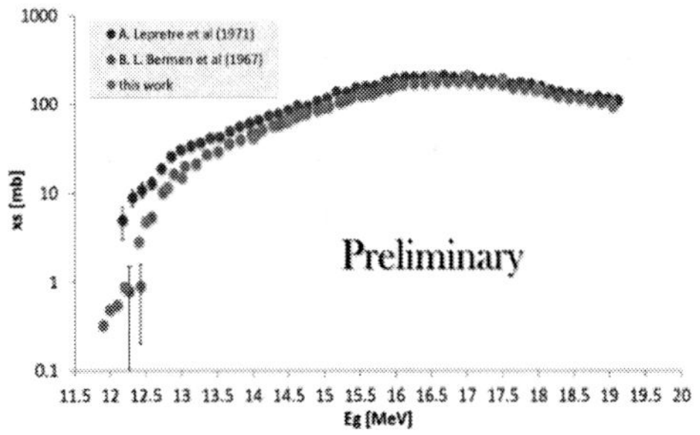

Fig. 3. Excitation function measured for the ^{90}Zr(γ,n) reaction (this work) illustrated in comparison with past work; data is presented in logarithmic scale.

This research was funded by the Research Corporation for Science Advancement – the Cottrell College Science Award.

References

1. The DOE/NSF Nuclear Science Advisory Committee, *"The Frontiers of Nuclear Science, A Long Range Plan"* Washington, December 2007; http://www.sc.doe.gov/np/nsac/
2. M. E. M. Burbidge, G. R. Burbidge, W. A. Fowler, F. Hoyle, Rev. Mod. Phys. **29**, 547 (1957)
3. David L. Lambert, The Astron. Astrophys. Rev. **3**, 201 (1992)
4. S. E. Woosley and W. M. Howard, The Astrophys. J. Suppl. S. **36**, 285 (1978)
5. M. Arnould and S. Goriely, Phys. Rep. **384**, 1 (2003)
6. T. Rauscher and F.-K. Thielemann, At. Data Nucl. Data Tables **75**, 1 (2000)
7. S. Goriely and E. Khan, Nucl. Phys. **A 706**, 217 (2002)
8. N. Tsoneva, S. Goriely, H. Lenske, and R. Schwengner, Phys. Rev. **C 91**, 044318 (2015)
9. H. Utsunomiya *et al.*, Phys. Rev. **C 80**, 055806 (2009).
10. H. Utsunomiya *et al.*, Phys. Rev. **C 88**, 015805 (2013)
11. D. M. Filipescu *et al.*, Phys. Rev. **C 90**, 064616 (2014)
12. H.-T. Nyhus *et al.*, Phys. Rev. **C 91**, 015808 (2015).
13. C. Travaglio, F. K. Röpke, R. Gallino, and W. Hillebrandt, Astroph. J. **739**, 93 (2011)
14. C. W. Arnold, T. B. Clegg, H. J. Karwowski, G. C. Rich, J. R. Tompkins, and C. R. Howell, Nucl. Instr. Meth. Phys. Res. **A 647**, 55 (2011)
15. R. Bernabei *et al.*, Phys. Rev. Lett. **57**, 1542 (1986)
16. Y. Birenbaum, S. Kahane, and R. Moreh, Phys. Rev. **C 32**, 1825 (1985)
17. M.A. Blackston, *"Precision Measurements of Deuteron Photodisintegration Using Linearly Polarized Photons of 14 and 16 MeV"*, Ph.D. Thesis, Duke University, 2007

Systematics of the electric dipole response in stable tin isotopes

Sergej Bassauer[1,*], *Peter* von Neumann-Cosel[1], and *Atsushi* Tamii[2]

[1] *Institut für Kernphysik, Technische Universität Darmstadt, 64289 Darmstadt, Germany*
[2] *Research Center for Nuclear Physics, Osaka University, Ibaraki, Osaka 567-0047, Japan*

Abstract. The electric dipole is an important property of heavy nuclei. Precise information on the electric dipole response provides information on the electric dipole polarisability which in turn allows to extract important constraints on neutron-skin thickness in heavy nuclei and parameters of the symmetry energy. The tin isotope chain is particularly suited for a systematic study of the dependence of the electric dipole response on neutron excess as it provides a wide mass range of accessible isotopes with little change of the underlying structure. Recently an inelastic proton scattering experiment under forward angles including 0° on ^{112,116,124}Sn was performed at the Research Centre for Nuclear Physics (RCNP), Japan with a focus on the low-energy dipole strength and the polarisability. First results are presented here. Using data from an earlier proton scattering experiment on ^{120}Sn the gamma strength function and level density are determined for this nucleus.

1 Introduction

The electric dipole response in heavy nuclei is dominated by the giant dipole resonance (GDR), a collective mode above the particle emission threshold usually described as an oscillation of protons against neutrons. Around the neutron threshold the electric dipole response shows itself as an oscillation of a neutron skin against an isospin saturated core. This mode is called pigmy dipole resonance (PDR). The GDR can be investigated using (γ, xn) experiments, whereas the investigation of the PDR is more complicated. Due to its occurrence around the neutron threshold, it can be accessed by (γ, γ') experiments below the neutron threshold and by (γ, xn) experiments above the neutron threshold. However, inelastic proton scattering experiments under 0° allow to investigate the GDR and PDR in a single experiment.

The electric dipole response provides information on the electric dipole polarisability which in turn allows to extract important constraints on the neutron-skin thickness [1–5] in heavy nuclei and parameters of the symmetry energy [2, 6–8]. Another important property of nuclei which can be extracted from the electric dipole response is the gamma strength function (GSF). GSFs describe the average gamma decay behaviour of a nucleus as a function of gamma energy. They serve as input in calculations of cross sections in astrophysics [9], reactor design [10] and waste transmutation [11] within statistical models. Using the so called fluctuation analysis [12] it is also possible to determine level densities in the GDR region [13].

*e-mail: sbassauer@ikp.tu-darmstadt.de

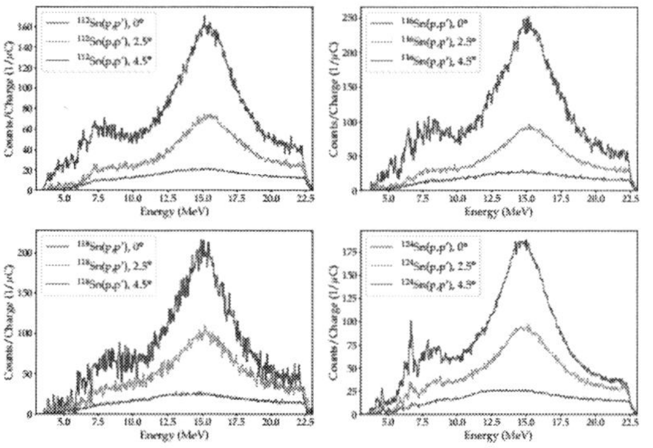

Figure 1. (Colour online) Inelastic proton scattering spectra under laboratory scattering angles of 0°, 2.5° and 4.5° normalised on the collected charge for ^{112,116,118,124}Sn.

2 Experiment and preliminary results

The proton scattering experiment was performed at the Research Centre for Nuclear Physics (RCNP), Japan. In this experiment unpolorised protons with a beam energy of 295 MeV and a beam current of up to 6 nA were used. An energy resolution of about 40 keV could be achieved. Data was taken at laboratory scattering angles of 0°, 2.5° and 4.5°. The main targets were ^{112}Sn, ^{116}Sn and ^{124}Sn. Additionally some data on ^{118}Sn could be taken. Details on the experimental set-up and data analysis can be found in [14].

In Fig. 1, inelastic proton scattering data for 0°, 2.5° and 4.5° normalised on the collected charge for ^{112,116,118,124}Sn can be seen. With increasing angle the typical decrease in the cross section due to the dominant

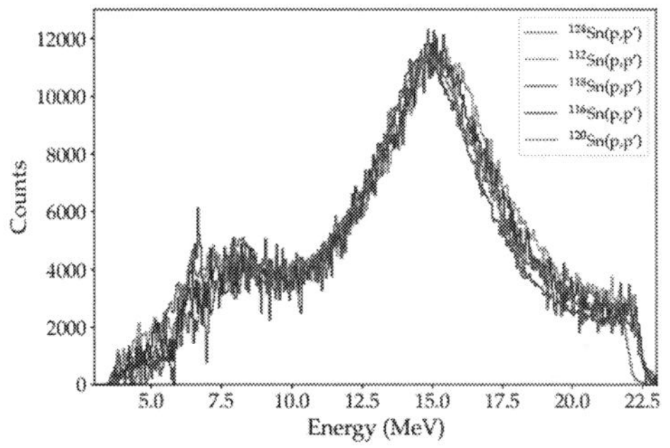

Figure 2. (Colour online) Comparison of the ^{112,116,118,120,124}Sn spectra (normalised on ^{124}Sn). The data for ^{120}Sn stems from [15, 16].

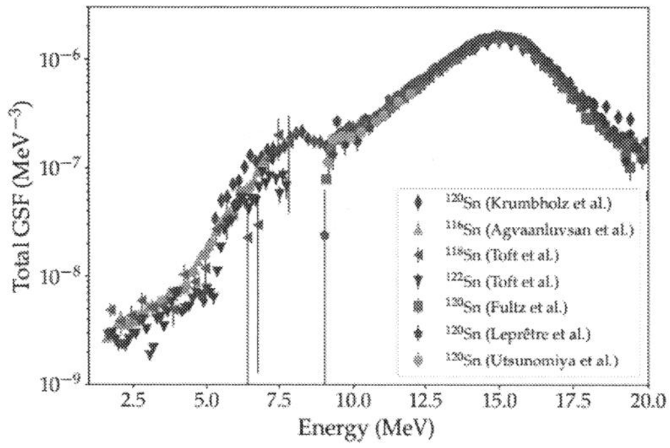

Figure 3. (Colour online) Total gamma strength function of ^{120}Sn in comparison to gamma strength functions from (γ, xn), (^{3}He,^{3} He$'\gamma$) and (^{3}He, $\alpha\gamma$) experiments [19–23].

Coulomb interaction under small angles can be observed. In all isotopes the GDR can be clearly identified around 15 MeV. Around the neutron threshold a bump can be seen. In ^{124}Sn strong structure is apparent in the low energy region.

Fig. 2 shows a comparison between the measured isotopes and ^{120}Sn from an earlier experiment [15]. For comparison all data was normalised on ^{124}Sn. As expected the centroid energy of the GDR is the lowest for ^{124}Sn. It is shifted to higher energies for lighter tin isotopes and is highest for ^{112}Sn. All spectra show a bump around 8 MeV and a structure around 6.5 MeV, most pronounced for ^{124}Sn.

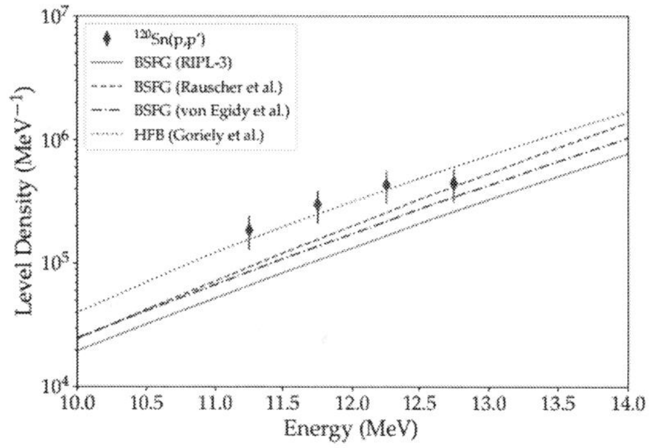

Figure 4. (Colour online) Level density of ^{120}Sn for 1^- states.

3 Gamma strength function and level density of ^{120}Sn

The double differential cross section for ^{120}Sn is available from an earlier proton scattering experiment [15, 16]. Using the so called virtual photon method the double differential cross section can be converted to photoabsorption cross section which allows to determine the E1 part of the GSF [17]. The M1 part of the GSF was extracted from M1 cross sections [15]. In Fig. 3 the total gamma strength function for ^{120}Sn is shown. In the GDR region the GSF from proton scattering is compared to GSFs from (γ, xn) experiments [19–21]. The agreement in this region is very good. In the low energy region gamma strength functions from (^{3}He,^{3} He$'\gamma$) and (^{3}He, $\alpha\gamma$) for ^{116,118,122}Sn are shown [22, 23]. Unfortunately no GSF for ^{120}Sn from this type of experiments is available. In comparison to ^{120}Sn from proton scattering the agreement is quite well in the low energy region. However, two bumps can be seen in the GSF from proton scattering around 6.5 MeV and 8.3 MeV. Around 8.3 MeV no direct comparison can be made since experimental data from gamma decay experiments are limited to lower energies. The bump around 6.5 MeV seems to be enhanced compared to (^{3}He,^{3} He$'\gamma$) and (^{3}He, $\alpha\gamma$) experiments (note the logarithmic scale of Fig. 3). This could be an indication of a violation of the Brink-Axel hypothesis (in contrast to a recent analysis in ^{96}Mo with the same technique [24]), which states that the GSF depends only on the gamma energy and is independent of the structure of the initial and final state [25, 26].

Using a fluctuation analysis [12] it was possible to extract level densities in the GDR region for 1^- states as shown in Fig 4. The results are compared to Back-shifted Fermi Gas Model (BSFG) calculations with different parametrisations and a Hartree-Fock-Bogoliubov Model (HBF) results [27–30]. From (^{3}He,^{3} He$'\gamma$) and (^{3}He, $\alpha\gamma$) experiments only total level densities can be extracted. Therefore the level densities from proton scattering for 1^- states must be converted to total level densities for a comparison. The procedure for this conversion is described in [31]. In Fig. 5 the total level density in ^{120}Sn from proton scattering is compared to the level densities from (^{3}He,^{3} He$'\gamma$) and (^{3}He, $\alpha\gamma$) experiments in neighbouring tin isotopes. In the low energy region the BSFG models reproduce the level densities quite well, whereas at higher energies the HFB model seems to describe the level densities more accurately.

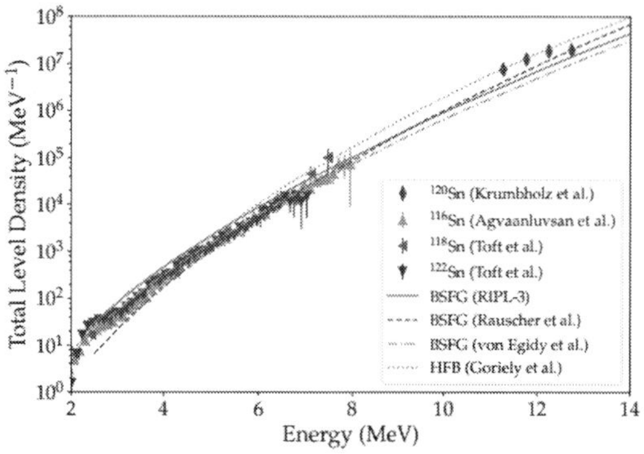

Figure 5. (Colour online) Total level density from proton scattering compared to the level densities from (^{3}He,^{3}He$'\gamma$) and (^{3}He,$\alpha\gamma$) experiments.

4 Summary and conclusions

First results from a recent inelastic proton scattering experiment on ^{112,116,118,124}Sn are presented. In all isotopes the GDR can be clearly identified. Around the neutron threshold a bump can be seen. Particularly in ^{124}Sn strong structure is visible at about 6.5 MeV. The expected shift of the centroid energy of the GDR can be observed in the tin isotope chain.

The gamma strength function and level density for ^{120}Sn was determined from the data of Refs. [15, 16]. The GSF agrees well in the GDR region with other experiments. In the low energy region an increase around 6.5 MeV can be seen in the GSF from proton scattering. This is in disagreement with the GSFs from (^{3}He,^{3}He$'\gamma$) and (^{3}He,$\alpha\gamma$) experiments (albeit not for ^{120}Sn but for neighbouring tin isotopes), which could be an indication of a Brink-Axel hypothesis violation.

The level densities agree well with the HFB model at higher energies. At lower energies however, a better agreement with the BSFG model is observed.

Acknowledgements

We thank all collaborators of the E316 and E422 experiments. This work was supported by the DFG under contract SFB 1245 and by the JSPS, KAKENHI Grant No. JP25105509.

References

[1] J. Piekarewicz, Phys. Rev. C **73**, 044325 (2006).
[2] A. Klimkiewicz *et al.*, Phys. Rev. C **76**, 051603(R) (2007).
[3] N. Tsoneva *et al.*, Phys. Rev. C **77**, 024321 (2008).
[4] J. Piekarewicz, Phys. Rev. C **83**, 034319 (2011).
[5] T. Inakura *et al.*, Phys. Rev. C **84**, 021302(R) (2011).
[6] A. Carbone *et al.*, Phys. Rev. C **81**, 041301(R) (2010).
[7] F.J. Fattoyev *et al.*, Phys. Rev. C **86**, 015802 (2012).
[8] M. B. Tsang *et al.*, Phys. Rev. C **86**, 015803 (2012).
[9] M. Arnould *et al.*, Phys. Rep. **450**, 97 (2007).
[10] M.B. Chadwick *et al.*, Nucl. Data Sheets **112**, 2887 (2011).
[11] M. Salvatore *et al.*, Prog. Part. Nucl. Phys. **66**, 144 (2011).
[12] P.G. Hansen *et al.*, Nucl. Phys. A **518**, 13 (1990).
[13] I. Poltoratska *et al.*, Phys. Rev. C **89**, 054322 (2014).
[14] A. Tamii *et al.*, Nucl. Instrum. Methods Phys. Res., Sect. A **605**, 3 (2009).
[15] A.M. Krumbholz, *et al.*, Phys. Lett. B **744**, 7 (2015).
[16] T. Hashimoto *et al.*, Phys. Rev. C **92**, 031305 (2015).
[17] C. Bertulani *et al.*, Phys. Rep. **163**, 299 (1988).
[18] J. Birkhan *et al.*, Phys. Rev. C **93**, 041302(R) (2016).
[19] A. Leprêtre *et al.*, Nucl. Phys. A **219**, 39 (1974).
[20] S.C. Fultz *et al.*, Phys. Rev. **186**, 1255 (1969).
[21] H. Utsunomiya *et al.*, Phys. Rev. C **84**, 055805 (2011).
[22] U. Agvaanluvsan *et al.*, Phys. Rev. C **79**, 014320 (2009).
[23] H.K. Toft *et al.*, Phys. Rev. C **83**, 044320 (2011).
[24] D. Martin *et al.*, Phys. Rev. Lett. **119**, 182503 (2017).
[25] D.M. Brink, Ph.D. thesis, Oxford University (1955).
[26] P. Axel, Phys. Rev. **126**, 671 (1962).
[27] R. Capote *et al.*, Nucl. Data Sheets **110**, 3107 (2009).
[28] T. Rauscher, Phys. Rev. C **78**, 032801 (2008).
[29] T. von Egidy *et al.*, Phys. Rev. C **80**, 054310 (2009).
[30] S. Goriely, *et al.*, Phys. Rev. C **78**, 064307 (2008).
[31] S. Bassauer, *et al.*, Phys. Rev. C **94**, 054313 (2016).

Selection of different reaction channels in ^{6}Li induced fusion reaction by a powerful combination of a charged particle array and a high-resolution gamma spectrometer

G. X. Zhang[1], *S. P. Hu*[2], *G. L. Zhang*[1,*], *H. Q. Zhang*[3], *Y. J. Yao*[1], *Z. Huang*[1], *M. L. Wang*[1], *H. B. Sun*[2], *J. J. Valiente-Dobòn*[4], *D. Testov*[5], *A. Goasduff*[4,5], *P. R. John*[6], *M. Siciliano*[5], *F. Galtarosa*[5], *R. Francesco*[5], *D. Mengoni*[5], *D. Bazzacco*[5], *E. T. Li*[2], and *X. Hao*[2]

[1] *School of Physics and Nuclear Energy Engineering, Beihang University, Beijing 100191, China*
[2] *School of Physics and Energy, Shenzhen University, Shenzhen 518060, China*
[3] *China Institute of Atomic Energy, Beijing 102413, China*
[4] *INFN, Laboratori Nazionali di Legnaro, Legnaro (Padova), Italy*
[5] *Dipartimento di Fisica and INFN, Sezione di Padova, Padova, Italy*
[6] *Institut für Kernphysik der Technischen Universität Darmstadt, Germany*

Abstract. Investigation of the breakup and transfer effect of weakly bound nuclei on the fusion process has been an interesting research topic in the past several years. In comparison with radioactive ion beam (RIB), the beam intensities of stable weakly bound nuclei such as ^{6,7}Li and ^{9}Be, which have significant breakup probability, are orders of magnitude higher. Precise fusion measurements induced by these nuclei have already been performed. However, the conclusion of reaction dynamics was not clear and has contradiction. In order to have a proper understanding of the influence of breakup and transfer of weakly bound projectiles on the fusion process, the ^{6}Li+^{89}Y experiment with incident energies of 22 MeV and 34 MeV was performed on Galileo array in combination with Si-ball EUCLIDES at Legnaro National Laboratory (LNL) in Italy. Using the coincidence by the charged particles and γ-rays, the different reaction channels can be clearly identified.

1 Introduction

Investigation of the breakup effect of weakly bound nuclei on the fusion process has been an interesting research topic in the past several years [1–3]. Several experiments and theoretical calculations have focused on this subject. The fusion reactions induced by these nuclei are more complicated with respect to tightly bound nuclei. Different processes can occur after the breakup of weakly bound nuclei (usually the projectiles). The whole projectile can also fuse with the target nucleus without breakup, and this process is named as direct complete fusion (DCF). When all the fragments fuse with target nucleus one by one, this process is called sequential complete fusion (SCF). Experimentally, the residues formed in SCF cannot be distinguished from ones from DCF due to the same compound nucleus. Therefore, in the experiment the complete fusion (CF) cross sections include the contributions of both SCF and DCF. When only part of the fragments fuses with the target nucleus, this process is called incomplete fusion (ICF). When any fragments are not captured, this process is called non-capture breakup (NCBU). The total fusion (TF) cross section is equal to the sum of CF and ICF, $\sigma_T = \sigma_{CF} + \sigma_{ICF}$. Besides the above processes, transfer process can also occur such as one neutron stripping and pickup. In ICF and transfer processes, the same residual nuclei can be formed. However, in these two processes the light charged particles with different energies can be emitted. In coincidence with the light charged particles, these two processes can be separated.

In comparison with radioactive ion beam (RIB), the beam intensities of stable weakly bound nuclei such as ^{6,7}Li and ^{9}Be, which have significant breakup probability, are orders of magnitude higher. Precise fusion measurements have already been performed, and the effect of breakup of those nuclei on the fusion process has been extensively studied. Experimental fusion cross sections induced by RIB on different target nuclei have been reported. In ^{8}He+^{197}Au experiment [4], by measuring γ rays which are emitted from the evaporated residual nuclei, fusion and transfer cross sections were obtained. In comparison with the cross sections of these two processes, it is indicated that the sub-barrier total reaction cross section is dominated by one and two neutron stripping. In the case of reactions with neutron-rich radioactive ion beams, the coupling to transfer is found to be important. Recently for the first time in ^{7}Li+^{198}Pt reaction [5] at energies near the Coulomb barrier, the α-capture cross sections were measured by in-beam exclusive measurements of prompt γ-rays from the heavy-residues with various light charged particles (α, t, d and p). The cross sections of the residues

* e-mail: zgl@buaa.edu.cn

resulting from the processes of fusion, *t*-capture and neutron transfer channels were measured by off-beam γ methods. The integrated cross sections for the above several channels were obtained. The different processes were distinguished including breakup fusion and the nucleon transfer. The result of a classical dynamic model [6–8] gave a good agreement with the experimental data and explained the critical role of different cluster structures of ^{7}Li in the dynamics of reaction mechanism for the first time. The present results are useful for the theoretical developments. It is important to carry out a systematic investigation on various targets especially for unstable weakly bound nuclei.

2 Experiments and results

The ^{6}Li+^{89}Y experiment was performed at the Tandem-XTU accelerator of LNL-INFN, Italy. A ^{6}Li^{3+} beam with E_{lab}=34 MeV and 22 MeV and an average beam intensity of 1.0 enA impinged on a ^{89}Y target of 550 μg/cm^2 backed with a 340 μg/cm^2 ^{12}C foil. The schematic of the experimental setup is shown in Fig. 1. The GALILEO array is composed of 25 Compton-Suppressed HPGe detectors, with 5 detectors placed at 152° , 5 at 129°, 5 at 119° and ten detectors at 90° with respect to the beam direction [9]. In Fig.1 the EUCLIDES is one of the key ancillary detectors of the GALILEO spectrometer [10–12] to detect the light-charged particles. It consists of 40 ΔE/E Si telescopes of different shapes with an angular coverage close to 4π. A single silicon telescope is usually assembled using a thin ΔE (130 μm) detector and a thick E (1 mm) detector. The most forward telescopes in the array are segmented into 4 sectors to sustain a higher counting rate. Each telescope is placed at distances around 6.5 cm from the target position. The solid angle of each telescope is about 0.2 sterad and the angular coverage is $\approx$11°.

An ^{89}Y target was mounted on the target holder and was put at the center of target chamber. An aluminum cylindrical absorber was inserted inside EUCLIDES and along the beam direction to protect the silicon detectors from the ions elastically scattered. Thickness of the absorber can be adjusted according with the beam-energy conditions and the reaction kinematics. In the commissioning experiment a ^{6}Li beam of 34 MeV and 22 MeV was sent on a ^{89}Y target. In that situation the thickness of the cylinder was 200 μm and the backward angles larger than 150° were unshielded by the absorber, see the Fig. 1 for further details of the setup.

3 Selection of reaction channels

For ^{6}Li+^{89}Y, many reaction channels can produce ^{90}Zr such as,

$$\begin{aligned} &\text{CF Process, } ^6\text{Li} + {}^{89}\text{Y} \rightarrow {}^{95}\text{Mo} \rightarrow {}^{90}\text{Zr} + \alpha + \text{n} \\ &\text{ICF Process, } ^2\text{H} + {}^{89}\text{Y} \rightarrow {}^{91}\text{Zr} \rightarrow {}^{90}\text{Zr} + \alpha + \text{n} \\ &\text{Transfer Process, } ^6\text{Li} + {}^{89}\text{Y} \rightarrow {}^{90}\text{Zr} + \alpha + \text{n} \end{aligned} \quad (1)$$

Since α particles can be from CF, ICF and transfer processes, even if one uses $\alpha - \gamma$ coincident method, it is impossible to distinguish these three reaction channels.

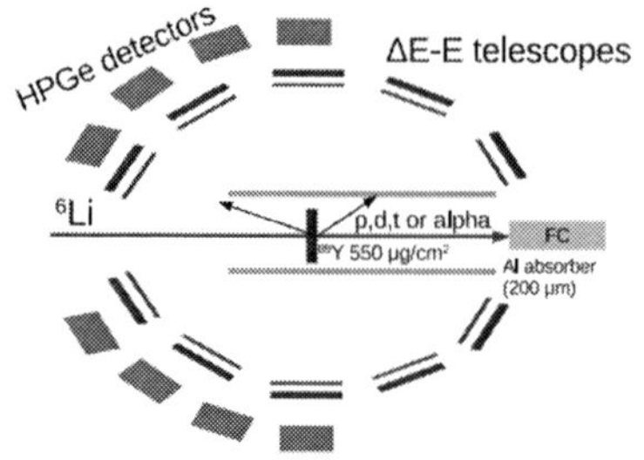

Figure 1. (color online). The schematic of the experimental setup (sectional view). HPGe detectors and ΔE-E telescopes present the schematic of a part of GALILEO and EUCLIDES arrays, respectively

So the reaction channels in fusion reactions induced by weakly bound nuclei cannot be separated by only using single γ energy spectra and even the charged particles-γ coincidence method without any limitation.

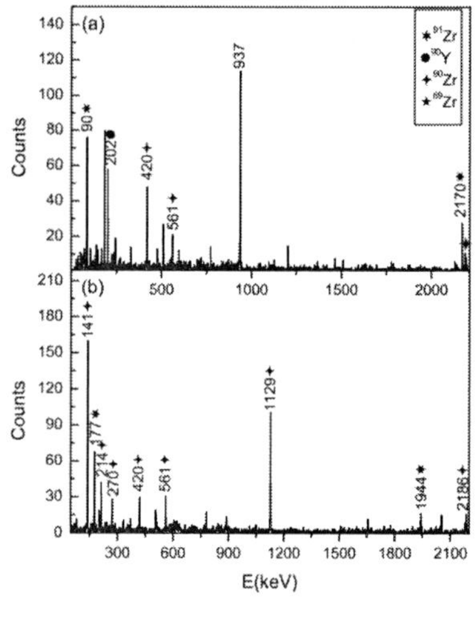

Figure 2. γ energy spectra in coincidence with the α particles of the angular ranges covered by aluminum absorber (a) and unshielded by aluminum absorber (b).

It is well known that the angular distributions and energy distributions of emitted charge particles from CF, ICF and transfer processes are different. To select the different channels and explore the different reaction mechanisms in fusion process induced by weakly bound nuclei, the method of γ energy spectra in coincidence with charged particles of different angular regions is applied. In Fig. 1 the aluminum absorber has a 200 μm thickness, and the ^{6}Li beam with an incident energy of 34 MeV or 22 MeV cannot pass it. In this setup, the α particles that have been observed in the telescope of angular ranges covered by the absorber were tentatively deemed to be mainly from transfer processes and the evaporated α particles cannot be detected. This conclusion is verified by the statistical evaporation model.

The γ-ray spectrum in coincidence with the α particles at angular ranges covered by the absorber is shown in Fig. 2 (a). The γ rays from ^{90}Zr ($1p$ stripping), ^{91}Zr ($1d$ stripping) and ^{90}Y ($1n$ stripping) are present. In the calculation of statistic evaporation model for complete fusion (CF) and incomplete fusion (ICF), ^{90}Y is not predicted and the yield of ^{91}Zr can be neglected. Therefore, ^{91}Zr and ^{90}Y are expected from $1d$ and $1n$ stripping reactions, respectively. In the angular range unshielded by the absorber the alpha-gated γ spectrum clearly shows the presence of ^{90}Zr and ^{89}Zr as shown in Fig. 2(b). In such uncovered ranges the α particles are mainly from evaporation processes from the compound nucleus in CF process. Therefore, they are related to the components of the fusion evaporation residues. In the two spectra of Fig. 2, the γ peaks of ^{90}Zr are visible. However, in the backward silicon range, in the alpha-gated γ spectrum the peaks related to the high-lying states in ^{90}Zr while in the covered range other states from ^{90}Zr are present. This indicates that the populated states of ^{90}Zr are coming from the different reaction kinematics. Therefore, according to the present experimental setup, the different reaction channels can be clearly identified by charged particles-γ coincident method.

4 summary

An experiment to explore the fusion reactions of weakly bound nuclei was performed with the GALILEO array in combination with the Si-ball EUCLIDES at the Legnaro National Laboratory (LNL), Italy. By the time coincidence measurement of light-charged particles and γ rays, the different channels can be separated. As the γ energy spectra in coincidence with the light-charged particles allow a clean selection of the reaction channels, we can derive the cross sections for the different reaction channels. The detailed analysis is still ongoing.

References

[1] L. F. Canto et al., Phys. Rep., **424**, 1 (2006)
[2] L. F. Canto et al., Phys. Rep., **596**, 1 (2015)
[3] N. Keeley et al., Prog. Part. Nucl. Phys., **59**, 579 (2007)
[4] A. Lemasson et al., Phys. Rev. Lett., **103**, 232701 (2009)
[5] A. Shrivastava et al., Phys. Lett. B., **718**, 931 (2013)
[6] A. Diaz-Torres et al., Phys. Rev. Lett., **98**, 152701 (2007)
[7] A. Diaz-Torres, J. Phys. G: Nucl. Part. Phys., **37**, 075109 (2010)
[8] A. Diaz-Torres, Comp. Phys. Comm., **182** 1100, (2011)
[9] J. J. Valiente-Dobòn *et al.*, INFN-LNL Annual Report, (2014).
[10] D. Testov *et al.*, INFN-LNL Annual Report, (2015).
[11] A. Gadea *et al.*, INFN-LNL Annual Report, 225 (1996).
[12] A. Gadea *et al.*, INFN-LNL Annual Report, 118 (1997).

The Gogny-HFB+QRPA dipole strength function and its application to radiative neutron capture cross section

Stephane Goriely[1,*], *Stephane* Hilaire[2], and *Sophie* Péru[2]

[1] *Institut d'Astronomie et d'Astrophysique, Université Libre de Bruxelles, CP-226, 1050 Brussels, Belgium*
[2] *CEA, DAM, DIF, F-91297 Arpajon, France*

Abstract. Valuable theoretical predictions of nuclear dipole excitations in the whole chart are of great interest for different nuclear applications, including in particular nuclear astrophysics. Here we extend our large-scale calculations of the $E1$ and $M1$ absorption γ-ray strength function obtained in the framework of the axially-symmetric deformed quasiparticle random phase approximation (QRPA) based on the finite-range D1M Gogny force to the determination of the de-excitation strength function. To do so, shell-model calculations of the de-excitation dipole strength function as well as experimental data are considered to provide insight in the low-energy limit and to complement the QRPA estimate phenomenologically. We compare our final prediction of the $E1$ and $M1$ strengths with available experimental data at low energies and show that a relatively good agreement can be obtained. Its impact on the average radiative width as well as radiative neutron capture cross section is discussed.

1 Introduction

Radiative neutron capture cross sections play a key role in almost all nuclear applications. Despite a huge effort to measure such radiative neutron capture cross sections, theoretical predictions are required to fill the gaps, both for nuclei for which measurements are not feasible at the present time, in particular for unstable targets, and for energies that cannot be reached in the laboratory. Some applications, such as nuclear astrophysics, also require the determination of radiative neutron capture cross sections for a large number of exotic neutron-rich nuclei [1]. In this case, large-scale calculations need to be performed on the basis of sound and accurate models to ensure a reliable extrapolation far away from the experimentally known region.

The neutron capture rates are commonly evaluated within the framework of the statistical model of Hauser-Feshbach, although the direct capture contribution plays an important role for very exotic nuclei [2]. The fundamental assumption of the Hauser-Feshbach model is that the capture goes through the intermediary formation of a compound nucleus in thermodynamic equilibrium. In this approach, the (n,γ) cross section strongly depends on the electromagnetic interaction, i.e the photon de-excitation probability. In turn, it is well known that the photon strength function is dominated by the dipole contribution. The various multipolarities of the γ-ray strength function are traditionally modeled by the phenomenological Lorentzian approximation or some of its energy-dependent variants [3]. The reliability of the γ-ray strength predictions can however be greatly improved by the use of microscopic or semi-microscopic models. Such an effort can be found in Refs. [4–8] where a complete set of $E1$ and $M1$ γ-ray strength functions was derived from mean field plus quasiparticle random phase approximation (QRPA) calculations. When compared with experimental data and considered for practical applications, all mean field plus QRPA calculations need however some phenomenological corrections. These include a broadening of the QRPA strength to take the neglected damping of collective motions into account as well as a shift of the strength to lower energies due to the contribution beyond the 1 particle - 1 hole excitations and the interaction between the single-particle and low-lying collective phonon degrees of freedom [9]. In addition, most of the mean field plus QRPA calculations assume spherical symmetries, so that phenomenological corrections need to be included in a way or another in order to properly describe the splitting of the giant dipole resonance in deformed nuclei [4–6].

Recently, axially-symmetric-deformed QRPA calculations based on Hartree-Fock-Bogoliubov (HFB) calculations using the finite-range Gogny interaction have been shown to provide rather satisfactory predictions of the $E1$ [7] and $M1$ strengths [8]. However, such QRPA calculations only describe the photoabsorption and it is well known that the de-excitation strength function may differ from the photoabsorption one, especially for low photon energies [3, 10]. In particular, a non-zero limit of the dipole strength has been observed experimentally [11, 12] and confirmed by shell-model calculations [13–15].

In the present paper, we complement our previous HFB+QRPA study of the $E1$ and $M1$ strengths [7, 8],

*e-mail: sgoriely@astro.ulb.ac.be

by correcting the low-energy limit on the basis of shell-model calculation. The paper is organized as follows. In Sect. 2, the $E1$ and $M1$ HFB+QRPA dipole photoabsorption strength functions are rapidly described together with the way corrections beyond QRPA are included phenomenologically. The de-excitation dipole strength function is discussed in Sect. 3 and low-energy corrections introduced to take into account the predictions of the shell model and available experimental data. In Sect 4, we compare measured average radiative width obtained with our D1M+QRPA strength and with Lorentzian-type models. Finally, the impact of the newly calculated γ-ray strength functions on the Hauser-Feshbach estimate of the radiative neutron capture cross section is discussed in Sect. 5. Conclusions are drawn in Sect. 6.

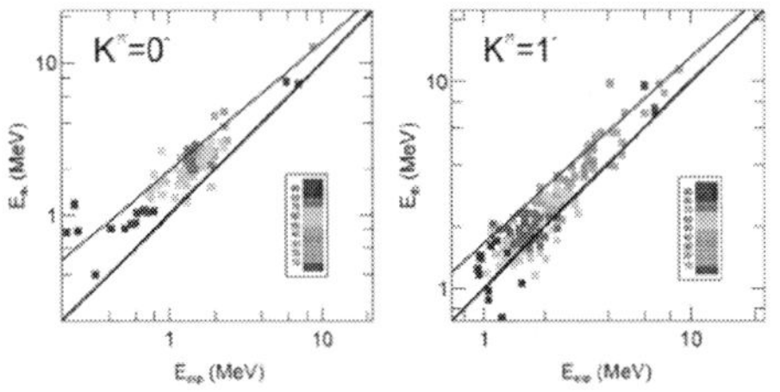

Figure 1. (Color online) Experimental – theoretical comparison of the energy of the low-energy vibrational states $K^\pi = 0^-$ (left panel) and $K^\pi = 1^-$ (right panel). The red curve corresponds to the energy shift applied to the HFB+QRPA strength. The colour code refers to the atomic mass range of the nuclei.

2 E1 and M1 photoabsorption strength functions

The QRPA formalism based on axially-symmetric-deformed HFB equations solved in a finite harmonic oscillator basis in cylindrical coordinates has been described in details in Refs. [7, 8, 16, 17]. In the present study, the D1M Gogny force [18] is used. Previous studies [7] have shown that the centroid energy of the $E1$ strength is systematically larger by typically 2.5 MeV than the Giant Dipole Resonance (GDR) energy determined experimentally. Similarly, a study of the QRPA low-energy vibrational states shows that the first experimental energies [19] are overestimated by typically 500 keV, as shown in Fig. 1. For this reason, the effects beyond the 1 particle - 1 hole QRPA can be empirically included by considering an energy shift that increases with energy. In comparison with our previous study of the $E1$ QRPA strength [7], we have now adopted a simpler treatment of the renormalization procedure to reproduce both the experimental GDR properties and low-energy vibrational states. More specifically, for the $E1$ and $M1$ QRPA strengths, we now apply an energy shift of $\Delta = 0.5$ MeV for $\varepsilon_\gamma \le 0.5$ MeV, $\Delta = 2.5$ MeV for $\varepsilon_\gamma = 18$ MeV and $\Delta = 5$ MeV for $\varepsilon_\gamma \ge 21$ MeV. For energies in the $0.5 \le \varepsilon_\gamma \le 21$ MeV range, the energy shift Δ is interpolated linearly between the anchor values at 0.5, 18 and 21 MeV.

Similarly, an empirical damping of the collective motions is introduced in the QRPA strength by folding each $E1$ strength by a standard Lorentzian (SLO) function of width Γ that has been adjusted on photoabsorption data and is assumed to be dependent on the atomic mass A only. We adopt the final expression Γ [MeV] $= 7 - A/45$ for $A \le 200$, and 2.5 MeV otherwise. For the $M1$ strength, a constant value of $\Gamma = 0.5$ MeV is adopted.

With such a systematic renormalization of the HFB+QRPA calculations, experimental strength functions based on photoabsorption measurements or nuclear resonance fluorescence can be rather well described, as already shown in Refs. [7, 8].

3 E1 and M1 de-excitation strength functions

When considering the de-excitation strength function, deviation from the photoabsorption strength can be expected, especially for γ-ray energies approaching the zero limit. In particular, a constant $E1$ strength of $f_{E1} \simeq 5\ 10^{-9}$ MeV^{-3} at energies $\varepsilon_\gamma \lesssim 5$ MeV was found by the shell model [15]. In addition, shell-model calculations [13–15] predict an exponential increase of the $M1$ de-excitation strength function at energies approaching zero. The upbend of the strength function observed experimentally [11, 12] has therefore been assumed to be of $M1$ nature, though no experimental evidence exists for the moment. Therefore, the de-excitation strength function can be consistently taken from the QRPA calculation (Sect. 2) provided corrections at low energies inspired from shell model predictions are included. These finally are expressed as

$$\overleftarrow{f_{E1}}(\varepsilon_\gamma) = f_{E1}^{QRPA}(\varepsilon_\gamma) + f_0 U/[1 + e^{(\varepsilon_\gamma - \varepsilon_0)}] \quad (1)$$

$$\overleftarrow{f_{M1}}(\varepsilon_\gamma) = f_{M1}^{QRPA}(\varepsilon_\gamma) + C e^{-\eta\varepsilon_\gamma}[1 + F\varepsilon_\gamma^3] \quad (2)$$

where f_{X1}^{QRPA} is the D1M+QRPA strength at the photon energy ε_γ, U (in MeV) is the excitation energy of the initial de-exciting state and $f_0 = 5\ 10^{-10}$ MeV^{-4}, $\varepsilon_0 = 5$ MeV, $C = 10^{-7}$ MeV^{-3}, $\eta = 2$ MeV^{-1} and $F = 0.4$ MeV^{-3} are free parameters that have been adjusted on the shell model results [15] and available experimental data [11, 12, 21]. The impact of the low-energy limit applied to the D1M+QRPA strength is illustrated in Fig. 2 for the $E1$ channel in ^{44}Ti and in Fig. 3 for both the $E1$ and $M1$ contributions in ^{57}Fe. The total dipole strength obtained with Eqs (1-2) is compared with experimental data extracted from photoabsorption measurements and the Oslo method in Fig. 4 for a sample of 4 nuclei. Note that the present $E1$ and $M1$ strengths are also found to be in rather good agreement with the data extracted from Average Resonance Capture (ARC) experiments, as shown in Ref. [20].

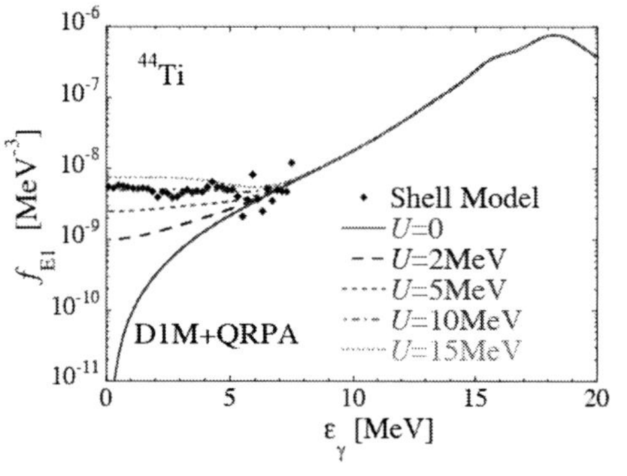

Figure 2. (Color online) Illustration for ^{44}Ti of the low-energy correction on the $E1$ strength function (Eq. 1) applied to the D1M+QRPA strength for various initial excitation energies U, as suggested by the shell-model calculations [15].

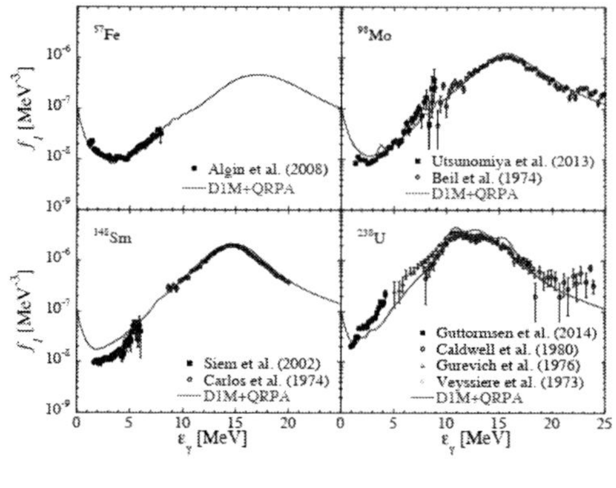

Figure 4. (Color online) The D1M+QRPA dipole strength function $f_1 = f_{E1} + f_{M1}$ given by Eqs. (1-2) is compared with experimental photodata and Oslo measurements for ^{57}Fe [21], ^{98}Mo [22, 23], ^{148}Sm [24, 25], and ^{238}U [26–29].

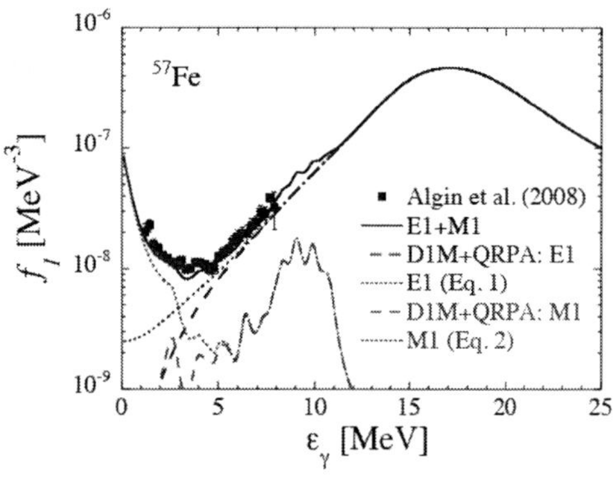

Figure 3. (Color online) Illustration for ^{57}Fe of the low-energy correction applied to the $E1$ and $M1$ D1M+QRPA strengths (Eqs. 1-2) and comparison with the low-energy data [21].

4 Average radiative width

Among the various experimental data, the average radiative width is known to play a key role in reaction modelling. The average radiative width is defined as [3]

$$\langle\Gamma_\gamma\rangle = \frac{D_0}{2\pi}\sum_{X,L,J,\pi}\int_0^{S_n+E_n} T_{XL}(\varepsilon_\gamma)\times\rho(S_n+E_n-\varepsilon_\gamma,J,\pi)d\varepsilon_\gamma \qquad (3)$$

where D_0 is the average resonance spacing for s-wave neutrons, S_n is the neutron separation energy, E_n is the neutron incident energy, $T_{XL} = 2\pi\varepsilon_\gamma^{2L+1}\overleftarrow{f_{XL}}(\varepsilon_\gamma)$ is the electromagnetic transmission coefficient ($X = M$ or E) for a multipolarity L and ρ is the energy-, spin- and parity-dependent nuclear level density.

It has been a long-standing problem that phenomenological SLO models tend to overestimate the average radiative width significantly, while its improved and widely used version, the so-called Generalized Lorentzian (GLO) model [10], underestimates $\langle\Gamma_\gamma\rangle$. For this reason, the SLO model also tends to overestimate the radiative neutron capture cross sections for low-energy (keV) neutrons, while the GLO model overestimates them (see also Ref. [30]). Such deviations are shown in Fig. 5 where the experimental average radiative widths are compared with predictions for SLO, GLO and the present corrected D1M+QRPA (Eqs. 1-2). It can be seen that in contrast to the SLO and GLO models, the present D1M+QRPA strengths (including both the $M1$ and $E1$ contributions) reproduce globally rather well the experimental average radiative width.

5 Radiative neutron captures

The radiative neutron capture cross sections and reaction rates of astrophysical interest have been calculated systematically on the basis of the Hauser-Feshbach statistical model described by the TALYS reaction code [31]. Both the widely used GLO model for the $E1$ strength function and the present D1M+QRPA models corrected at low energies are considered here. The HFB plus combinatorial model of nuclear level densities [32] is adopted for the cross section calculations. Fig. 6 shows the ratio of the Maxwellian-averaged neutron capture cross sections (MACS) at a temperature $T = 10^9$K obtained with the present corrected D1M+QRPA to those obtained with the GLO $E1$ model and SLO $M1$ model recommended in Refs. [3, 10].

When approaching the neutron dripline ($S_n = 0$), the MACS calculated with the corrected D1M+QRPA strength is seen to be a factor 100 to 1000 larger than the one obtained with the traditional GLO model. The most significant effects responsible for such an increase of the MACS are *i)* the low-energy $E1$ strength predicted for neutron-rich nuclei by the HFB+QRPA approach (as already discussed in Ref. [7]) and *ii)* the low-energy $M1$

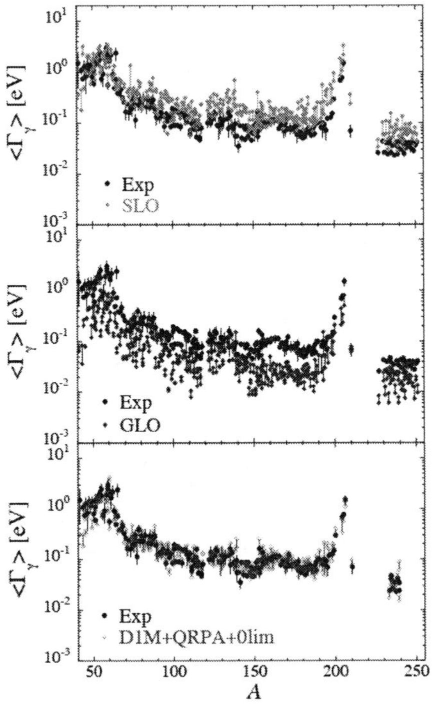

Figure 5. (Color online) Comparison between experimental (black circles) [3] and theoretical (coloured diamonds) average radiative width $\langle\Gamma_\gamma\rangle$ as a function of A. The upper panel is obtained with the SLO model [3] for both the $E1$ and $M1$ strengths and the middle panel with the GLO model for the $E1$ and the SLO model for the $M1$ [3, 10]. The lower panel corresponds to the present D1M+QRPA corrected by the low-energy limits (Eqs. 1-2). The error bars illustrate the uncertainties associated with the use of different nuclear level density models.

upbend introduced in Eq. (2). In particular, the $M1$ upbend can affect the MACS of exotic neutron-rich nuclei by a factor up to 100. In contrast, the low-energy limit of the $E1$ strength function included in Eq. (1) only affects the MACS by 20 to 50% due to the ε_γ^3 dependence weighting the γ-ray strength function in the calculation of the transmission coefficient T_{E1} (see Sect. 4). This is not the case for the $M1$ upbend, where the fast increase of the $M1$ strength at decreasing energies approaching zero is capable of counterbalancing the ε_γ^3 effect in the calculation of T_{M1}. As far as the HFB+QRPA prediction of the $M1$ spin flip is concerned, it is almost always insignificant with respect to the stronger $E1$ contribution at the relevant energies of 8-9 MeV, so that its impact on the MACS is reduced to typically 10% in comparison with the MACS obtained with a SLO $M1$ strength [3]. However, the low-energy $M1$ scissors mode predicted in deformed nuclei by the HFB+QRPA can impact the MACS by a factor of 2. This extra strength is never included in cross section calculations except recently through a phenomenological approximation [30]. It affects the average radiative width as well as the neutron capture cross section of deformed nuclei in a non-negligible way.

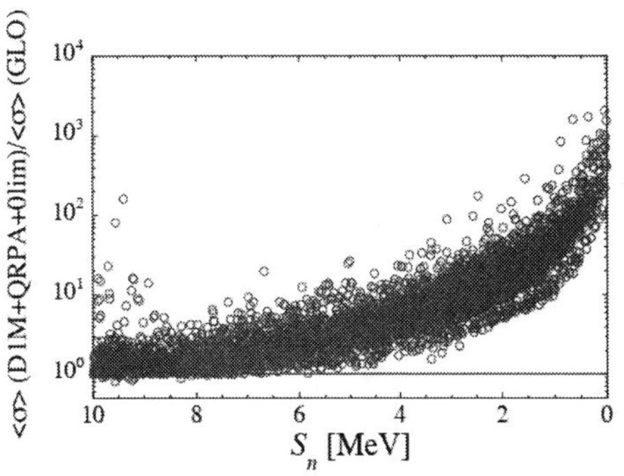

Figure 6. (Color online) Ratio of the MACS at $T = 10^9$K obtained with the present D1M+QRPA corrected by the low-energy limits (Eqs. 1-2) to the one obtained with the GLO model for the $E1$ and SLO model for the $M1$ [3, 10]. The ratio is given for all nuclei with $8 \leq Z \leq 94$ lying between the proton and neutron driplines as a function of the neutron separation energy S_n.

6 Conclusions

HFB+QRPA models are now available for applications and have shown their capacity to predict photoabsorption γ-ray strength functions. Contributions beyond the 1 particle - 1 hole excitations and due to the interaction between the single-particle and low-lying collective phonon degrees of freedom can be empirically taken into account by renormalizing the HFB+QRPA strength through an energy shift and a damping width. Here we extend our HFB+QRPA calculations to the determination of the de-excitation strength function. To do so, shell-model calculations of the de-excitation dipole strength function as well as experimental data are considered to provide insight in the low-energy limit. We showed that, in a first approximation, the HFB+QRPA strength can be complemented by simple analytical expressions to account for the missing strength at low energies. These corrections have been shown to affect the calculation of the average radiative width as well as radiative neutron capture cross sections, so that future work will need to estimate them more systematically on the basis of microscopic models. The present QRPA prediction provides an alternative accurate and reliable way to estimate the dipole $M1$ and $E1$ strength for a large set of nuclei with respect to the phenomenological and approximate Lorentzian-type description traditionally used in reaction codes.

Acknowledgments: We thank K. Sieja for valuable exchanges. SG acknowledges the support of the FRS-FNRS.

This work was performed within the IAEA CRP on "Updating the Photonuclear data Library and generating a Reference Database for Photon Strength Functions" (F410 32).

References

[1] M. Arnould, S. Goriely, K. Takahashi, Phys. Repts. **450**, 97 (2007)
[2] Y. Xu, S. Goriely, A. Koning, S. Hilaire, Phys. Rev. C **90**, 024604 (2014)
[3] R. Capote, M. Herman, P. Oblozinsky, P. Young, S. Goriely, T. Belgya, A. Ignatyuk, A. Koning, S. Hilaire, V. Plujko et al., Nuclear Data Sheets **110**, 3107 (2009)
[4] S. Goriely, E. Khan, Nucl. Phys. A **706**, 217 (2002)
[5] S. Goriely, E. Khan, M. Samyn, Nucl. Phys. A **739**, 331 (2004)
[6] I. Daoutidis, S. Goriely, Phys. Rev. C **86** (2012)
[7] M. Martini, S. Péru, S. Hilaire, S. Goriely, F. Lechaftois, Phys. Rev. C **94**, 014304 (2016)
[8] S. Goriely, S. Hilaire, S. Péru, M. Martini, I. Deloncle, F. Lechaftois, Phys. Rev. C **94**, 044306 (2016)
[9] S. Kamerdzhiev, V. Tkachev, Phys. Lett. B **142**, 225 (1984)
[10] J. Kopecky, M. Uhl, Phys. Rev. C **41**, 1941 (1990)
[11] A. Voinov, E. Algin, U. Agvaanluvsan, T. Belgya, R. Chankova, M. Guttormsen, G. Mitchell, J. Rekstad, A. Schiller, S. Siem, Phys. Rev. Lett. **93**, 142504 (2004)
[12] M. Guttormsen, R. Chankova, U. Agvaanluvsan, E. Algin, L.A. Bernstein, F. Ingebretsen, T. Lönnroth, S. Messelt, G.E. Mitchell, J. Rekstad et al., Phys. Rev. C **71**, 044307 (2005)
[13] R. Schwengner, S. Frauendorf, A.C. Larsen, Phys. Rev. Lett. **111**, 232504 (2013)
[14] B.A. Brown, A.C. Larsen, Phys. Rev. Lett. **113** (2014)
[15] K. Sieja, Phys. Rev. Lett. **119**, 052502 (2017)
[16] S. Peru, H. Goutte, Phys. Rev. C **77**, 044313 (2008)
[17] S. Peru, M. Martini, Eur. Phys. J. A **50**, 88 (2014)
[18] S. Goriely, S. Hilaire, M. Girod, S. Péru, Phys. Rev. Lett. **102**, 242501 (2009)
[19] S. Hilaire, M. Girod, S. Goriely, A. Koning, Phys. Rev. C **86**, 064317 (2012)
[20] J. Kopecky, S. Goriely, S. Peru, S. Hilaire, M. Martini, Phys. Rev. C **95**, 054317 (2017)
[21] E. Algin, U. Agvaanluvsan, M. Guttormsen, A.C. Larsen, G.E. Mitchell, J. Rekstad, A. Schiller, S. Siem, A. Voinov, Phys. Rev. C **78**, 054321 (2008)
[22] H. Utsunomiya, S. Goriely, T. Kondo, C. Iwamoto, H. Akimune, T. Yamagata, H. Toyokawa, H. Harada, F. Kitatani, Y.W. Lui et al., Physical review. C, Nuclear physics **88**, 015805 (2013)
[23] H. Beil, R. Bergere, P. Carlos, A. Lepretre, A.D. Miniac, A. Veyssiere, Nucl. Phys. A **227**, 427 (1974)
[24] S. Siem, M. Guttormsen, K. Ingeberg, E. Melby, J. Rekstad, A. Schiller, A. Voinov, Phys. Rev. C **65**, 0044318 (2002)
[25] P. Carlos, H. Beil, R. Bergere, A. Lepretre, A.D. Miniac, A. Veyssière, Nucl. Phys. A **225**, 171 (1974)
[26] M. Guttormsen, L.A. Bernstein, A. Görgen, B. Jurado, S. Siem, M. Aiche, Q. Ducasse, F. Giacoppo, F. Gunsing, T.W. Hagen et al., Phys. Rev. C **89**, 014302 (2014)
[27] J.T. Caldwell, E.J. Dowdy, B.L. Berman, R.A. Alvarez, P. Meyer, Phys. Rev. C **21**, 1215 (1980)
[28] G. Gurevich, L. Lazareva, V. Mazur, G. Solodukhov, B. Tulupov, Nucl. Phys. A **273**, 326 (1976)
[29] A. Veyssiere, H. Beil, R. Bergere, P. Carlos, . Lepretre, K. Kernbath, Nucl. Phys. A **199**, 45 (1973)
[30] M.R. Mumpower, T. Kawano, J.L. Ullmann, M. Krtička, T.M. Sprouse, Phys. Rev. C **96**, 024612 (2017)
[31] A.J. Koning, D. Rochman, Nucl. Data Sheets **113**, 2841 (2012)
[32] S. Goriely, S. Hilaire, A. Koning, Phys. Rev. C **78**, 064307 (2008)

Masses and lifetimes for *r*-process nucleosynthesis: FRIB outlook

Rebecca Surman[1,*] and *Matthew* Mumpower[2,**]

[1]Department of Physics, University of Notre Dame, Notre Dame, IN USA
[2]Theoretical Division, Los Alamos National Laboratory, Los Alamos, NM USA

Abstract. Nuclear masses and lifetimes are key inputs for calculations of rapid neutron capture (*r*-process) nucleosynthesis. Masses and half-lives for thousands of nuclei from the valley of stability to the neutron drip line are required and only a fraction have been experimentally measured. Here we examine the promise of the Facility for Rare Isotope Beams, now under construction at Michigan State University, to dramatically reduce uncertainties in *r*-process abundance patterns due to uncertain masses and half-lives.

GW170817 [1] has ushered in a new era of *r*-process studies. Observations of the GW170817 neutron star merger event in the optical and infrared are interpreted to be a kilonova: an electromagnetic transient powered by the radioactive decay of heavy nuclei freshly produced via rapid neutron capture (*r*-process) nucleosynthesis [2, 3]. This is the first clear evidence of *r*-process production and may even suggest that most of the galactic *r*-process material originates in merger events [4]. Still many questions remain. Are mergers the only site of *r*-process nucleosynthesis, or do multiple sites contribute? Can we understand neutron star merger nucleosynthesis from first principles? At what level of detail can we interpret this and future kilonova observations?

Nuclear physics plays a key role in all of the above questions. The nuclear network calculations used to estimate the nuclei produced in a merger and their associated radioactive heating rely on nuclear physics information—masses, reaction rates, β-decay and fission properties—for thousands of nuclei from the valley of stability to the neutron drip line [5, 6]. Only a fraction of the needed data has be measured experimentally, leading to large uncertainties in predicted abundance patterns [4, 6–9] and radioactive heating estimates [10].

The next generation of radioactive beam facilities will have unprecedented reach into exotic neutron-rich regions of the nuclear chart. In the United States, the Facility for Rare Isotope Beams (FRIB) is currently under construction at Michigan State University and first beam is scheduled for 2020. Here we look forward to the capabilities of FRIB and examine its potential for reducing uncertainties in simulated *r*-process abundance patterns.

Nuclear masses are of fundamental importance for *r*-process simulations. Neutron separation energies S_n appear explicitly in the evaluation of photodissociation rates via detailed balance. S_n set the location of the *r*-process path and shape the decay to stability, for as long as the temperature is high enough for photodissociation to compete with neutron capture and β decay. Masses also set the Q-values for neutron capture and β decay, thus influencing calculations of these quantities where experimental data is not available.

*e-mail: rsurman@nd.edu
**e-mail: matthew@mumpower.net

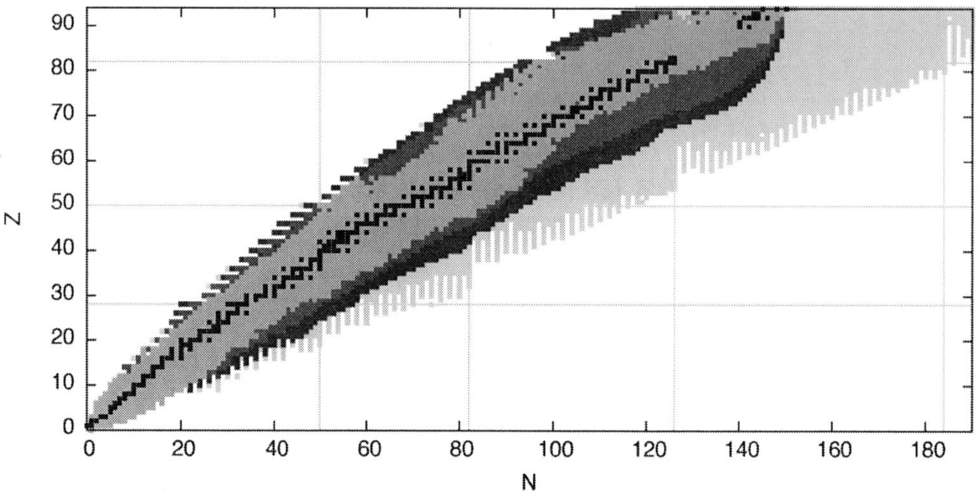

Figure 1. Nuclear chart highlighting the anticipated FRIB reach, showing stable nuclei (black), nuclei with measured masses included in the 2016 Atomic Mass Evaluation [11] (light green squares), nuclei with masses estimated to be measurable at FRIB on Day 1 (green squares), and nuclei with masses estimated to be measurable once FRIB reaches its designed beam intensity (dark green squares). FRIB rates are from [16]. Here we take a mass to be measurable if the stopped beam rate is anticipated to be greater than 10^{-3} pps and the predicted half-life is greater than 10 ms. The grey shaded region indicates where $S_n > 0$ and $S_p > 0$ as calculated using FRDM2012 nuclear masses.

The current extent of mass measurements is shown in Fig. 1, where the isotopes with experimental mass values appearing in the 2016 Atomic Mass Evaluation [11] are highlighted in light green. Outside of this region, *r*-process simulations rely on theoretical mass models. Different models tend to agree fairly well where data is available and diverge markedly away from this region (see, e.g, Fig. 6 of [6]). Separation energy predictions among different models can disagree by MeV (see, e.g., Fig. 3 of [12]). The extent of positive values of neutron separation energies for one popular mass model, the Finite Range Droplet Model (FRDM2012) [15], is shown as the light grey shaded region in Fig. 1. All of the nuclei on the neutron-rich side of this region can potentially participate in a neutron star merger *r*-process.

Fig. 1 also highlights the upcoming mass measurement capabilities anticipated for FRIB [16]. Beam requirements depend on the nature of the mass measurement technique. Here we estimate a mass to be measurable if FRIB can achieve a stopped beam intensity of approximately 10^{-3} pps, and if the isotope has a estimated or measured half-life of at least 10 μs. When FRIB begins operations in 2020, the planned Day 1 beam intensity of 10 kW puts the experimental reach for mass measurements as shown in the green shaded region of Fig. 1. The ultimate design intensity of FRIB at 400 kW results in a predicted reach given by the dark green shaded region of Fig. 1.

It is clear from Fig. 1 that dedicated experimental campaigns at FRIB can result in an unprecedented number of new mass measurements. This has the potential to radically transform the capability of *r*-process simulations to provide precision estimates of nucleosynthetic abundances and heating rates. To roughly illustrate this, we perform Monte Carlo variations of uncertain nuclear masses in an *r*-process simulation. We choose astrophysical conditions similar to those expected for a heated neutron star merger accretion disk wind, with entropy $s/k \sim 87$, dynamical timescale of 80 ms, and initial electron fraction $Y_e = 0.26$. These con-

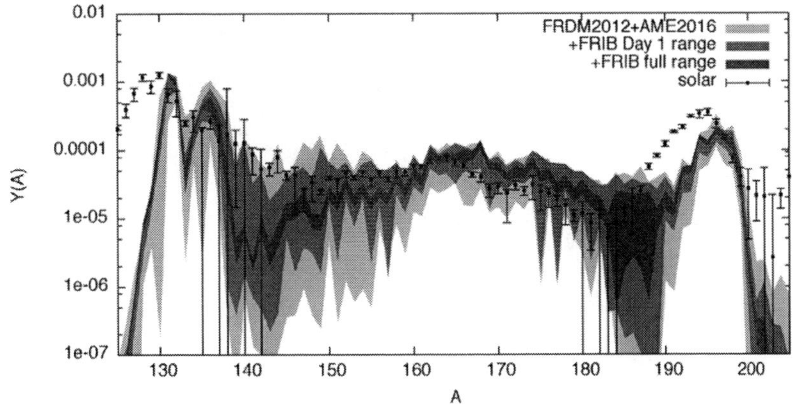

Figure 2. Abundance pattern 1σ uncertainty bands obtained from 10,000 sets of uncorrelated, random variations of individual nuclear masses outside the range of: the AME2016 (light green band), the FRIB Day 1 anticipated reach (green band), and the FRIB design reach (dark green band), for hot wind astrophysical conditions ($s/k = 87$, $\tau = 80$ ms, $Y_e = 0.26$). Points show scaled solar r residuals [5].

ditions produce a hot, (n, γ)-(γ, n) equilibrium r process, where the influence of the masses is predominantly through the neutron separation energies [13]. Neutron separation energies set the abundances along each isotopic chain while equilibrium holds and influence the individual neutron capture rates and photodissociation rates that are important once equilibrium fails. The r-process calculations proceed as in [6, 8]. The baseline simulation uses AME2016 masses with FRDM2012 masses where experimental values are not available, NUBASE 2016 [14] β-decay rates with rates from [17] where experimental values are not available, and NONSMOKER [18] neutron capture rates. Fission is treated as in [6], though the conditions chosen do not result in appreciable fission cycling.

The base simulation is repeated 10,000 times, with each step including random, uncorrelated variations of all masses assumed to be unknown. The random mass variations are pulled from Gaussian distributions of width 0.5 MeV. The latter choice is guided by the typical σ_{RMS} values of the mass models most commonly used in r-process calculations, $\sigma_{RMS} \sim 0.4 - 1.0$ MeV. For the chosen astrophysical conditions the greatest influence of the mass is through the photodissociation rates, which set the equilibrium abundances and compete with neutron capture and β-decay in freezeout. Photodissociation rates are calculated by detailed balance and thus have an exponential dependence on separation energy; mass changes of ~ 0.5 MeV can result in changes to photodissociation rates as large as two orders of magnitude. In contrast, the same mass changes produce changes to neutron capture rates of factors of two to five and β-decay rates of factors of two to four [13]. Thus to capture the primary influence of masses for these astrophysical conditions, we update the neutron separation energies and photodissociation rates for each step as in [8]. The light green shaded region of Fig. 2 shows the 1-σ variation in the resulting 10,000 abundance patterns when all of the masses outside of the AME2016 are randomly varied in this way. The width of the shaded band is similar to that found in past studies of uncorrelated mass variations [22] and in recent studies of systematic mass variations [4]. To estimate the improvements expected from dedicated FRIB measurement campaigns, we repeat the full set of 10,000 simulations twice, where we restrict the range of random nuclear mass variations to (1) outside the Day 1 FRIB reach and (2) outside the full FRIB reach. The resulting abundance pattern variations are shown in the green and dark green bands, respectively, of Fig. 2. Note significant improvements to the abundance

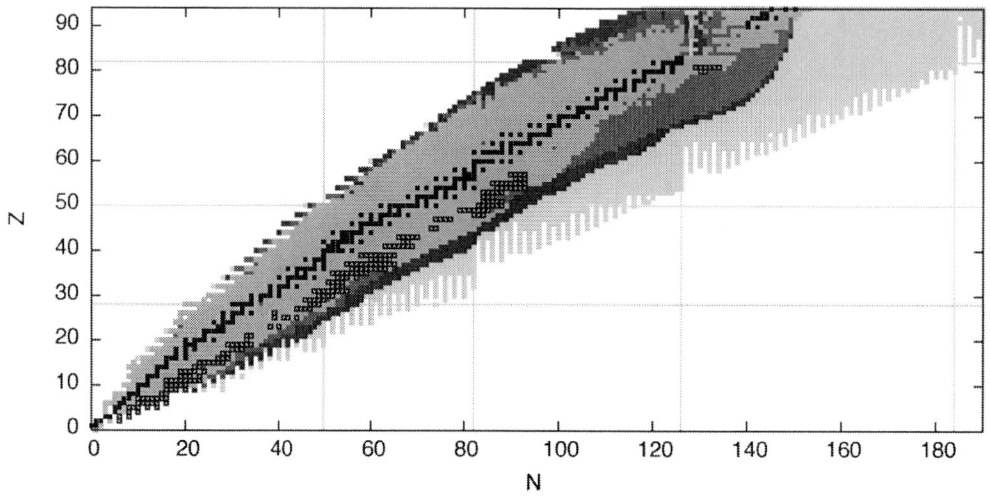

Figure 3. Nuclear chart highlighting the anticipated FRIB reach, showing stable nuclei (black), nuclei with measured half-lives from the NUBASE 2016 compilation (light blue squares), nuclei with measured P_n from NUBASE 2016 (open black squares), nuclei with half-lives estimated to be measurable at FRIB on Day 1 (blue squares), and nuclei with half-lives estimated to be measurable once FRIB reaches its designed beam intensity (dark blue squares). FRIB rates are from [16]. Here we take a half-life to be measurable if the stopped beam rate is anticipated to be greater than 10^{-5} pps. The grey shaded region is defined as in Fig. 1.

pattern uncertainties can be realized with FRIB at 10 kW intensity, and the potential improvement with FRIB at full intensity is dramatic. Fine details of the abundance pattern that are currently obscured by uncertainties in nuclear masses can potentially be clearly resolved.

The calculations presented in Fig. 2 consider only the role of nuclear masses in a hot, (n,γ)-(γ,n) equilibrium r process. For a cold or very neutron rich r-process, photodissociation plays a more limited role and thus the full uncertainties in r-process abundances due to nuclear masses can only be captured when mass changes are propagated through to the neutron capture rates and β-decay lifetimes [13]. Monte Carlo studies for these conditions are presented in [12]. Additional uncertainties come from the uncertainty in nuclear structure of neutron-rich nuclei, which shapes the level densities and γ-strength functions required for neutron capture rate calculations and the β-strength functions needed for β-decay lifetimes and β-delayed neutron emission estimates. FRIB has the potential to dramatically reduce these uncertainties as well.

Lifetimes for β decay determine the relative abundances of the different isotopic chains for an r-process in (n,γ)-(γ,n) equilibrium and govern the decay back to stability once equilibrium fails. Thanks in part to recent measurement campaigns at RIKEN [19, 20] more experimental values are available for β-decay lifetimes than any other set of nuclear data. This is illustrated in Fig. 3, where the nuclei with experimental half-lives compiled in NUBASE 2016 are shown in light blue. Outside this region, r-process simulations use lifetimes calculated from gross theory or with some application of the quasiparticle random phase approximation (QRPA), e.g., [9, 17, 21]. Interestingly, the variations among different approaches which reach orders of magnitude for long-lived nuclei narrow far from stability to approximately a factor of two [9], as Q-values increase and the lifetimes become less sensitive to details of the estimated nuclear structure. Monte Carlo determinations of r-process abundance pattern uncertainties due to this factor of two uncertainty in β-decay rates are presented in Fig. 17 in [9].

The reach of FRIB for new lifetime measurements is also shown in Fig. 3, for the Day 1 and design beam intensities. The improvements to r-process uncertainties due to extending lifetime measurements to the FRIB full range are not anticipated to be as dramatic as in the case of nuclear masses for the reasons mentioned above; the current half-life data extends further into neutron-rich regions than the mass data and the theoretical model disparities are not as large. Instead the importance shifts to providing new measurements of lifetimes near the $N \sim 126$ closed shell region, reducing the errors on existing measurements, and measuring the probabilities P_n that β decay is followed by the emission of one or more neutrons. Also of key importance is the determination of nuclear structure information via the β-decay strength functions [23]. Nuclei close to the drip line will always be out of reach of experiment, and lifetimes alone are insufficient to constrain theory. P_n measurements and improvements to lifetimes beyond a factor of two are required to study the fine details of the final abundance pattern, for example, in the rare earth peak [22], see also Fig. 14 of [9].

Precision predictions of r-process abundance patterns and careful interpretation of kilonova signals require improvements in our understanding of nuclear physics far from stability. Here we examined the part that the Facility for Rare Isotope Beams is poised to play in realizing these improvements. For nuclear masses, dedicated experimental campaigns for direct measurements alone have the potential to result in dramatic improvements in r-process abundance predictions. For β decay, the greatest need is to go beyond lifetime measurements to investigate β-strength functions and β-delayed neutron emission probabilities [24]. Also important to note is the promise of FRIB to contribute to unknown neutron capture rates through transfer reaction [25] and β-Oslo [7] measurements. We encourage FRIB to fully embrace its potential role as the next-generation r-process machine in planning for Day 1 experiments and beyond.

Acknowledgements This work was supported in part by U.S. Department of Energy under contract No. DE-SC0013039 (RS) and by the National Nuclear Security Administration of the U.S. Department of Energy at Los Alamos National Laboratory under Contract No. DE-AC52-06NA25396 (MM).

References

[1] B.P. Abbott *et al.* Phys. Rev. Lett. **119**, 161101, (2017)
[2] P.S. Cowperthwaite *et al.*, Astrophys. J. Lett. **848**, L17 (2017).
[3] D. Kasen, B. Metzger, J. Barnes, E. Quataert, E. Ramirez-Ruiz, Nature **551**, 80 (2017)
[4] B. Côté, C.L. Fryer, K. Belczynski, O. Korobkin, M. Chruślińska, N. Vassh, M.R. Mumpower, J. Lippuner, T. M. Sprouse, R. Surman, R. Wollaeger, accepted in Astrophys. J., arXiv:1710.05875 (2018)
[5] M. Arnould, S. Goriely, K. Takahashi, Phys. Rep. **450**, 97 (2007)
[6] M.R. Mumpower, R. Surman, G.C. McLaughlin, A. Aprahamian *Prog. Part. Nucl. Phys.* **86**, 86 (2016)
[7] S.N. Liddick *et al.*, Phys. Rev. Lett. **116**, 242502 (2016)
[8] R. Surman, M. Mumpower, A. Aprahamian, Acta Phys. Polonica B **47**, 673 (2016)
[9] T. Shafer, J. Engel, C. Fröhlich, G.C. McLaughlin, M. Mumpower, R. Surman, Phys. Rev. C **94**, 055802 (2016)
[10] J. Barnes, D. Kasen, M.-R. Wu, G. Martínez-Pinedo, Astrophys. J. **829**, 110 (2016)
[11] M. Wang, G. Audi, F. G. Kondev, W. J. Huang, S. Naimi, X. Xu, Chinese Phys. C **41** 030003 (2017)

[12] R. Surman, M. Mumpower, G.C. McLaughlin, 14th International Symposium on Nuclei in the Cosmos (NIC2016) 010612 (2017)
[13] M. Mumpower, R. Surman, D.-L. Fang, M. Beard, P. Möller, T. Kawano, A. Aprahamian, Phys. Rev. C **92**, 035807 (2015)
[14] G. Audi, F.G. Kondev, M. Wang, W.J. Huang, S. Naimi, Chinese Phys. C **41**,030001 (2017)
[15] P. Möller, A.J. Sierk, T. Ichikawa, H. Sagawa, At. Data Nucl. Data Tables **109**, 1 (2016)
[16] https://groups.nscl.msu.edu/frib/rates/fribrates106.html
[17] P. Möller, B. Pfeiffer, K.-L. Kratz, Phys. Rev. C **67**, 055802 (2003)
[18] T. Rauscher, F.-K. Thielemann, At. Data Nucl. Data Tables **75**, 1 (2000)
[19] G. Lorusso *et al.*, Phys. Rev. Lett. **114**, 192501 (2015)
[20] J. Wu *et al.*, Phys. Rev. Lett. **118**, 072701 (2017)
[21] T. Marketin, L. Huther, G. Martínez-Pinedo, Phys. Rev. C **93**, 025805 (2016)
[22] M.R. Mumpower, G.C. McLaughlin, R. Surman, A.W. Steiner, J. Phys. G **44**, 034003 (2017)
[23] A. Spyrou *et al.*, Phys. Rev. Lett. **117**, 142701 (2016)
[24] R. Grzywacz *et al.*, FRIB Decay Station white paper, in preparation
[25] J.A. Cizewski, A. Ratkiewicz, J.E. Escher, A.Lepailleur, S.D. Pain, G. Potel, EPJ Web of Conferences **165**, 01013 (2018)

EPJ Web of Conferences **178**, 04003 (2018) https://doi.org/10.1051/epjconf/201817804003
CGS16

Astrophysical relevance of the low-energy dipole strength of ^{206}Pb

A.P. Tonchev[1,2], N. Tsoneva[3], S. Goriely[4], C. Bhatia[2,5], C.W. Arnold[6], S.L. Hammond[7,5], J.H. Kelley[8,5], E. Kwan[9], H. Lenske[3], J. Piekarewicz[10], R. Raut[11], G. Rusev[6], T. Shizuma[12], and W. Tornow[2,5]

[1]Nuclear and Chemical Sciences Division, Lawrence Livermore National Laboratory, Livermore, California 94550, USA
[2]Duke University, Department of Physics, Box 90308, Durham, North Carolina 27708-0308, USA
[3]Institut für Theoretische Physik, Universität Gießen, Gießen, D-35392, Germany
[4]Institut d'Astronomie et d'Astrophysique, Université Libre de Bruxelles, Campus de la Plaine, CP-226, 1050 Brussels, Belgium
[5]Triangle Universities Nuclear Laboratory, Durham, North Carolina 27708-0308, USA
[6]Chemistry Division, Los Alamos National Laboratory, Los Alamos, New Mexico 87545, USA
[7]Department of Physics and Astronomy, University of North Carolina at Chapel Hill, Chapel Hill, North Carolina 27599-3255, USA
[8]North Carolina State University, Department of Physics, Box 8202, Raleigh, North Carolina 27695-8202, USA
[9]National Superconducting Cyclotron Laboratory, East Lansing, Michigan 48824, USA
[10]Department of Physics, Florida State University, Tallahassee, Florida 32306-4350, USA
[11]UGC-DAE Consortium for Scientic Research, Kolkata Centre LB-8 Sector-III, Bidhannagar, Kolkata 700098, India
[12]Quantum Beam Science Center, Japan Atomic Energy Agency, Tokaimura, Ibaraki 319-1184, Japan

Abstract. The dipole strength of ^{206}Pb was studied below the neutron separation energy using photon scattering experiments at the HIGS facility. Utilizing the technique of nuclear resonance fluorescence with 100% linearly-polarized photon beams, the spins, parities, branching ratios and decay widths of excited states in ^{206}Pb from 4.9 - 8.1 MeV have been measured. The new experimental information is used to reliably predict the neutron capture cross section of ^{205}Pb, an important branch point nucleus along the s-process path of nucleosynthesis.

Introduction

Extinct ^{205}Pb isotopes can be used as powerful cosmochemical tracers in the early solar system [1]. The nuclide ^{205}Pb, which decays to ^{205}Tl with a half-life of 15.3 Myr is of particular astrophysical importance as it is the only short-lived nuclide produced solely by *s*-process nucleosynthesis. As can be seen in Fig. 1, ^{205}Pb is screened from the *r*-process by the stable ^{205}Tl isotope and provides highly important chronometric information about the time span between the last *s*-process nucleosynthetic events that have modified the composition of the solar nebula and the formation of solid bodies in the solar system. The abundance of ^{205}Pb in the early solar system inferred from carbonaceous chondrites data can also be used to assess whether Asymptotic Giant Branch (AGB) stars and massive Wolf-Rayet stars are the most likely sites of the *s*-process [4,5]. However, this investigation remains sensitive to the amount of freshly produced *s*-process material, including the survival of ^{205}Pb in AGB stars [6] through both neutron capture and weak interaction processes. Thus, one needs to explain the initial absolute *s*-process abundance of stable lead isotopes which depend on the neutron capture rates. At present the value of the cross section used for calculating abundances in stellar models are purely based on theoretical Hauser-Feshbash predictions [7]. There is an upper limit for the ^{205}Pb - ^{205}Tl abundance ratio of 9x10^{-5} from meteorites [8]. However, no experimental information is available. Due to its astrophysical importance, the Maxwellian-averaged capture cross section on ^{205}Pb needs to be known accurately, and if not measurable directly, it should at least be experimentally constrained. Accurate capture cross-section measurements on branching-point nuclei such as ^{205}Pb and ^{204}Tl could be used to provide information on the mean neutron flux during the *s*-process.

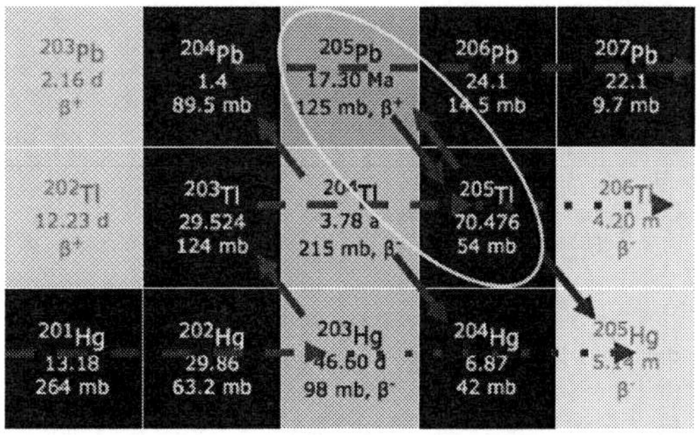

Fig. 1. *s*-process path throughout the thallium-lead isotopes. The green arrows show the s-process path, while the blue and red arrows show β^- and β^+ decays respectively.

* Corresponding author: tonchev2@llnl.gov

Furthermore, the presence of additional Pygmy Dipole Resonance (PDR) strength below and near the neutron separation threshold has an impact on neutron-capture rates of astrophysical importance [9, 10]. These J =1⁻ states forming the PDR act as wide open doorway states in the neutron capture process. The short lifetimes of these states permit for fast depopulation of the capture levels towards lower-lying bound states, and stabilize the newly formed nucleus against subsequent neutron emission. Applying detailed balance in the photon bath formed in astrophysical *r*-process event suggests the PDR levels near the neutron threshold will enhance the reverse process of (γ,n) photodissociation.

Experimental technique

The Nuclear Resonance Fluorescence (NRF) technique [11] was used to measure the nuclear dipole response of ^{206}Pb at the High-Intensity Gamma-Ray Source (HIGS) [12]. Linearly-polarized photons were produced via Compton backscattering of a relativistic electrons from free-electron laser photons. The resultant high intensity photon beam travels 53 m in a high-vacuum tube before passing through a lead collimator of length 30.5 cm with a cylindrical hole of 1.27 cm diameter located 4 m upstream of the ^{206}Pb target. The energy distribution of the photon flux was measured with a large volume High Purity Germanium (HPGe) detector, of efficiency 123% relative to a 7.62 cm x 7.62 cm NaI scintillator, placed in the incident beam. For this measurement, the beam was attenuated by copper absorbers mounted upstream. The true energy spectrum from 6500 keV incident photon beam, shown in green, is given in Fig. 2 (a). This spectrum was obtained after removing the response of the HPGe detector (yellow) [13,16]. The energy range of interest was scanned, with beam centroid energies, incrementing in steps of approximately the FWHM of the beam (~200 keV), from 4.8 MeV up to 8.3 MeV. These spectra are shown in Fig. 2 (b).

The sample was 4000 mg of 99.3% enriched ^{206}Pb metal powder. The scattered rays from the lead target were measured by four HPGe detectors, with 60% relative efficiency, positioned around the ^{206}Pb target at (θ,φ) = (90°, 0°), (90°, 90°), (90°, 180°), and (90°, 270°), where θ is the polar angle with respect to the horizontally polarized incoming photon beam and φ is the azimuthal angle measured from the polarization plane. A fifth detector, of 25%, relative efficiency was placed at (θ,φ) = (135°, 0°) to distinguish the spins of positive parity states. The distance between the center of the target to the front surface of the 90° detectors was 10 cm. All detectors had passive shielding consisting of 3 mm copper and 2 cm thick lead cylinders. Lead and copper absorbers of thickness 5 and 3 mm, respectively, covered the openings of the detectors to reduce the low-energy part of the scattered spectrum. The absolute photopeak efficiencies of the HPGe detectors were measured using a set of γ-ray sources placed in the target position. The measured efficiency was also simulated using MCNP6 [17, 49].

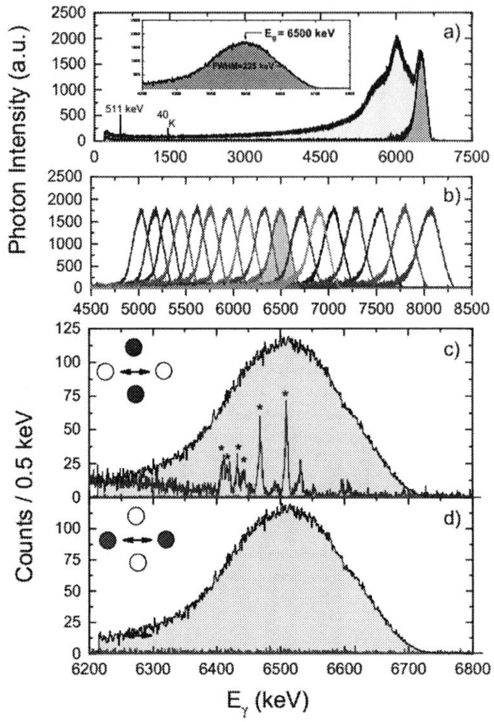

Fig. 2. (a) A photon beam with energy peaked at 6500 keV generated the γ-ray spectra. (b) The photon energy spectra from all beam energies used in this experiment. (c) The 90 up/down spectra, indicating E1 transitions. (d) The 90 left/right spectra, indicating possible M1 transitions.

Results

The primary results from photon scattering on ^{206}Pb was published in Ref. [18]. To calculate the partial decay width Γ_0^x for the newly observed state we compare to the partial decay width of the known states, Γ_0^r, observed in Refs. [14, 15], by:

$$\Gamma_0^x = \frac{N^x F^r (E_\gamma^x)^2}{N^r F^x (E_\gamma^r)^2} \Gamma_0^r$$

where N^x (N^r) is the observed number of counts in an elastic peak in the unknown (reference) state, corrected for the detector efficiency and, if necessary, resonant self-absorption. The normalization factor F^x (F^r) depends on the unknown (reference) states position in the beam energy distribution, corresponding to the relative flux while E_γ^x (E_γ^r) are the transition γ-ray energy for the unknown (reference) state. This method was validated using ^{11}B resonance states [16]. A portion of the of NRF γ-ray spectra from the photon beam with peak energy of 6500 keV is presented in Fig. 2 (c) and (d). In this particular energy interval, the γ-ray transitions are only steaming from the vertical detectors, showing unambiguous distinction between E1 and M1 transitions [21, 29]. More detailed information about the NRF technique used at the HIGS facility can be found in Refs. [19, 20, 22-31].

With higher photon fluxes available at the HIGS facility over the entire energy range, many new states of interest were observed comprised to Refs. [14, 15]. For example, in Fig. 2 (c) the know-resonance states around 6.5 MeV excitation energy from Refs. [14, 15] are labeled with stars. Hence, some relatively weak transitions (with $\Gamma_0^x < 100$ meV) are left undetected in the bremsstrahlung NRF experiments. One hundred 1^- states were observed. Twenty-six M1 excited states were also measured. However, it is clear that in this energy region the dipole response is predominantly electric in nature. The M1 strength is localized in two groups around 6.0 and 7.5 MeV which may be attributed to the M1 spin- resonance [29, 30].

The measured elastic, inelastic, and total photoabsorption cross sections of ^{206}Pb, integrated over 250 keV bins, are shown in Fig. 3. As can be seen for excitation energies lower than ~ 6 MeV the total photoabsorption cross section is determined by the elastic cross section. Near the neutron separation energy of 8.1 MeV, the elastic and inelastic cross sections become equal. There is a sudden drop of the elastic (and the corresponding total) cross section at about 6.25 MeV.

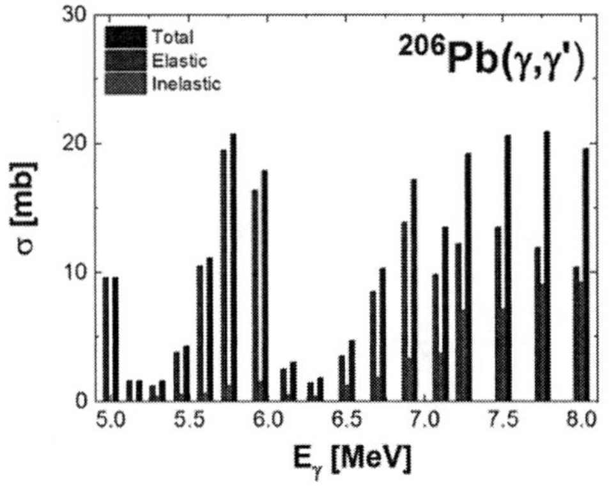

Fig. 3. Elastic, inelastic, and total photoabsorption cross section of ^{206}Pb below the one-neutron separation energy (S_n= 8.1 MeV) averaged over 250 keV energy bins.

Astrophysical relevance

In order to constrain the capture cross section on ^{205}Pb we carried out statistical model calculations using the TALYS-1.8 code [32]. The results are shown in Fig. 4 a) where the "Recommended" curve corresponds to the radiative cross section obtained with the presently measured E1 and M1 γ-ray strength, complemented by the energy–density functional plus quasiparticle–phonon model (EDF+QPM) predictions outside the experimental energy range. As shown in Fig. 4 (a), the QPM supplemented with the QRPA calculations lead to a cross section in excellent agreement with the experimentally constrained recommended one. The uncertainty band in Fig. 4 (a) stems from the experimental uncertainties associated with the γ-ray strength but also the use of different nuclear level density models [33, 34]. The recommended cross section was obtained with the combinatorial model of Ref. [35]. Note that level densities are constrained by the cumulative number of low-lying levels, including the J^π=1 states, reliably measured in the present experiment up to about 6.5 MeV. The contribution of the M1 strength as well as the PDR plus core polarization contributions to the cross section is illustrated in Fig. 4 (b), where the M1 component is seen to be rather insignificant (less than 5%), but the PDR plus core polarization contributions are crucial for a proper cross-section description.

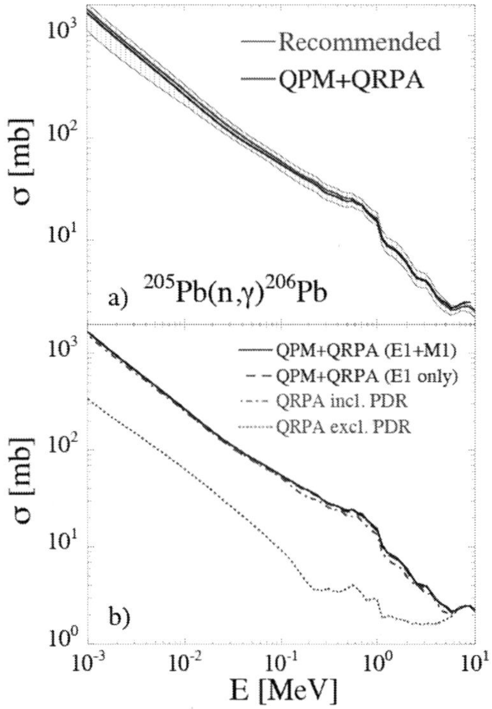

Fig. 4. (a) Radiative capture cross section ^{205}Pb(n,γ)^{206}Pb using as input the experimental E1 and M1 dipole strength (red curve) and three-phonon EDF+QPM plus EDF+QRPA predictions (blue curve). (b) Same (a) but now assessing the impact of the various energy regions and multipolarities to the EDF+QPM plus EDF+QRPA calculations. See text for further details.

The experimentally constrained, Maxwellian-averaged cross section at 30 keV is estimated to be 130±25 mb, close to the theoretical value of 125±22 mb recommended in Ref. [36] and traditionally used in *s*-process calculations. With this updated cross section, the AGB origin of ^{205}Pb on the basis of its early solar system abundance and present nucleosynthesis models can be put on more solid ground, despite the uncertainties still affecting the distraction of ^{205}Pb by weak processes and the *s*-process model in AGB stars.

Summary

In conclusion, experimental high-resolution studies of the PDR mode and EDF+ QPM predictions of the electromagnetic response of ^{206}Pb permit separation of the PDR from the tail of the GDR and multi-phonon excitations due to core polarization effects. Our findings suggest that the low-energy dipole strength is predominantly electric in character and mainly due to a PDR skin oscillation. However, a substantial contribution from both the low-energy tail of the GDR and multi-phonon states to the total E1 strength responsible for the fragmentation pattern of low-energy dipole states is also observed. Moreover, the EDF+QPM theory successfully reproduces the low-energy M1 spectral distribution, suggesting that this distribution is mostly due to spin-flip excitations.

In the context of stellar nucleosynthesis, an updated, experimentally-constrained, Maxwellian-averaged radiative capture cross section for the ^{205}Pb(n,γ)^{206}Pb reaction is obtained. The study illustrates the vital and ever increasing role that measurements of exotic modes of excitation in neutron-rich nuclei are playing in the determination of observables of critical astrophysical importance.

Acknowledgments

This work was performed under the auspices of US DOE by LLNL under contract DE-AC52-07NA27344 and the DOE grants DE-FG02-97ER41033, DE-FG02-97ER41041, DE-FG02-97ER41042, DE-FG52 06NA26155, DFG grant Le439/6, and the U.S. Department of Energy, Office of Nuclear Physics, under Grant No. DE-FG02- ER41033.

References

1. C. Iliadis. Nuclear Physics of Starts. Wiley-VCH Verlag GmbH&Co. KGaA (2015).
2. S.G. Nielsen, M., Rehkämper, A.N. Halliday, Geochim. Cosmochim. Acta **70**, 2643 (2006).
3. R.G.A. Baker, M. Schönbächler, M. Rehkämper, H.M. Williams, A.N. Halliday, Earth Planet. Sci. Lett. **291**, 39 (2010).
4. M. Arnould, G. Paulus, G. Meynet, Astron. Astrophys. **321**, 452 (1997).
5. G.J. Wasserburg, M. Busso, R. Gallino, K.M. Nollett, Nucl. Phys. A **777**, 5 (2006).
6. N, Mowlavi, S. Goriely, M. Arnould, Astron. Astrophys. **330**, 206 (1985).
7. KADoNiS website at http://www.kadonis.org.
8. K. Yokoi et al., Astron. Astrophys. **145**, 339 (1985).
9. S. Goriely, Phys. Lett. B **436**, 10 (1998).
10. N. Tsoneva et. al., Phys. Rev. C **91**, 044319 (2015).
11. U. Kneissl et al., Prog. Part. Nucl. Phys. **34**, 285 (1995).
12. H. Weller et al., Prog. Part. Nucl. Phys. **62**, 257 (2009).
13. C. Sun, Y. K.Wu, J. Li, G. Rusev, and A. P. Tonchev, Phys. Rev. Spec. Top. **12**, 062801 (2009).
14. T. Chapuran, R. Vodhanel, M.K. Brusssel, Phys. Rev. C **2**, 1420 (1980).
15. J. Enders et al., Nucl. Phys. A **741**, 3 (2004).
16. G. Rusev et al., Phys. Rev. C **79**, 061302 (2009).
17. S. Carson et al. Nuc. Instr. Meth. Phys. Res. A **618**, 190 (2010).
18. A.P. Tonchev et al., Phys. Lett. B **773**, 20 (2017).
19. R. Massarczyk et al., Phys. Rev. Lett. **112**, 072501 (2014).
20. N. Pietralla et al., Phys. Rev. Lett. **88**, 012502 (2002).
21. A.P. Tonchev et al., Phys. Rev. Lett. **104**, 072501 (2010).
22. P.M. Goddard et al., Phys. Rev. C **88**, 064308 (2013).
23. C. Romig et al., Phys. Rev. C **88**, 044331 (2013).
24. M. Scheck et al., Phys. Rev. C **87**, 051304(R) (2013).
25. R. Schwengner at al., Phys. Rev. C. **87**, 024306 (2013).
26. S. Volz et al., Nucl. Phys. A **779**, 1 (2006).
27. H. Ohgaki, et al. Nucl. Phys. A **649**, 73c (1999).
28. R. Schwengner et al., Phys. Rev. C **78**, 064314 (2008).
29. G. Rusev et al., Phys. Rev. Lett. **104**, 072501 (2013).
30. Krishichayan et al., Phys. Rev. C **91**, 044328 (2015).
31. B. Leher et al., Phys. Lett. B **756**, 72 (2016).
32. A.J. Koning, D. Rochman, Nuclear Data Sheets **113**, 2841 (2012).
33. A.J. Koning, S. Hilaire, S. Goriely, Nucl. Phys. A **810**, 13 (2008).
34. R. Capote, M. Herman, P. Oblozinsky, et al., Nuclear Data Sheets **110**, 3107 (2009).
35. S. Goriely, S. Hilaire, and A.J. Koning, Phys. Rev. C **78**, 064307 (2008)
36. Z. Y. Bao et al., Atomic Data and Nuclear Data Tables **76**, 70 (2000).

The Great Wall: Urca Cooling Layers in the Accreted NS Crust

Zach Meisel[1,*]

[1]Department of Physics and Astronomy, Ohio University, 45701 Athens, Ohio USA

Abstract. Accreting neutron stars host a number of astronomical observables which can be used to infer the properties of the underlying dense matter. These observables are sensitive to the heating and cooling processes taking place in the accreted neutron star (NS) crust. Within the past few years it has become apparent that electron-capture/beta-decay (urca) cycles can operate within the NS crust at high temperatures. Layers of nuclei undergoing urca cycling can create a thermal barrier, or Great Wall, between heating occurring deep in the crust and the regions above the urca layers. This paper briefly reviews the urca process and the implications for observables from accreting neutron stars.

1 Introduction

Describing how matter behaves across the range of temperatures and densities occurring in nature is the ultimate goal of much of modern physics research. The only avenue presently available to study matter at the highest densities, near and above those of an atomic nucleus, at relatively low temperatures is the study of neutron stars. Neutron stars accreting material from a binary companion are particularly useful in this regard, as their observable phenomena provide clues as to the underlying structure of the neutron star (NS).

Gleaning information from such observables requires comparisons between the observed data and results from astrophysics model calculations. However, successful reproduction of the observed data requires high quality physics input. As such, the model-observation comparison process is inherently iterative, where the model input physics is adjusted until agreement is achieved. Unfortunately, many degeneracies exist and so efforts must be made to reduce the number of free parameters in models by constraining the physics input as best as possible.

While much of the physics of dense matter is beyond the reach of terrestrial experiments, and so only theoretical constraints can be obtained, the nuclear physics of relevance for the NS outer layers can often be measured. In particular, the nuclear reactions of the outer crust and above can be constrained using indirect measurements at present stable and radioactive ion-beam facilities.

The nuclear physics of particular interest for this work is the phenomenon of electron-capture/β^--decay cycling, known as urca cooling, which was recently found to be active in the crusts of accreting neutron stars. Prior to describing urca cooling and its impact, a brief discussion of reactions in the outer NS (shown schematically in Figure 1) will help provide context.

2 Journey of a Nucleus in the Outer NS

Low mass X-ray binary systems which host observables such as type-I X-ray bursts, X-ray superbursts, and decaying X-ray emission from NS crust cooling, are generally thought to involve accretion of hydrogen and/or helium-rich material onto the NS surface. The H/He fuel builds-up on the NS surface, bringing in roughly 200 MeV per accreted nucleon. The combination of heat and fuel results in nuclear burning on and near the NS surface which produces a variety of heavier nuclei. A number of burning processes produce iron-rich ashes, though a wide variety of nuclei up to the tin region can be made, including special burning regimes which enrich the NS ocean in carbon [1–4].

Subsequent accretion buries the surface burning ashes to deeper, denser conditions. The electron Fermi energy $E_{e^-,Fermi}$ rises with increasing depth, so that buried nuclei eventually encounter a region where the degenerate electron gas can pay the energy debt required for electron capture Q_{EC}. Near $E_{e^-,Fermi} \approx Q_{EC}$, electron-capture EC proceeds, transforming a proton within a nucleus into neutron. Therefore, in the upper regions of the outer crust, where EC is the dominant reaction mechanism, buried ashes maintain their original abundance distribution in terms of mass number A, but are transformed into more and more neutron rich isobars.

Eventually, EC has driven the composition near to the neutron-drip line, so that neutron emission via (EC,*x*n), where *x* is some integer, can occur. The neutron-drip point marks the transition between the outer and inner crust. At similar and deeper densities, the zero-point vibrations of nuclei are sufficient to tunnel through the energy barrier between them, so that density driven (pycnonuclear) fusion proceeds. Ultimately the nuclei that were born on and near the NS surface will dissolve into the core once buried to a depth near nuclear

* Email: meisel@ohio.edu

saturation density. The entire journey takes on the order of a million years for an NS continually accreting near the Eddington accretion rate (The rate at which the force of radiation emitted from accretion is balanced with the inward force of gravity).

The combination of EC and fusion reactions just described dramatically transforms the outer NS composition and drives it from thermal equilibrium [5].

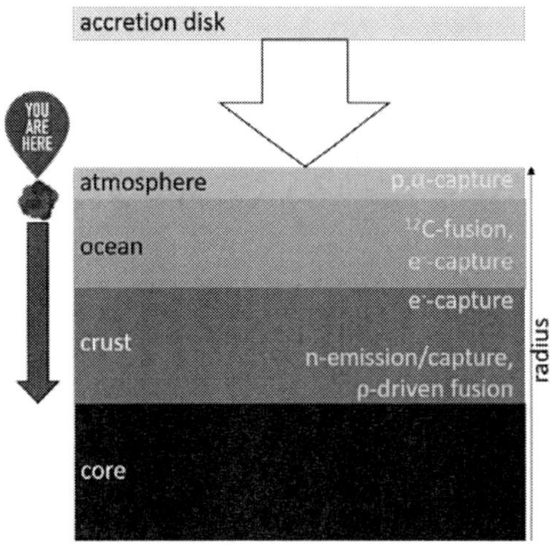

Fig. 1. Nuclear processes relevant for a nucleus as it is buried by accretion to various depths in the NS outer layers.

3 Urca Cycling and the Great Wall

For most of the time since EC reactions in the NS crust were hypothesized, they were considered to be heat sources. In particular, EC reactions to excited states, which are likely prevalent in two-step EC reactions that occur for even-A isobars, can release several-hundred keV per EC [6]. The reverse process of β^--decay went mostly ignored since the electron-degenerate environment implies most of the phase-space is blocked for an electron to re-enter the environment. However, recent calculations of the thermal and compositional evolution of an accreting crust employing a multi-species nuclear reaction network showed that β^--decay is likely possible for a large number of nuclei [7]. Once β^--decay proceeds for such cases, EC is still energetically favorable and so the process can repeat cyclically. This is generally limited to odd-A nuclides.

EC/β^--decay cycling was proposed for dense stellar interiors by Gamow and Schoenberg, who coined it the urca process [8,9]. The cycling occurs as shown in Figure 2, where EC proceeds near Q_{EC} to the ground or a low-lying excited state of the EC daughter. Low-lying states of the EC parent can also participate due to thermal population. In each case, "low-lying" means on the order of k_BT, where k_B is the Boltzmann constant and T is the environment temperature. The neutrino luminosity L_ν for isotopes undergoing urca cycling, referred to as an urca pair, is proportional to the weak transition rate between the nuclides, as quantified by the comparative half-life *ft*; the phase-space for the weak transition, which is sensitive to Q_{EC}; and the thickness of the urca layer, which is proportional to k_BT and inversely proportional to the local acceleration due to gravity *g*.

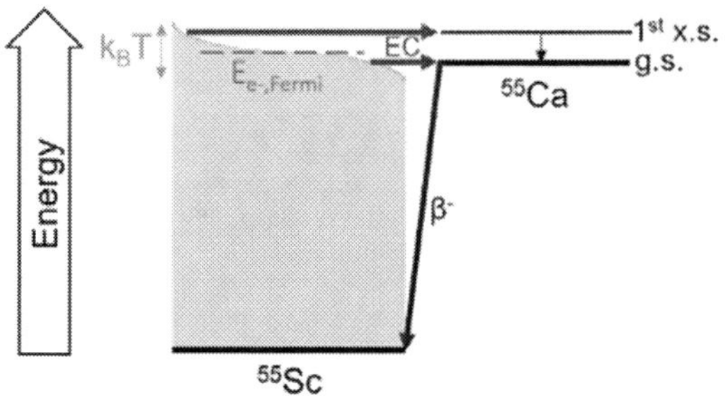

Fig. 2. Urca cycling between ^{55}Sc and ^{55}Ca at the depth in the NS crust where $E_{e^-,Fermi} = Q_{EC}(^{55}Sc)$, including a hypothetical low-lying excited state in ^{55}Ca which is only included for illustrative purposes.

Quantitatively, for an EC parent ^AZ [10,11]: $L_\nu(Z,A) \approx L_{34}(Z,A)X(A)T_9^5(2/g_{14})(R/10km)^2\cdot 10^{34}$ erg/s, where X(A) is the mass fraction of A, T_9 is the local temperature in gigakelvin, g_{14} is $g/10^{14}$ cm/s^2, and R is the NS radius in kilometers. $L_{34}(Z,A)$ is the intrinsic cooling strength for an urca pair with EC parent Z,A: $L_{34} = 0.87(10^6 s/ft)(56/A)(Q_{EC}(Z,A)/4MeV)^5(2\langle F\rangle^*)$, where $\langle F\rangle^* = (\langle F\rangle^+\langle F\rangle^-)/(\langle F\rangle^+ + \langle F\rangle^-)$, the Coulomb factor $\langle F\rangle^\pm \approx 2\pi\alpha/|1-\exp(\mp 2\pi\alpha Z)|$, and α is the fine structure constant. For this equation *ft* is the average *ft*-value for the EC and β^--decay, which can differ slightly due to different state degeneracies.

For context, typical crustal heating is, depending on the heat deposited per accreted nucleon and the accretion rate, on the order of a few $\times 10^{34}$–10^{36} erg/s [7]. In practice, urca cooling impacts the local environment temperature significantly enough to impact observables (see the following section) when the temperature is around ~1GK and $X(A)L_{34}(Z,A) \gtrsim 1$. For example, if X(A) ~ 1%, this would be roughly satisfied if Q_{EC} ~ 8 MeV and log(*ft*) ~ 5, or if Q_{EC} ~ 15 MeV and log(*ft*) ~ 6.5. Of course, combinations of larger Q_{EC} and smaller log(*ft*) result in larger L_ν.

Estimates of L_ν are sensitive to nuclear physics input through Q_{EC}, *ft*, and X(A). For the majority of nuclei in the NS crust, Q_{EC} is known to a few percent, though theoretical predictions for unmeasured cases can differ by more than 10%. On the other hand, *ft* is generally unknown and is often uncertain by several orders of magnitude. X(A) relies on results from model calculations of surface nuclear burning processes, which have many astrophysics uncertainties, but even the uncertainties in nuclear reaction rates can impact X(A) by orders of magnitude [12].

As an example, consider the urca pair ^{55}Sc–^{55}Ca. The atomic mass excess ME of ^{55}Sc is known to roughly half an MeV [13] and the mass of ^{55}Ca is unmeasured, though the disagreement between theoretical predictions in this region can be on the order of an MeV. As such, the uncertainty for Q_{EC} (= ME(Z,A) - ME(Z-1,A)) in this case is roughly 13%, resulting in around a factor of 2 uncertainty in L_ν. The value of *ft* has not been measured for the ground-state to ground-state transition between ^{55}Sc and ^{55}Ca, so an estimate must be made. QRPA and shell-model calculations for this transition predict log(*ft*) = 4.2 and 3.9, respectively [7,14]. Alternatively, a data-based systematics approach [15] can be used (which avoids costly, large-scale calculations). From a compilation of log(*ft*) for transitions with a known change in spin *J* and parity *π*, log(*ft*) can be adopted based on a transition's $\Delta J^{\Delta\pi}$. For the ^{55}Sc–^{55}Ca ground-state to ground-state transition, $\Delta J^{\Delta\pi} = 1^+$, according to tentative J^π assignments [16]. From a compilation [17], it is apparent that the range of log(*ft*) for $\Delta J^{\Delta\pi} = 1^+$ for ~700 odd-A nuclei is between 3.5–9.1, where the distribution is centered about log(*ft*) = 5.9 ± 1. Thus, a conservative estimate for the uncertainty contribution of *ft* to L_ν is a factor of 10. Estimating the uncertainty in X(A) is very difficult, though an empirical upper-limit for odd-A nuclides appears to be somewhere near X(A) < 10%. For X-ray bursts it is likely more than an order of magnitude due to nuclear reaction rate uncertainties alone [12]. For X-ray superbursts, the X(A) distribution is mostly determined by nuclear statistical equilibrium [18] and is therefore determined by the relatively well known ME for nuclei around stable iron. However, the uncertainty contribution to late-time H/He capture reactions on superburst ashes has not been explored. Therefore, it is clear that a number of nuclear physics experiments are needed to adequately constrain L_ν for this case and many more like it (see section 5).

The consequence of including an urca layer in the NS crust is that a high-temperature thermostat exists at that depth. The T^5 dependence ensures that any significant heat flux encountering the urca layer will be bled away by neutrinos. This amounts to a thermal barrier, or Great Wall, between the deep crust, where (EC,xn) and pycnonuclear fusion occur, and the ocean and atmosphere where H/He and C burning drive bursts. As an example, Figure 3 shows the impact of an urca layer for an NS crust heated to over 1 GK by accretion.

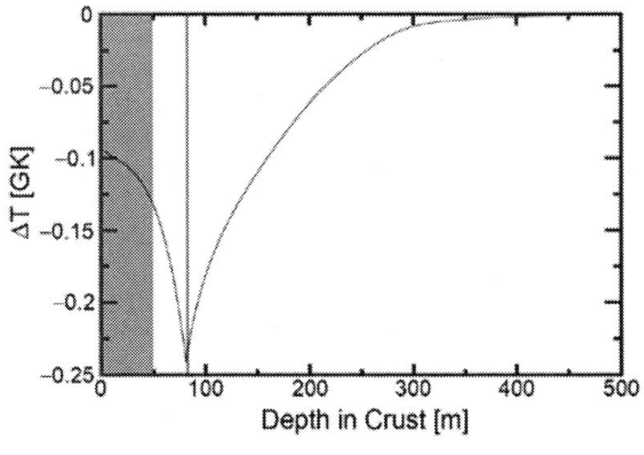

Fig. 3. Residual for the NS crust temperature after 480 days of accretion at 1.3×10^{18} g/s, assuming a shallow heating (red region) of 8 MeV per accreted nucleon in a crust with impurity parameter (variance in the nuclear charge) $Q_{imp} = (1/n_{ion})\sum_j n_j(Z - \langle Z\rangle)^2 = 1$, between `dStar` [19] models with and without Urca cooling. The cooler model adopted an urca layer (blue line) mimicking ^{55}Sc–^{55}Ca cycling assuming log(*ft*)=4.9 and X(55) =1.8%.

4 Implications for Observables

A number of accreting NS observables are sensitive to the thermal properties of the NS outer layers. The ignition of X-ray bursts and superbursts are sensitive to the heat flux from the underlying NS, often referred to as base heating. Cooling transients directly probe the NS composition and temperature as a function of depth, since the X-ray flux at any given point in the cooling light curve originates from the depth with which the surface has reached thermal equilibrium. It is then clear that urca cooling is of interest for accreting NS model-observation comparisons.

For X-ray superbursts, an unphysical urca cooling L_ν would be required to alter the light curve. However, $X(A)L_{34}(Z,A) \gtrsim 10$ is sufficient to modify the depth of the carbon ignition which powers superbursts. Within estimated uncertainties for log(*ft*), the superburst ignition depth predicted from model calculations could be lowered to more than an order of magnitude larger pressure [11]. Interestingly, the ignition depth inferred from observations is already an order of magnitude shallower than found in models [20]. An as-yet unidentified shallow heating source has been invoked to reconcile the discrepancy (and is also required for adequate reproduction of most cooling transients [21] and some type-I X-ray bursters [22]). As A=55 ashes are robustly produced in calculations of superburst ashes, present nuclear physics estimates would require the shallow heat source be located above the ^{55}Sc EC depth, as deeper heat sources would be thermally isolated from heating the NS ocean due to ^{55}Sc–^{55}Ca urca cooling.

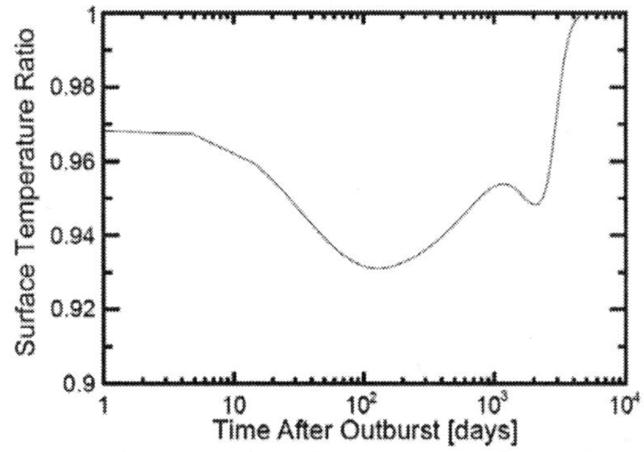

Fig. 4. Ratio between the surface temperatures for the `dStar` models described in Figure 3 for the time after accretion has ceased. The hundred-day dip is directly due to the inclusion of the urca layer, whereas the feature near one thousand days is due to the change in absolute time when the surface reaches equilibrium with deep crustal heating when an urca layer is included. Note that typical observational uncertainties are on the percent level [23,24].

The impact on cooling transients is limited to NS crusts which have been heated out of equilibrium to a temperature of ~1 GK, as is thought to be the case for the source MAXI J0556-332 for its earliest observed crust cooling phase [23,24]. For such a hot NS crust, an urca layer with $X(A)L_{34}(Z,A) \gtrsim 1$ is sufficient to appreciably alter the outer NS crust temperature profile [15] (Figure

3 shows an example with $X(A)L_{34}(Z,A) \approx 44$). Due to the associated modifications in thermal gradients in the NS crust, heat transport time-scales are modified and therefore the rate at which the surface reaches thermal equilibrium with particular depths is changed. The result is a dip that is introduced in the crust cooling light curve around ~100 days after accretion ceases (see Figure 4). Though a large number of poorly constrained astrophysics inputs contribute to uncertainties in models of cooling transient light curves [21], and many inputs are degenerate in their impacts, no other known mechanism exists to mimic the hundred-day dip. Therefore, a positive identification of such a dip would be direct evidence for urca cooling in the crust, possibly validating surface nucleosynthesis models.

Though urca cooling has gone undetected, one can use the negative result for MAXI J0556-332 to place interesting constraints on this source. As the urca mechanism is robust when using a wide range of nuclear physics and astrophysics model assumptions, the lack of an urca cooling signature in the cooling transient light curve for an NS with a hot crust implies that urca nuclei are not present at $E_{e^-,Fermi} \approx Q_{EC}$. This then implies that these nuclei were not produced on or near the NS surface in the past, where the exact time constraint depends on the assumed past accretion rate and duty cycle. Since both X-ray burst and superburst ashes are thought to produce urca nuclei with significant associated L_ν, we can conclude that MAXI J0556-332 did not exhibit type-I X-ray bursts or superbursts in the past centuries to millenia (or at least not for sustained periods) [15]. This source presently exhibits stable burning near the Eddington accretion rate, which is not thought to yield significant abundances of urca nuclides [1,15]. Taken with the previous conclusion, this implies MAXI J0556-332 has not reduced its accretion rate to the roughly tenth of Eddington rate needed to have bursting and superbursting behaviour.

Since high accretion rates are needed to reach gigakelvin crust temperatures and stable burning proceeds at these rates [1], it becomes clear that a special sort of NS source would be required to result in a detectable urca signature in its cooling transient light curve. The source must accrete at a low enough rate to exhibit bursts and/or superbursts for centuries so that a full urca layer can be built-up (whose width in $E_{e^-,Fermi}$ is ~k_BT), then accrete at a near-Eddington accretion rate for at least a year in order to heat the crust to ~gigakelvin temperatures, and then cease accretion for several hundred days or more so that the crust cooling phase can be observed. Though seemingly contrived, a source that appears to almost fulfill this criterion was already observed, though KS1731-260 appears to host relatively modest shallow heating [25,26] and therefore too low of an NS crust temperature to exhibit significant urca cooling luminosities. The observation of cooling transients is less than two decades old, with most model-observation comparisons taking place in the last decade, so the existence of a source fulfilling the previously stipulated conditions is a distinct possibility.

5 Directions for Future Research

As demonstrated, urca cooling has significant implications for our interpretation of observables from accreting neutron stars. Positive identification of an urca signature would provide unique evidence of nuclear reactions in the NS crust, whereas the lack of an urca signature where it is expected would require a change in our picture of the accreting NS outer layers. However, observational data for phenomena sensitive to urca cooling are limited, systematic studies of degeneracies between urca cooling signatures and other astrophysical effects have not been performed, and nuclear physics inputs that determine L_ν are mostly poorly constrained. This provides opportunities for continued work in experiment, theory, and observation.

The only observable phenomena known where a direct signature of urca cooling is expected is crust cooling from transiently accreting systems. However, less than ten such systems are known and only one appears to have had a crust hot enough for L_ν to be significant [27]. While this may be cause for concern, the research focus of cooling transient observations is in its relative infancy (e.g. compare to nearly half a century of type-I X-ray burst observations) and several groups are actively working to identify new objects and to monitor additional cooling events from known sources. Should another source as hot as the first observed outburst of MAXI J0556-332 be observed, it will be imperative for repeated observations to occur in the several tens to several hundreds of days after accretion ceases in order to definitively characterize the cooling light curve around the possible hundred-day dip.

The possibility of an urca signature in cooling transient light curves has only recently been identified and so few model calculations have investigated this phenomenon [15,28]. While it has been shown that an observable urca signature is possible, the range of signatures for different surface burning ashes and different accretion conditions has not been explored. Such systematics studies will be necessary to maximize the information obtained from model-observation comparisons should a transient be observed with the anticipated hundred-day dip in the light curve. Ideally, advances will be made in deducing surface burning conditions from model-observation comparisons of bursts on the NS surface. This would enable more reliable predictions for ash production on the surface of particular sources, and therefore remove degrees of freedom when modelling cooling transients. Indeed, such multi-observable modelling will likely be key to remove degeneracies that are present in the impact of astrophysics inputs on model calculation results.

The orders of magnitude uncertainties in L_ν from nuclear physics inputs (see section 3) will require a considerable experimental effort spanning much of the nuclear chart below A~100, as indicated by Figure 5.

Studies of surface burning processes, such as the rapid proton-capture (*rp*-) process that powers type-I X-ray bursts, must be sure to include investigations of odd-A nuclide production. While the majority of studies to date have focused on the impact of nuclear physics uncertainties on the X-ray burst light curve, this work shows that understanding odd-A nuclide production is just as critical. In particular, odd-A isobars that are suspected to have neutron-rich isotopes with relatively low log(*ft*) are of interest. These tend to be located in regions between nuclear closed shells, where there is appreciable deformation [7].

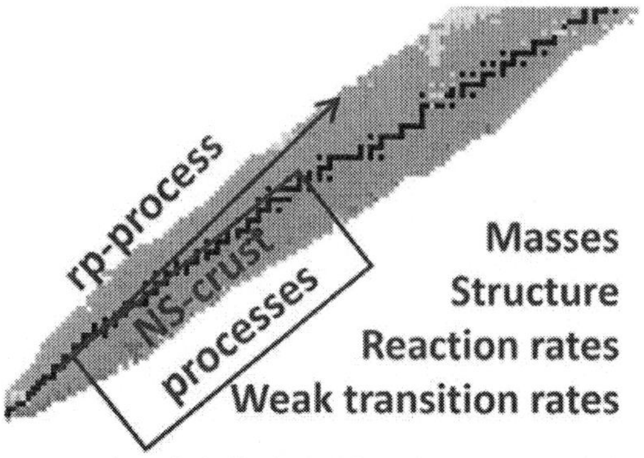

Fig. 5. Schematic indicating nuclear physics properties that need to be constrained to better estimate the L_ν associated with urca cooling in the NS crust.

Nuclear mass measurements are required to better constrain Q_{EC}, which impacts the strength and determines the location of urca layers. When low-lying excited states are suspected to play a role, higher precision masses ($\delta Q_{EC} \lesssim k_BT$) are necessary to determine the extent to which they can participate in urca cycling. This is also true for any even-A cases with anomalously small odd-even mass-staggering [13].

Weak transition rates, as quantified by *ft*, are arguably the most uncertain input in present estimates of L_ν. Precise constraints will require β^--decay measurements far off of stability which carefully account for all decay branches, including delayed neutron emitting channels. In their absence, spectroscopy measurements to determine J^π of ground and low-lying states can be used to roughly estimate *ft* from known systematics related to $\Delta J^{\Delta\pi}$.

The above discussion leaves a long to-do list for nuclear physics experimentalists at both stable and radioactive ion beam facilities. Stable beam facilities will be able to map nuclear structure properties influencing nuclear reaction rates near the valley of stability, whereas radioactive ion beam facilities will be required to do most of the rest of the work. Of course, any improvements in nuclear theory would be welcome to provide best-guesses for any nuclear data that presently defies measurements.

These are exciting times for those interested in the physics of the accreting NS outer layers and the associated observables. The possible presence of Great Walls of urca cooling provides new challenges in experiment, theory and observation, which are sure to keep the field busy for some time.

Acknowledgements

This work resulted from collaborations with Edward Brown, Andrew Cumming, Alex Deibel, and Hendrik Schatz. The author is supported in part by the U.S. Department of Energy through grant № DE-FG02-88ER40387. This material is based on work supported by the National Science Foundation through grant № PHY-1430152 (Joint Institute for Nuclear Astrophysics – Center for the Evolution of the Elements).

References

[1] H. Schatz *et al.*, Astrophys. J. **524**, 1014 (1999)
[2] A. Parikh *et al.*, Prog. Part. Nuc. Phys. **69**, 225 (2013)
[3] J. Stevens *et al.*, Astrophys. J. **791**, 106 (2014)
[4] L. Keek *et al.*, Mon. Not. R. Astron. Soc. **454**, 3559 (2015)
[5] K. Sato, Prog. Theor. Phys. **62**, 957 (1979)
[6] S. Gupta *et al.*, Astrophys. J. **662**, 1188 (2007)
[7] H. Schatz *et al.*, Nature **505**, 62 (2014)
[8] G. Gamow & M. Schoenberg, Phys. Rev. **58**, 1117 (1940)
[9] G. Gamow & M. Schoenberg, Phys. Rev. **59**, 539 (1941)
[10] S. Tsuruta & A. Cameron, Astro. Space Sci. **7**, 374 (1970)
[11] A. Deibel *et al.*, Astrophys. J. **831**, 13 (2016)
[12] R. Cyburt *et al.*, Astrophys. J. **830**, 55 (2016)
[13] Z. Meisel *et al.*, Phys. Rev. Lett. **115**, 162501 (2015)
[14] A. Cole *et al.*, Phys. Rev. C **86**, 015809 (2012)
[15] Z. Meisel & A. Deibel, Astrophys. J. **837**, 73 (2017)
[16] H. Junde, Nuc. Data Sheets **109**, 787 (2008)
[17] B. Singh *et al.*, Nuc. Data Sheets **84**, 487 (1998)
[18] H. Schatz *et al.*, Astrophys. J. Lett. **583**, L87 (2003)
[19] E. Brown, Astro. Source Code Lib., ascl :1505.034 (2015)
[20] L. Keek & A. Heger, Astrophys. J. **743**, 189 (2011)
[21] A. Turlione *et al.*, Astron. Astrophys. **577**, A5 (2015)
[22] L. Keek & A. Heger, Astrophys. J. **842**, 113 (2017)
[23] J. Homan *et al.*, Astrophys. J. **795**, 131 (2014)
[24] A. Parikh *et al.*, Astrophys. J. Lett. **851**, L28 (2017)
[25] R. Rutledge *et al.*, Astrophys. J. **580**, 413 (2002)
[26] R. Merritt *et al.* Astrophys. J. **833**, 186 (2016)
[27] R. Wijnands *et al.*, J. Astro. Astron. **38**, 49 (2017)
[28] A. Deibel *et al.*, Astrophys. J. Lett. **809**, L31 (2015)

Nuclear Neutrino Spectra in Late Stellar Evolution

G. Wendell Misch[1,2,*], *Yang* Sun[1,2,3,**], and *George* Fuller[4,***]

[1]School of Physics and Astronomy, Shanghai Jiao Tong University, Shanghai 200240, China
[2]Collaborative Innovation Center of IFSA, Shanghai Jiao Tong University, Shanghai 200240, China
[3]Institute of Modern Physics, Chinese Academy of Sciences, Lanzhou 730000, China
[4]Department of Physics, University of California, San Diego La Jolla, CA 92093, USA

Abstract. Neutrinos are the principle carriers of energy in massive stars, beginning from core carbon burning and continuing through core collapse and after the core bounce. In fact, it may be possible to detect neutrinos from nearby pre-supernova stars. Therefore, it is of great interest to understand the neutrino energy spectra from these stars. Leading up to core collapse, beginning around core silicon burning, nuclei become dominant producers of neutrinos, particularly at high neutrino energy, so a systematic study of nuclear neutrino spectra is desirable. We have done such a study, and we present our *sd*-shell model calculations of nuclear neutrino energy spectra for nuclei in the mass number range $A = 21 - 35$. Our study includes neutrinos produced by charged lepton capture, charged lepton emission, and neutral current nuclear deexcitation. Previous authors have tabulated the rates of charged current nuclear weak interactions in astrophysical conditions, but the present work expands on this not only by providing neutrino energy spectra, but also by including the heretofore untabulated neutral current de-excitation neutrino pairs.

1 Introduction

Neutrinos from nearby (<1 kpc) presupernova stars, of which there are several candidates, may soon be detectable with Earth-based neutrino detection experiments [1–3]. Therefore, there has been recent work to produce realistic neutrino spectra from presupernova cores. In the final phases before collapse, when the star begins to fuse silicon into iron, nuclei become the dominant sources of neutrinos, particularly at high energy [4, 5].

Reference [5] presented expected presupernova neutrino spectra and detector signals from simulation results. However, they used the technique of reference [6] to compute the nuclear component of the spectra from published nuclear weak rate tables. This method consists of treating each nucleus at each temperature-density point as having a single transition from parent to daughter with an effective Q-value chosen to reproduce the published average neutrino energy. However, reference [7] showed that nuclear structure must be properly considered to obtain accurate nuclear neutrino spectra; electron/positron captures and decays can in principle proceed from any state in the parent nucleus to any energetically-allowed state in the daughter nucleus. These two approaches are shown schematically in figure 1.

*e-mail: wendell.misch@gmail.com
**e-mail: sunyang@sjtu.edu.cn
***e-mail: gfuller@ucsd.edu

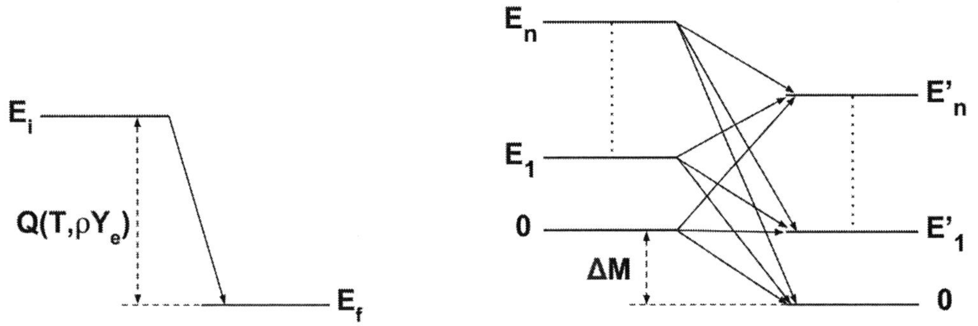

Figure 1. Two approaches to computing the nuclear contribution to neutrino spectra. Left: Single effective Q-value method. For each nucleus at each temperature and density, assume a single effective transition from the parent to the daughter with a Q-value chosen to reproduce the average neutrino energy in published rate tables. Right: Full structure method. Consider all transitions between thermally populated initial parent states and final daughter states.

We also show in section 2 that neutral current nuclear de-excitation, wherein an excited nucleus relaxes by emitting a virtual Z^0 boson that decays to a ν-$\bar{\nu}$ pair, is likely a dominant source of high-energy antineutrinos late in silicon burning and in early collapse, but it is generally not included in stellar models.

2 Results

Because nuclei are the dominant source of detectable neutrinos and because nuclear structure is an important component of producing realistic spectra, we have produced a catalog of neutrino spectra for 70 *sd*-shell nuclei on the Fuller-Fowler-Newman [8] temperature-density grid. We present here a small sample of spectra from our tables.

Figure 2 shows neutrino spectra from electron capture and positron decay of ^{27}Si at a temperature and density appropriate for a silicon-burning stellar core. Structure effects are evident as features in the spectra, particularly near ~5 MeV. The features are due to transitions proceeding through different parent and/or daughter nuclear states.

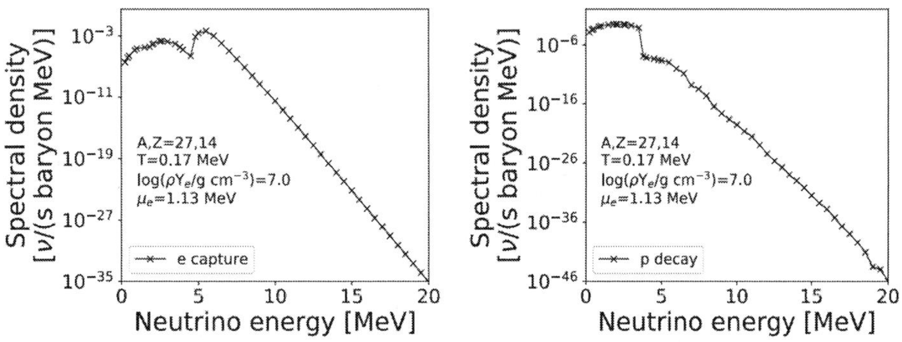

Figure 2. Neutrino spectra from electron capture (left) and positron decay (right) in ^{27}Si at temperature $T = 0.17$ MeV and $\rho Y_e = 10^7$ g/cm^3. Nuclear structure manifests as features in the spectra.

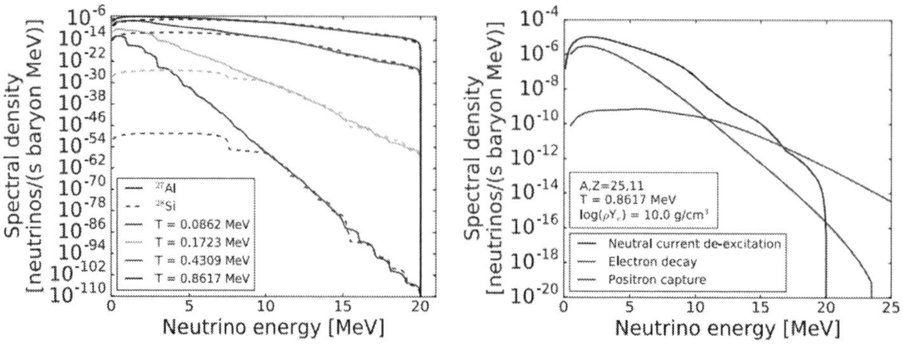

Figure 3. Left panel: Neutral current de-excitation neutrino spectra for ^{27}Al (solid) and ^{28}Si (dashed) at various temperatures leading up to collapse. Right panel: Antineutrino spectra from ^{25}Na in early collapse conditions. Neutral current de-excitation is extremely sensitive to temperature, and in early collapse conditions, it is a dominant source of high energy antineutrinos.

Figure 3 shows neutrino spectra from de-excitation of ^{27}Al and ^{28}Si at a wide range of temperatures along with a comparison of antineutrino spectra from three processes in ^{25}Na under early collapse conditions. Neutral current de-excitation is incredibly sensitive to temperature, increasing by orders of magnitude with a factor of 2 increase in temperature. Because no charged leptons are involved, this process is independent of density.

^{25}Na will not be abundant in early collapse, but it has the correct electron-to-baryon ratio Y_e, so it serves as a proxy for more realistic nuclei. In these conditions, neutral current de-excitation antineutrinos dominate both positron capture and electron decay up to at least 17 MeV neutrino energy. Because we do not include initial states above ~20 MeV in our calculations, the neutral current spectra cut off sharply at 20 MeV. Therefore, it may be that this process continues to dominate at even higher neutrino energies. Regardless, it is clear that a realistic estimate of pre-supernova neutrino spectra must include neutral current nuclear de-excitation.

The techniques used to produce these spectra are detailed in reference [9]. Our complete nuclear neutrino spectra tables are available for download and use at http://www.jinaweb.org/html/mischnuspectra.html.

References

[1] A. Odrzywolek, M. Misiaszek, M. Kutschera, Acta Physica Polonica B **35**, 1981 (2004)
[2] A. Odrzywolek, M. Misiaszek, M. Kutschera, Astroparticle Physics **21**, 303 (2004)
[3] K. Asakura, A. Gando, Y. Gando, T. Hachiya, S. Hayashida, H. Ikeda, K. Inoue, K. Ishidoshiro, T. Ishikawa, S. Ishio et al., ApJ **818**, 91 (2016)
[4] A. Odrzywolek, A. Heger, Acta Physica Polonica B **41**, 1611 (2010)
[5] K.M. Patton, C. Lunardini, R.J. Farmer, The Astrophysical Journal **840**, 2 (2017)
[6] K. Langanke, G. Martínez-Pinedo, J.M. Sampaio, Phys. Rev. C **64**, 055801 (2001)
[7] G.W. Misch, G.M. Fuller, Phys. Rev. C **94**, 055808 (2016)
[8] G.M. Fuller, W.A. Fowler, M.J. Newman, ApJ **252**, 715 (1982)
[9] G.W. Misch, Y. Sun, G.M. Fuller, ApJ **852**, 43 (2018)

Systematic Studies on the β-decay Half-lives of r-process Nuclei

Zhihong Li*, *Yong* Zhou, *XinYue* Li, *Youbao* Wang, *Bing* Guo, *Ding* Nan, *Weiping* Liu

[1]China institute of atomic energy, Beijing 102413, China

Abstract. Based on the accurate macroscopic-microscopic mass formula and the experimental data of β-decay half-lives of the nuclei with atomic number ranging from 20 to 190, a systematic formula has been proposed to calculate β-decay half-lives of neutron-rich nuclei. The formula is proved to reproduce the experimental β-decay half-lives of neutron-rich nuclei very well, and then is used to study the r-process nucleosynthesis in models of high-entropy mass outflows. The calculated abundances show a good agreement with the solar r-abundances around the third peak and the rare earth mass region.

1 Introduction

The rapid neutron-capture process (r-process) of stellar nucleosynthesis is believed to explain the production of about half of the nuclides heavier than iron in the universe [1, 2]. A large amount of nuclear data, such as mass, decay and reaction rates, are necessary in the studies of r-process. Among them, the β-decay half-lives of a large range of extremely neutron-rich nuclei are one of the most important inputs in the r-process model calculations. However, for the lack of experimental data, the identification of the realistic site and the exact path of the r-process has proved to be very difficult.

In this work, we have systematically investigated the variation of β-decay half-lives with the decay energy Q and nucleon number (Z, N) based on the experimental data. A systematic formula has been proposed to calculate the β-decay half-lives of that neutron-rich nuclei after taking into account the shell effects and pairing effects. The formula is tested for the resent experimental data, and then used to predict the β-decay half-lives of some nuclei far from stability including the ones on the possible r-process path.

2 Systematic formula for β-decay half-lives

According to Fermi's golden rule [3], the decay constant (or transition probability) of β-decay is defined by

$$\lambda = \frac{ln2}{T_{1/2}} = \frac{m_e^5 c^4 g^2 |M_{if}|^2}{2\pi^3 \hbar^7} f(Z, E_m). \qquad (1)$$

To obtain the decay constant, the transition matrix element M_{if} and phase-space factor $f(Z, E_m)$ should be worked out firstly. $f(Z, E_m)$ is a function of the radius of the nucleus R =

*e-mail: zhli@ciae.ac.cn

$r_0 A^{1/3}$, nuclear charge Z, and the energies of the emitted electron E_e. A formula for β-decay half-lives $\ln(T_{1/2})$ should include these three terms: $\alpha^2 Z^2 \ln(2R)$, $(\alpha^2 Z^2 - 5)\ln Q$ and $-\alpha Z\pi$. The term $(\alpha^2 Z^2 - 5)\ln Q$ is essential for the generalized Sargent law. The atomic masses are the key values to calculate Q-values, thus we select the very recent atomic mass table given by Wang [4] as our input data. The deviations between experimental and calculated Q-values are about 5%.

The nuclear matrix element contains the information of nuclear structure, the shell effects and pairing effects. The pairing effects on β-decay half-lives versus Q-values can be well described by Eq. (2).

$$\delta = (-1)^N + (-1)^Z. \tag{2}$$

The shell effects become strong near the neutron and proton magic-numbers, which can be well characterized by

$$\begin{aligned} S(Z,N) =& a_1 e^{-\frac{(N-28)^2+(Z-20)^2}{12}} + a_2 e^{-\frac{(N-50)^2+(Z-38)^2}{43}} + a_3 e^{-\frac{(N-82)^2+(Z-50)^2}{13}} \\ &+ a_4 e^{-\frac{(N-82)^2+(Z-58)^2}{24}} + a_5 e^{-\frac{(N-110)^2+(Z-70)^2}{244}}. \end{aligned} \tag{3}$$

It must be pointed out that a complete correction of shell effects should include all major shell as well as sub-shell closures. However, when taking account of the range of available experimental data, only portions of magic-numbers which have the most remarkable effects are included in Eq. (3).

Consequently, a systematic formula for β-decay half-lives of neuron-rich nuclei far from stability line can be derived from the relationship between β-decay half-lives and Q-values, nucleon numbers (Z,N), neutron excess $(N - Z)/A$, with corrections of pairing effects and shell effects. The new formula is expressed as

$$\begin{aligned} \ln(T_{1/2}) =& a_6 + (\alpha^2 Z^2 - 5 - a_7 \frac{N-Z}{A}) \ln(Q - a_8 \delta) + a_9 \alpha^2 Z^2 \\ &+ \frac{1}{3}\alpha^2 Z^2 \ln A - \alpha Z \pi + S(Z,N), \end{aligned} \tag{4}$$

with the parameters $a_i (i = 1, 2, \cdots, 9)$ determined according to all available experimental data of β-decay half-lives of nuclei far from stability line ($\delta N > 5$). The most important term $(\alpha^2 Z^2 - 5 - a_7 \frac{N-Z}{A}) \ln(Q - a_8\delta)$ has a vital effect on the calculation of β-decay half-lives. The latter three terms in Eq. (4) have relatively small contribution to the total value and have minor fluctuation versus (Z, N). As for the shell correction $S(Z,N)$, it has obvious contribution only near the nucleon magic-numbers but contributes nothing far from nucleon magic-numbers.

Through a least-square fitting to all available experimental data of 350 nuclei far from the β-stability line, we obtain all parameters in Eq. (4): $a_i (i = 1, 2, \cdots, 9)$ = 3.016, 3.879, 1.322, 6.030, 1.669, 11.09, 1.07, -0.935, -5.398, respectively. The average ratio between calculated β-decay half-lives and experimental ones is about 1.69. It reproduces 308 (88%) nuclei of 350 available experimental data within a factor of 3, and 261 (74.6%) nuclei within a factor of 2. To demonstrate the extrapolating capacity of the systematic formula, we have also calculated the β-decay half-lives of the isotopic chains such as Zr, Rb, Sr, Y and compare them with the experimental data [5]. The results are shown in Fig. 1. One can see that the agreements are very well. The ratios between calculated β-decay half-lives and experimental ones of these 17 nuclei are within a factor of 1.15. It is interesting to see from Fig. 1 that extrapolations of these isotopic chains have smaller deviations than the other reproduced β-decay half-lives.

The half-lives deduced with the current systematic formula has been used to study the r-process nucleosynthesis in models of high-entropy mass outflows. The calculated abundances show a good agreement with the solar r-abundances around the third peak and the rare earth mass region.

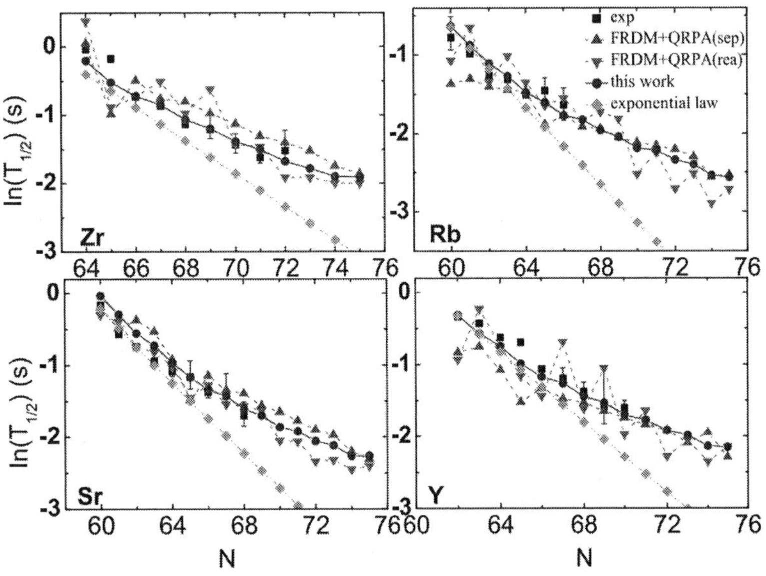

Figure 1. (Color online)A comparison among the calculated logarithm of half-lives from the exponential law [6] (diamonds), this work (dots), microscopic approaches [7, 8] (triangles) and measured ones [5] for Zr, Rb, Sr, Y isotopes.

3 Summary

Based on the Fermi's theory of β-decay and experimental data, the variation of β-decay half-lives with the decay energy Q and nucleon number (Z, N) has been systematically investigated. A empirical formula has been proposed for the calculation of β-decay half-lives of neutron-rich nuclei. The present formula works well in extrapolating the experimentally unknown r-process nuclei and in the model calculations of r-process nucleosynthesis.

This work was supported by the National Natural Science Foundation of China (Grant Nos. 11490563 and 11375269), the 973 program of China under Grant No: 2013CB834406 and the National key Research and Development Program of China under Grant No. 2016YFA0400502.

References

[1] J. J. Cowan, F. K. Thielemann, J. W. Truran, Phys. Rep., 1991, 208: 267-394
[2] M. Arnould , S. Goriely, K. Takahashi, Phys. Rep., 2007, 450(4): 97-213
[3] E. Fermi. Z. Phys., 1934, 88: 161
[4] N. Wang, M. Liu, X. Z. Wu, J. Meng, Phys. Lett. B 2014, 734: 215-219
[5] G. Lorusso et al., Phys. Rev. Lett., 2015, 114(19): 192501
[6] X. P. Zhang, Z. Z. Ren, Q. J. Zhi, Q. Zheng, J. Phys. G 2007, 34(12): 2611
[7] P. Möller, B. Pfeiffer, K.-L. Kratz, Phys. Rev. C 2003, 67(5): 055802
[8] D.L. Fang, B. A. Brown, T. Suzuki, Phys. Rev. C, 2013, 88(2): 024314

Capture and photonuclear reaction rates involving charged-particles: Impacts of nuclear ingredients and future measurement on ELI-NP

Y. Xu[1,*], *S. Goriely*[2], *D.L. Balabanski*[1], *S. Chesnevskaya*[1], *G.L. Guardo*[1], *M. La Cognata*[3], *H.Y. Lan*[4], *D. Lattuada*[1], *W. Luo*[4], and *C. Matei*[1]

[1] *Extreme Light Infrastructure-Nuclear Physics, RO-077125, Magurele, Ilfov, Romania*
[2] *Institut d'Astronomie et d'Astrophysique, CP-226, Universite Libre de Bruxelles, 1050, Brussels, Belgium*
[3] *INFN, Laboratori Nazionali del Sud, Via Santa Sofia 62, 95123, Catania, Italy*
[4] *School of Nuclear Science and Technology, University of South China, 421001, Hengyang, China*

Abstract. The astrophysical p-process is an important way of nucleosynthesis to produce the stable and proton-rich nuclei beyond Fe which can not be reached by the s- and r-processes. In the present study, the impact of nuclear ingredients, especially the nuclear potential, level density and strength function, to the astrophysical reaction rates of (p,γ), (α,γ), (γ,p), and (γ,α) reactions are systematically studied. The calculations are performed basad on the modern reaction code TALYS for about 3000 stable and proton-rich nuclei with $12 \leq Z \leq 110$. In particular, both of the Wood-Saxon potential and the microscopic folding potential are taken into account. It is found that both the capture and photonuclear reaction rates are very sensitive to the nuclear potential, thus the better determination of nuclear potential would be important to reduce the uncertainties of reaction rates. Meanwhile, the Extreme Light Infrastructure-Nuclear Physics (ELI-NP) facility is being developed, which will provide the great opportunity to experimentally study the photonuclear reactions in p-process. Simulations of the experimental setup for the measurements of the photonuclear reactions ^{96}Ru(γ,p) and ^{96}Ru(γ,α) are performed. It is shown that the experiments of photonuclear reactions in p-process based on ELI-NP are quite promising.

1 Introduction

In massive star evolution and stellar explosive site, the astrophysical p-process is an important way of nucleosynthesis to produce the stable and proton-rich nuclei beyond Fe which cannot be reached by the s- and r-processes. The common picture is that p-nuclei are synthesized by photodisintegration of pre-existing s- and r-processes nuclei via the capture and photo-disintegration reaction channels. To completely determine the nucleosynthesis of p-process, accurate knowledge of capture and photodisintegration reaction rates for the stable and proton-rich nuclei is essential. However, available experimental information for the reactions involved in p-process is quite limited [1]. For the reactions whose experimental data are not available yet, the state-of-the-art nuclear reaction models and the nuclear structure knowledge deduced by microscopic models should be taken into account for the evaluations of reaction rates.

In the present study, we perform the systematic computations of (p,γ), (α,γ), (γ,p), and (γ,α) reaction rates on about 3000 nuclei based on the TALYS software with various sets of nuclear structure ingredients. Furthermore, simulations of the experimental setup for the measurements of photonuclear reactions of astrophysics interest in p-process based on the Extreme Light Infrastructure-Nuclear Physics (ELI-NP) facility are performed, and the preliminarily results are correspondingly presented.

2 Systematic study on the impacts of nuclear ingredients to capture and photonuclear reaction rates

The main purpose of the present study is to systematically study how the nuclear structure ingredients, especially the nuclear level density, strength function, and optical model, impact the cross sections and reaction rates of capture and photo-disintegration reactions involving charged particles. Therefore, calculations of (p,γ), (α,γ), (γ,p), and (γ,α) reactions on about 3000 stable and proton-rich nuclei are performed with TALYS software[2, 3], by varying the specific nuclear structure ingredients. TALYS provides a complete description of all reaction channels and observables, and takes into account all types of direct, pre-equilibrium, and compound mechanisms to estimate the total reaction probability as well as the competition between the various open channels.

In particular, 6 sets of nuclear level density (NLD) and 5 sets of strength function (SF) are available in TALYS. The present calculation is performed each time by combining any set of level density models and strength function,

*e-mail: yi.xu@eli-np.ro

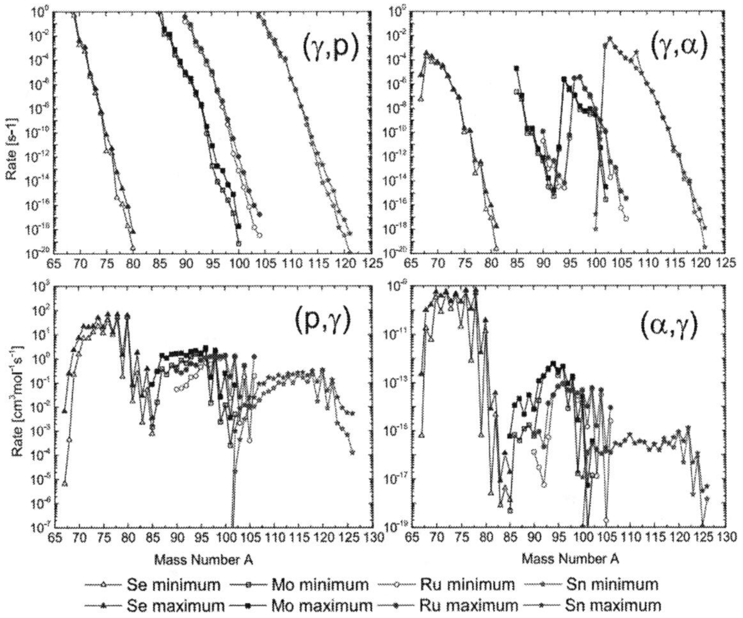

Figure 1. Comparisons of the maximal and minimal values among the reaction rates at T_9=2.5 calculated by all the available inputs of NLDs and SFs in TALYS for 4 isotopic targets of Se, Mo, Ru and Sn.

thus eventually 30 sets of calculated results are obtained. Among these results, the maximal and minimal values are chosen for comparison. The obtained maximal and minimal reaction rates at T_9=2.5 for 4 isotopic targets (Se, Mo, Ru and Sn) are shown in figure 1 corresponding to the 4 kinds of reactions. It is found that for the capture reactions, the differences between the maximal and minimal reaction rates are large, which means the capture rate is sensitive to NLD and SF. However, the photonuclear reaction rates are hardly changed by varying the NLD and SF.

Furthermore, calculations of reaction rates for about 3000 stable and proton-rich nuclei using the folding potential (JLM [4] for proton and M3Y [5] for α-particle) and the phenomenological Wood-Saxon potential [6, 7] are performed. At T_9=2.5, the ratio of the photonuclear reaction rates calculated by the 2 sets of potentials are shown on the (N,Z) plane in figure 2. It is shown that for the reaction involving proton, the ratios are smaller than a factor of 10 for most nuclei. However, the ratios of reaction rates involving α-particle are larger than a factor of 100 for most nuclei, which means the rates are very sensitive to the α-particle optical model. Therefore, it is proposed to experimentally study the photonuclear reactions based on the ELI-NP facility, especially for the (γ,α) channel. The (γ,α) reaction rates could be better determined using the improved α-particle potential that can be extracted from the measurements of (γ,α) reactions.

3 Simulations of the experimental setup for the measurements of photonuclear reactions based on ELI-NP facility

A very brilliant gamma beam, produced via Compton backscattering of a laser beam on a relativistic electron beam, will be available at Extreme Light Infrastructure - Nuclear Physics (ELI-NP) [8]. The gamma beam is characterized by a narrow bandwidth (<0.5% bdw) and tunable energy of up to almost 20 MeV. The brilliance of the gamma beam at the peak energy could reach 10^{23}/s mm^2 mrad^2 with 0.1% bdw. The research of nuclear astrophysics based on gamma-induced reactions with very low cross sections will largely benefit form the use of the high intensity gamma beams at ELI-NP. Meanwhile, for the detection system of charged particles, silicon detectors represent one of the best solution for their detection because they can guarantee the exceptional energy resolution and they have essentially 100% efficiency in charged particle detection. In the case of photonuclear reactions of astrophysical relevance, the emitted charged-particles have low energies, ranging from few hundreds keV to few MeV.

A silicon strip detector array ELISSA, will be realized in a common effort by ELI-NP and INFN Laboratori Nazionali del Sud (INFN-LNS) in Catania, Italy. The prototype of ELISSA was built and tested at INFN-LNS [9]. On this occasion, the simulations of the experimental setup for the measurements of (γ,α) and (γ,p) reactions on ^{96}Ru target are performed with the help of GEANT4. In the simulation, the configuration of ELISSA is taken

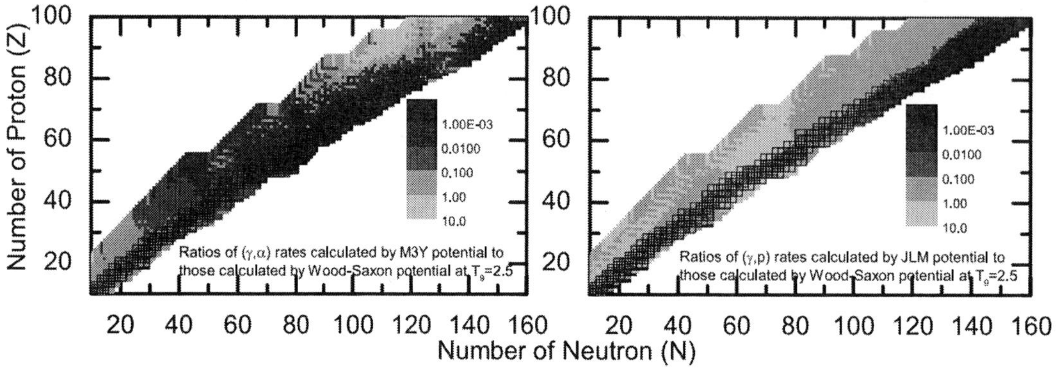

Figure 2. Representation in the (N,Z) plane of the ratios of (γ,α) and (γ,p) rates calculated by folding potential to those calculated by Wood-Saxon potential at T_9=2.5 for about 3000 stable and proton-rich nuclei.

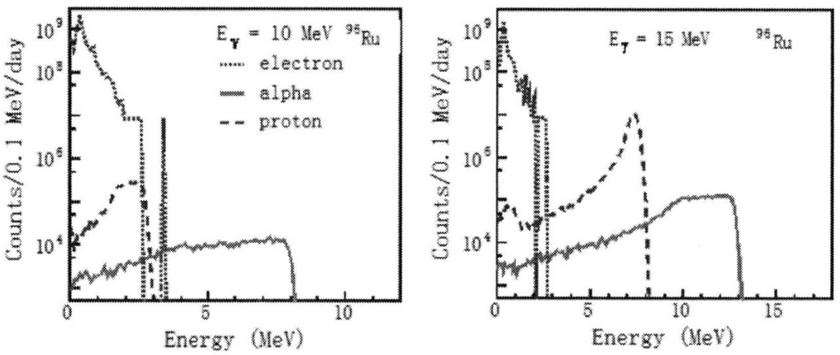

Figure 3. Energy spectrum of proton and alpha particles for ^{96}Ru(γ,p) and ^{96}Ru(γ,α) reactions obtained by GEANT4 simulations. The incident gamma-ray energies of 10 MeV and 15 MeV are used. The target thickness of Ruthenium is 10 μm.

into account, and the thickness of ^{96}Ru target is 10 μm. For the incident gamma-ray energies of 10 MeV and 15 MeV, the energy spectrum of proton and α-particle for ^{96}Ru(γ,p) and ^{96}Ru(γ,α) reactions are shown in figure 3. One can find that 10^4 counts per day can be obtained at E_γ = 10 MeV for the experimental yield of α-particle, and 10^5 counts per day can be obtained at E_γ = 15 MeV. We believe that the experiment of the simulated photonuclear reaction is quite promising, and it would be helpful to constraint the α-particle potential used for the determination of astrophysical reaction rates.

References

[1] T. Szücs et al., Nuclear Data Sheets **120**, 191 (2014)

[2] A. J. Koning, S. Hilaire, and M. C. Duijvestijn, NRG-report 21297/04.62741/P (2004)

[3] Y. Xu, S. Goriely, A. J. Koning and S. Hilaire, Phys. Rev. C **90**, 024604 (2014)

[4] E. Bauge, J. P. Delaroche, and M. Girod, Phys. Rev. C **58**, 1118 (1998)

[5] P. Demetriou, C. Grama and S. Goriely, Nucl. Phys. A **707**, 253 (2002)

[6] A. J. Koning and J.P. Delaroche, Nucl. Phys. A **713**, 231 (2003)

[7] V. Avrigeanu, M. Avrigeanu, and C. Manailescu, Phys. Rev. C **90**, 044612 (2014)

[8] C. A. Ur et al., Nucl. Inst. Meth. B **355**, 198 (2015)

[9] M. La Cognata et al., JINST **12**, C03079 (2017)

Experimental measurement of the $^{12}C+^{16}O$ fusion cross sections at astrophysical energies

X. Fang[1,⋆], *W. P. Tan*[2], *M. Beard*[2], *R. J. deBoer*[2], *G. Gilardy*[2], *H. Jung*[2], *Q. Liu*[2], *S. Lyons*[2], *D. Robertson*[2], *K. Setoodehnia*[2], *C. Seymour*[2], *E. Stech*[2], *B. Vande Kolk*[2], *M. Wiescher*[2], *R. de Souza*[3], *S. Hudan*[3], *V. Singh*[3], *X. D. Tang*[4], and *E. Uberseder*[5]

[1] *Sino-French Institute of Nuclear Engineering and Technology, Sun Yat-Sen University, Zhuhai, Guangdong 519082, China*
[2] *Institute for Structure and Nuclear Astrophysics, Joint Institute for Nuclear Astrophysics, University of Notre Dame, Notre Dame, Indiana 46556, USA*
[3] *Department of Chemistry, Indiana University Bloomington, Bloomington, Indiana 47405, USA*
[4] *Institute of Modern Physics, Chinese Academy of Science, Lanzhou, Gansu 730000, China*
[5] *Texas A&M University, College Station, Texas 77843, USA*

Abstract. The total cross sections of the $^{12}C+^{16}O$ fusion have been experimentally determined at low energies to investigate the role of this reaction during late stellar evolution burning phases. A high-intensity oxygen beam was produced by the 5MV pelletron accelerator at the University of Notre Dame impinging on a thick ultra-pure graphite target. Protons and γ-rays were measured simultaneously in the center-of-mass energy range from 3.64 to 5.01 MeV, using strip silicon and HPGe detectors. Statistical model calculations were employed to interpret the experimental results. A new broad resonance-like structure is observed for the $^{12}C+^{16}O$ reaction, and a decreasing trend of its S-factor towards low energies is found.

1 Introduction

The $^{12}C+^{16}O$ reaction is widely considered to play a critical role in early stellar evolution, in explosive carbon burning and explosive and hydrostatic oxygen burning[1, 2]. The $^{12}C+^{16}O$ reaction is relatively unimportant due to its high Coulomb barrier when $^{12}C+^{12}C$ is still the dominant reaction. It becomes more and more important when temperature and density of stars increase during oxygen burning stage, and the $^{16}O(\gamma,\alpha)^{12}C$ reaction is a supply of ^{12}C nuclei. The produced ^{12}C will be consumed either by reacting with itself or with ^{16}O, and the final abundances of remain elements after oxygen burning can be greatly affected by them. Type Ia supernovae are interpreted as the consequence of explosive carbon burning ignited near the core of white dwarf stars. The importance of $^{12}C+^{12}C$ used to be emphasized as ignition process, however, $^{12}C+^{16}O$ reaction may also plays a significant role. Recent studies showed that the $^{12}C+^{16}O$ rate is expected to have an unusually large effect on the calcium and sulfur yields in type Ia supernovae. The higher $^{12}C+^{16}O$ rate suppresses the alpha-particle abundance, which in turn decreases the Ca/S mass ratio in supernovae remnants [3].

⋆e-mail: fangx26@mail.sysu.edu.cn

The corresponding stellar thermal energy range of the $^{12}C+^{16}O$ fusion spans from 3.0 MeV to 7.2 MeV, where the major competing exit modes are the $^{24}Mg+\alpha$ (Q=6.772 MeV) and $^{27}Al+p$ (Q=5.170 MeV) channels. The contribution of $^{27}Si+n$ (Q=-0.424 MeV) channel is only a small percentage of the total due to low Q-value. The cross section of the fusion is governed by possibilities of penetrating through the Coulomb Barrier, and it thus rapidly drops when the reaction energy decreases under the barrier. This results to extremely low cross section values near and within its astrophysical interested energies, therefore, the direct experimental measurement at stellar energies is always difficult. The $^{12}C+^{16}O$ fusion has been widely studied at stellar energies, by measuring either γ rays[4, 5] or charged particles[6]. *Christensen*[5] *et al.* extended studied energies down to $E_{c.m.}$=3.9 MeV, and *Čujec*[4] *et al.* extrapolated the total cross section down to $E_{c.m.}$=3.0 and 3.5 MeV. It is generally accepted that the compound nucleus mechanism dominates the reaction cross section at low bombarding energies. The selective population of ^{24}Mg, ^{27}Al and ^{27}Si excited states can be explained by the compound nucleus mechanism. Hauser-Feshbach (HF) statistical-model calculations have been performed to explain the branching ratio of different channels and the relative population possibilities of excited states. Nuclear level density is critical input for HF theory, thus essential for determining the total cross section.

2 Measurement procedure

This work measured simultaneously both charged particles (mainly protons) and γ-rays emitted by the $^{12}C+^{16}O$ fusion, using silicon and HPGe detectors. $^{16}O^{5+}$, $^{16}O^{3+}$ ions produced by the Santa Ana 5MeV accelerator were applied to bombard a thick pure graphite target. The Santa Ana accelerator is a new electrostatic Pelletron accelerator at the Nuclear Science Laboratory (NSL) at the University of Notre Dame. This accelerator provides voltage up to 5 MV and high intensity (~10s particle μA) heavy ion beams. The target is highly oriented pyrolytic graphite (HOPG) with dimension of 20mm×20mm×1mm, which has a layered structure and is made of many graphene layers.

The schematic of experimental setup can be found in the reference [7].The silicon-detector array consists of six YY1-type 16-strip silicon detectors and one S2-type silicon detector, covering angles from 102° to 146° and 151° to 170°. HPGe detector was applied to measure γ-rays. The absolute efficiencies for typical γ-rays are 1.74% for 844 keV, 1.65% for 1014 keV and 1.46% for 1368 keV. The data was collected by ASIC (Application Specific Integrated Circuit) data acquisition system at NSL. During the measurement, $^{16}O^{3+}$ beams of energies from 8.5 MeV to 11.7 MeV in step of 100 or 200 keV, equivalent to 3.64 MeV$\leqslant E_{c.m.}\leqslant$5.01 MeV, were applied to bombard the target. $^{16}O^{5+}$ beams were used at 15.0 and 16.0 MeV for testing purpose. A measurement was also taken at E_{beam}= 7.0 MeV ($E_{c.m.}$=3.0 MeV) to study beam induced background.

3 Results and conclusion

Based on the measurement introduced above, the partial cross sections for various exit channels in the $^{12}C+^{16}O$ fusion were obtained from the thick-target yields for γ transitions associated with the different decay channels, protons, proton and γ coincident events, individually. An advantage of current measurement is that coincidence between protons and γ rays can be performed. This is the first time the coincidence method applied to $^{12}C+^{16}O$ measurement. To provide total cross sections from experimental data, the ratios of α, p and n and typical γ rays partial cross sections were provided by statistical calculations performed with the code SAPPHIRE [8]. Therefore, two sets of total cross sections and thus S(E) factors were accomplished from p and γ singles and p-γ coincidence datas, being displayed in figure 1, which shows the S(E) factor of the $^{12}C+^{16}O$ fusion reaction as a function of center-of-mass energy, comparing with other results[4–6]. The two data sets show good agreement with each other within the experimental uncertainties. An R-matrix fit using the code AZURE2 [9, 10] was also obtained and could be found in the reference [7]. The fit is based on the analysis of the different proton channels, normalized to the total S(E) factors. A decreasing trend of its S-factor towards low energies is predicted. However, present data do not evidently support the hindrance effect model [12] due to huge uncertainties.

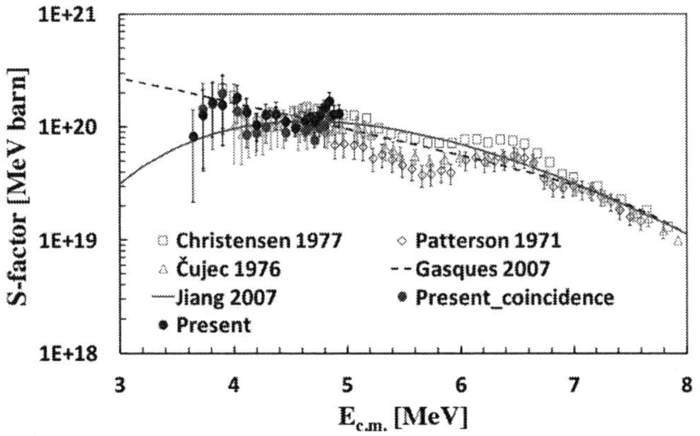

Figure 1. The S(E) factors of $^{12}C+^{16}O$ fusion are shown, including the present work (solid symbols) and previous data (open symbols) [4–6]. Reference [6] does not include the *n* channel. The Sao Paulo potential calculation (dashed line) [11] and hindrance model fit (solid line) [12] are also displayed.

In conclusion, the total cross sections for the $^{12}C+^{16}O$ fusion are determined by this work, and evidences for the existence of wide resonance-like structures from $E_{c.m.}$=3.9 MeV to 4.0 MeV in the $^{12}C+^{16}O$ fusion are observed. The decreasing trend of S-factor towards low energies could not be conclusive due to the huge uncertainties in the lower energy range.

This work was supported by the National Science Foundation through Grants No. PHY- 1068192 and No. PHY-1419765, the Joint Institute for Nuclear Astrophysics (JINA) under Grant No. PHY-0822648, the Joint Institute for Nuclear Astrophysics–Center for the Evolution of Elements (JINA-CEE) under Grant No. PHY-1430152, the National Natural Science Foundation of China under Grant No. 11775316, and the US Department of Energy under Grant No. DE-FG02-88ER-40404.

References

[1] M. Wiescher, F. Käppeler, and K. Langanke, Annu. Rev. Astron. Astrophys. **50**, 165 (2012)
[2] S. E. Woosley, W. D. Arnett, and D. D. Clayton, Phys. Rev. Lett. **27**, 213 (1971)
[3] H. Martínez-Rodríguez, C. Badenes, H. Yamaguchi *et al.*, Astrophys. J. **843**, 35 (2017)
[4] B. Čujec and C. Barnes, Nucl. Phys. A **266**, 461 (1976)
[5] P. R. Christensen, Z. E. Switkowskiw, and R. A. Dayras, Nucl. Phys. A **280**, 189 (1977)
[6] J. R. Patterson, B. Nagorcka, G. Symons, and W. Zuk, Nucl. Phys. A **165**, 545 (1971)
[7] X. Fang, W. P. Tan, M. Beard, R. J. deBoer *et al.*, Phys. Rev. C **96**, 045804 (2017)
[8] M. Beard, E. Uberseder, R. Crowter, M. Wiescher, Phys. Rev. C **90**, 034619 (2014)
[9] R. E. Azuma, E. Uberseder, E. C. Simpson, C. R. Brune *et al.*, Phys. Rev. C **81**, 045805 (2010)
[10] AZURE2 R-matrix code, https://azure.nd.edu
[11] L. R. Gasques, E. F. Brown, A. Chieffi, C. L. Jiang *et al.*, Phys. Rev. C **76**, 035802 (2007)
[12] C. L. Jiang, K. E. Rehm, B. B. Back, and R. V. F. Janssens, Phys. Rev. C **75**, 015803 (2007)

Production of neutron-rich nuclei approaching r-process by gamma-induced fission of ^{238}U at ELI-NP

Bo Mei[1,2,⋆], *Dimiter* Balabanski[1], *Paul* Constantin[1], *Tuan Anh* Le[1,3], and *Phan Viet* Cuong[1,4]

[1] *Extreme Light Infrastructure Nuclear Physics, "Horia Hulubei" National Institute for Physics and Nuclear Engineering, Str. Reactorului 30, 077125 Bucharest Magurele, Romania*
[2] *Institute of Modern Physics, Chinese Academy of Sciences, Lanzhou 730000, China*
[3] *Graduate University of Science and Technology, Vietnam Academy of Science and Technology, Hanoi, Vietnam*
[4] *Centre of Nuclear Physics, Institute of Physics, Vietnam Academy of Science and Technology, Hanoi, Vietnam*

Abstract. The investigation of neutron-rich exotic nuclei is crucial not only for nuclear physics but also for nuclear astrophysics. Experimentally, only few neutron-rich nuclei near the stability have been studied, however, most neutron-rich nuclei have not been measured due to their small production cross sections as well as short half-lives. At ELI-NP, gamma beams with high intensities will open new opportunities to investigate very neutron-rich fragments produced by photofission of ^{238}U targets in a gas cell. Based on some simulations, a novel gas cell has been designed to produce, stop and extract ^{238}U photofission fragments. The extraction time and efficiency of photofission fragments have been optimized by using SIMION simulations. According to these simulations, a high extraction efficiency and a short extraction time can be achieved for ^{238}U photofission fragments in the gas cell, which will allow one to measure very neutron-rich fragments with short half-lives by using the IGISOL facility proposed at ELI-NP.

1 Introduction

The production and measurement of very neutron-rich nuclei are of significant importance for nuclear physics as well as the r-process in nuclear astrophysics. However, so far, only few neutron-rich nuclei near the stability line have been measured, while most neutron-rich exotic nuclei have not been experimentally investigated because of their small production cross sections and short half-lives. Therefore, more experimental studies are required for these neutron-rich nuclei away from the stability line.

Very brilliant and intense gamma beams obtained by laser Compton backscattering at ELI-NP will provide new chances to investigate very neutron-rich nuclei, including those close to the r-process path, produced by photofission of ^{238}U targets placed at the center of a gas cell [1]. An IGISOL (Ion Guide Isotope Separator On-Line) facility with a gas cell is proposed at ELI-NP. This facility is mainly aimed at studying neutron-rich exotic nuclei, especially refractory isotopes [1–3], which are difficult to be investigated at traditional ISOL (Isotope Separator On-Line) facilities.

In this work, the production cross sections and rates of neutron-rich fragments produced by photofission of ^{238}U are reliably calculated for the planned IGISOL facility at ELI-NP. Furthermore, some SIMION simulations are performed for the optimization of the gas cell of the ELI-NP IGISOL facility, used to produce and extract the neutron-rich fragments.

⋆e-mail: bo.mei@eli-np.ro

2 Calculations for the production of neutron-rich exotic nuclei at ELI-NP

Accurate calculations for production cross sections of fragments produced by ^{238}U photofission at low energies over the whole energy region of the giant dipole resonance (GDR) are required for estimating the production yields of neutron-rich fragments and optimizing nuclear physics experiments at the planned IGISOL facility at ELI-NP. The gamma beam used at this IGISOL facility will cover a wide energy range between about 10 and 18.5 MeV [1–3].

Recently, a reliable empirical parametrization, called GIF^{238}U (Gamma Induced Fission of ^{238}U) [4], has been proposed for accurately calculating the production cross sections (rates) of neutron-rich fragments produced by photofission of ^{238}U at low energies (E_γ < 30 MeV). This parametrization is based on some measured ^{238}U photofission data and has been validated by comparing with many elemental, mass, and isotopic yields measured in ^{238}U photofission experiments at different energies (see Ref. [4] for details).

Figure 1 presents the production cross sections of various fragments produced by ^{238}U photofission at 14 MeV near the GDR peak, which are calculated by the GIF^{238}U parametrization [4]. For many neutron-rich nuclei with

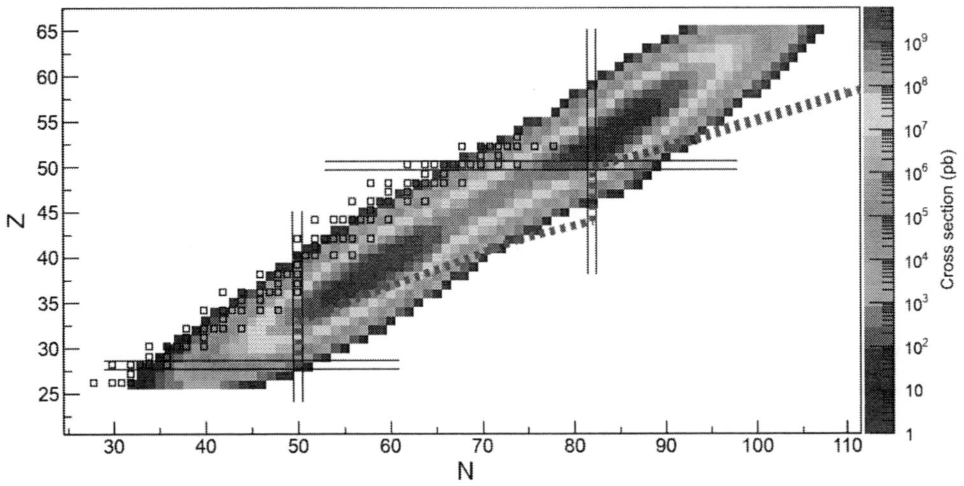

Figure 1. (Color online) Isotopic cross sections of fragments produced by photofission of ^{238}U at 14 MeV, calculated by the GIF^{238}U parametrization (see Ref. [4] for details). The stable nuclei are donated by open squares. The dotted lines show the position of the r-process path [5]. The proton and neutron closed shells are indicated with the black solid lines.

$31 \leqslant Z \leqslant 62$ and $80 \leqslant A \leqslant 160$, which are on or close to the r-process path [5], their production cross sections of the order around 1 mb can be achieved by using ^{238}U photofission, and thus ^{238}U photofission is very suitable for producing many neutron-rich nuclei away from the stability. For example, production cross sections of $^{80-82}$Ge and $^{85-87}$Se around the closed shell $N = 50$ are roughly 0.17 and 1.1 mb, respectively, while production cross sections of ^{132}Sn and ^{134}Te around the closed shell $N = 82$ are about 0.66 and 6.4 mb, respectively.

The above cross sections calculated with the GIF^{238}U parametrization can be used to estimate the production rates of neutron-rich nuclei produced by ^{238}U photofission at the ELI-NP IGISOL facility. As an example, production rates of ^{132}Sn and ^{134}Te in the ^{238}U targets with a total thickness of 251 μm are about 2×10^4 and 2×10^5 ions/s, respectively, according to calculations by the GIF^{238}U parametrization for the IGISOL facility at ELI-NP. At the gamma production point, the total intensity of the gamma beam with a broad energy distribution between 10 and 18.5 MeV is conservatively estimated to be 5×10^{10} γ/s for the above calculations of day-one experiments, while it can be improved to be around 1×10^{12} γ/s for future experiments [1–3]. The ^{238}U target in the gas cell is placed around 7 m after this gamma production point.

3 Simulations for ^{238}U photofission fragments in the ELI-NP gas cell

According to the design of the ELI-NP IGISOL facility, the ^{238}U target is sliced into roughly 33 foils and tilted at 15° for a high release rate of produced fragments [2, 3]. The fragments released from ^{238}U foils are then stopped by the helium buffer gas filled in the gas cell. To maximize the rate of the released and stopped photofission

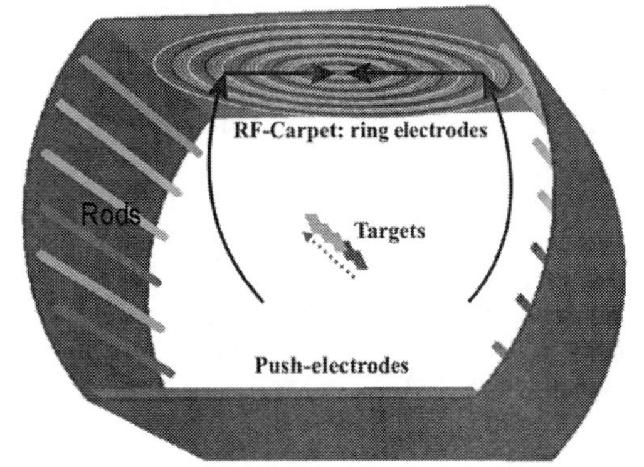

Figure 2. (Color online) Schematic drawing of the gas cell for the proposed IGISOL facility at ELI-NP. The RF carpet is composed of a stack of concentric ring electrodes. The direction of the gamma beam is indicated with the dotted arrow. Examples of trajectories of two photofission fragments in the gas cell are shown by the solid arrows.

fragments, the geometry of the gas cell for the ELI-NP IGISOL facility has been optimized by simulations using the Geant4 toolkit [6] (see Ref. [2] for detailed Geant4 simulations). After the stopping process, the stopped fragments are transported by the DC as well as RF fields to the extraction nozzle positioned at the center of the RF carpet, namely, the innermost ring electrode. Figure 2 shows the preliminary design for the gas cell of the planned IGISOL facility at ELI-NP. As shown in Fig. 2, the fragment extraction is performed in the perpendicular direction with respect to the gamma beam, which leads to a very short extraction path and time [7]. This perpendicular extraction method is particularly useful for the investigation of very exotic and short-lived nuclei. Typical trajectories of two fragments in the gas cell are shown with the black solid arrows in Fig. 2.

The DC and RF fields in the gas cell have been optimized by simulations using the SIMION software [8]. The DC field produced by the push-electrodes pushes the produced fragments towards the RF carpet. Voltages on the right and left-side rods are set in order to constrain fragment trajectories and prevent fragments from hitting the wall of the gas cell. The final position and velocity of stopped fragments in the stopping process are from Geant4 simulations [2] and are used as inputs for SIMION simulations. The DC drag field produced by the RF carpet is used to drag fragments from other positions to the center nozzle, where fragments are finally extracted, and the RF field of the carpet is applied to prevent the ions from hitting the RF carpet.

According to many SIMION simulations, an extraction efficiency of about 50% and a short extraction time (average time around 15 ms) can be reached for photofission fragments after the DC and RF fields have been optimized. Thus, the total rate of fragments extracted from the gas cell is around 5×10^5 ions/s. Final rates of ^{132}Sn and ^{134}Te extracted from the gas cell are about 1×10^3 and 1×10^4 ions/s, respectively. In these calculations, a release efficiency of about 15% is used for the produced fragments, according to Geant4 simulations in Ref. [2]. More details about the final design of the optimized gas cell will be discussed in future works.

4 Summary

In order to study neutron-rich exotic nuclei, an IGISOL facility including a gas cell is proposed at ELI-NP. The production yields (cross sections) of ^{238}U photofission fragments in the gas cell have been calculated by a reliable empirical parametrization GIF^{238}U, which is based on ^{238}U photofission experimental data. Calculations indicate that a production rate between about 10^4 and 10^5 ions/s can be obtained for many neutron-rich nuclei with $31 \leqslant Z \leqslant 62$ and $80 \leqslant A \leqslant 160$.

Furthermore, the preliminary design for the gas cell of the planned IGISOL facility at ELI-NP is discussed. Both the DC and RF fields have been optimized by using SIMION simulations. According to these SIMION simulations, a high extraction efficiency of about 50% and a short extraction time (average time around 15 ms) can be obtained for the optimized gas cell, which demonstrates that this gas cell is very suitable for the investigation of the short-lived neutron-rich nuclei. The total rate of all fragments extracted from the gas cell is estimated to be around 5×10^5 ions/s, while the rate of extracted isotopes is of the order around 10^3 ions/s for many neutron-rich ones.

This research was supported by the Extreme Light Infrastructure Nuclear Physics (ELI-NP) Phase II, a project co-financed by the Romanian Government and the European Union through the European Regional Development Fund - the Competitiveness Operational Programme (1/07.07.2016, COP, ID 1334). P. V. Cuong and L. T. Anh would like to acknowledge the support from the Vietnam Academy of Science and Technology under Grant No. VAST. CTVL.03/17-18.

References

[1] D. L. Balabanski et al., Rom. Rep. Phys. **68**, S621 (2016)

[2] P. Constantin, D. L. Balabanski, L. T. Anh, P. V. Cuong, and B. Mei, Nucl. Instrum. Methods Phys. Res., Sect. B **397**, 1 (2017)

[3] P. Constantin, D. L. Balabanski, and P. V. Cuong, Nucl. Instrum. Methods Phys. Res., Sect. B **372**, 78 (2016)

[4] B. Mei, D. L. Balabanski, P. Constantin, L. T. Anh, and P. V. Cuong, Phys. Rev. C, **96**, 064610 (2017)

[5] H. Schatz and K. Blaum, Europhys. News **37**, 16 (2006)

[6] S. Agostinelli et al., Nucl. Instrum. Methods Phys. Res., Sect. A **506**, 250 (2003)

[7] T. Dickel et al., Nucl. Instrum. Methods Phys. Res., Sect. B **376**, 1 (2016)

[8] http://simion.com

Analogous Gamow-Teller and $M1$ Transitions in $T_z = \pm 1/2$ Mirror Nuclei and in $T_z = \pm 1, 0$ Triplet Nuclei relevant to Low-energy Super GT state

Yoshitaka FUJITA[1,2,★], *Hirohiko* FUJITA[1], and *Yusuke* Tanumura[3]

[1] *Research Center for Nuclear Physics, Osaka University, Ibaraki, Osaka 567-0047, Japan*
[2] *Department of Physics, Osaka University, Toyonaka, Osaka 560-0043, Japan*
[3] *Department of Physics, Tohoku University, Sendai 980-8578, Japan*

Abstract. Nuclei have spin- and isospin-degrees of freedom. Therefore, Gamow-Teller (GT) transitions caused by the $\sigma\tau$ operator (spin-isospin operator) are unique tools for the studies of nuclear structure as well as nuclear interactions. They can be studied in β decays as well as charge-exchange (CE) reactions. Similarly, $M1$ γ decays are mainly caused by the $\sigma\tau$ operator. Combined studies of these transitions caused by Weak, Strong, and Electro-Magnetic interactions provide us a deeper understanding of nuclear spin-isospin-type transitions. We first compare the strengths of analogous GT and $M1$ transitions in the $A = 27$, $T_z = \pm 1/2$ mirror nuclei ^{27}Al and ^{27}Si. The comparison is extended to the $T_z = \pm 1, 0$ nuclei. The strength of GT transition from the ground state (g.s.) of ^{42}Ca to the 0.611 MeV first $J^\pi = 1^+$ state in ^{42}Sc is compared with that of the analogous $M1$ transition from the 0.611 MeV state to the $T = 1$, 0^+ g.s. (isobaric analog state: IAS) in ^{42}Sc. The 0.611 MeV state has the property of Low-energy Super GT (LeSGT) state, because it carries the main part of the GT strength of all available transitions from the g.s. of ^{42}Ca (and ^{42}Ti) to the $J^\pi = 1^+$ GT states in ^{42}Sc.

1 Introduction: GT and $M1$ operators

Gamow-Teller (GT) transitions caused by the $\sigma\tau$-type operator are "allowed transitions" in β decays [1]. They are also selectively excited in charge-exchange (CE) reactions, such as (p, n) or (^{3}He, t) reactions, performed at intermediate incident energies of more than 100 MeV/nucleon and 0° [2–4]. Note that $\sigma\tau$-type nucleon-nucleon (N-N) interaction is strong under such conditions [5].

The $M1$ transitions studied in γ decays or (e, e') reactions are the simplest "magnetic-type transitions" caused by the electro-magnetic (EM) interaction [1, 4, 6]. In spite of the name $M1$, the operator that causes $M1$ transition is similar to that of GT in the sense that the main contribution is expected from the $\sigma\tau$ operator.

Our aim here is to compare the strengths of analogous GT and $M1$ transitions and derive one-step deeper structure information that are not obtained in individual studies of GT and $M1$ transitions.

2 Reduced GT and $M1$ transition strengths

The strengths of GT and $M1$ transitions can be compared by means of the "reduced transition strength (probability)" B(GT) and $B(M1)$. The expression for these values reduced in spin (J) can be found in text books (see *e.g.* [6, 7]). In order to compare transition strengths for analogous GT and/or $M1$ transitions, however, it is important to reduce the matrix element in terms of isospin as well [8, 9]. Therefore, we start here with the reduced matrix elements in spin (J) but not in isospin and follow the convention of Edmonds [10].

2.1 Reduced GT transition strength

The reduced GT transition strength B(GT) for the transition from the initial state with spin J_i, isospin T_i, and z-component of isospin T_{zi} to the final state with J_f, T_f, and T_{zf} is given by [11]

$$B^{\pm1}(\mathrm{GT}) = \frac{1}{2J_i+1}\left|\langle J_f T_f T_{zf} \| \frac{1}{\sqrt{2}}\sum_{j=1}^{A}(\sigma_j \tau_j^{\pm1}) \| J_i T_i T_{zi}\rangle\right|^2, \tag{1}$$

where $\tau^{\pm1} = \mp(1/\sqrt{2})(\tau_x \pm i\tau_y)$ and transforms as a tensor of rank one, and $T_z = (N-Z)/2$. By applying the Wigner-Eckart theorem in isospin space, we get

$$\begin{aligned} B(\mathrm{GT}) &= \frac{1}{2J_i+1}\,\frac{1}{2}\,\frac{C_{\mathrm{GT}}^2}{2T_f+1}\,\left|\langle J_f T_f ||| \sum_{j=1}^{A}(\sigma_j\tau_j) ||| J_i T_i\rangle\right|^2 \\ &= \frac{1}{2J_i+1}\,\frac{1}{2}\,\frac{C_{\mathrm{GT}}^2}{2T_f+1}\left[M_{\mathrm{GT}}(\sigma\tau)\right]^2, \end{aligned} \tag{2}$$

where C_{GT} is the isospin Clebsch-Gordan (CG) coefficient $(T_i T_{zi} 1 \pm 1 | T_f T_{zf})$, and the $M_{\mathrm{GT}}(\sigma\tau)$ is the isovector (IV) spin-type GT matrix element.

From this expression for the "reduced" GT transition strength, we see that B(GT) consists of the squared value of the matrix element of the IV spin operator $M_{\mathrm{GT}}(\sigma\tau)$ and

★e-mail: fujita@rcnp.osaka-u.ac.jp

spin and isospin geometrical factors. Therefore, even if the initial and final states are common, transitions in reversed directions have different B(GT) values. For example, a GT transition from a state having $|J\ T\ T_z\rangle$ of $|0\ T_0\ T_0\rangle$ to the $|1\ T_0-1\ T_0-1\rangle$ state has three times larger B(GT) than that in the reverse direction.

2.2 Reduced $M1$ transition strength

The operator μ for $M1$ transitions and magnetic moments consists of an orbital part $g_\ell \boldsymbol{\ell}$ and a spin part $g_s \boldsymbol{s}$ [= $(1/2)g_s\boldsymbol{\sigma}$]. It can be rewritten as the sum of isoscalar (IS) and IV terms (for example, see [6]) as,

$$\mu = \left[\sum_{j=1}^{A} g_\ell \boldsymbol{\ell}_j + g_s \boldsymbol{s}_j\right]\mu_N$$
$$= \left[\sum_{j=1}^{A}\left(g_\ell^{\mathrm{IS}}\boldsymbol{\ell}_j + g_s^{\mathrm{IS}}\boldsymbol{s}_j\right) - \sum_{j=1}^{A}\left(g_\ell^{\mathrm{IV}}\boldsymbol{\ell}_j + g_s^{\mathrm{IV}}\boldsymbol{s}_j\right)\tau_{zj}\right]\mu_N, \quad (3)$$

where μ_N is the nuclear magneton, the z-component of the isospin operator $\tau_{zj} = 1$ for neutrons and -1 for protons, and τ_{zj} is τ_j^0. The coefficients g^{IS} and g^{IV} are the IS and IV combinations of gyromagnetic factors (g factors): $g_\ell^{\mathrm{IS}} = \frac{1}{2}(g_\ell^\pi + g_\ell^\nu)$, $g_s^{\mathrm{IS}} = \frac{1}{2}(g_s^\pi + g_s^\nu)$, $g_\ell^{\mathrm{IV}} = \frac{1}{2}(g_\ell^\pi - g_\ell^\nu)$, and $g_s^{\mathrm{IV}} = \frac{1}{2}(g_s^\pi - g_s^\nu)$. For bare protons and neutrons, the orbital and spin g factors are $g_l^\pi = 1$ and $g_l^\nu = 0$, and $g_s^\pi = 5.586$ and $g_s^\nu = -3.826$, respectively. Therefore, we get $g_\ell^{\mathrm{IS}} = 0.5$, $g_s^{\mathrm{IS}} = 0.880$, $g_\ell^{\mathrm{IV}} = 0.5$, and $g_s^{\mathrm{IV}} = 4.706$.

Starting from the reduced matrix elements in spin but not in isospin, and following the convention of Edmonds [10], the reduced $M1$ transition strength B(M1) can be written as [11]

$$B(M1) = \frac{1}{2J_i+1}\frac{3}{4\pi}\left|\langle J_f T_f T_{zf}\|\mu\|J_i T_i T_{zi}\rangle\right|^2$$
$$= \frac{1}{2J_i+1}\frac{3}{4\pi}\mu_N^2\left|\langle J_f T_f T_{zf}\|\sum_{j=1}^{A}\left(g_\ell^{\mathrm{IS}}\boldsymbol{\ell}_j + g_s^{\mathrm{IS}}\frac{1}{2}\boldsymbol{\sigma}_j\right)\right.$$
$$\left. - \sum_{j=1}^{A}\left(g_\ell^{\mathrm{IV}}\boldsymbol{\ell}_j + g_s^{\mathrm{IV}}\frac{1}{2}\boldsymbol{\sigma}_j\right)\tau_j^0\|J_i T_i T_{zi}\rangle\right|^2, \quad (4)$$

where $T_z = T_{zf} = T_{zi}$ for $M1$ transitions. By again applying the Wigner-Eckart theorem in isospin space, we get

$$B(M1) = \frac{1}{2J_i+1}\frac{3}{4\pi}\mu_N^2\left[\left(g_\ell^{\mathrm{IS}}M_{M1}(\ell) + g_s^{\mathrm{IS}}\frac{1}{2}M_{M1}(\sigma)\right)\right.$$
$$\left. - \frac{C_{M1}}{\sqrt{2T_f+1}}\left(g_\ell^{\mathrm{IV}}M_{M1}(\ell\tau) + g_s^{\mathrm{IV}}\frac{1}{2}M_{M1}(\sigma\tau)\right)\right]^2. \quad (5)$$

The isospin CG coefficient $C_{M1} = (T_i T_{zi} 10|T_f T_{zf})$ comes out explicitly by the use of reduced matrix elements, where $T_{zf} = T_{zi} = T_z$. The matrix elements $M_{M1}(\ell)$, $M_{M1}(\sigma)$, $M_{M1}(\ell\tau)$, $M_{M1}(\sigma\tau)$ correspond to the four terms of the μ operator, respectively.

Since the coefficient g_s^{IV} (= 4.706) is the largest, the contribution from the term of IV spin matrix $M_{M1}(\sigma\tau)$ is often the largest [7, 12]. On the other hand, it is expected that the contributions of the IS terms are small [12].

2.3 Meson exchange current

In the comparison of GT transitions with the analogous $M1$ transitions, a simple relationship is obtained for the transition strengths if the IV spin term is dominant in the $M1$ transitions. From the comparison of Eq. (2) and Eq. (5), the "quasi proportionality" between B(GT) and $B(M1)$ is expressed as

$$B(M1) \approx \frac{3}{8\pi}(g_s^{\mathrm{IV}})^2\mu_N^2\frac{C_{M1}^2}{C_{\mathrm{GT}}^2}B(\mathrm{GT}). \quad (6)$$

However, it is known that the contributions from meson-exchange currents (MEC) enhance $B(M1)$ strength over the corresponding B(GT) strength [8, 9]. The enhancement is traced back to larger and additive contributions of the vector MEC over the axial-vector MEC which are active in $M1$ and GT transitions, respectively [13]. Taking the enhancement factor R_{MEC} into account, we have now the relationship

$$B(M1) \approx \frac{3}{8\pi}(g_s^{\mathrm{IV}})^2\mu_N^2\frac{C_{M1}^2}{C_{\mathrm{GT}}^2}R_{\mathrm{MEC}}B(\mathrm{GT}). \quad (7)$$

In order to remove the constructive and destructive contributions of both IS and orbital terms, it was proposed to sum up the strengths over a wide range of excitation energy [14] and such cancellation of orbital contributions was predicted in shell-model calculations [15]. The analyses performed for the even-even self-conjugate nuclei ^{24}Mg and ^{28}Si suggested that the R_{MEC} value is in the range of 1.20 - 1.85 [14, 16, 17].

Further detailed analysis has been performed for the analogous $B(M1)$ and B(GT) values in the $A = 27$ mirror nuclei ^{27}Al and ^{27}Si [8, 9] (also see Sec. 3). The cumulative sums of $B(M1)$ and B(GT) values were calculated for the states with reliable B(GT) values and good correspondence in the region up to the excitation energy $E_x = 8.2$ MeV. From the ratio of these cumulative sums, an enhancement factor $R_{\mathrm{MEC}} = 1.4$ has been derived [8, 9]. Since the sum is for a limited region in excitation energy, it is not appropriate to extract a definite conclusion for the value of R_{MEC}. We simply mention that this value 1.4 is consistent with the previous results.

2.4 Isoscalar and orbital contributions in $M1$ transitions

From Eq. (7), we now see that the value

$$B(M1_{\sigma\tau}) = 2.644\mu_N^2\frac{C_{M1}^2}{C_{\mathrm{GT}}^2}R_{\mathrm{MEC}}B(\mathrm{GT}). \quad (8)$$

is the $\sigma\tau$ contribution expected in the $M1$ transition.

The interference of IS and IV orbital terms with the IV spin term in an $M1$ transition can be studied for the jth analogous $M1$ and GT transitions in isobars by examining the ratio defined by

$$R_{\mathrm{ISO}}^j = \frac{B^j(M1)}{B^j(M1_{\sigma\tau})}. \quad (9)$$

By comparing Eq. (2) and Eq. (5), it is seen that $R_{\rm ISO} > 1$ usually indicates that the IS term and/or the IV orbital term make a constructive contribution to the IV spin term, while $R_{\rm ISO} < 1$ shows a destructive contribution. As discussed in [9, 18], the contribution of the IS term is usually minor. Therefore, it is expected that the deviation of $R_{\rm ISO}$ from unity mainly shows a contribution of the IV orbital term in each $M1$ transition.

3 Orbital contribution in $M1$ transitions in $T = 1/2$, $A = 27$ nuclei

We examine the $M1$ γ decays in the $T_z = 1/2$, ^{27}Al nucleus and compare the $M1$ transition strengths with the GT strengths obtained from the ^{27}Al(^{3}He, t)^{27}Si reaction. The jth state with $J^\pi_j = 3/2^+, 5/2^+$, and $7/2^+$ can decay directly to the $J^\pi_0 = 5/2^+$ g.s. by $M1$ transitions.

3.1 Analog states and analogous transitions

A pair of isospin $T = 1/2$ mirror nuclei is characterized by $T_z = \pm 1/2$. All other quantum numbers of corresponding states are the same. Thus, with the assumption that isospin is a good quantum number, for every state in one of the mirror nuclei, an analog state should be found in the other nucleus (see Fig. 1). The energy levels should be almost identical in the pair nuclei, although small differences are expected from the state-dependent differences in the Coulomb displacement energies. In addition, the Coulomb displacement energy itself allows the g.s. of the $T_z = -1/2$ nucleus to undergo β decay to the g.s. as well as to several low-lying states of the $T_z = +1/2$ nucleus.

As a result of the analogous nature of corresponding states in the mirror nuclei, following analogous transitions are expected. They are transitions from an initial to a final state in the $T_z = +1/2$ and -1/2 nuclei and all other transitions in which initial and/or the final state is replaced by the respective analog state. Transitions reversing the initial and final states are also possible. Analogous $M1$ and GT transitions from or to the g.s. with spin value $J_{\rm g.s.} \neq 0$ and with isospin $T = 1/2$ are schematically shown in Fig. 1.

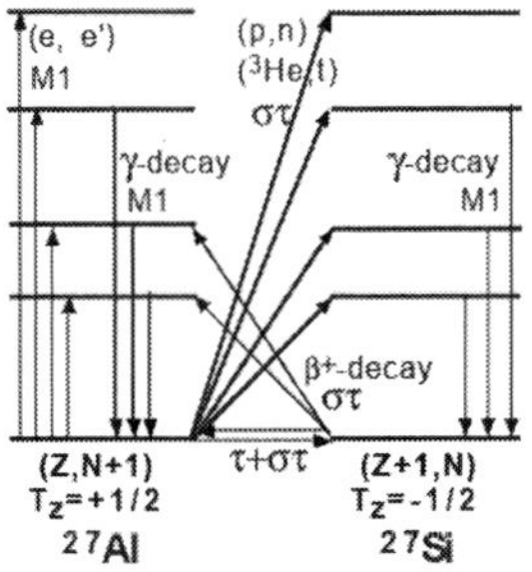

Figure 1. Schematic level schemes are shown for $T = 1/2$, odd-A mirror nuclei. Analogous transitions connecting the ground states in the mirror nuclei with excited states in the same nucleus and the conjugate nucleus are indicated. The type of reaction or decay and the relevant interactions causing each transition are shown along the arrows indicating the transitions.

3.2 Comparison of $M1$ and GT transition strengths

By comparing the analogous $B(M1)$ and B(GT) strengths in $T = 1/2$ mirror nuclei, it is possible to extract the combined contribution of the IS term and the IV orbital term in the $B(M1)$ strength using Eq. (9). Since the effect of MEC should be independent of the wave function of the individual state, the value $R_{\rm MEC} = 1.4$ derived from the analysis of the $A = 27$ system is used (for details, see Ref. [8]). The $R_{\rm ISO}$ ratios were calculated for the fourteen pairs of $M1$ and GT transitions up to the excitation energy of 9.7 MeV in ^{27}Al, where pairs well corresponding in E_x values were selected.

The results are shown in Fig. 2 for the $M1$ transitions in the $T_z = +1/2$ ^{27}Al nucleus as a function of $B(M1)\uparrow$ value [the $B(M1)$ value from the g.s. to the jth final excited state]. It is interesting to note that the $R_{\rm ISO}$ value tends to deviate from unity by more than a factor of two when the $B(M1)\uparrow$ is less than approximately 0.1. This shows that the "combined IS and orbital contribution" is rather large in weaker transitions and the quasi proportionality [Eq. (7)] of the $B(M1)$ values for $\Delta T = 0$ $M1$ transitions and the analogous B(GT) values is lost. This finding is interpreted as follows; since the IS term is small and the IV orbital term cannot be large for the s or d orbits with smaller orbital angular momentum ℓ, the dominance of the IV spin term of the $M1$ operator is guaranteed if the transitions are at least of average strength. However, the contribution of the IV spin term can also be small. Then the relative contribution of the IS term and the IV orbital term becomes significant although the transition itself is weak [8, 12, 18, 19]. A similar discussion applies to the $M1$ transitions in $T_z = -1/2$ mirror nuclei [9].

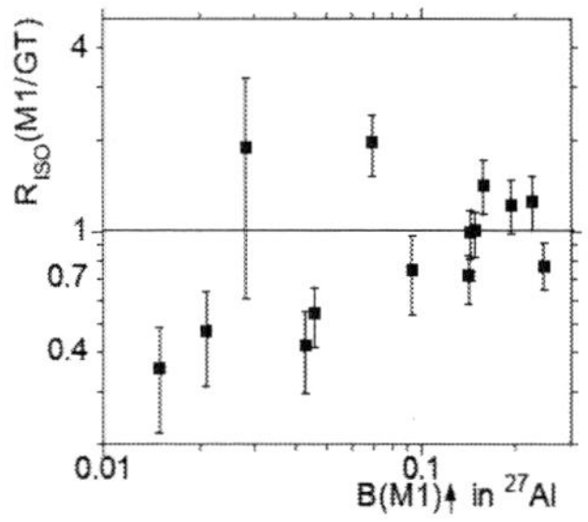

Figure 2. The ratio $R_{\rm ISO}$ for the $M1$ transitions in ^{27}Al as a function of $B(M1)\uparrow$. (Note the log-log scale of the figure.) The ratio is sensitive to the combined contribution of IS term and IV orbital term to each $M1$ transition. Values of $R_{\rm ISO} > 1 (< 1)$ suggest constructive (destructive) interference of these terms with the IV spin term. For the definition of $R_{\rm ISO}$, see text.

4 Comparison of analogous $M1$ and GT transition strengths in $A = 42$ nuclei

In a simple shell model (SM) picture, GT transitions are allowed among the LS-partner orbits $j_>$ and $j_<$. As a result, GT excitations are expected in the low-energy region (nominally at $E_x = 0$) and the region around $E_x = 3 - 6$ MeV (the energy difference of the LS partner orbits).

Table 1. The matrix elements M_{GT} for the configurations contributing to the GT transitions exciting $T = 0$, $J^{\pi} = 1^{+}$ low-lying states in ^{42}Sc from the 0^{+} g.s. of ^{42}Ca. The *pf*-shell SM calculation was performed [23] using the GXPF1J interaction [27]. The notation $f7 \rightarrow f7$, for example, stands for the transition with the $\nu f_{7/2} \rightarrow \pi f_{7/2}$ type. The summed value of the matrix elements is denoted by ΣM_{GT} and its squared value is the *B*(GT), where the *B*(GT) values do not include the quenching factor of the SM calculation.

States in ^{42}Sc		Configurations						Transition Strengths	
E_x (MeV)	T	$f7 \rightarrow f7$	$f7 \rightarrow f5$	$f5 \rightarrow f7$	$p3 \rightarrow p3$	$p3 \rightarrow p1$	$p1 \rightarrow p3$	ΣM_{GT}	B(GT)
0.33	0	1.383	0.548	0.063	0.031	0.024	0.016	2.07	4.28
4.41	0	0.719	-0.742	-0.085	-0.079	-0.073	-0.048	-0.31	0.09
7.41	0	0.193	-0.788	-0.090	0.142	0.060	0.040	-0.44	0.19

Note that the overview of GT responses can be studied by CE reactions [3, 4]. Against the above expectation for the GT strength distribution, these studies showed that the distribution in each nucleus can be largely different and dependent on the specific nuclear structure.

4.1 Low-energy Super Gamow-Teller transition observed in ^{42}Ca(^{3}He, *t*)^{42}Sc reaction

The most famous structure formed by GT excitation is the GT resonance (GTR) situated in high E_x regions of 9-15 MeV. In 1980s, GTRs were intensively studied by pioneering (p, n) reactions [3]. They were observed in almost all nuclei with mass $A > 60$ and $N > Z$. It was found that GTRs consume $\approx$ 60% of the GT strength predicted by the Ikeda sum rule [20].

Various random-phase-approximation (RPA) calculations can reproduce the high E_x values and the concentration of the available GT strengths in GTRs by introducing IV-type effective residual interactions (ERIs). The "repulsive" IV-type ERIs push GT strengths up in energy than is expected in a simple SM picture and concentrate the GT strengths in GTRs. In addition, available configurations contributing to the excitation of GTRs are "in phase" [21], showing the collective nature of them. It is known that IV-type ERIs are active in configurations with particle–hole (p–h) nature in GT excitations. Note that this condition is always realized in $N >> Z$ nuclei.

On the other hand, high resolution (^{3}He, *t*) study on ^{42}Ca [22, 23], and recently on ^{18}O [24], showed that the GT strength can also be concentrated in the lowest-energy GT state. As we see in Fig. 3, the 1^{+}, 0.611 MeV GT state in ^{42}Sc is dominant in the 0-degree spectrum. A total *B*(GT) value of 2.7(4) is in discrete states up to 13 MeV [about half of the Ikeda sum-rule value of $3(N - Z) = 6$], and $\approx$ 80% of that [*B*(GT) = 2.17] is concentrated in the lowest 0.611 MeV GT state. We call it the Low-energy Super GT state (LeSGT state) [22, 23]. In RPA calculations, it was found that LeSGT states are formed by the contribution of IS-type ERIs [22, 23, 25] that are active in proton particle–neutron particle (πp–νp) configurations [25, 26] on top of the *LS*-closed core of ^{40}Ca.

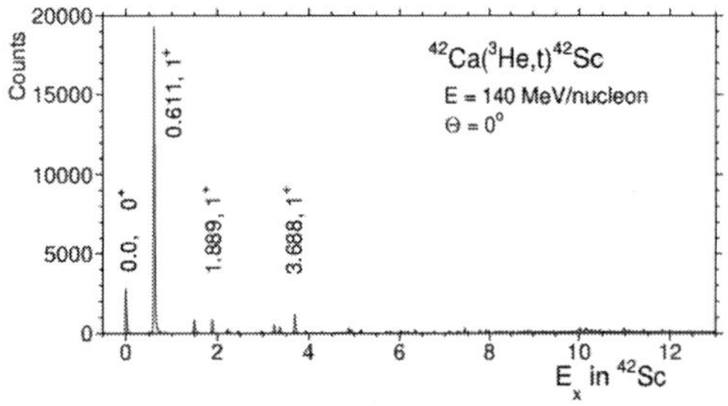

Figure 3. The 0°, ^{42}Ca(^{3}He, *t*)^{42}Sc spectrum. Two prominent peaks observed in the low-energy region are the $J^{\pi} = 0^{+}$ IAS (g.s.) and the 0.611 MeV, 1^{+} LeSGT state excited with $\Delta L = 0$ transitions. No GTR structure, usually expected at $\approx$ 10 MeV, is observed.

It was found that the SM calculation for ^{42}Ca $\rightarrow ^{42}$Sc GT transitions using the GXPF1J interaction [23, 27] reproduces the concentration of the GT transition strength to the lowest $T = 0$, 1^{+} state (see Table 1). Note that not only the main configurations, i.e., $\nu f_{7/2} \rightarrow \pi f_{7/2}$ and $\nu f_{7/2} \rightarrow \pi f_{5/2}$, but also *p*-shell configurations contribute all in phase to the excitation of the lowest 1^{+} state at 0.33 MeV, i.e., the LeSGT state. On the other hand, contributions of the configurations are out of phase in the other two 1^{+} states.

4.2 Analogous *M*1 and GT transitions from and to the LeSGT state

Figure 4 shows the analogous *M*1 and GT transitions in the low-energy region of the $A = 42$ isobars. The $T_z = 0$ nucleus ^{42}Sc is unique in the sense that it has the $T = 1$ g.s. (the IAS of the g.s. of ^{42}Ca and ^{42}Ti). We notice that GT transitions from the ground states of ^{42}Ca and ^{42}Ti to the 0.611 MeV $T = 0$, 1^{+} state in ^{42}Sc are analogous. Using the experimental data evaluated in Ref. [28], *B*(GT) = 2.17 (5), the value mentioned earlier,

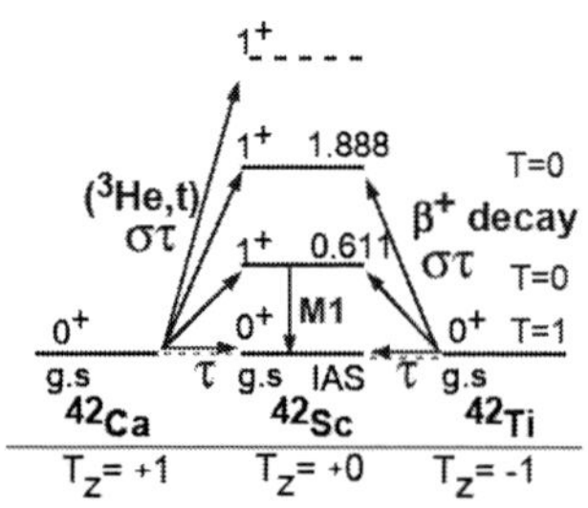

Figure 4. Gamow-Teller and *M*1 transitions in $A = 42$, $T_z = 1$, 0, -1 isobar nuclei ^{42}Ca, ^{42}Sc, and ^{42}Ti, where ^{42}Ca and ^{42}Ti are the mirror nuclei. The mirror GT^{-} and GT^{+} transitions that can be studied by the ^{42}Ca(^{3}He, *t*)^{42}Sc reaction and ^{42}Ti β^{+} decay, respectively, are caused by the $\sigma\tau$ operator. On the other hand, *M*1 γ transition in ^{42}Sc is caused by the *M*1 (μ) operator.

is obtained from the ^{42}Ti β^+ decay and the same value is expected in the ^{42}Ca(^{3}He, t)^{42}Sc reaction [23] on the basis of isospin symmetry between the mirror transitions. This large B(GT) value, as discussed, is due to the in-phase nature of all available f- and p-shell configurations contributing to this GT transition.

We also notice that the $M1$ γ decay from the 1^+, 0.611 MeV state in ^{42}Sc to the 0^+ g.s. is analogous to the GT transitions (see Fig. 4). For this $M1$ transition, a large $B(M1)$ of 6.2 (26) μ_N^2 [$B(M1)$ (W.u.) = 3.4 (15)] has been derived [28]. Large uncertainty is due to that in the half-life measurement of the 0.611 MeV state [$T_{1/2}$ = 28 (12) fs]. From the B(GT) = 2.17 (5) of the analogous GT transition and using Eq. (8), a value $B(M1_{\sigma\tau})\uparrow$ = 8.1 (24) μ_N^2, i.e., the $B(M1_{\sigma\tau})$ value from the T = 1 g.s. to the 1^+, 0.611 MeV state, is obtained, where $R_{\rm MEC}$ = 1.4 is assumed. The value $B(M1_{\sigma\tau})\downarrow$ that can be directly compared with the γ-decay $B(M1)$ is derived taking the differences of spin and isospin CG coefficients into account. We obtained a rather small value of 2.7 (8) μ_N^2. Relatively large uncertainty comes from that of the value of $R_{\rm MEC}$ = 1.2 - 1.85 that is not well known for pf-shell nuclei [14].

As discussed, under the assumption that strong $M1$ transitions are mainly caused by the $\sigma\tau$-part of the $M1$ operator, it is expected that analogous GT and $M1$ transitions have corresponding transition strengths. However, here, we hesitate to come to this simplified conclusion. Note that the contribution of the IV orbital term in Eq. (4) can be larger for larger ℓ orbits (ℓ = 3 for the $f_{7/2}$ and $f_{5/2}$ orbits). Looking for a clue to answer this question, we performed a three-body model calculation assuming valence p and n particles on top of a ^{40}Ca core [29], where both IS and IV pairing correlations were included. The calculation showed a $B(M1)$ value of 6.8 μ_N^2. It also suggested that contributions of the spin and orbital terms are constructive. The ratio of these contributions to the transition amplitude is approximately 1 : 0.47, which suggests that the $B(M1)$ can be approximately twice larger than the $B(M1_{\sigma\tau})$ value.

As we have seen, the combined studies of GT and $M1$ transitions for the analogous transitions allow us to access "one-step deeper fields" in nuclear physics such as orbital and isoscalar contributions in $M1$ transitions or meson-exchange currents in $M1$ and GT transitions. Steady efforts for quantitative measurements, such as life-time measurements of excited states [30], and combined efforts of γ- and β-decay studies as well as studies using CE reactions are needed to approach such quantities.

Acknowledgments. YF acknowledges the support by JSPS KAKENHI, Japan, under Grants No.15540274, No.18540270 and No.JP15K05104. YT acknowledges the support from the Graduate Program on Physics for the Universe (GP-PU), Tohoku University.

References

[1] A. Bohr and B.R. Mottelson, *Nuclear Structure* (Benjamin, New York, 1969), Vol. 1.
[2] F. Osterfeld, Rev. Mod. Phys. **64**, 491 (1992).
[3] J. Rapaport and E. Sugarbaker, Annu. Rev. Nucl. Part. Sci. **44**, 109 (1994).
[4] Y. Fujita, B. Rubio, and W. Gelletly, Prog. Part. Nucl. Phys. **66**, 549 (2011).
[5] W.G. Love and M.A. Franey, Phys. Rev. C **24**, 1073 (1981).
[6] H. Morinaga and T. Yamazaki, *In Beam Gamma-Ray Spectroscopy* (North-Holland, Amsterdam, 1976).
[7] S.S. Hanna, in *Isospin in Nuclear Physics*, edited by D. H. Wilkinson (North-Holland, Amsterdam, 1969), Chap. 12.
[8] Y. Fujita, *et al.*, Phys. Rev. C **59**, 90 (1999).
[9] Y. Fujita, *et al.*, Phys. Rev. C **62**, 044314 (2000).
[10] A. R. Edmonds, *Angular Momentum in Quantum Mechanics* (Princeton University Press, Princeton, 1960).
[11] L. Zamick and D. C. Zheng, Phys. Rev. C **37**, 1675 (1988).
[12] E. K. Warburton and J. Weneser, in *Isospin in Nuclear Physics*, ed. D. H. Wilkinson, (North-Holland, Amsterdam, 1969) Chap. 5, and references therein.
[13] I. S. Towner and F. C. Khanna, Nucl. Phys. **A399**, 334 (1983).
[14] A. Richter, A. Weiss, O. Häusser, and B. A. Brown, Phys. Rev. Lett. **65**, 2519 (1990).
[15] M. Hino, K. Muto, and T. Oda, J. Phys. G **13**, 1119 (1987).
[16] Y. Fujita, *et al.*, Phys. Rev. C **55**, 1137 (1997).
[17] P. von Neumann-Cosel, A. Richter, Y. Fujita, and B. D. Anderson, Phys. Rev. C **55**, 532 (1997).
[18] Y. Fujita *et al.*, Phys. Rev. C **66**, 044313 (2002).
[19] G. Morpurgo, Phys. Rev. **110**, 721 (1958); *ibid.* **114**, 1075 (1959).
[20] K. Ikeda, S. Fujii, J.I. Fujita, Phys. Lett. **3**, 271 (1963).
[21] M.N. Harakeh, A. van der Woude, *Giant Resonances* Oxford Studies in Nucl. Phys. **24**, (Oxford University Press, Oxford, 2001).
[22] Y. Fujita, *et al.*, Phys. Rev. Lett. **112**, 112502 (2014).
[23] Y. Fujita, *et al.*, Phys. Rev. C **91**, 064316 (2015).
[24] H. Fujita, *et al.*, private communications.
[25] C.L. Bai, *et al.*, Phys. Rev. C **90**, 054335 (2014).
[26] *Fifty Years of Nuclear BCS, Pairing in Finite Systems* edited by R.A. Broglia and V. Zelevinsky (World Scientific, Singapore, 2013).
[27] M. Honma, T. Otsuka, T. Mizusaki, M. Hjorth-Jensen, and B.A. Brown, J. Phys.: Conf. Ser. **20**, 7 (2005).
[28] Jun Chen and Balraj Singh, Nucl. Data Sheets, **135**, 1 (2016).
[29] Y. Tanimura, H. Sagawa, and K. Hagino, Prog. Theor. Exp. Phys. **2015**, 053D02.
[30] A. Dewald, O. Möller, and P. Petkov, Prog. Part. Nucl. Phys. **67**, 786 (2012).

A geometry for the shell model

P. Van Isacker[1,a]

[1] *Grand Accélérateur National d'Ions Lourds, CEA/DRF–CNRS/IN2P3*
Bd Henri Becquerel, B.P. 55027, F-14076 Caen, France

Abstract. A geometric interpretation is given of matrix elements of a short-range interaction between states that are written in terms of aligned neutron–proton pairs.

1 Introduction

In this contribution matrix elements of the nucleon–nucleon interaction between shell-model states are studied from a geometric point of view. An approach of this kind goes back to the seminal study of Schiffer [1]. The interaction between two nucleons can be expressed in terms of the angle between the angular momenta of the nucleons. Furthermore, the strength of the nucleon–nucleon interaction has a universal dependence on this angle—universality that can be shown to result from the short-range character of the effective interaction [2]. This geometric interpretation of the nucleon–nucleon interaction is by now an accepted feature of nuclear structure; a discussion of it can be found in textbooks on nuclear physics, see e.g. Sect. 4.2 of Ref. [3].

The geometry of the shell model can be extended from two- to four-nucleon configurations [4]. In this contribution the geometric interpretation of the neutron–proton (np) interaction is examined for a four-nucleon configuration when written in terms of aligned np pairs.

2 Matrix elements of the neutron-proton interaction in a 2n–2p basis

States for two neutrons with angular momentum J_ν and two protons with angular momentum J_π, coupled to total angular momentum J, can be written as $|J_\nu J_\pi; J\rangle \equiv |j_\nu j_\nu(J_\nu), j_\pi j_\pi(J_\pi); J\rangle$, which, for all even values of J_ν and J_π allowed by angular-momentum coupling, form a complete and orthogonal basis. Matrix elements of the np interaction in the 2n–2p basis are

$$\langle J'_\nu J'_\pi; J|\hat{V}_{\nu\pi}|J''_\nu J''_\pi; J\rangle = 4[J'_\nu][J''_\nu][J'_\pi][J''_\pi]\sum_R (2R+1)\begin{bmatrix} j_\nu & & j_\pi & & J'_\pi & & J'_\nu & \\ & R & & j_\pi & & J & & j_\nu \\ j_\nu & & j_\pi & & J''_\pi & & J''_\nu & \end{bmatrix} V^R_{\nu\pi}, \qquad (1)$$

with $[x] \equiv \sqrt{2x+1}$, and where $V^R_{\nu\pi} \equiv \langle j_\nu j_\pi; R|\hat{V}_{\nu\pi}|j_\nu j_\pi; R\rangle$ are the np two-body matrix elements and the symbol in square brackets is a $12j$ coefficient of the second kind, which is known as a sum of products of four $6j$ coefficients [5].

[a] e-mail: isacker@ganil.fr

The expression (1) is used in Ref. [6] to demonstrate the crucial role played by the np matrix element $V^R_{\nu\pi}$ with aligned neutron and proton angular momenta, that is, for $R = j_\nu + j_\pi$. This generic feature of the np interaction results from angular momentum coupling [i.e., from the $12j$ coefficient in Eq. (1)] as well as from the matrix element in the aligned configuration, which is moderately to strongly attractive for any reasonable np interaction. The importance of the aligned np matrix element raises the question whether a shell-model approximation can be formulated in terms of aligned np pairs, as recently proposed in Ref. [7]. (See Sect. 4.3 of Ref. [8] for a review of this debate.) This question is also addressed in Ref. [6] for a 2n–2p system, and more generally for kn–kp systems in $N = Z$ nuclei in Ref. [9], with the conclusion that the wave functions of many, but certainly not all, yrast states have a dominant component in terms of aligned np pairs. Here I examine whether a description in terms of aligned np pairs gives rise to a geometry of the shell model.

3 Matrix elements of the neutron-proton interaction in a np–np basis

An alternative basis exists in terms of np pairs of the form $|J_1J_2;J\rangle\!\rangle \equiv |j_\nu j_\pi(J_1), j_\nu j_\pi(J_2); J\rangle\!\rangle$. If J_1 and J_2 acquire all values allowed by angular-momentum coupling, this defines an overcomplete basis of states, which are not necessarily normalized, as indicated by the double bracket. Matric elements in the np–np basis can be obtained with the help of the transformation

$$|J_1J_2;J\rangle\!\rangle = -\sum_{J_\nu\ \mathrm{even}}\sum_{J_\pi\ \mathrm{even}} \begin{bmatrix} j_\nu & j_\pi & J_1 \\ j_\nu & j_\pi & J_2 \\ J_\nu & J_\pi & J \end{bmatrix} |J_\nu J_\pi; J\rangle, \tag{2}$$

where the symbol in square brackets is a unitary $9j$ coefficient [10]. Note that the basis states on the right-hand side are orthogonal and normalized while those on the left-hand side are not. In the application of interest one constructs states in terms of a single np pair such that $J_1 = J_2 \equiv \bar{J}$, in which case the overlap matrix element is

$$\langle\!\langle \bar{J}^2; J|\bar{J}^2; J\rangle\!\rangle = \frac{1+(-)^J}{4}\left(1 - \begin{bmatrix} j_\nu & j_\pi & \bar{J} \\ j_\pi & j_\nu & \bar{J} \\ \bar{J} & \bar{J} & J \end{bmatrix}\right), \tag{3}$$

which is zero for odd values of J. This shows that the np pair behaves as a boson since two identical np pairs can only coupled to even J. Henceforth it is assumed that J is even. The matrix elements of the np interaction in the np–np basis are

$$\begin{aligned} \langle\!\langle \bar{J}^2; J|\hat{V}_{\nu\pi}|\bar{J}^2; J\rangle\!\rangle &= V^{\bar{J}}_{\nu\pi} - 2\begin{bmatrix} j_\nu & j_\pi & \bar{J} \\ j_\pi & j_\nu & \bar{J} \\ \bar{J} & \bar{J} & J \end{bmatrix} V^{\bar{J}}_{\nu\pi} \\ &\quad + (2\bar{J}+1)^2\sum_R (2R+1)\begin{bmatrix} j_\nu & & j_\pi & & \bar{J} & & \bar{J} & \\ & R & & j_\nu & & J & & j_\pi \\ j_\nu & & j_\pi & & \bar{J} & & \bar{J} & \end{bmatrix} V^R_{\nu\pi}. \end{aligned} \tag{4}$$

For aligned np pairs, $J_{\max} = j_\nu + j_\pi$, the unitary $9j$ coefficient in the overlap (3) and in the matrix element (4) is much smaller than 1 and can be neglected. [For $J = 2J_{\max}$ the unitary $9j$ coefficient equals 1 but this angular momentum is excluded by the Pauli principle—see Eq. (3).] The sum over the $12j$ coefficients, however, cannot be neglected. To a good approximation the matrix elements of the np interaction between normalized states are therefore

$$\langle J^2_{\max}; J|\hat{V}_{\nu\pi}|J^2_{\max}; J\rangle \approx 2V^{J_{\max}}_{\nu\pi} + 2(2J_{\max}+1)^2\sum_R (2R+1)\begin{bmatrix} j_\nu & & j_\pi & & J_{\max} & & J_{\max} & \\ & R & & j_\nu & & J & & j_\pi \\ j_\nu & & j_\pi & & J_{\max} & & J_{\max} & \end{bmatrix} V^R_{\nu\pi}. \tag{5}$$

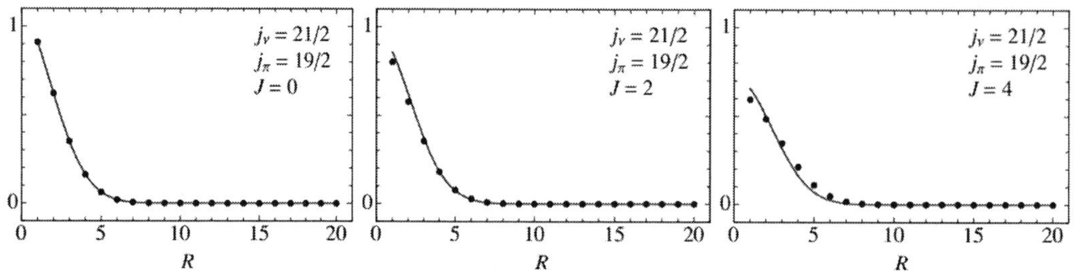

Figure 1. Illustration of the $12j$-coefficient approximation for $j_\nu = 21/2$, $j_\pi = 19/2$ and $J = 0, 2, 4$, as a function of R. The dots are the exact expression on the left-hand side and the curve is its approximation on the right-hand side of Eq. (8).

Not surprisingly, one finds that, in a basis constructed out of aligned np pairs, the main contribution to the 2n–2p configuration stems from the matrix element $V_{\nu\pi}^{J_{\max}}$. There are however corrections to this dominant contribution, which depend on the total angular momentum J and on the strengths of the np interaction with $R \neq J_{\max}$.

4 Expressions in the high-J and the low-J limits

For a schematic short-range interaction the expressions in the np–np basis simplify. Specifically, for a surface delta interaction (SDI) classical approximations to the sum in Eq. (5) can be worked out for aligned np pairs, $J_{\max} = j_\nu + j_\pi$. With the method of Ref. [12], based on the classical limit of $3nj$ coefficients [11], the matrix element (5) can be reduced to

$$\langle J_{\max}^2; J|\hat{V}_{\nu\pi}^{\rm SDI}|J_{\max}^2; J\rangle \approx 2V_{\nu\pi}^{J_{\max}} - \frac{3a_0 + a_1}{\pi \sin\theta} + (-)^{\ell_\nu + j_\nu + \ell_\pi + j_\pi} \frac{a_0 - a_1}{\pi \tan\theta}, \tag{6}$$

where a_0 (a_1) is the isoscalar (isovector) strength of the np interaction and θ is the angle between the angular momenta $J_{\max}$ of the two np pairs that couple to total angular momentum J,

$$\theta = \arccos \frac{J(J+1) - 2J_{\max}(J_{\max}+1)}{2J_{\max}(J_{\max}+1)}. \tag{7}$$

The classical approximation (6) is reasonable except for low angular momentum J, in particular for $J = 0$ or $\theta = \pi$ when it diverges. Unlike in Ref. [12] one is interested in the case $J = 0$ since it corresponds to the energy of the ground state. The following approximate expression can be derived for the $12j$ coefficient under the assumption that J is low and that the np pair is aligned, $J_{\max} = j_\nu + j_\pi$:

$$\begin{aligned}(J_{\max}+1)^2(2J_{\max}+1)&\begin{bmatrix} j_\nu & & j_\pi & & J_{\max} & & J_{\max} & \\ & R & & j_\nu & & J & & j_\pi \\ j_\nu & & j_\pi & & J_{\max} & & J_{\max} & \end{bmatrix} \\ &\approx \exp\left[\frac{-4R(R+1) - J(J+1) + \sqrt{R(R+1)J(J+1)} + 4(j_\nu - j_\pi)^2}{2(J_{\max}+1)}\right]. \end{aligned} \tag{8}$$

This approximation is accurate for $J = 0$ but deteriorates rapidly for increasing values of J, unless the single-particle angular momenta are very large (see Fig. 1). The low-J and high-J approximations to the matrix element (5) are illustrated in Fig. 2.

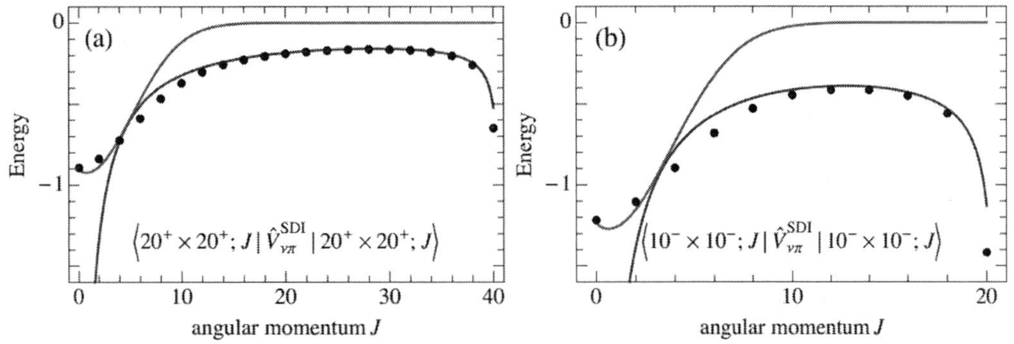

Figure 2. The exact expression for the 2n–2p matrix element (5) of the np interaction (dots) compared with the high-J (red) and low-J (blue) approximations of Eqs. (6) and (8), respectively. The comparison is carried out for (a) a SDI with $a_0 = a_1 = 0.25$ and for $j_\nu = 21/2$ and $j_\pi = 19/2$ and ℓ_ν and ℓ_π even, and (b) a SDI with $a_0 = 0.75$ and $a_1 = 0.25$, and for $j_\nu = h_{11/2}$ and $j_\pi = g_{9/2}$. The constant contribution $2V_{\nu\pi}^{J_{\max}}$ is not included.

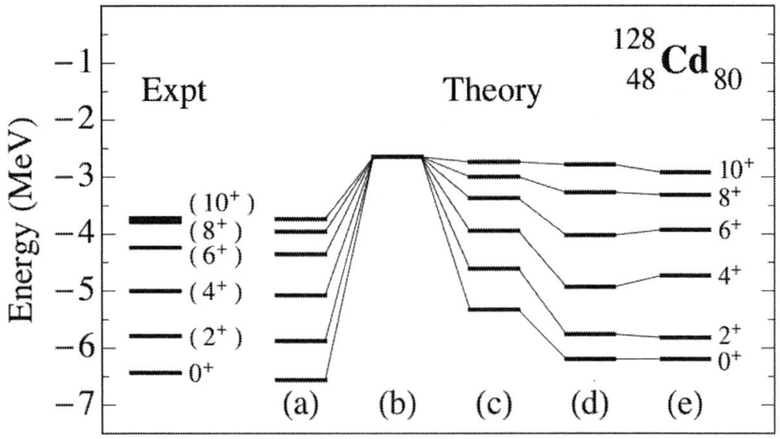

Figure 3. The experimental spectrum of ^{128}Cd compared to the shell-model approximations (a)–(e) discussed in the text.

5 The example of ^{128}Cd

Let me now illustrate the different approximations with the example of the nucleus ^{128}Cd, with two neutron (proton) holes in $\nu 0h_{11/2}$ ($\pi 0g_{9/2}$) with respect to the core ^{132}Sn. The relevant two-body np matrix elements are derived from the realistic interaction jj45pna [13], and the nn and pp interactions are taken from ^{130}Sn and^{130}Cd, respectively. The results of a shell-model calculation with this interaction in the $(\nu 0h_{11/2})^{-2}(\pi 0g_{9/2})^{-2}$ model space are shown in Fig. 3(a), and are seen to agree with the observed energies. The jj45pna interaction can therefore be considered as a realistic basis for further approximations. The approximation (b) consists of keeping only the np matrix element $V_{\nu\pi}^{10^-}$ and putting all others (nn, pp and np) to zero. As already pointed out by Zamick and Escuderos [14], this clearly is inadequate. It is essential to keep the nn and pp interactions and therefore, in order to the test the influence of the np interaction, in the approximations (c)–(e) the realistic nn and pp interactions are taken without modification. In Fig. 3(c) are shown the results obtained in the

$(\nu 0h_{11/2})^{-2}(\pi 0g_{9/2})^{-2}$ model space with the exact nn and pp interactions but with a single non-zero matrix element $V_{\nu\pi}^{10^-}$. This is the approximation advocated in Ref. [6]; it is indeed remarkable that with a *single* component of $V_{\nu\pi}^{R}$ that much of the np correlations can be accounted for. The results (d) are obtained with Eq. (5) and equivalent expressions for the nn and pp interactions (not given here). This is the aligned-np-pair approximation [9]: take the full, realistic interaction within the restricted space constructed out of aligned np pairs, which in this example consists of a single state. Comparison of (a) and (d) shows that the aligned-np-pair approximation is adequate for low J but fails for $J = 8$ and 10. Finally, the results (e) are obtained with the np interaction in the low-J approximation (8) for $J \leq 4$ and the classical approximation (6) for $J \geq 4$ with $a_0 = 0.5$ and $a_1 = 0.13$, strengths obtained from a fit of the SDI to the jj45pna interaction. The results for $J = 4$ are essentially indistinguishable in the two approximations.

6 Concluding remark

It is important to distinguish between the following two approximations: (i) The full shell-model interaction is considered in the restricted space constructed out of aligned np pairs; (ii) Only the component of the np interaction in the aligned configuration is taken and diagonalized in the full shell-model space. The two approximations are *not* equivalent and have their respective merits and problems. If approximation (i) is made for a short-range interaction, a simple geometric interpretation can be obtained, as argued in this contribution. Approximation (ii), on the other hand, can be shown to give rise to a partially solvable problem for the four-nucleon system [15].

Acknowledgement

I acknowledge fruitful discussions with Yung Hee Kim, Maurycy Rejmund and Antoine Lemasson.

References

[1] J. P. Schiffer, Ann. Phys. (N.Y.) **66** (1971) 798.
[2] A. Molinari, M. B. Johnson, H. A. Bethe and W. M. Alberico, Nucl. Phys. A **239** (1975) 45.
[3] R. F. Casten, *Nuclear Structure from a Simple Perspective* (Oxford University Press, Oxford, 2000).
[4] P. Van Isacker, EPJ Web Conf. **78** (2014) 03004.
[5] A. P. Yutsis, I. B. Levinson and V. V. Vanagas, *The Theory of Angular Momentum* (Israel Program for Scientific Translations, Jerusalem, 1962).
[6] Y. H. Kim, M. Rejmund, P. Van Isacker and A. Lemasson, to be published.
[7] B. Cederwall *et al.*, Nature **469** (2011) 68.
[8] S. Frauendorf and A. O. Macchiavelli, Prog. Part. Nucl. Phys. **78** (2014) 24.
[9] P. Van Isacker, Int. J. Mod. Phys. E **20** (2011) 191.
[10] I. Talmi, *Simple Models of Complex Nuclei. The Shell Model and the Interacting Boson Model* (Harwood, Chur, 1993)
[11] E. P. Wigner, *Group Theory and Its Application to the Quantum Mechanics of Atomic Spectra* (Academic Press, New York, 1959).
[12] P. Van Isacker and A. O. Macchiavelli, Phys. Rev. C **87** (2013) 061301(R).
[13] M. Hjorth-Jensen, T. T. Kuo and E. Osnes, Phys. Reports **261** (1995) 125.
[14] L. Zamick and A. Escuderos, Phys. Rev. C **87** (2013) 044302.
[15] P. Van Isacker, unpublished.

Partial dynamical symmetry and the vibrational structure of Cd isotopes

A. Leviatan[1,*], *N.* Gavrielov[1,**], *J.E.* García-Ramos[2,***], and *P.* Van Isacker[3,****]

[1]Racah Institute of Physics, The Hebrew University, Jerusalem 91904, Israel
[2]Departamento de Ciencias Integradas, Universidad de Huelva, 21071 Huelva, Spain
[3]Grand Accélérateur National d'Ions Lourds, CEA/DRF-CNRS/IN2P3, Bvd Henri Becquerel, B.P. 55027, F-14076 Caen, France

Abstract. The recently reported deviations of selected non-yrast states in ^{110}Cd from the expected spherical-vibrator behaviour, is addressed by means of an Hamiltonian with U(5) partial dynamical symmetry. The latter preserves the U(5) symmetry in a segment of the spectrum and breaks it in other states. The effect of intruder states is treated in the framework of the interacting boson model with configuration mixing.

The Cd isotopes have been traditionally considered to be a prime example of spherical vibrators. Recently, advanced experimental studies have reported significant deviations from this behaviour in selected two- and three-phonon states, along the Cd chain (A=108-126) [1–3]. These observations have led to claims for the "breakdown of the vibrational motion" in these isotopes and the need for a paradigm shift [1, 2]. In the present contribution, we examine an alternative explanation for the structure of the Cd isotopes, in terms of U(5) partial dynamical symmetry (PDS) [4].

A convenient starting point for describing spherical nuclei is the U(5) limit of the interacting boson model (IBM) [5], corresponding to the chain of nested algebras,

$$\mathrm{U}(6) \supset \mathrm{U}(5) \supset \mathrm{SO}(5) \supset \mathrm{SO}(3)\,. \quad (1)$$

The basis states $|[N], n_d, \tau, n_\Delta, L\rangle$ have quantum numbers which are the labels of irreducible representations of the algebras in the chain. Here N is the total number of monopole (s) and quadrupole (d) bosons, n_d and τ are the d-boson number and seniority, respectively, and L is the angular momentum. The multiplicity label n_Δ counts the maximum number of d-boson triplets coupled to $L = 0$. The dynamical symmetry (DS) Hamiltonian has the form

$$\hat{H}_{\mathrm{DS}} = t_1\,\hat{n}_d + t_2\,\hat{n}_d^2 + t_3\,\hat{C}_{\mathrm{SO(5)}} + t_4\,\hat{C}_{\mathrm{SO(3)}}\,, \quad (2)$$

where $\hat{C}_{\mathrm{G}}$ is the Casimir operator of G, and $\hat{n}_d = \sum_m d_m^\dagger d_m = \hat{C}_{\mathrm{U(5)}}$. $\hat{H}_{\mathrm{DS}}$ is completely solvable with eigenstates $|[N], n_d, \tau, n_\Delta, L\rangle$ and eigen-energies

$$E_{\mathrm{DS}} = t_1\,n_d + t_2\,n_d^2 + t_3\,\tau(\tau+3) + t_4\,L(L+1)\,. \quad (3)$$

A typical U(5)-DS spectrum exhibits n_d-multiplets of a spherical vibrator, with a two-phonon ($n_d = 2$) triplet of states ($L = 4, 2, 0$) at an energy $E(n_d = 2) \approx 2E(n_d = 1)$ above the ground state ($n_d = L = 0$), and a three-phonon ($n_d = 3$) quintuplet of states ($L = 6, 4, 3, 0, 2$) at $E(n_d = 3) \approx 3E(n_d = 1)$. A quadrupole operator proportional to

$$\hat{Q} = d^\dagger s + s^\dagger \tilde{d}\,, \quad (4)$$

enforces strong ($n_d + 1 \rightarrow n_d$) E2 transitions with particular ratios, *e.g.*, $\frac{B(E2;\,n_d=2,L=0,2,4\rightarrow n_d=1,L=2)}{B(E2;\,n_d=1,L=2\rightarrow n_d=0,L=0)} = 2\frac{(N-1)}{N}$.

The empirical spectrum of ^{110}Cd consists of both normal levels [shown in Fig. 1(a)], and intruder levels [shown in Fig. 2(a)] based on 2p-4h proton excitations across the Z=50 closed shell. The experimental energies and E2 rates in Fig. 1(a), demonstrate that most normal states have good spherical vibrator properties, and conform well with the U(5)-DS calculation shown in Fig. 1(b). However, the measured rates for E2 decays from the non-yrast states, 0_3^+, ($n_d = 2$) and [0_4^+, 2_5^+ ($n_d = 3$)], reveal marked deviations from this behaviour. In particular, $B(E2;\,0_3^+ \rightarrow 2_1^+) < 7.9$, $B(E2;\,2_5^+ \rightarrow 4_1^+) < 5$, $B(E2;\,2_5^+ \rightarrow 2_2^+) < 0.7^{+0.5}_{-0.6}$ W.u., are extremely small compared to the U(5)-DS values: 46.29, 11.02, 19.84 W.u., respectively. Absolute B(E2) values for transitions from the 0_4^+ state are not known, but its branching ratio to 2_2^+ is small.

Attempts to explain the above deviations in terms of strong mixing between the normal spherical [U(5)-like] states and intruder deformed [SO(6)-like] states have been shown to be unsatisfactory [1, 2]. The reasons are twofold. (i) The strong mixing required for an adequate description of the two-phonon 0_3^+ state, results in serious disagreements with the observed decay pattern of three-phonon yrast states. (ii) The discrepancy in the decays of the non-yrast two- and three-phonon states persists also in the heavier ^{A}Cd isotopes (A=110-126), even though the energy of intruder states rises away from neutron mid-shell, and the mixing is reduced. These observations have led to the conclusion that the normal-intruder strong-mixing scenario needs to be rejected, and have raised serious questions on the validity of the multi-phonon interpretation [1, 2]. In what follows, we consider a possible

*e-mail: ami@phys.huji.ac.il
**e-mail: noam.gavrielov@mail.huji.ac.il
***e-mail: enrique.ramos@dfaie.uhu.es
****e-mail: isacker@ganil.fr

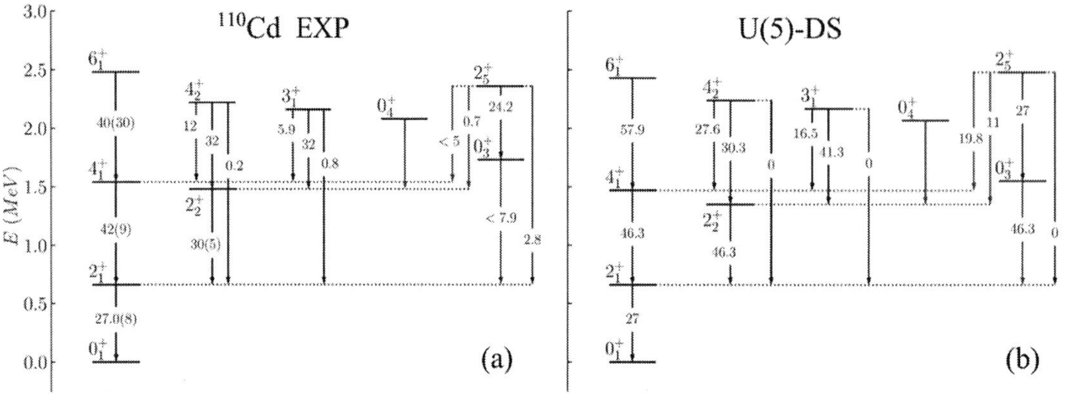

Figure 1. (a) Experimental spectrum and representative E2 rates [2, 6] (in W.u.) of normal levels in ^{110}Cd. (b) U(5)-DS spectrum obtained from $\hat{H}_{\rm DS}$ (2) with parameters $t_1 = 641.5$, $t_2 = 67.9$, $t_3 = -26.1$, $t_4 = 8.7$ keV and $N = 7$. The E2 operator is $e_B\,\hat{Q}$, Eq. (4), with $e_B = 1.96$ (W.u.)$^{1/2}$. For additional experimental error bars on B(E2) values, not shown in Figs. 1(a) and 2 (a), see [2, 6].

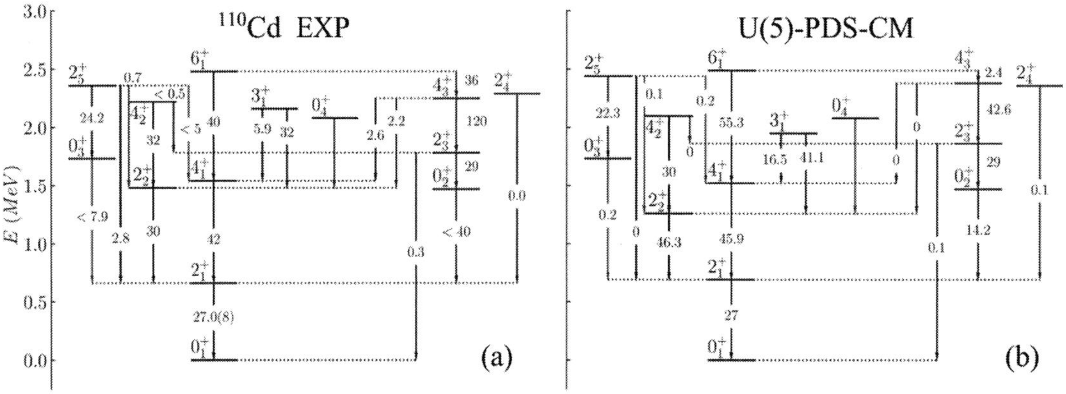

Figure 2. (a) Experimental spectrum and representative E2 rates [2, 6] (in W.u.) of normal [as in Fig. 1(a)] and intruder levels (0^+_2, 2^+_3, 4^+_3, 2^+_4) in ^{110}Cd. (b) U(5)-PDS-CM spectrum, obtained from $\hat{H}$ (11) with parameters $t_1 = 767.8$, $t_2 = -t_3 = 73.6$, $t_4 = 18.5$, $r_0 = 2.1$, $e_0 = -6.9$, $\kappa = -72.7$, $\Delta/2 = 4989.5$, $\alpha = -42.8$ keV and $N = 7$. The E2 operator (12) used has $e_B^{(N)} = 1.96$, $e_B^{(N+2)} = 1.19$ (W.u.)$^{1/2}$.

explanation for the "Cd problem", based on a partial dynamical symmetry (PDS). The latter corresponds to a situation in which a given symmetry is obeyed by only a subset of states and is broken in other states. Such a notion has been previously employed in nuclear spectroscopy in conjunction with the SU(3)-DS [7–9] and SO(6)-DS [10–12] chains of the IBM. In the present contribution, we show the relevance of U(5)-PDS to the Cd problem.

The lowest normal levels comprise three classes of states,

$$
\begin{aligned}
&\text{Class A}: && n_d = \tau = 0, 1, 2, 3 && (n_\Delta = 0)\,, && \text{(5a)}\\
&\text{Class B}: && n_d = \tau + 2 = 2, 3 && (n_\Delta = 0)\,, && \text{(5b)}\\
&\text{Class C}: && n_d = \tau = 3 && (n_\Delta = 1)\,. && \text{(5c)}
\end{aligned}
$$

In the U(5)-DS calculation of Fig 1(b), the "problematic" states [0^+_3 ($n_d = 2$) and 2^+_5 ($n_d = 3$)] belong to class B, and 0^+_4 ($n_d = 3$) belongs to Class C. The remaining "good" spherical-vibrator states [0^+_1 ($n_d = 0$); 2^+_1 ($n_d = 1$); $4^+_1, 2^+_2$ ($n_d = 2$); $6^+_1, 4^+_2, 3^+_1$ ($n_d = 3$)] belong to Class A. As mentioned, the spherical-vibrator interpretation is valid for most states in Fig. 1(a), but not all. We are thus confronted with a situation in which some states in the spectrum (assigned to Class A) obey the predictions of U(5)-DS, while other states (assigned to Classes B and C) do not. These empirical findings signal the presence of a partial dynamical symmetry, U(5)-PDS.

The construction of Hamiltonians with U(5)-PDS follows the general algorithm [12, 13], by identifying operators which annihilate particular sets of U(5) basis states. In the present case, we consider the following interaction,

$$\hat{V}_0 \;=\; r_0\, G^\dagger_0 G_0 + e_0 \left(G^\dagger_0 K_0 + K^\dagger_0 G_0 \right) , \qquad (6)$$

where $G^\dagger_0 = [(d^\dagger d^\dagger)^{(2)} d^\dagger]^{(0)}$, $K^\dagger_0 = s^\dagger (d^\dagger d^\dagger)^{(0)}$ and standard notation of angular momentum coupling is used. $\hat{V}_0$ of Eq. (6) is in normal-ordered form and satisfies

$$\hat{V}_0 |[N], n_d = \tau, \tau, n_\Delta = 0, L\rangle = 0\,, \qquad (7)$$

with $L = \tau, \tau+1, \ldots, 2\tau-2, 2\tau$. Eq. (7) follows from the fact that the indicated states have $n_d = \tau$ and $n_\Delta = 0$, hence do not contain a pair or a triplet of d-bosons coupled to $L = 0$ and, as such, are annihilated by K_0 [5] and G_0 [14].

The states of Eq. (7), which include those of Class A, form a subset of U(5) basis states, hence remain solvable eigenstates of the following Hamiltonian

$$\hat{H}_{\rm PDS} = \hat{H}_{\rm DS} + \hat{V}_0 \,, \tag{8}$$

with good U(5) symmetry and energies given in Eq. (3) with $n_d = \tau$. It should be noted that while $\hat{H}_{\rm DS}$ (2) is diagonal in the U(5)-DS chain (1), the r_0-term (e_0-term) in $\hat{V}_0$ connects states with $\Delta n_d = 0$ and $\Delta\tau = 0, \pm2, \pm4, \pm6$ ($\Delta n_d = \pm1$ and $\Delta\tau = \pm1, \pm3$). Accordingly, the remaining eigenstates of $\hat{H}_{\rm PDS}$ (8), in particular those of classes B and C, are mixed with respect to U(5) and SO(5). The U(5)-DS is therefore preserved in a subset of eigenstates but is broken in other states. By definition, $\hat{H}_{\rm PDS}$ exhibits U(5)-PDS. Cubic terms of the type present in $\hat{V}_0$, Eq. (6), are frequently encountered in PDS Hamiltonians, *e.g.*, in conjunction with signature splitting [9], band structure [10–12], and shape-coexistence [15, 16] in nuclei.

The effect of intruder levels can be studied in the framework of the interacting boson model with configuration mixing (IBM-CM) [17]. The latter involves the space of normal states described by a system of N bosons representing valence nucleon pairs, and the space of intruder states described by a system of $N+2$ bosons, accounting for particle-hole shell model excitations. This procedure has been used extensively in describing coexistence phenomena in nuclei [18–21]. In the present study of ^{110}Cd, the Hamiltonian in the normal sector is taken to be $\hat{H}_{\rm PDS}$ of Eq. (8), acting in a space of $N = 7$ bosons. The Hamiltonian in the intruder sector is taken to be of SO(6)-type [18],

$$\hat{H}_{\rm intrud} = \kappa \hat{Q}\cdot\hat{Q} + \Delta \,, \tag{9}$$

acting in a space of $N = 9$ bosons, with $\hat{Q}$ given in Eq. (4). A mixing term between the $[N]$ and $[N+2]$ boson spaces is defined as [18–21],

$$\hat{V}_{\rm mix} = \alpha\left[(s^\dagger)^2 + (d^\dagger d^\dagger)^{(0)}\right] + {\rm H.c}, \tag{10}$$

where H.c. means Hermitian conjugate. The combined Hamiltonian for the two configurations has the form

$$\hat{H} = \hat{H}^{(N)}_{\rm PDS} + \hat{H}^{(N+2)}_{\rm intrud} + \hat{V}^{(N,N+2)}_{\rm mix} \,. \tag{11}$$

Here $\hat{O}^{(N)} = \hat{P}^\dagger_N \hat{O}\,\hat{P}_N$ and $\hat{O}^{(N,N')} = \hat{P}^\dagger_N \hat{O}\,\hat{P}_{N'}$ for an operator $\hat{O}$, with $\hat{P}_N$ a projection operator onto the $[N]$ boson space. Similarly, the E2 operator is defined as,

$$\hat{T}(E2) = e_B^{(N)}\,\hat{Q}^{(N)} + e_B^{(N+2)}\,\hat{Q}^{(N+2)} \,, \tag{12}$$

with boson effective charges, $e_B^{(N)}$ and $e_B^{(N+2)}$.

The experimental energies and E2 rates for both normal and intruder levels in ^{110}Cd, are shown in Fig. 2(a). They are well reproduced by an IBM-PDS-CM calculation, shown in Fig. 2(b), employing the Hamiltonian of Eq. (11) and the E2 operator of Eq. (12). The mixing between the intruder states and normal states of class A is weak. The latter states retain a high degree of purity and good U(5) quantum numbers. This is reflected in their E2 decay properties, which are essentially the same as those of the U(5)-DS shown in Fig. 1(b). In contrast, the states in classes B and C, whose decay properties show marked deviations from the U(5)-DS limit, are mixed with other normal and intruder states. The resulting calculated values: $B(E2;\, 0^+_3 \to 2^+_1) = 0.25$, $B(E2;\, 2^+_5 \to 4^+_1) = 0.19$, $B(E2;\, 2^+_5 \to 2^+_2) = 0.12$ W.u., are consistent with the measured upper limits: 7.9, 5, $0.7^{+0.5}_{-0.6}$ W.u., respectively.

In summary, we have considered the vibrational structure of ^{110}Cd, by means of U(5)-PDS. The PDS Hamiltonian retains good U(5) symmetry for yrast states, but breaks it in particular non-yrast states. The mixing with the intruder levels is weak, and affects mainly the broken U(5)-DS states. Most low-lying normal states maintain the vibrational character and only specific states exhibit a departure from this behaviour, in line with the empirical data. Calculations are underway to see if this approach can be implemented in other neutron-rich Cd isotopes.

This work is supported in part (A.L. and N.G.) by the Israel Science Foundation (Grant 586/16).

References

[1] P.E. Garrett, K.L. Green and J.L. Wood, Phys. Rev. C **78**, 044307 (2008)

[2] P.E. Garrett *et al.*, Phys. Rev. C **86**, 044304 (2012)

[3] J.C. Batchelder *et al.*, Phys. Rev. C **89**, 054321 (2014)

[4] A. Leviatan, Prog. Part. Nucl. Phys. **66**, 93 (2011)

[5] F. Iachello and A. Arima, *The Interacting Boson Model* (Cambridge Univ. Press, Cambridge, 1987)

[6] G. Gürdal and F.G. Kondev, NDS **113**, 1315 (2012)

[7] A. Leviatan, Phys. Rev. Lett. **77**, 818 (1996)

[8] A. Leviatan *et al.*, Phys. Rev. C **60**, 061301(R) (1999)

[9] A. Leviatan, J.E. García-Ramos and P. Van Isacker, Phys. Rev. C **87**, 021302(R) (2013)

[10] P. Van Isacker, Phys. Rev. Lett. **83**, 4269 (1999)

[11] A. Leviatan and P. Van Isacker, Phys. Rev. Lett. **89**, 222501 (2002)

[12] J.E. García-Ramos, A. Leviatan and P. Van Isacker, Phys. Rev. Lett. **102**, 112502 (2009)

[13] Y. Alhassid and A. Leviatan J. Phys. A **25**, L1265 (1992)

[14] I. Talmi, in *Symmetries in Nuclear Structure*, (A. Vitturi and R.F. Casten Eds.), World Scientific p. 10 (2004)

[15] A. Leviatan and D. Shapira, Phys. Rev. C **93**, 051302(R) (2016)

[16] A. Leviatan and N. Gavrielov, Phys. Scr. **92**, 114005 (2017)

[17] P.D. Duval and B.R. Barrett, Nucl. Phys. A **376**, 213 (1982)

[18] K. Heyde *et al.*, Nucl. Phys. A **586**, 1 (1995)

[19] R. Fossion, K. Heyde, G. Thiamova and P. Van Isacker, Phys. Rev. C **67**, 024306 (2003)

[20] J.E. García-Ramos, V. Hellemans and K. Heyde, Phys. Rev. C **84**, 014331 (2011)

[21] J.E. García-Ramos and K. Heyde, Phys. Rev. C **89**, 014306 (2014); Phys. Rev. C **92**, 034309 (2015)

Shape phase transition and shape coexistence in the Bose-Fermi system

Yu Zhang[1,⋆,⋆⋆], *Wenting* Dong[1], and *He* Jiang[1]

[1] *Department of Physics, Liaoning Normal University, Dalian, 116029, China*

Abstract. A classical analysis of shape phase transition and shape coexistence in odd-even nuclei has been carried within the interacting boson-fermion model. The results indicate that shape coexistence can be taken as a signature of shape phase transition in odd-even nuclei.

1 Introduction

Shape phase transitions (SPTs) in odd-even nuclei have received some attentions in recent years [1–10]. Evidences of the spherical to prolate SPT and the associated shape/phase coexistence have been clearly identified in the odd Sm nuclei [8]. A theoretical tool suitable for analyzing the SPTs in odd-even nuclei is the interacting boson-fermion model (IBFM) [11], in which odd-even nucleus is approximately considered as the Bose-Fermi system with an even-even core (boson) coupled to a single particle (fermion). Previous investigations reveal that the effects of single particle may influence different type of SPT in different way [4–6]. In this work, we will present a classical analysis of the effects of single particle on the spherical to prolate SPT within the framework of IBFM. Particularly, we will emphasize how the presence of an odd particle can influence the shape/phase coexistence in this type of SPT.

2 Classical analysis of the spherical to prolate SPTs

In the following discussion, an IBFM Hamiltonian describing the system with the even-even core coupled to a single fermion via quadrupole-quadrupole interaction is adopted and given as

$$\hat{H} = \varepsilon\left[(1-\eta)\hat{n}_d - \frac{\eta}{4N}\hat{Q}_{BF}\cdot\hat{Q}_{BF}\right], \tag{1}$$

where $\hat{n}_d \equiv \sum_u d_u^\dagger d_u$ and $\hat{Q}_{BF} \equiv \hat{Q}_B^\chi + \hat{q}_F$ with $\hat{Q}_B^\chi = (d^\dagger s + s^\dagger \tilde{d})^{(2)} + \chi(d^\dagger\tilde{d})^{(2)}$ and $\hat{q}_F = (a_j^\dagger \tilde{a}_j)^{(2)}$. Besides, η and χ are the control parameters with $\eta \in [0,1]$ and $\chi \in [-\sqrt{7}/2, 0]$, and ε is a scale factor which is set as 1 for convenience. One can prove that the spherical to prolate SPTs may occur at $\eta_c = \frac{28N}{56(N-1)+\chi^2(5+2N)}$ with $\chi \in [-\sqrt{7}/2, 0)$ if only the even-even core corresponding to $\hat{Q}_{BF} = \hat{Q}_B^\chi$ in

⋆Support from National Natural Science Foundation of China (11375005)
⋆⋆e-mail: dlzhangyu_physics@163.com

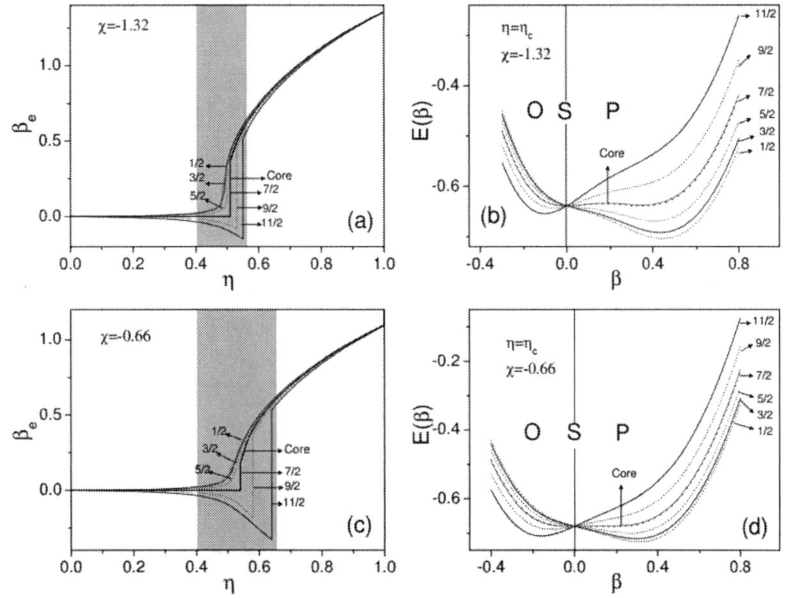

Figure 1. Evolution of the classical order parameter as a function of η and the potential energy surface at the corresponding critical point in the cases with $\chi = -1.32$ and $\chi = -0.66$, respectively. P, O and S represent prolate, oblate and spherical deformation.

(1) is taken into account. Since a single-j shell is considered for the odd fermion, the pure fermion part contributes only an additional constant to the ground state energy [4].

To this end, we take the single fermion with $j = 11/2$ as an example. Specifically, we adopt the intrinsic state for the coupled system [4, 5]

$$|\beta, \gamma, N; j, K\rangle_{BF} = \frac{1}{\sqrt{N!(1+\beta^2)^N}}[s^\dagger + \beta\cos\gamma\, d_0^\dagger + \frac{1}{\sqrt{2}}\beta\sin\gamma(d_2^\dagger + d_{-2}^\dagger)]^N|0\rangle_B \otimes a_{jK}^\dagger|0\rangle_F\,.$$

to diagonalize the boson-fermion Hamiltonian (1) in the basis with $K = 1/2,\ 3/2, ..., j$ in order to get the energy eigenvalues $E(\beta, \gamma)$. Here, we consider, for simplicity, the case $\gamma = 0°$ for which the quantum K would be a good quantum number [4, 5]. If the effects of single fermion is analogous as an external field to the boson core, here $\gamma = 0°$ means the field with an axial symmetry [5]. Minimization of $E(\beta)_K$ with respect to β for given χ and η gives the optimal values β_e with $\beta_e = 0$, $\beta_e > 0$, and $\beta_e < 0$ denoting the spherical, prolate, and oblate, respectively. Then, β_e is considered as the classical order parameter to indicate order and type of the SPT. For a 1st order SPT, β_e as a function of η is not continuous. For a 2nd order SPT, β_e is continuous but $\frac{\partial\beta_e}{\partial\eta}$ is not. To perform a concrete analysis, the results calculated for $\chi = -1.32$ and $\chi = -0.66$ are shown in Fig. 1. As seen in Fig. 1(a), the results indicate that the effect of single fermion may weaken the phase transitional features for states with $K = 1/2,\ 3/2,\ 5/2$ but enhance them for states with $K = 9/2,\ 11/2$, when comparing the cases for the even-even core. In contrast, the state for $K = 7/2$ presents almost the same transitional feature as the one for the even-even core. It should be noted that these features were firstly revealed in [5, 6]. More interestingly, one can find that a critical region (grey area in Fig. 1(a)), where β_e for different K suddenly changes with η, has been developed around the critical point η_c. It means that the critical behavior in the odd-even system can persist within a wider range of parameters in contrast to the adjacent even-even system. As seen from the Fig. 1(b), the potentials at η_c for different K have different configurations thus indicating different types of deformation or shape. Specifically, the cases for $K = 1/2,\ 3/2,\ 5/2$ favor prolate deformation corresponding to $\beta_e > 0$, while the cases for $K = 9/2,\ 11/2$ favor weakly oblate deformation corresponding to $\beta_e < 0$. In contrast, the case for $K = 7/2$ has a configuration as same as the one for even-even core, with two degenerate minima at $\beta = 0$ and $\beta > 0$, respectively. As further seen from Fig. 1(a), all the cases favor either the spherical ($\beta_e = 0$) or the prolate ($\beta_e > 0$) when η runs outside of the critical region. It is thus concluded that the Bose-Fermi system may allow coexistence of different shapes only in the critical region, which in turn suggests that shape coexistence could be considered as a signature of the SPT in odd-even nuclei. Here, shape coexistence means that different types of deformation induced by both single-particle excitation and collective excitation may coexist in the low-energy structure at the same point (η, χ). In fact, it has been revealed that phenomenon of shape/phase coexistence indeed occurs in the low-lying spectrum of ^{151}Sm [8], which can be taken as the critical point nucleus in the spherical to prolate SPT [7]. Furthermore, one can find in Fig. 1(c) and (d) that the transitional features in the case with $\chi = -0.66$ are similar to those corresponding to $\chi = -1, 32$ shown in Fig. 1(a) and (d), except that the critical region in the former is enlarged and shape coexistence becomes more pronounced due to the involvement of the O(6) components. Anyway, it suggests that there could be many more chances to identify phenomenon of shape coexistence in the Bose-Fermi system undergoing SPT.

3 Summary

In summary, a classical analysis of the shape phase transitions and the associated shape coexistence in odd-even nuclei have been performed within the framework of the IBFM. Specifically, we have analyzed the evolution of the deformations and the critical potential configurations in the case with the even-even core coupled to an single-fermion moving in the $j = 11/2$ shell. The results indicate that the effects of odd particle may induce more pronounced shape coexistence in odd-even nuclei than in even-even nuclei around the critical point. Since only the quadrupole-quadrupole Bose-Fermi interaction term is considered in the IBFM Hamiltonian in the present discussion, other Bose-Fermi terms such as the exchange term and monopole term, which may potentially affect the shape phase transitional behaviors [6], should be included in the Hamiltonian to further prove the above conclusion.

References

[1] J. Jolie, S. Heinze, P. Van Isacker, and R. F. Casten, Phys. Rev. C **70**, 011305(R) (2004).
[2] F. Iachello, Phys. Rev. Lett. **95**, 052503 (2005).
[3] C. E. Alonso, J. M. Arias, L. Fortunato, and A. Vitturi, Phys. Rev. Lett. **98**, 052501 (2007).
[4] M. Böyükata, C. E. Alonso, J. M. Arias, L. Fortunato, and A. Vitturi, Phys. Rev. C **82**, 014317 (2010).
[5] F. Iachello, A. Leviatan and D. Petrellis, Phys. Lett. B **705**, 379 (2011).
[6] D. Petrellis, A. Leviatan and F. Iachello, Ann. Phys. **326**, 926 (2011).
[7] Y. Zhang, L. N. Bao, X. Guan, F. Pan, and J. P. Draayer, Phys. Rev. C **88**, 064305 (2013).
[8] Y. Zhang, F. Pan, Y. X. Liu, Y. A. Luo, and J. P. Draayer, Phys. Rev. C **88**, 014304 (2013).
[9] D. Bucurescu and N. V. Zamfir, Phys. Rev. C **95**, 014329 (2017).
[10] K. Nomura, T. Nikšić, and D. Vretenar, Phys. Rev. C **96**, 014304 (2017).
[11] F. Iachello and P. Van Isacker, *The Interacting Boson-Fermion Model* (Cambridge University Press, Cambridge, 1991)

On the nature of the shape coexistence and the quantum phase transition phenomena: lead region and Zr isotopes

José-Enrique García-Ramos[1,2,*] and *Kris* Heyde[3,**]

[1] *Departamento de Ciencias Integradas y CEAFMC Universidad de Huelva, 21071 Huelva*
[2] *Instituto Carlos I de Física Teórica y Computacional, Universidad de Granada, Fuentenueva s/n, 18071 Granada, Spain*
[3] *Department of Physics and Astronomy, Ghent University, Proeftuinstraat, 86 B-9000 Gent, Belgium*

Abstract. The goal of this contribution is to analyze the connection between shape coexistence and quantum phase transition, two seemingly unrelated phenomena that share common aspects, namely, the rapid change in the ground state structure along an isotope chain or the presence of several minima at the mean-field level. To illustrate the similarities and differences between both phenomena, we will focus in the Pb region, in particular in Pt and Hg isotopes, as well as in Zr isotopes.

1 Introduction

The atomic nuclei can present different shapes, namely, spherical, axially deformed, either prolate or oblate, or even triaxial. In general, the shape evolves in a gradual way as one passes from a nucleus to its neighbor, but in certain cases the change can be very abrupt. This is the key element for the appearance of a quantum phase transition (QPT) [1]. On the other hand, in a given nucleus different shapes can coexist in the energy spectrum, leading to the appearance of shape coexistence [2]. The presence of extra configurations can be easily singled out in the very conspicuous parabolic behaviour in the energy systematics centered around the mid-shell. Moreover, another relevant indicator of the presence of intruder configurations is the lowering of the first excited 0^+ state that presents a minimum also around the mid-shell.

Along this contribution we try to disentangle which are the relationships and differences between both phenomena. To this end, we analyze three different chains of nuclei, Pt, Hg, and Zr which are ideal cases to see the competition between QPT and shape coexistence phenomena.

2 Shape coexistence and quantum phase transition in a nutshell

Shape coexistence in nuclear physics was first proposed by Morinaga in the 1950's and since them it has given rise to a property of atomic nuclei that appears throughout the entire nuclear landscape, especially to those nuclei at or near shell or sub-shell closures [2]. Shape coexistence presents certain distinct experimental features: a U shape in the energy systematics of certain excited bands, lowering of certain excited 0^+ states, a rapid change in the value of spectroscopic quadrupole moments, and the existence of strong E0 transitions. All of them are enhanced and present an almost symmetric behaviour with respect to the corresponding mid-shell. The appearance of intruder configurations reflects the competition between the energy gap, that tends to maintain spherical shapes, and the residual interaction, that favours the deformation of the nucleus and lower, in some cases considerably, the excitation energy of the intruder states [2].

Shape coexistence can be understood in terms of two major theoretical approaches, namely the spherical shell model and the mean field. According to the shell model approach, the nucleus is described accordingly as an inert core and a set of valence nucleons that occupy certain orbits and interact amongst them through a residual two-body interaction. The promotion of pairs of nucleons across the shell gaps at shell and subshell closures is the mechanism responsible of the creation of intruder configurations. From the point of view of mean-field theory, using the self-consistent Hartree-Fock-Bogoliubov (HFB) theory, one obtains an energy surface depending on certain deformation parameters. In this framework, the ground band will correspond to the deepest minimum, but the others can be interpreted as intruder configurations, that will appear higher in energy and corresponding with larger deformations.

The concept of QPT refers to the sudden change of the atomic nucleus ground state structure as a function of a control parameter. Such control parameter can be, e.g., the neutron number and, therefore, a QPT can appear in an isotopic chain where the ground state deformation varies in an abrupt way when passing from an isotope to its neighbor. Taking into account that the two-neutron separation energy (S_{2n}) is somehow proportional to the derivative of the energy, a first-order QPT will involve a discontinu-

*e-mail: enrique.ramos@dfaie.uhu.es
**e-mail: kris.heyde@ugent.be

ity in S_{2n} while a second-order one a discontinuity in the derivative of S_{2n}. In this contribution we are mainly interested in the case of first-order QPT because it allows the coexistence of two phases around the phase transition region.

An ideal framework to deal with QPTs are the algebraic models, such as the interacting boson model (IBM) [3]. In such a framework a QPT can be modeled in terms of a Hamiltonian that is a combination of two given symmetries of the system, combined through a control parameter, $H = x\,H_{\rm sym1} + (1-x)\,H_{\rm sym2}$. The onset of a QPT is denoted by the existence of a critical value of the control parameter, x_c, for which the structure of the system passes from one phase with *symmetry 1* to another phase with *symmetry 2*.

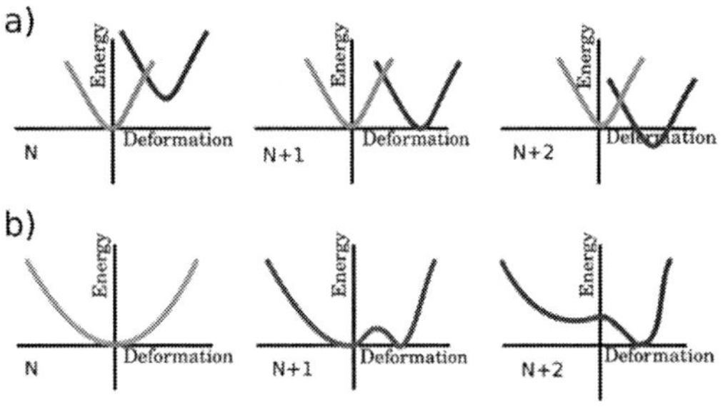

Figure 1. Schematic representation (both scales in arbitrary units) of the evolution of the energy surface for shape coexistence (a) and QPT (b) cases, as a function of the neutron number, *N*.

Are shape coexistence and QPT related in any way? On one hand, both phenomena involve a rapid change in the structure of a certain set of states, either ground or excited states. On the other hand, in both cases, at the mean-field level, several minima coexist. In a pictorial way, in Fig. 1 we depict the evolution of two minima as a function of the neutron number. In panel a) we treat the shape coexistence situation where the relative position of the regular (green) and intruder (red) minima changes as a function of the neutron number. In this case, we assume that intruder and regular minima become degenerated and even that the intruder describes the ground state of the system, although this situation is not general and in many cases the intruder band never becomes the ground state. In panel b) we consider the QPT case, for which in the left most panel only one spherical minimum exists, in the central one an energy surface with two minima develops and finally in the right most panel the spherical minimum disappears and only the deformed one remains. The case where the two minima coexist leads to the existence of a QPT. Obviously both situations present clear similarities, namely, around the region of interest two minima coexist and the value of the deformation changes abruptly in the central panels. However, at the quantum level one expects to observe clear differences. For example, the height of the barrier between both configurations should be a clear source of differences because the states of the bands with corresponding different wells will effectively interact for low barriers while remain unperturbed for high ones. In the figure, we depicted the case of a high barrier with no interacting configurations for the case of shape coexistence, while a low one for QPT. Note that other situations can also exist.

3 Zr, Pt, and Hg

The Pb region is, probably, the best example of shape coexistence along the mass table and, indeed, in Pb up to three different configurations have been identified. In the Hg and Pt nuclei, which are the nuclei we are interested in, two kind of configurations show up with a clear presence of low lying 0^+ states, combined with a parabolic shape in the excitation energy systematics in Hg, whereas not so obvious for Pt. On the other hand, Zr is known to exhibit the faster change in deformation ever observed, with ^{98}Zr almost spherical and ^{100}Zr a very well deformed nucleus.

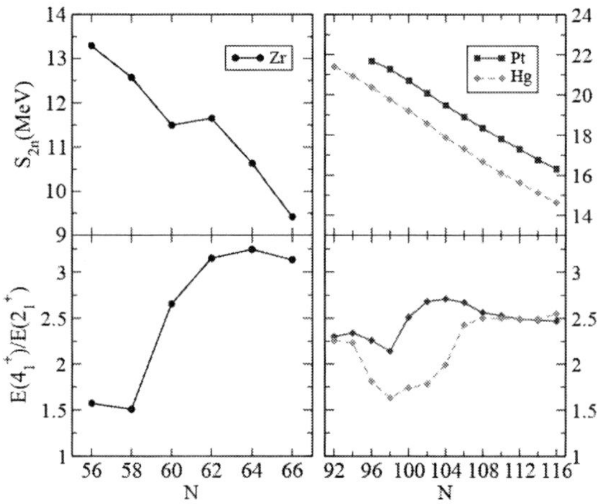

Figure 2. S_{2n} and $E(4_1^+)/E(2_1^+)$ as a function of the neutron number,*N*, for Zr, Pt and Hg.

From the point of view of a QPT, there are two key observables that can be used as indicators of the onset of a QPT. The first one is the S_{2n} that will experience a discontinuity at the first-order transition point, and the second one is the ratio between the excitation energies of the states 2_1^+ and 4_1^+, $E(4_1^+)/E(2_1^+)$, which is related with the deformation or, in other words, with the value of the order parameter of the QPT. In Fig. 2 we present the systematics of S_{2n} and $E(4_1^+)/E(2_1^+)$, as well, for Zr and Pt and Hg. Regarding the value of S_{2n}, one notices a clear discontinuity in Zr, but a fully linear behavior in Hg and Pt. On the other hand, $E(4_1^+)/E(2_1^+)$ shows for Zr the typical rapid change that is observed in a QPT, passing from a value close to 2 (even 1.5 for $^{96-98}$Zr) to another close to 3.3. However, in Pt and Hg the interpretation of the systematics is not so obvious. While in Pt it partially resembles what is observed in a QPT, in Hg there is a noticeable dropping of the ratio around mid-shell. In the latter two cases the systematics can be understood in terms of the coexistence of two configurations that cross or approach very closely.

To shed some more light on the problem, in Fig. 3 we show results corresponding to IBM calculations that were presented elsewhere ([4] for Pt, [5] for Hg and [6] for Zr). In particular, the results correspond to excitation energies of selected states, both regular and intruder, for which the interaction term between intruder and regular sectors of the Hamiltonian has been switched off. In these three cases one observes a rather flat energy systematics for the intruder states while the conspicuous parabolic behaviour for the intruder ones centered around the mid-shell ($A = 106, 182$, and 184 for Zr, Pt, and Hg, respectively). However, several distinct features are worth to be mentioned. First, in Zr and Pt nuclei, a intruder state becomes the ground state around the mid-shell, while it never happens for Hg. That means that the structure of the ground state changes abruptly around the mid-shell in Zr and Pt, but not in Hg. In Hg a change in the structure is observed in the 2^+ states. Note that the more realistic calculation involves a certain degree of mixing between the regular and the intruder states, which is rather small in Zr [6] and in Hg [5], but much larger in Pt [4], which means that the observed behavior in Pt is smoothed out in the more realistic calculation, but remains essentially the same in Zr and Hg.

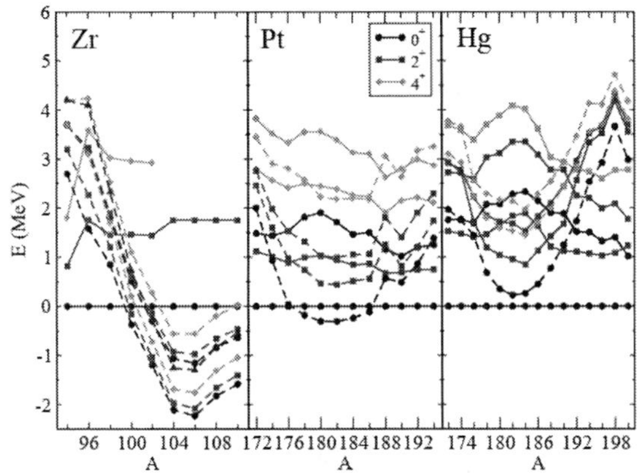

Figure 3. Theoretical IBM excitation energies for selected states in Zr, Pt, and Hg and Hamiltonians without interactions between the regular and the intruder sectors, as a function of A. Full lines stand for regular states and dashed ones for intruder ones.

The above features have direct consequences in the systematics of S_{2n} and $E(4_1^+)/E(2_1^+)$. In the case of Zr, as observed in Fig. 2 can be easily understood in terms of the crossing of configurations depicted in Fig. 3. Indeed, the change in the nature of the ground state produces a sudden change in the slope of the energy systematics and therefore a discontinuity in the value of S_{2n}. On the other hand, because the crossing is so fast, not only the ground state presents an intruder character, but also the rest of low-lying members of the yrast band involved in the energy ratio $E(4_1^+)/E(2_1^+)$, which makes that for $A < 100$ all the states present a vibrational character, while for $A \geq 100$ a rotational one appears. On the other hand, in the case of Hg, because the ground state exhibits a regular character and, moreover, there is no interaction between intruder and regular sector, the linear tendency of S_{2n} even at the mid-shell can be understood. Besides, the crossing between 2_1^+ and 4_1^+ regular and intruder states can explain the drop of the ratio $E(4_1^+)/E(2_1^+)$ around the mid-shell. Finally, to explain the systematics of Pt in Fig.2 is a challenge. First, the crossing of regular and intruder 0^+ energies would suggest a discontinuity in S_{2n} as the observed one in Zr. However experimental values show a fully linear systematics. The reason for such a behaviour is the strong mixing between the regular and the intruder sectors (see [4]). On the other hand, the energy ratio $E(4_1^+)/E(2_1^+)$ seems to present the precursors of a QPT, but they are hindered. Most probably, this too is due to the strong mixing between the regular and intruder states. Note that according to [7] the value of the deformation, both in the IBM and in the HFB calculations, changes as it should be in a QPT. However the strong mixing somehow hides its effect in both the S_{2n} and $E(4_1^+)/E(2_1^+)$ observables.

4 Conclusions

We have analyzed the interplay between shape coexistence and QPT in three chain of isotopes, namely Zr, Pt, and Hg, that seem to be the most promising candidates to disentangle the relationship between both phenomena. In the case of Zr, the QPT that appears around $A = 100$ is easily explained in terms of the existence of two families of states that cross at the phase transition point. In Hg, because the ground state never becomes of intruder character, no indications of QPT are shown, neither for S_{2n} not for $E(4_1^+)/E(2_1^+)$. Finally, in Pt, the crossing of regular and intruder configurations could induce a QPT, but it is not observed in S_{2n} and only partially in $E(4_1^+)/E(2_1^+)$. The reason for that is most probably the strong mixing between regular and intruder configurations. Therefore, we have found a mechanism for inducing a QPT from shape coexistence, though probably is not the only one.

Financial support from the Interuniversity Attraction Poles Program of the Belgian State-Federal Office for Scientific, Technical and Cultural Affairs (IAP Grant No. P7/12) and the Spanish MINECO and FEDER under Project No. FIS2014-53448-C2-2-P are acknowledged.

References

[1] P. Cejnar, J. Jolie, and R.F. Casten, Rev. Mod. Phys. **82**, 2155 (2010).

[2] K. Heyde and J.L. Wood, Rev. Mod. Phys. **83**, 1467 (2011).

[3] F. Iachello and A. Arima, *The Interacting Boson Model*, Cambridge University Press (1987).

[4] J.E. García-Ramos and K. Heyde, Nucl. Phys. **A 825**, 39 (2009); J.E. García-Ramos, V. Hellemans, and K. Heyde, Phys. Rev. C **84**, 014331 (2011).

[5] J.E. García-Ramos, and K. Heyde, Phys. Rev. C **89**, 014306 (2014).

[6] J.E. García-Ramos and K. Heyde, work in progress.

[7] J.E. García-Ramos, K. Heyde, L.M. Robledo, and R. Rodriguez-Guzmán, Phys. Rev. C **89**, 034313 (2014).

Isospin Non-Conservation in Shell Model Approach and Applications*

Yi Hua Lam[1,**] and *Nadezda A.* Smirnova[2,***]

[1]Key Laboratory of High Precision Nuclear Spectroscopy, Institute of Modern Physics, Chinese Academy of Sciences, Lanzhou 730000, China
[2]CENBG, CNRS/IN2P3 and Université de Bordeaux, Chemin du Solarium, 33175 Gradignan cedex, France

Abstract. Up to now, empirical shell-model Hamiltonians for valence space calculations provide the most accurate description of the low-energy spectra and individual transitions of *sd*- and *pf*-shell nuclei. These features made them of particular importance for the description of the isospin-symmetry-breaking phenomena, such as energy splitting of the isobaric multiplets or isospin-forbidden transition rates. In this contribution, we demonstrate the applications of a recently constructed isospin non-conserving (INC) Hamiltonian in *sd* shell [Lam *et al.* Phys. Rev. C **87**, 054304 (2013)]. First, we explore the partial decay scheme of ^{24}Si and discuss the states affected by the Thomas-Ehrman shift. Second, we show the accuracy of the INC Hamiltonian for the description of the mirror energy differences.

1 Introduction

The concept of isospin symmetry significantly simplifies nuclear many-body calculations and enables us to understand the structure of nuclei in the vicinity of the $N = Z$ line. The isospin symmetry implies that a nuclear many-body Hamiltonian is invariant under all rotations in isospin space. However, it is only an approximate symmetry of atomic nuclei due to the Coulomb interaction between protons and a slight neutron-proton mass difference. Modern experiments at radioactive ion beam facilities are capable to bring accurate information on the isospin-symmetry breaking in various phenomena, ranging from the splittings of the isobaric multiplets to rates of isospin-forbidden decays. It is therefore a challenge for a microscopic nuclear theory to describe the experimental observations, making thus a link to the isospin properties of the nucleon-nucleon interaction.

In this contribution, we use an *sd*-shell INC Hamiltonian [1], which is based on the isospin-conserving USDB Hamiltonian [2] and includes (i) a two-body Coulomb potential, normalized for the short-range correlations by the unitary correlation operator method [3], (ii) an effective charge-dependent two-body term of nuclear origin, and (iii) isovector single-particle energies. The strengths of those charge-dependent terms have been determined by

*The work is supported by the National Natural Science Foundation of China under Grant No. 11775277, the financial support to the "Théorie-ISB" project (IN2P3/CNRS, France), and the Ministry of Foreign Affairs and International Development of France in the framework of PHC Xu GuangQi 2015 under project No. 34457VA.

**e-mail: lamyihua@gmail.com
***e-mail: smirnova@cenbg.in2p3.fr

a least-squares fit to reproduce the experimental isovector and isotensor coefficients of the isobaric-multiplet-mass equation (IMME) [4]. The INC Hamiltonian is named cdUSDB.

Two applications are discussed below. One is related to a beta (β)-delayed proton decay of ^{24}Si, while the second part is devoted to the description of the mirror energy differences in *sd*-shell $T = 1/2$ nuclei. All calculations have been performed with the NuShellX@MSU shell-model code [5]. The results are compared with the existing experimental data.

2 Partial-decay scheme of 24Si.

The β-delayed particle-radioactivity is a powerful tool in studies of proton-rich nuclei [6], since it provides us with direct information on decay properties of excited states in exotic nuclei. In the context of the isospin-symmetry violation, we are particularly interested in the decay of the isobaric analogue state (IAS) in a daughter nucleus, which is often characterized by competing electromagnetic and isospin-forbidden particle decay channels, important for theoretical understanding [7, 8].

The decay of ^{24}Si has been a subject of intensive experimental and theoretical studies in recent decades [9–12]. The β_+ decay of ^{24}Si populates, besides the IAS, 0^+, $T = 2$, a number of 1^+, $T = 1$ states in ^{24}Al. All these states manifest themselves by gamma decay and/or by proton emission, populating the low-lying $T = 1/2$ states in ^{23}Mg. Obviously, the proton emission from the IAS is forbidden by the isospin symmetry. Thus, the measurements of the proton branching ratio give information on the existence of the isospin-mixing in the IAS. Besides, the analysis of β-γ coincidences in the decay of ^{24}Si [11, 12] has provided information on the Gamow-Teller (GT) branches to the two lowest 1^+ states. The interest is due to the fact that the second excited 1^+ state is known to be strongly shifted compared to its mirror state in ^{24}Na, suggesting the important role of the weakly bound $1s_{1/2}$ orbital (Thomas-Ehrman effect).

The purpose of the present work is to reproduce the main features of the decay scheme within the shell model. Our preliminary results are summarized in two tables inserted in Fig.1 near the experimental partial decay scheme. In addition to cdUSDB, we use the INC term established by Ormand and Brown [13], on top of the USDB Hamiltonian (OBUSDB). The GT operator is quenched by a standard factor $q = 0.776$ [14]. The half-life of ^{24}Si is about 20% larger than the experimental value and the distribution of the GT strength is in robust agreement with the experimental data. It is seen from the tables on Fig.1, that both interactions predict very similar GT branching ratios as it is the isospin-invariant USDB Hamiltonian which is mainly responsible for that. The theoretical results reproduce very well the GT feeding of the two lowest 1^+ states, however, the numbers for the next two 1^+ states seem to be inverted with respect to the experiment.

We also employed these two INC Hamiltonians to describe the ratio of the $B(GT)^+$ to $B(GT)^-$ strengths to the two lowest 1^+ states in mirror nuclei ^{24}Al and ^{24}Na, as well as the ratio of $B(GT)^+$ to the 1^+_1 and 1^+_2 states in ^{24}Al, as shown in Table 1. It is remarkable that the theoretical B(GT)-values and their ratios are very close to the experimental values. As we mentioned in the introduction, the second 1^+ state in ^{24}Al manifests a noticeable Thomas-Ehrman effect: its excitation energy is 258 keV lower than of its mirror in ^{24}Na. Indeed, the $1s_{1/2}$ orbital occupation is about twice larger than that in the 1^+_1 state. The detailed analysis of the mirror (a)symmetry will be published elsewhere [15].

In addition, we have analyzed the β-delayed proton emission from ^{24}Si. The branching ratios of the isospin-forbidden proton decay of the IAS are presented in the table inserted in Fig. 1 (left part). It is evident that two Hamiltonians produce different branching ratios, due to the different parameterizations of the INC term used. The branching ratios from OBUSDB are in a slightly better agreement with the recent values published by Ichikawa *et al* [11].

Both interactions predict about 11-12% branch to the lowest $5/2^+$ state in ^{23}Mg, which has not yet been seen experimentally. It would be interesting to have more refined experimental data to be able to compare with the theory in more detail.

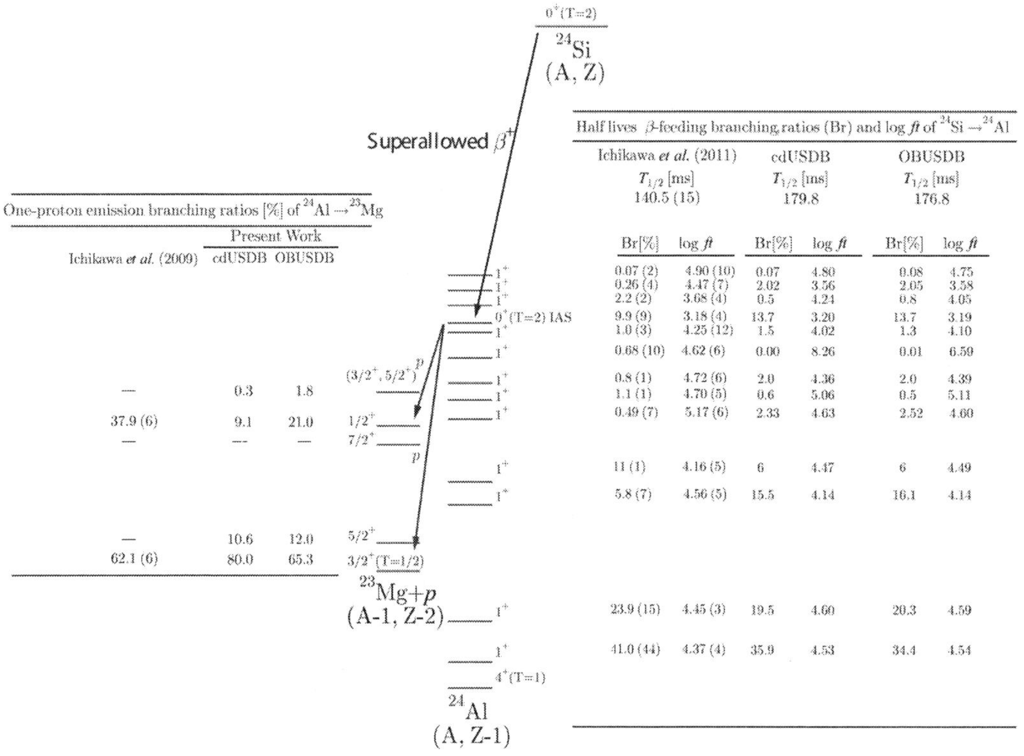

Figure 1. Partial-decay scheme of ^{24}Si.

Table 1. Gamow-Teller strengths of $^{24}\text{Si} \rightarrow ^{24}\text{Al}$ and $^{24}\text{Ne} \rightarrow ^{24}\text{Na}$ for the 1_1^+ and 1_2^+ states.

	$^{24}\text{Si} \rightarrow ^{24}\text{Al}$			$^{24}\text{Ne} \rightarrow ^{24}\text{Na}$			$B(\text{GT}^+)/B(\text{GT}^-)$	
	$B(\text{GT}^+)^{1_1^+}$	$B(\text{GT}^+)^{1_2^+}$	ratio	$B(\text{GT}^-)^{1_1^+}$	$B(\text{GT}^-)^{1_2^+}$	ratio	1_1^+	1_2^+
Ref. [12]	0.17(2)	0.14(1)	1.21(17)	0.167(4)	0.155(9)	1.077(68)	1.02(12)	0.90(8)
OBUSDB	0.1631	0.1463	1.12	0.1832	0.1586	1.155	0.89	0.92
cdUSDB	0.1936	0.1665	1.16	0.2116	0.1815	1.166	0.91	0.92

3 Mirror energy difference

The other study is focused on excitation-energy differences between excited states of mirror nuclei, known as mirror energy difference (MED). In the last decades, a good amount of accurate experimental information has become available on the MEDs in *sd*- and *pf*-shell nuclei and it has been recognized that the MEDs are a very sensitive probe to the details of the INC Hamiltonian (for review see Refs. [16, 17] and references therein). Each MED is caused by the Coulomb interactions between protons and by a smaller contribution from effective charge-dependent interaction of nuclear origin. Analyzed as a function of the excitation energy and of the angular momentum, the MEDs contain valuable information on the dynamical changes in the structure of the excited states and allow us to understand the role of the broken isospin symmetry in quantum many-body systems.

The MED is defined as a difference in excitation energies of the mirror states and may be considered as a function of the angular momentum J and isospin T [16, 17],

$$\mathrm{MED}_{J,T} = E^*_{J,T,T_z=-T} - E^*_{J,T,T_z=T}\,, \tag{1}$$

where $T = |T_z|$ as we only study a set of excited states of the lowest isospin. For $T = 1/2$ multiplets, the MED can be related to the differences of the isovector (b) coefficient of IMME, $\mathrm{MED}_{J,1/2} = -\Delta b_J$, where Δb_J is the difference of the b coefficients between the excited state of spin J and the ground state. The magnitude of these energy differences is rather small compared to the total contribution of the Coulomb energy to the nuclear mass of a given nucleus, and to the splitting of IASs. Nevertheless, this energy difference carries important information on a subtle charge asymmetry of the nuclear interaction. The measured MED vary typically from a few tens of keV and up to about 120 keV for sd-shell $T = 1/2$ doublets. Figures 2 and 3 show the available experimental data on MEDs of $T = 1/2$ yrast states in $A = 21$ [18] and $A = 23$ [19, 20].

It was found by Zuker *et al.* [21], see also [16], that an addition of a specific J-dependent matrix element is necessary to reproduce the MED trends in pf-shell nuclei. In the present study, we show that with the INC Hamiltonians described above we reasonably reconstruct the MEDs of yrast states of sd-shell $T = 1/2$ multiplets. The results from the cdUSDB and OBUSDB interactions for $A = 21$ and $A = 23$ are shown in Figs. 2 and 3. We notice that in $A = 21$, the OBUSDB results are in very good agreement with the experiment, except for $7/2^+$ state. The results obtained with cdUSDB exhibit some more discrepancies for high-spin states. Contrary, in the $A = 23$ case, both interactions well reproduce the experimental trend with cdUSDB being the best. We remark that the MED trends of a few others sd-shell mirror pairs are robustly reproduced as well, e.g., in $T = 1/2$, $A = 23$, 29, 31, and 35. Nevertheless, further work is needed to tune the INC terms of the Hamiltonian to improve the predictive power for MEDs of $T = 1/2$, $A = 21$, 25, 27, and 33. A detailed study of the whole series of multiplets and a further analysis of the agreement will be published elsewhere. We hope that this work will shed more light on the structure of the INC Hamiltonian, in particular, on the most optimal form of the charge-dependent terms of nuclear origin.

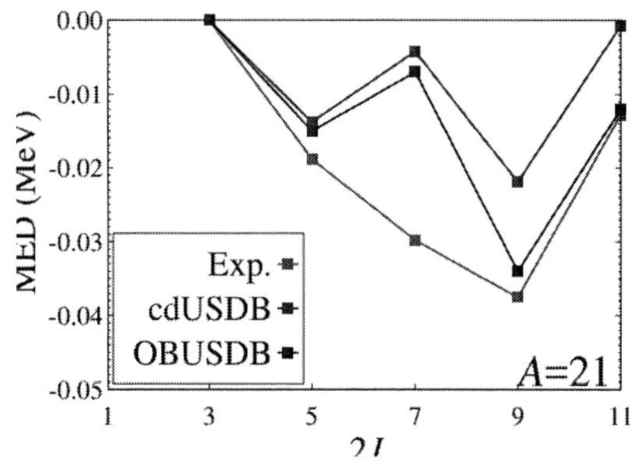

Figure 2. The MED of mass $A = 21$. Experimental data are deduced from NNDC [18].

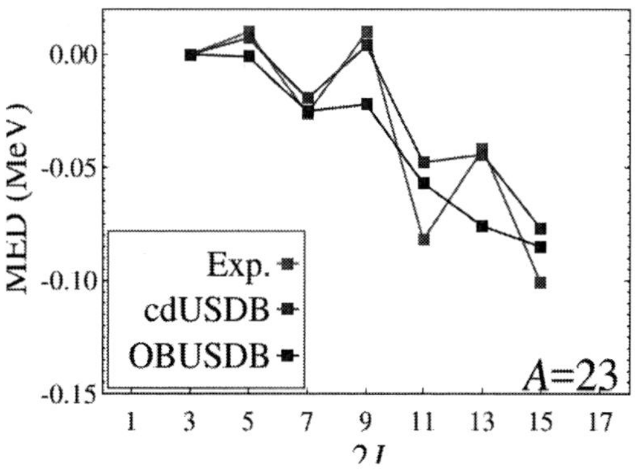

Figure 3. The MED of mass $A = 23$. Experimental data are quoted from Refs. [19, 20].

4 Conclusions

We have used a recently constructed INC *sd*-shell Hamiltonian based on USDB to describe a partial-decay scheme of ^{24}Si and MEDs in $T = 1/2$ mirror $A = 21$ and $A = 23$ nuclei. The results are compared with another INC Hamiltonian, OBUSDB, comprised from an earlier parameterization of charge-dependent terms on top of the USDB interaction. We found that while the USDB interaction well describes the β-decay lifetime and the GT strength distribution, the INC part robustly reproduces the measured isospin-forbidden proton branching ratios. In addition, we find that both Hamiltonians can reconstruct reasonably well the trends of MEDs in a few *sd*-shell $T = 1/2$ isobaric mirror pairs, while MEDs for some states stay a challenge. It would be interesting and important to pursue this study in order make a link to the necessity of a J-dependent matrix element.

References

[1] Y. H. Lam, N. A. Smirnova, E. Caurier, Phys. Rev. C **87**, 054304 (2013).
[2] B. A. Brown and W. A. Richter, Phys. Rev. C **74**, 034315 (2006).
[3] R. Roth and T. Neff and H. Feldmeier, Prog. Part. Nucl. Phys. **65**, 50 (2010).
[4] Y. H. Lam *et al.*, At. Data Nucl. Data Tables **99**, 680 (2013).
[5] B. A. Brown and W. D. M. Rae, Nucl. Data Sheets **120**, 115 (2014).
[6] B. Blank and M. J. G. Borge, Prog. Part. Nucl. Phys. **60**, 403 (2008).
[7] B. A. Brown, Phys. Rev. Lett. **65**, 2753 (1990).
[8] N. A. Smirnova *et al.*, Phys. Rev. C **93**, 044305 (2016).
[9] S. Czajkowski *et al.*, Nucl. Phys. A **628**, 537 (1998).
[10] V. Banerjee *et al.*, Phys. Rev. C **63**, 024307 (2001).
[11] Y. Ichikawa *et al.*, Phys. Rev. C **80**, 044302 (2009).
[12] Y. Ichikawa *et al.*, J. Phys.: Conf. Ser. **312**, 092031 (2011).
[13] W. E. Ormand and B. A. Brown, Nucl. Phys. A **491**, 1 (1989).
[14] W. A. Richter, S. Mkhize, B. A. Brown, Phys. Rev. C **78**, 064302 (2008).
[15] Y. H. Lam, L. Xayavong, N. A. Smirnova, B. Blank, *in preparation.*
[16] M. A. Bentley, S. M. Lenzi, Prog. Part. Nucl. Phys. **59**, 497 (2007).
[17] S. M. Lenzi and M. A. Bentley, *Test of Isospin Symmetry Along the $N = Z$ Line*, in: J. S. Al-Khalili, E. Roeckl (eds) The Euroschool Lectures on Physics with Exotic Beams, Vol. III, Lect. Notes Phys. **764**, 57 (2009), Springer Berlin, Heidelberg.
[18] National Nuclear Data Center (NNDC) online, http://www.nndc.bnl.gov/ensdf
[19] D. G. Jenkins *et al.*, Phys. Rev. C **87**, 064301 (2013).
[20] A. Boso *et al.*, Acta Phys. Pol. B **48**, 313 (2017).
[21] A. P. Zuker *et al.*, Phys. Rev. Lett. **89**, 142502 (2002).

Isospin Conservation in Neutron Rich Systems of Heavy Nuclei

Ashok Kumar Jain[1,*] and *Swati* Garg[1]

[1]Department of Physics, Indian Institute of Technology Roorkee, Roorkee-247667, India

Abstract. It is generally believed that isospin would diminish in its importance as we go towards heavy mass region due to isospin mixing caused by the growing Coulomb forces. However, it was realized quite early that isospin could become an important and useful quantum number for all nuclei including heavy nuclei due to neutron richness of the systems [1]. Lane and Soper [2] also showed in a theoretical calculation that isospin indeed remains quite good in heavy mass neutron rich systems. In this paper, we present isospin based calculations [3, 4] for the fission fragment distributions obtained from heavy-ion fusion fission reactions. We discuss in detail the procedure adopted to assign the isospin values and the role of neutron multiplicity data in obtaining the total fission fragment distributions. We show that the observed fragment distributions can be reproduced rather reasonably well by the calculations based on the idea of conservation of isospin. This is a direct experimental evidence of the validity of isospin in heavy nuclei, which arises largely due to the neutron-rich nature of heavy nuclei and their fragments. This result may eventually become useful for the theories of nuclear fission and also in other practical applications.

1 Introduction

Isospin is a useful and fundamental quantum number in nuclear and particle physics. In nuclei, we consider neutron and proton to be the two isospin states of a common entity called a nucleon with isospin $T = 1/2$. The third component of isospin is different for both the states, $T_3 = +1/2$ for neutron and $T_3 = -1/2$ for proton. This convention is just opposite to what is normally used in particle physics.

We plot a graph similar to nuclear landscape in Fig. 1, but in terms of isospin T_3 on one axis. We note that $T_3 = (N - Z/2)$ for a given nucleus and is a well defined quantity. The line in the middle (black in color) shows the line of β-stability. We have the neutron rich nuclei on the left of the line of stability and the neutron deficient nuclei on the right, plotted on the basis of the presently known experimental data. Theoretical predictions for the drip lines are shown by the wavy lines, while the presently known experimental limits are shown by joining the known data points. We can see that experimentally known neutron-rich nuclei are quite far from the theoretical predictions. On the other hand, the experimental data touch the theoretical predictions for the proton drip line at least up to $A = 200$. Around $A = 240$, the experimentally known neutron rich and neutron deficient nuclei merge into the near-stability line, suggesting that a large number of neutron-rich and neutron-deficient isotopes are yet to be found.

There are two main features in Fig. 1 that we would like to emphasize. Firstly, as we go from light to heavy nuclei, the isospin increases due to the fact that $N > Z$ in heavier nuclei. Secondly, in a given chain of isotopes, the range of isospin values could be very large. For example, in the chain of Pb isotopes, the isospin values could range from near zero in the neutron deficient isotopes to nearly 50 in the neutron rich isotopes. In the most stable lead isotope ^{208}Pb, the ground state isospin would be $T = 22$ as $T_3 = 22$. It is thus obvious that very large T values are involved in dealing with heavy nuclei. These two fac-

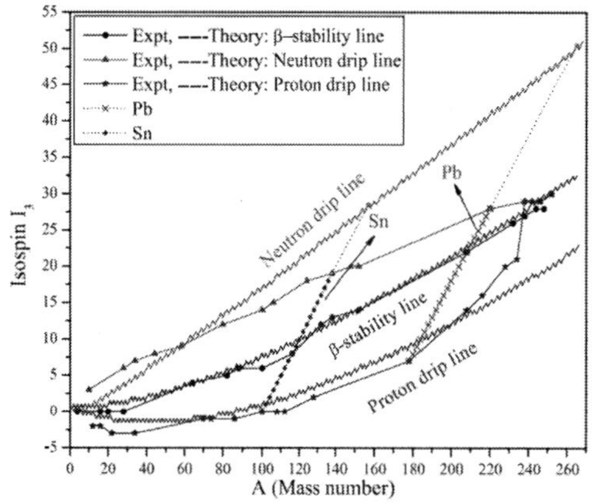

Figure 1. Experimental neutron-rich nuclei (upper solid line), β-stable nuclei (mid-line) and neutron deficient nuclei (lower solid line), plotted from the known experimental data. The wavy lines show the theoretical predictions for the drip line nuclei.

[*]e-mail: ajainfph@iitr.ac.in

tors make the heavy nuclei very interesting to study. We also note that the fission fragments coming out from such sources will be even more neutron-rich. Therefore, the experimental data on fission fragment distributions of heavy nuclei is a good testing ground for verifying the conservation of isospin in heavy nuclei.

However, the data where this could be directly tested is very scarce. Fission fragment distribution data are available in plenty but the mass resolution is about 3-4 units. This means that the Z and A of each fragment is not known precisely in most of the cases. In many situations, particularly the thermal neutron fission, the fragment mass distribution of heavy fragments is still not known. Only recently, more precise fragment distribution data of HI induced fission are becoming available where gamma ray spectroscopy of fragments are being used to identify each fragment, although only in even-even nuclei so far. This in itself presents a huge opportunity for experimnetalists so that good fragment distribution data become available for all the partitions and also for odd-A fragments eventually.

For the first time in 1962, Lane and Soper [2] theoretically demonstrated that isospin may remain as a good quantum number in heavy nuclei as in light nuclei. In simple terms, if we consider a nucleus having N neutron and Z protons, we may look upon it as made up of a core ($N = Z$) nucleus having isospin zero and ($N - Z$) valence neutrons. These excess neutrons act in a way so as to reduce the isospin impurity in the nucleus. As the number of excess neutrons rises, the isospin tends to become more pure quantum number. Sliv and Kharitonov (1965) [5] calculated the isospin admixture in light ($N = Z$) nuclei and heavy ($N > Z$) nuclei by using harmonic oscillator shell model wave functions and showed that the isospin admixture in the ground state of ^{16}O is nearly same as in ^{208}Pb. A detailed discussion of some these developments may be found in the review by Auerbach [6].

In the backdrop of this discussion, we analyze the heavy ion induced fission-fusion reaction ^{208}Pb(^{18}O, f) in which neutron-rich fission fragments are emitted. We treat the isospin to be a conserved quantity and follow the isospin algebra. We calculate the fission fragment mass distribution using this concept of goodness of isospin. An important conjecture given by Kelson [7] is quite helpful in this respect. Kelson has given a theoretical explanation of how n-emission in fission leads to the formation of Isobaric Analog states (IAS) in final fission products. Kelson's ideas help in resolving our problem of assigning isospin values to various fission fragments. We find that our calculated values match experimental data reasonably well. There are a some deviations which may be due to the presence of shell closure or the presence of isomers.

2 Formalism

Consider a projectile Y with isospin $T_Y = T_{3_Y}$ incident upon a target with $T_X = T_{3_X}$, leading to a compound nucleus (CN) with $T_{CN} =\mid T_X - T_Y \mid, ..., T_X + T_Y$. However, from isospin algebra which behaves quite similar to SU(2) algebra of spin while dealing with neutrons and protons, $T_{CN} \geqslant T_{3_{CN}}$ and $T_{3_{CN}} = T_X + T_Y$. We assume that only ground or, low lying states of target and projectile are involved. Therefore, the CN always has only one possible value of isospin, $T_{CN} = T_{3_{CN}} = T_X + T_Y$. For the reaction under consideration in present work ^{208}Pb (^{18}O, f), we have $T_X(^{208}Pb) = T_{3_X} = 22$ and $T_Y(^{18}O) = T_{3_Y} = 1$. The isospin of CN (^{226}Th) can have values $T_{CN} = 21, 22$ or 23 but since $T_{3_{CN}} = 23$, there is only one allowed value of $T_{CN} = 23$. This CN further fissions to give two fragments F_1 ($T_{F1}, T_{3_{F1}}$) and F_2 ($T_{F2}, T_{3_{F2}}$) with the emission of n number of neutrons. This is many body problem and to simplify it to two-body problem, we invoke the concept of residual compound nucleus (RCN) formed after the emission of n number of neutrons and has a isospin T_{RCN}. Now, our complete reaction will look like,

$$\begin{aligned} Y(T_Y, T_{3_Y}) + X(T_X, T_{3_X}) &\rightarrow CN(T_{CN}, T_{3_{CN}}) \\ &\rightarrow RCN(T_{RCN}, T_{3_{RCN}}) + n \\ &\rightarrow F_1(T_{F1}, T_{3_{F1}}) + F_2(T_{F2}, T_{3_{F2}}) + n \end{aligned}$$

and isospin of RCN should satisfy the two conditions,

$$\mid T_{CN} - n/2 \mid \leq T_{RCN} \leq (T_{CN} + n/2)$$
$$\text{and} \quad \mid T_{F1} - T_{F2} \mid \leq T_{RCN} \leq (T_{F1} + T_{F2})$$

Now, we would like to assign isospin values to various fission fragments emitted in different partitions, which is not so straight forward. We formulate two conjectures based on the ideas put forth by Kelson [7] in assigning the isospin values to fission fragments. First conjecture states that neutron emission from CN leads to formation of highly excited states with $T > T_3$. Using this conjecture, we fix the isospin of RCN, $T_{RCN} = T_{F1} + T_{F2}$. Kelson's second conjecture states that the fission fragments are preferably emitted in isobaric analog states (IAS). We thus assign isospin to fission fragments on the basis of this second conjecture. We make three isobars of each mass number having T_3 values as T_3, $T_3 - 2$ and $T_3 - 4$. Then, we assign $T = T_3$ to that particular mass number since this is the minimum value of isospin required to generate all the members of isobaric multiplet formed corresponding to that mass number. For ^{208}Pb(^{18}O, f), we have six experimentally observed partitions. We consider eight nuclides on the lighter side and eight on the heavier side of a partition and for each mass number, we make three isobars. Now, we assign isospin to each mass number using Kelson's second conjecture as described above. For example, for $A = 112$, we have three isobars namely ^{112}Ru with $T_3 = 12$, ^{112}Pd with $T_3 = 10$ and ^{112}Cd with $T_3 = 8$. We assign maximum of three T_3 values which is 12 to $A = 112$ as it can generate all the three isobars ^{112}Ru, ^{112}Pd and ^{112}Pd.

Once we have assigned the isospin values, we proceed to calculate the relative yields of fragments by using isospin part of the wave function. For a particular pair of fragments emitted in a given n-emission channel, the isospin wave function for RCN can be written as,

$$\begin{aligned} \mid T_{RCN}, T_{3RCN}\rangle_n = \langle T_{F1} T_{F2} T_{3_{F1}} T_{3_{F2}} \mid T_{RCN} T_{3_{RCN}}\rangle \\ \mid T_{F1}, T_{3_{F1}}\rangle \mid T_{F2}, T_{3_{F2}}\rangle \end{aligned} \quad (1)$$

The first part of the equation on R.H.S. gives us the isospin C.G. coefficient (CGC). The intensity can be calculated by taking the square of CGC,

$$I_n = \langle CGC \rangle^2 = \langle T_{F1} T_{F2} T_{3_{F1}} T_{3_{F2}} \mid T_{RCN} T_{3_{RCN}} \rangle^2 \quad (2)$$

To calculate the relative yield of any fragments, we multiply intensity by weight factor (w_n) of that particular n-emission channel. The weight factors are obtained by relative normalization with respect to n-emission channel having maximum number of counts in the corresponding partition. Therefore, the final yield of any fragment is,

$$I = \sum_n I_n \times w_n = \sum_n \langle CGC \rangle^2 \times w_n \quad (3)$$

where we take summation over all the experimentally known n-emission channels. In the same way, we calculate the yields of all the lighter and heavier fragments in a partition. As we are not calculating the exact yields and want only relative yields, we normalize the yields of all the fragments with respect to the maximum yield. We perform this normalization separately for lighter side and heavier side of a partition. Similarly, we can have values of relative yields of fragments in all the partitions. Our calculations do not give the absolute yields at any point.

We also calculate the total mass distribution of fragments although only relative yields can be calculated. For this calculation, we use the same procedure of assigning the isospin values to fragments and then, in the same way, we calculate CGC of pair of fragments emitted in a n-emission channel in a partition. Since, now we are looking at complete mass distribution, we have to take into account the weightage of each partition simultaneously. For this, we normalize the weight factors (w'_n) of each n-emission channel of all the six partitions with respect to n-emission channel having maximum number of counts among all the partitions. Therefore, our equation for calculating the intensity of a fragment is,

$$I' = \sum_n I_n \times w'_n = \sum_n \langle CGC \rangle^2 \times w'_n \quad (4)$$

After calculating yields of all the fragments in individual partitions, we simply add the yields of all the three members of an isobaric multiplet to have yield corresponding to that particular mass number. Then we again normalize all the yields of mass numbers with respect to mass number with maximum yields. This gives us the relative yields of all the fragments present in the total mass distribution.

3 Results and Discussion

We have plotted in Fig. 2, the calculated relative yields of fragments for each partition separately and compare them with the available experimental data of Bogachev *et al.* [8]. There is good agreement between the calculated and experimental data in most of the partitions. This agreement is not so good for Zr-Sn and Kr-Xe partitions. One reason for this deviation could be the presence of closed shell configuration at $A = 124$ ($Z = 50$) and $A = 136$ ($N = 82$) and their complementary fragments at $A = 84$ and $A = 92$ [9].

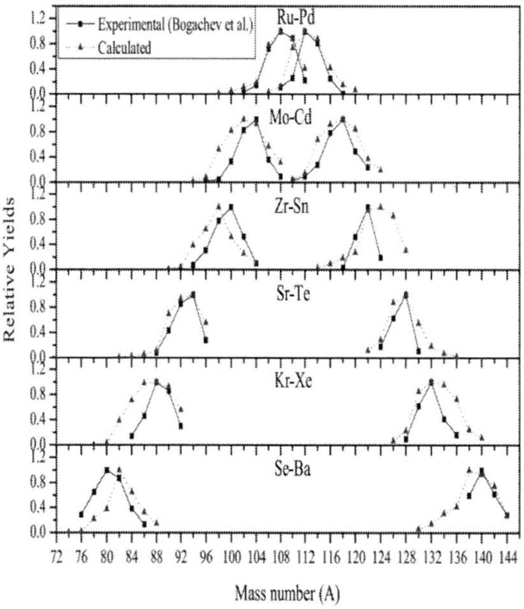

Figure 2. Comparison of calculated relative yields of fragments emitted in all the six partitions with the experimental data from Bogachev et al. [8].

Another possibility is the presence of isomers at some of the points. Also, Bogachev *et al.* [8] estimated an error of 10-30% in the experimental data. Considering these factors, we feel that our calculations reproduce the experimental trends and data reasonably well.

Next, we have plotted in Fig. 3 the relative total yield of fission fragments and compare it with the fission fragment mass distribution given in Bogachev *et al.* [8]. The two curves look very similar to each other with the exceptions at the shell closures as discussed above. These results confirm that isospin and its conservation seems to be valid in heavy neutron-rich nuclei.

4 Conclusion

We have calculated the relative yields of fission fragments partition-wise and also total mass distribution of fragments. In doing so we have assumed that the basic concept of isospin and its conservation remains valid in heavy nuclei. We assign the isospin values by using Kelson's conjectures which are based on sound physics arguments. We then follow the isospin algebra and find that the fission fragments are preferably formed in isobaric analogue states forming isospin multiplets. A given isospin multiplet is assigned a T value corresponding to the maximum isospin projection in the isobaric multiplet. A reasonable agreement of the calculated results with the observed fission fragment distribution provides a direct experimental evidence of the validity of isospin in heavy nuclei. Isospin, therefore, seems to emerge as a powerful tool in neutron-rich heavy nuclei and may play an important role in many phenomena and applications.

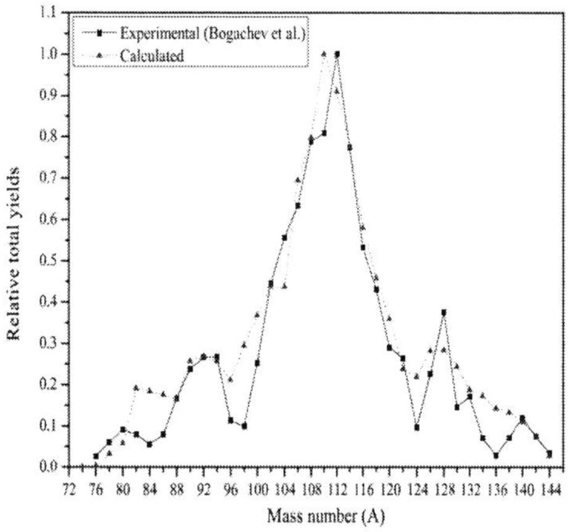

Figure 3. Comparison of calculated relative total yields of fragments with the experimental data from Bogachev et al. [8].

Acknowledgment

Support from Ministry of Human Resource Development (Government of India) to SG in the form of a fellowship is gratefully acknowledged. The authors also acknowledge the financial support in the form of travel grant from IIT Roorkee Alumni funds and IIT Roorkee.

References

[1] D. Robson, Science **179**, 133 (1973).
[2] A. M. Lane and J. M. Soper, Nucl. Phys. **37**, 663 (1962).
[3] A. K. Jain, D. Choudhary, and B. Maheshwari, Nuclear Data Sheets **120**, 123 (2014).
[4] S. Garg and A. K. Jain, Phys. Scr. **92**, 094001 (2017).
[5] L. A. Sliv and Yu. I. Kharitononv, Phys. Lett. **16**, 176 (1965).
[6] N. Auerbach, Phys. Rep. **98**, 273 (1983).
[7] I. Kelson, Proc. of the Conference on "Nuclear Isospin", edited by D. Anderson, S.D. Bloom, J. Cerny and W.W. True, Academic Press (1969) Pg. 781.
[8] A. Bogachev *et al.*, Eur. Phys. J. A **34**, 23 (2007).
[9] L. S. Danu *et al.*, Phys. Rev. C **81**, 014311 (2010).

Does Compound Nucleus remember its Isospin- An Evidence from the Fission Widths

Swati Garg[1,*] and *Ashok Kumar* Jain[1]

[1]Department of Physics, Indian Institute of Technology Roorkee, Roorkee-247667, India

Abstract. We present an evidence of isospin effects in nuclear fission by comparing the fission widths for reactions involving different isospin states of the same compound nucleus (CN). Yadrovsky [1] suggested this possibility in 1975. Yadrovsky obtained the fission widths for two reaction data sets, namely ^{206}Pb(α,f) and ^{209}Bi(p,f), both leading to same CN, and concluded that "a nucleus remembers the isospin value of the nuclear states leading to fission". We obtain the fission decay widths for both the $T_0 + \frac{1}{2}$ and $T_0 - \frac{1}{2}$ states of CN by using two appropriate reaction data sets. We then compare the fission widths for the two isospin states of CN. More specifically, we have chosen the combination of ^{206}Pb(α,f) and ^{209}Bi(p,f) same as presented in Yadrovsky's paper [1] in this study. A significant difference between the ratios of fission decay widths to total decay widths for different isospin values suggests that isospin plays an important role in fission.

1 Introduction

Isospin is a very simple but useful concept in nuclear physics. As the name suggests, it behaves in the same manner as spin. It is important from both structural and reaction point of view [2]. In the early days since its discovery, it was considered to be useful only for light nuclei because it was considered to be pure for these nuclei. As we go towards heavy nuclei, isospin mixing starts increasing and it no longer remains a good quantum number. However, in 1962, Lane and Soper [3] argued on theoretical basis that isospin impurity decreases as we go towards neutron-rich nuclei. The extra neutrons dilute the isospin impurity and isospin again becomes a nearly good quantum number in neutron-rich nuclei. We have used this concept of isospin conservation in reproducing the yields of neutron-rich fragments [4, 5].

In 1975, Yadrovsky [1] calculated the fission branching ratios for two possible isospin states $T_0 + \frac{1}{2}$ and $T_0 - \frac{1}{2}$ of CN for a combination of ^{206}Pb(α,f) and ^{209}Bi(p,f) reactions, both leading to same CN. It was concluded that the CN remembers isospin of initial nuclei. However, the author in Ref. [1] was not certain whether this could lead to isospin conservation in fission or not. In the present work, we try to test Yadrovsky idea. We also calculate the same quantity for the combination taken by Yadrovsky but with a different set of experimental data. Our calculations provide a strong support to the concept that isospin is actively involved in fission.

*e-mail: swat90.garg@gmail.com

2 Formalism

We, first, consider a combination of proton and alpha induced reactions giving same CN. Suppose a proton with $T = \frac{1}{2}, T_3 = \frac{-1}{2}$ is incident on a target with $T = T_3 = T_0$ forming a CN. From isospin algebra, the possible values for isospin of CN are, $T_{CN} = T_0 + \frac{1}{2}, T_0 - \frac{1}{2}$ and $T_3 = T_0 - \frac{1}{2}$. Similarly, for α induced reaction on a target with $T' = T'_3 = T_0 - \frac{1}{2}$, there is only one possible isospin state of CN, $T_{CN} = T_{3_{CN}} = T_0 - \frac{1}{2}$ since α particle has isospin $T = T_3 = 0$. The formalism, we are going to use is summarized by Yadrovsky in his paper [1]. Yadrovsky deduced the following expressions for ratio of fission width (Γ_f) to total decay width (Γ_{total}) for $T_0 - \frac{1}{2}$,

$$\frac{\Gamma_f^{T_0-1/2}}{\Gamma_{total}^{T_0-1/2}} = \frac{\langle\sigma_{\alpha,f}\rangle}{\sigma_\alpha} \tag{1}$$

and $T_0 + \frac{1}{2}$ states of CN,

$$\frac{\Gamma_f^{T_0+1/2}}{\Gamma_{total}^{T_0+1/2}} = (2T_0 + 1)[\frac{\langle\sigma_{p,f}\rangle}{\sigma_p} - \frac{\langle\sigma_{\alpha,f}\rangle}{\sigma_\alpha}] \tag{2}$$

From Eq. (1) and (2), we can calculate the fission branching ratios for $T_0 - \frac{1}{2}$ and $T_0 + \frac{1}{2}$ states, respectively. For ^{209}Bi(p, f) and ^{206}Pb(α, f) datasets, the value of $T_0 = 21.5$. Yadrovsky has taken the experimental data of the ratio of cross-sections $\frac{\langle\sigma_{p,f}\rangle}{\sigma_p}$ and $\frac{\langle\sigma_{\alpha,f}\rangle}{\sigma_\alpha}$ for ^{209}Bi(p, f) and ^{206}Pb(α, f), respectively from Gadioli *et al.* [6].

Yadrovsky, further, calculated the fission widths by considering total decay width to be equal to the sum of neutron decay width, proton decay width and fission width and then, obtained the proton and neutron decay width using optical model. But in our work, we limit ourselves to fission branching ratios for two isospin states of CN by using a different experimental dataset.

3 Results and Discussion

3.1 ^{209}Bi(p, f) and ^{206}Pb(α, f)

First, we look for the experimental data of fission cross-sections for ^{209}Bi (p,f) and ^{206}Pb (α,f) at different excitation energies of CN, ^{210}Po. For ^{209}Bi(p, f), we extract the data of fission cross-sections from Fig. 1 of Prokofiev [7]. According to Prokofiev, the method of data normalization done by Gadioli *et al.* [6] is not clear and, therefore, unreliable. Therefore, by using different experimental data sets from EXFOR, Prokofiev analyzed and obtained the data for cross sections for many proton induced reactions at different incident energies as shown in Fig. 1 of Ref. [7]. For ^{206}Pb(α, f), we use the fission cross-section data from Huizenga *et al.* [8]. Huizenga *et al.* [8] reported experimental X-sections at five different projectile energies. We fit this curve exponentially so that we can have fission X-sections at excitation energies of CN same as obtained for ^{209}Bi(p, f). We calculate the CN fusion cross-sections σ_p and σ_α for ^{209}Bi (p,f) and ^{206}Pb (α,f) using PACE4 [9] at different energies of projectiles. After getting all the required quantities, we put these values in Eq. (1) and (2) to calculate the fission widths for $T_0 - \frac{1}{2}$ and $T_0 + \frac{1}{2}$ states of CN, table 1. In the table 1, E_p and E_α represent the incident energies of projectiles, proton and alpha and E^*_{CN} represents the excitation energy of CN.

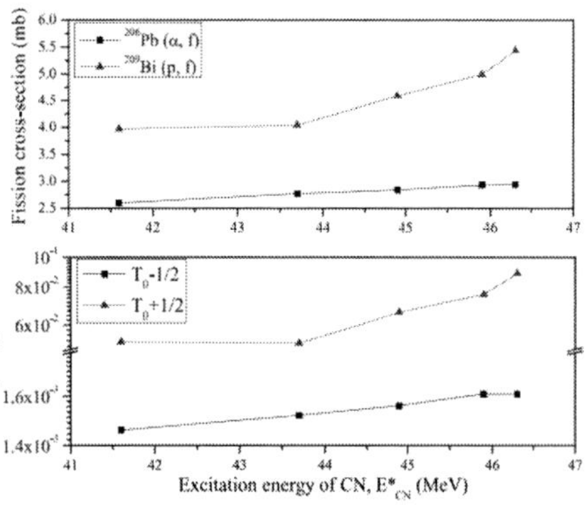

Figure 1. Comparison of fission cross sections $\sigma_{p,f}$ and $\sigma_{\alpha,f}$ for ^{209}Bi (p,f) and ^{206}Pb (α,f), respectively (upper panel). Comparison of fission branching ratios for two isospin states of CN, $T_0 + \frac{1}{2}$ and $T_0 - \frac{1}{2}$ (lower panel).

In Fig. 1, we compare the fission cross-section for ^{209}Bi (p,f) and ^{206}Pb (α,f) and fission branching ratios for $T_0 + \frac{1}{2}$ and $T_0 - \frac{1}{2}$ states of CN, ^{210}Po using table 1. We can see that there is a reasonable difference between the fission branching ratios for two isospin states of CN. The fission branching ratio for $T_0 + \frac{1}{2}$ is greater by an order than for $T_0 - \frac{1}{2}$ state. This supports Yadrovsky's claim that CN remembers the isospin value during fission.

4 Conclusion

We have calculated the ratio of fission width to total decay width or fission branching ratio for $T_0 + \frac{1}{2}$ and $T_0 - \frac{1}{2}$ states of CN at different excitation energies of CN using Yadrovsky's idea. Yadrovsky has done the calculation for a combination of ^{209}Bi (p,f) and ^{206}Pb (α,f) giving a CN ^{210}Po and found that there is a noticeable difference of the order of 10^2-10^3 between fission branching ratios for two isospin states of CN. For fission decay width, this difference goes upto 10^5. We have done the calculation of fission branching ratios for the same combination of datasets taken by Yadrovsky but with different experimental data. We also use PACE4 [9] to calculate CN formation cross-section at different projectile energies for different sets of reactions. There is reasonable difference of fission branching ratios for two different isospin states of CN. This reinforces the claim made by Yadrovsky that a nucleus remembers isospin of the initial states leading to fission. This also provides further support to our claim of isospin conservation in fission as presented in [4, 5].

Acknowledgment

Support from Ministry of Human Resource Development (Government of India) to SG in the form of a fellowship is gratefully acknowledged. The authors also acknowledge the financial support in the form of travel grant from IIT Roorkee Alumni funds.

References

[1] E. L. Yadrovsky, Phys. Lett. **55B**, 157 (1975).
[2] D. Robson, Ann. Rev. Nucl. Sci. **16**, 119 (1966).
[3] A. M. Lane and J. M. Soper, Nucl. Phys. **37**, 663 (1962).
[4] A. K. Jain, D. Choudhary, and B. Maheshwari, Nuclear Data Sheets **120**, 123 (2014).
[5] S. Garg and A. K. Jain, Phys. Scr. **92**, 094001 (2017).
[6] E. Gadioli *et al.*, Nucl. Phys. A **151**,16 (1970).
[7] A. V. Prokofiev, Nucl. Instr. Meth. Phys. Res. A **463**, 557 (2001).
[8] J. R. Huizenga *et al.*, Phys. Rev. **126**, 210 (1962).
[9] O. B. Tarasov, D.Bazin, NIM B **266**, 4657-4664 (2008); A. Gavron, Phys. Rev. C **21**, 230-236 (1980); http://lise.nscl.msu.edu/pace4.

 https://doi.org/10.1051/epjconf/201817805008

Table 1. Ratio of fission width to total decay width for $T_0 + \frac{1}{2}$ and $T_0 - \frac{1}{2}$ states of CN at different E^*_{CN} by using experimental data of fission cross-section for ^{209}Bi(p,f) and ^{206}Pb(α,f) datsets.

E_p (MeV)	E_α (MeV)	E^*_{CN} (MeV)	$\sigma_{p,f}$ (mb)	$\sigma_{\alpha,f}$ (mb)	σ_p (mb)	σ_α (mb)	$\frac{\Gamma_f^{T_0+1/2}}{\Gamma_{total}^{T_0+1/2}}$	$\frac{\Gamma_f^{T_0-1/2}}{\Gamma_{total}^{T_0-1/2}}$
36.7	47.9	41.6	3.97	2.6	1.49×10^3	1.78×10^3	5.28×10^{-2}	1.46×10^{-3}
38.9	50.1	43.7	4.04	2.77	1.49×10^3	1.82×10^3	5.24×10^{-2}	1.52×10^{-3}
40.1	51.3	44.9	4.59	2.84	1.5×10^3	1.82×10^3	6.6×10^{-2}	1.56×10^{-3}
41.1	52.3	45.9	4.99	2.93	1.5×10^3	1.82×10^3	7.55×10^{-2}	1.61×10^{-3}
41.5	52.7	46.3	5.44	2.94	1.5×10^3	1.82×10^3	8.87×10^{-2}	1.61×10^{-3}

EPJ Web of Conferences **178**, 06001 (2018) https://doi.org/10.1051/epjconf/201817806001
CGS16

Gamma-widths, lifetimes and fluctuations in the nuclear quasi-continuum

M. Guttormsen[1,⋆], *A. C.* Larsen[1,⋆⋆], *J. E.* Midtbø[1], *L.* Crespo Campo[1], *A.* Görgen[1], *V. W.* Ingeberg[1], *T.* Renstrøm[1], *S.* Siem[1], *G. M.* Tveten[1], *F.* Zeiser[1], and *L. E.* Kirsch[2]

[1]*Department of Physics, University of Oslo, N-0316 Oslo, Norway*
[2]*Lawrence Berkeley National Laboratory, Berkeley, CA 94720, USA*

Abstract. Statistical γ-decay from highly excited states is determined by the nuclear level density (NLD) and the γ-ray strength function (γSF). These average quantities have been measured for several nuclei using the Oslo method. For the first time, we exploit the NLD and γSF to evaluate the γ-width in the energy region below the neutron binding energy, often called the quasi-continuum region. The lifetimes of states in the quasi-continuum are important benchmarks for a theoretical description of nuclear structure and dynamics at high temperature. The lifetimes may also have impact on reaction rates for the rapid neutron-capture process, now demonstrated to take place in neutron star mergers.

1 Introduction

Nature displays a huge span of lifetimes, from the birth and death of stars to the population and decay of states in the micro-cosmos. In the world of quantum physics, unstable states are associated with an energy width Γ, which is related to the lifetime through $\tau\Gamma = \hbar$. Both quantities depend on available final states and the γ strength into these states.

The nuclear level density (NLD) is an exponentially increasing function of excitation energy. When the number of states reaches 100-1000 levels per MeV, detailed spectroscopy becomes almost impossible and less useful. In this quasi-continuum region, the NLD and the average γ-ray strength function (γSF) become fruitful concepts. These two quantities replace the accurate position of initial and final states and the transition probabilities between them in conventional discrete spectroscopy.

The Oslo method has provided NLDs and γSFs for many nuclei in the vicinity of the β-stability line[1]. From these observables, lifetimes, γ widths, and fluctuations can be explored in the quasi-continuum. In this work, we will demonstrate the wealth of information that is hidden in these data.

The present study deals with the properties of levels in the quasi-continuum excitation region below the neutron separation energy S_n. The level density is ranging from around 10^3 to 10^7 levels per MeV at S_n, when going from nuclear masses of $A \sim 50$ to 240. For γ energies around 3 MeV, the corresponding increase in γ strength is only one order of magnitude. This makes sense, because the NLD is fundamentally a combinatorial problem of the number of active quasi-particles, while the electric-dipole γ strength scales linearly with the number of protons.

⋆e-mail: magne.guttormsen@fys.uio.no
⋆⋆e-mail: a.c.larsen@fys.uio.no

[1]Published data on NLDs and γSFs measured with the Oslo method are avaliable at *http://ocl.uio.no/compilation/*

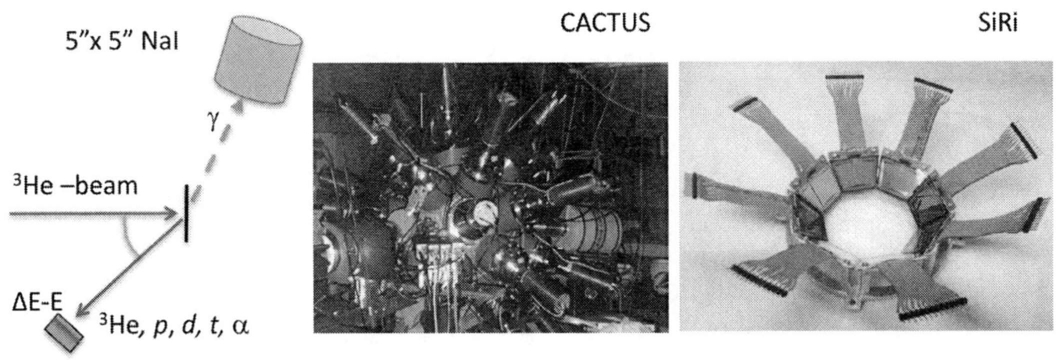

Figure 1. (Color online) Typical particle-γ coincidence set-up for the Oslo method. The 64 silicon particle telescopes of SiRi are placed in the vacuum chamber at the center of CACTUS.

2 The Oslo Method

In this section we give a short review of the Oslo method [1] for which the starting point is a set of γ-ray spectra measured as a function of initial excitation energy. The γ rays are measured in coincidence with the charged ejectile from light ion reactions such as $(d, p\gamma)$, $(p, p'\gamma)$ and $(^3\text{He}, \alpha\gamma)$, where the ejectile determines the initial excitation energy of each γ spectrum. Typical beam energies for the three reactions are 12 MeV, 16 MeV and 30 MeV, respectively.

Figure 1 shows a schematic drawing of the set-up. A silicon particle detection system (SiRi) [2], which consists of 64 telescopes, is used for the selection of a certain ejectile types and to determine their energies. The front ΔE and back E detectors have thicknesses of 130 μm and 1550 μm, respectively. Coincidences with γ rays are performed with the CACTUS array [3], consisting of 26 collimated 5" × 5" NaI(Tl) detectors with a total efficiency of 14.1% at $E_\gamma = 1.33$ MeV.

With the raw γ-ray spectra at hand, we arrange these into a particle-γ matrix $R(E, E_\gamma)$. Then, for all initial excitation energies E, the γ spectra are unfolded with the NaI response functions giving the matrix $U(E, E_\gamma)$ [4]. The procedure is iterative and stops when the folding $\mathcal{F}$ of the unfolded matrix equals the raw matrix within the statistical fluctuations, i.e. when $\mathcal{F}(U) \approx R$.

In the next step the primary γ-ray spectra are extracted from the unfolded matrix U. This is obtained by subtracting a weighted sum of $U(E', E_\gamma)$ spectra below excitation energy E:

$$P(E, E_\gamma) = U(E, E_\gamma) - \sum_{E'<E} w(E, E')U(E', E_\gamma). \tag{1}$$

The weighting coefficients $w(E, E')$ are determined in an iterative way described in Ref. [5]. After a few iterations, $w(E, E')$ converges to $P(E, E_\gamma)$, where we have normalized each γ spectrum by $\sum_{E_\gamma} P(E, E_\gamma) = 1$. This conversion of $w \to P$ is exactly what is expected, namely that the weighting function should equal the primary γ-ray spectrum. We rely on the fact that quasi-continuum decay is dominated by dipole transitions [9, 13], and consider only $E1$ and $M1$ transitions in the following. It should be noted that the validity of the procedure rests on the assumption that the γ-energy distribution is the same whether the levels were populated directly by the nuclear reaction or by γ decay from higher-lying states.

To extract the level density and the γ-ray strength function, we exploit a part of the primary $P(E, E_\gamma)$ matrix where the level density is high, typically well above 2Δ (the pairing gap), and where no single γ lines dominate. This statistical part of the matrix is described by the product of two vectors:

$$P(E, E_\gamma) \propto \rho(E - E_\gamma)\mathcal{T}(E_\gamma), \tag{2}$$

where the decay probability is assumed to be proportional to the NLD at the final energy $\rho(E - E_\gamma)$ according to Fermi's golden rule [6, 7]. The decay is also proportional to the γ-ray transmission coefficient $\mathcal{T}$, which according to a generalized version of the Brink hypothesis [8] is independent of spin and excitation energy; only the transitional energy E_γ plays a role. The γSF can be calculated from our measured transmission coefficient through [9]

$$f_{XL}(E_\gamma) = \frac{1}{2\pi}\frac{\mathcal{T}_{XL}(E_\gamma)}{E_\gamma^{2L+1}}. \tag{3}$$

It remains to normalize ρ and $\mathcal{T}$ to known experimental information from other experiments. The normalization procedures and the precisions obtained depend on available external data. Further description and tests of the Oslo method and the normalization procedures are given in Refs. [1, 10].

One could argue that the level densities and γSFs would depend on the light-ion reaction used. However, although these reactions are very selective, the γ decay appears much later and thus from a thermalized, compound-like system. This has been demonstrated by the Oslo group for many reactions. As an example, the (^{3}He, $\alpha\gamma$) and (^{3}He, ^{3}He'γ) reactions have been studied populating the same final nuclei, ^{96}Mo and ^{97}Mo [11, 12]. Also the very different reactions $(p, p'\gamma)$ and (^{3}He, $\alpha\gamma$) into ^{56}Fe confirm the independence of the reaction [13]. Minor differences may appear which probably are due to deviations in the spin distributions populated by the various reactions.

3 The evaluation of γ width and lifetime

The γ width (Γ) and lifetime (τ) can be evaluated from the measured NLD and γSF obtained with the Oslo method. However, one should note some differences when comparing with neutron capture data. First of all, significantly more levels are populated in the charged-particle reaction, giving a large spin distribution of typically $J \approx 0-6$ and populations of both parities. Secondly, the initial excitation bin is much larger (100-200 keV) than for neutron capture data that may even select only one ore a few resonances. These conditions ensures that the Oslo type of data represent an averaging over a broader initial excitation energy region and spins and parities.

The γ-decay strength function for γ-ray emission of multipole XL from levels of spin J and parity π at E_x is defined by Bartholomew *et al.* [14] as

$$f_{XL}^{J\pi}(E_\gamma, E_x) = \frac{\langle\Gamma_{XL}^{E_\gamma}(E_x, J, \pi)\rangle\rho(E_x, J, \pi)}{E_\gamma^{2L+1}}, \tag{4}$$

where $\langle\Gamma_{XL}^{E_\gamma}(E_x, J, \pi)\rangle$ is the *partial* γ width for the transition $E_x \to E_x - E_\gamma$. In the equation, it is assumed that E_x takes a fixed value while E_γ takes variable values, i.e. the final excitation energy $E_x - E_\gamma$ varies. We now apply Eq. (3) with the assumption that $f_{XL}(E_\gamma, E_x)$ is independent of E_x [8] and find

$$\langle\Gamma_{XL}^{E_\gamma}(E_x, J, \pi)\rangle = \frac{1}{2\pi\rho(E_x, J, \pi)}\mathcal{T}_{XL}(E_\gamma). \tag{5}$$

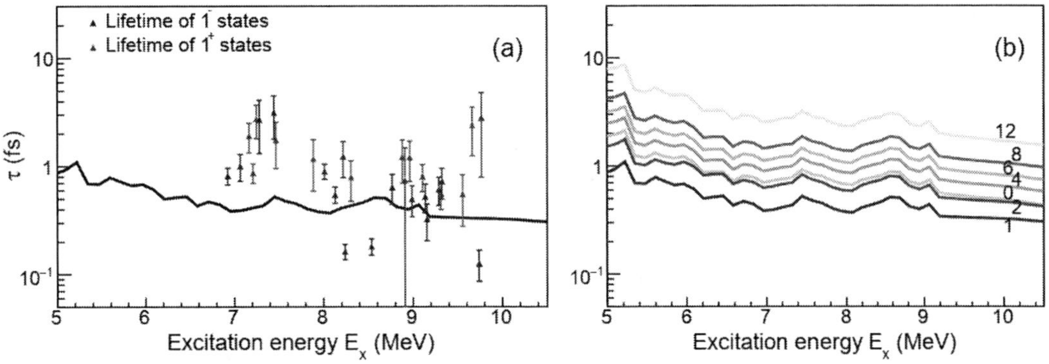

Figure 2. (Color online) Experimental lifetimes of $J^\pi = 1^+$ and 1^- states in ^{56}Fe [15, 16] compared with the estimates from the Oslo method (black line) (a). The predictions for various spins are shown in panel (b).

In order to obtain the the average *total* γ width of levels with excitation energy E_x, spin J and parity π, we sum up the strength for all possible primary transitions below E_x as prescribed by Kopecky and Uhl [9]:

$$\langle\Gamma(E_x, J, \pi)\rangle = \frac{1}{2\pi\rho(E_x, J, \pi)} \sum_{J_f,\pi_f} \int_0^{E_x} \mathrm{d}E_\gamma \mathcal{T}_{XL}(E_\gamma) \times \rho(E_x - E_\gamma, J_f, \pi_f), \tag{6}$$

where the summation run over all final levels with spin J_f and parity π_f that are accessible by XL transitions with energy E_γ. It should be noted that for the normalization of $\mathcal{T}$, we apply Eq. (6) using the initial spin(s) populated in neutron capture to reproduce the total γ width $\langle\Gamma(S_n)\rangle$. If we for simplicity assume that all levels within the initial energy bin are populated in the charged particle reaction, we obtain the average total γ width by

$$\langle\Gamma(E_x)\rangle = \sum_{J\pi} g(E_x, J)\mathcal{P}(E_x, \pi)\langle\Gamma(E_x, J, \pi)\rangle, \tag{7}$$

where g and $\mathcal{P}$ are the spin and parity distributions, respectively. From $\langle\Gamma(E_x)\rangle$, we get the lifetime in the quasi-continuum by

$$\langle\tau(E_x)\rangle = \frac{\hbar}{\langle\Gamma(E_x)\rangle} = \frac{658.2}{\langle\Gamma(E_x)\rangle}\text{fs}, \tag{8}$$

where the γ width is in units of meV.

4 Results and discussion

The γ widths and lifetimes of $J^\pi = 1^+$ and 1^- states in the quasi-continuum of ^{56}Fe have been measured by photon-scattering experiments [15, 16]. In Fig. 2 (a) we show the measured lifetimes, and compare with the corresponding estimates of lifetimes evaluated from the NLD and γSF obtained from the Oslo method [13]. We observe that the experimental data fluctuate up to a factor of ten, which is a result of the random structural overlap with few final states. Assuming Porter-Thomas fluctuations [17], the relative fluctuations are $\sqrt{2/\nu}$ where the degree of freedom ν can be estimated by the number of primary transitions from the excited state. Figure 2 (b) shows that the spin 1 states

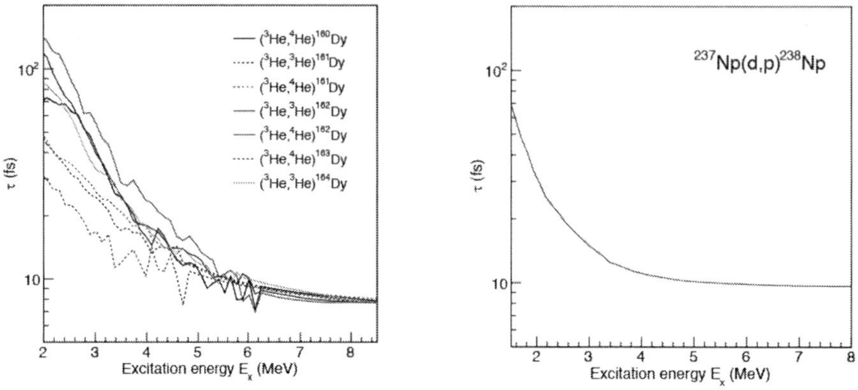

Figure 3. (Color online) Lifetimes in the quasi-continuum region for $^{160-164}$Dy (a) and ^{238}Np (b).

represent the fastest dipole transitions in the quasi-continuum, which can be explained by their direct decay to the 0^+ ground and first excited 2^+ states.

The number of levels in odd-mass dysprosiums are about a factor of seven larger than for the even-even neighbours. This reduces the decay time for the odd-mass isotopes as seen in Fig. 3 (a). However, at higher excitations ($E_x \geq 5$ MeV) the lifetimes for all isotopes converge to a common value. Furthermore, Fig. 3 (b) demonstrates that the lifetimes of ^{238}Np seem to flatten out for $E_x \geq 4$ MeV. The saturation in τ seen for the two cases can be understood from Eq. (6): the decay by higher γ energies for higher E_x is suppressed by a factor $\exp(-E_\gamma/T)$ where T is the nuclear temperature. It is surprising that the saturation in lifetimes is the same (8 – 10 fs) for a large range of mass numbers.

References

[1] A. Schiller *et al.*, Nucl. Instrum. Methods Phys. Res. A **447** 494 (2000).
[2] M. Guttormsen *et al.*, Nucl. Instrum. Methods Phys. Res. A **648**, 168 (2011).
[3] M. Guttormsen *et al.*, Phys. Scr. **T 32**, 54 (1990).
[4] M. Guttormsen *et al.*, Nucl. Instrum. Methods Phys. Res. A **374**, 371 (1996).
[5] M. Guttormsen *et al.*, Nucl. Instrum. Methods Phys. Res. A **255**, 518 (1987).
[6] P. A. M. Dirac, Proc. R. Soc. Lond. A 1927 114, 243-265.
[7] E. Fermi, Nuclear Physics. University of Chicago Press (1950).
[8] D. M. Brink, Doctoral thesis (unpublished), Oxford University, 1955.
[9] J. Kopecky and M. Uhl, Phys. Rev. C **41**, 1941 (1990).
[10] A. C. Larsen *et al.*, Phys. Rev. C **83**, 034315 (2011).
[11] M. Guttormsen *et al.*, Phys. Rev. C **71**, 044307 (2005).
[12] R. Chankova *et al.*, Phys. Rev. C **73**, 034311 (2006).
[13] A. C. Larsen *et al.*, Phys. Rev. Lett. **111**, 242504 (2013).
[14] G. A. Bartholomew *et al.*, Adv. Nucl. Phys. **7**, 229 (1973).
[15] F. Bauwens *et al.*, Phys. Rev. C **62**, 024302 (2000).
[16] T. Shizuma *et al.*, Phys. Rev. C **87**, 024301 (2013).
[17] T. Porter and R. G. Thomas, Phys. Rev. **104**, 483 (1956).

Gamma Strength Functions and Level Densities from High-Resolution Proton Scattering under 0°

Peter von Neumann-Cosel[1,a], Sergej Bassauer[1], Dirk Martin[1], and Atsushi Tamii[2]

[1] *Institut für Kernphysik, Technische Universität Darmstadt, 64289 Darmstadt, Germany*
[2] *Research Center for Nuclear Physics, Osaka University, Ibaraki, Osaka 567-0047, Japan*

Abstract. Inelastic proton scattering at energies of a few 100 MeV and forward angles including 0° provides a novel method to measure gamma strength functions (GSF) in nuclei in an energy range of about 5 – 20 MeV. The experiments provide not only the E1 but also the M1 part of the GSF. The latter is poorly known in heavy nuclei. Comparison with gamma decay data (e.g. from the Oslo method) allows to test the generalised Brink-Axel (BA) hypothesis in the energy region of the pygmy dipole resonance (PDR) crucial for the modelling of (n,γ) and (γ,n) reactions in astrophysical reaction networks. From the two test cases studied, ^{208}Pb remains inconclusive in the energy region of the PDR because of large Porter-Thomas fluctuations due to the small level density (LD), while the BA hypothesis seems to hold in case of ^{96}Mo. A fluctuation analysis of the high-resolution data also provides a direct measure of the LD in the energy region of the isovector giant dipole resonance (IVGDR) well above the neutron threshold, where hardly any experimental information is available. This permits an independent test of the decomposition of GSF and LD in Oslo-type experiments.

1 Introduction

The gamma strength function describes the average γ decay behavior of a nucleus. Their knowledge is required for applications of statistical nuclear theory in astrophysics [1], reactor design [2], and waste transmutation [3]. The GSF is dominated by E1 radiation with smaller contributions from M1 strength, while higher multipoles contribute little. Above particle threshold it is governed by the IVGDR but at lower excitation energies one observes In nuclei with neutron excess the formation of a PDR [4] sitting on the low-energy tail of the IVGDR. This low-energy E1 strength impacts on astrophysical reaction rates and the resulting abundances in the *r* process [5–7].

Most applications imply an environment of finite temperature, notably in stellar scenarios [8], and thus reactions on initially excited states become relevant. Their contributions to the reaction rates are usually estimated applying the generalized Brink-Axel (BA) hypothesis [9, 10], which states that the GSF is independent of the properties of the initial and final states and thus should be the same in γ emission and absorption experiments. Although historically formulated for the IVGDR, where it seems to hold approximately for not too high temperatures [11], this is nowadays a commonly used assumption to calculate the low-energy E1 and M1 strength functions. Recent theoretical studies [12, 13] put that into question demonstrating that the strength functions of collective modes built on excited states do show an energy dependence. However, numerical results for E1 strength functions obtained in Ref. [12] showed an approximate constancy as a function of excitation energy consistent with the BA hypothesis.

Recent work utilizing compound nucleus γ decay with the so-called Oslo method [14] has demonstrated independence of the GSF from excitation energies and spins of initial and final states in a given nucleus in accordance with the BA hypothesis once the LD is sufficiently large to suppress strong intensity fluctuations of individual transitions [15]. However, there are a number of experimental results which indicate violations of the BA hypothesis in the low-energy region. For example, the GSF in heavy deformed nuclei at excitation energies of 2 – 3 MeV is dominated by the orbital M1 scissors mode [16] and potentially large differences in B(M1) strengths are observed between γ emission and absorption experiments [17–19]. At very low energies (< 2 MeV) an increase of GSFs is observed in Oslo-type experiments [20, 21], which for even-even nuclei cannot have a counterpart in ground state absorption experiments because of the pairing gap.

For the low-energy E1 strength in the region of the PDR, the question is far from clear when comparing results from the Oslo method with photoabsorption data. Below particle thresholds most information on the GSF stems from nuclear resonance fluorescence (NRF) experiments. A striking example of disagreement is the GSF of ^{96}Mo, where the results from the Oslo method [22, 23] and from NRF [24] differ by factors 2 – 3 at excitation energies between 4 and 7 MeV. A problem with the NRF method are the experimentally unobserved braching ratios to excited states. While many older analyses of NRF data assume

[a]e-mail: vnc@ikp.tu-darmstadt.de

that these are negligible, in Ref. [24] they are included by a Hauser-Feshbach calculation assuming statistical decay. The resulting corrections are sizable, reaching a factor of five close to the neutron threshold. On the other hand, there are indications of non-statistical decay behavior of the PDR from recent measurements [25, 26]. Violation of the BA hypothesis was also claimed in a simultaneous study of the (γ, γ') reaction and average ground state branching ratios [27] in ^{142}Nd (see, however, Ref. [28]). Clearly, information on the GSF – in particular in the energy region of the PDR – from independent experiments is of high interest.

A new method for the measurement of complete E1 strength distributions – and thereby the E1 part of the GSFs – in nuclei from about 5 to 20 MeV has been developed using relativistic Coulomb excitation in polarized inelastic proton scattering at energies of a few hundred MeV and scattering angles close to 0° [29–33]. These data allow the dipole polarizability to be determined which provides important constraints on the neutron skin of nuclei and the poorly known parameters of the symmetry energy [34]. They also permit extraction of the M1 part of the GSF due to spinflip excitations [35], which energetically overlaps with the PDR strength. Furthermore, when performed with good energy resolution, the level density can be extracted in the excitation region of the IVGDR from the giant resonance fine structure independent of the GSF [36]. This allows an important test of the model-dependent decomposition of LD and GSF in the Oslo method [14] .

2 Experiments and and multipole decomposition

By way of example we explain the the experimental methods and major steps of the data analysis with the data for ^{96}Mo [37]. Further details can be found in Ref. [38]. Figure 1(a) displays the spectra taken at spectrometer angles 0°, 3°, and 4.5°. Besides discrete transitions at low excitation energies, a resonance-like structure around 8 MeV and the prominent IVGDR centered at $E_x \approx 16$ MeV are observed. The cross sections show a strong forward-angle peaking indicating the dominance of Coulomb excitation. The total spin transfer

$$\Sigma = \frac{3 - (2D_{NN'} + D_{LL'})}{4} \tag{1}$$

shown in Fig. 1(b) can be extracted from the measured polarization transfer observables. It takes a value of one for spinflip ($\Delta S = 1$) and zero for non-spinflip ($\Delta S = 0$) transitions at 0° [39], interpreted as M1 and E1 cross sections parts, respectively. Values in between 0 and 1 result from the summation over finite energy bins (200 keV up to an excitation energy of 10 MeV and 500 keV above). The polarization transfer analysis (PTA) reveals the presence of a few spinflip transitions between 5 and 7.5 MeV and a concentration of spinflip strength in the energy region 7.5 – 10 MeV identified as the spin-M1 resonance in ^{96}Mo. Cross sections above 10 MeV are dominantly of $\Delta S = 0$ character as expected for Coulomb excitation. These findings

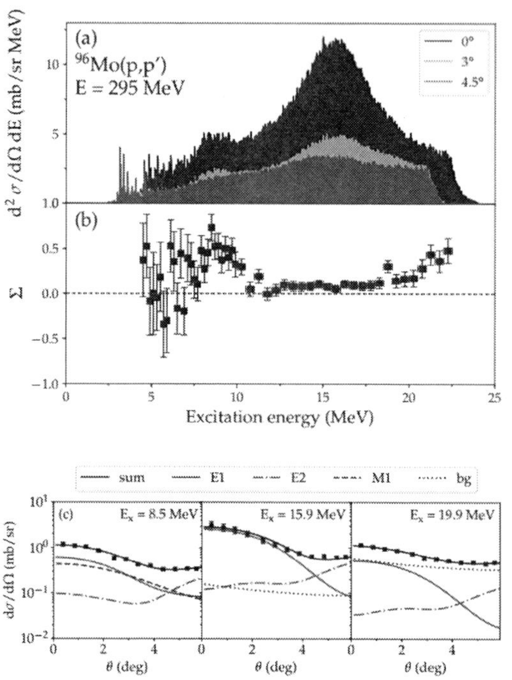

Figure 1. (a) Spectra of the ^{96}Mo($\vec{p}, \vec{p}'$) reaction at $E_p = 295$ MeV with the spectrometer placed at 0° (blue), 3° (yellow), and 4.5° (red). (b) Total spin transfer Σ, Eq. (1). (c) Examples of the MDA for selected energy bins. From Ref. [37].

are consistent with the results in ^{208}Pb [29] and ^{120}Sn [32]. The $\Delta S = 1$ strength observed at high E_x can be understood to arise from quasi-free scattering [40].

Alternatively, a multipole decomposition analysis (MDA) was performed for angular distributions of the cross sections in the PDR and GDR regions. For this purpose, angular cuts were applied to the spectra of Fig. 1(a) providing 4 data points each. The MDA followed closely the approach described in Refs. [29, 30]. For the nuclear background the empirical parametrization found for ^{208}Pb [36] was adopted assuming the same momentum transfer dependence. Figure 1(c) presents representative examples of the MDA for 200 keV excitation energy bins at different excitation energies. They demonstrate that in the low-energy bump M1 contributions are sizable while E1 dominates in the energy region of the IVGDR. At even higher energies, the nuclear background becomes dominant.

3 Gamma strength functions of ^{96}Mo and ^{208}Pb

After the decomposition of E1 and M1 cross sections described above, both can be converted into photoabsorption cross sections, or equivalently, the GSF. For the E1 strength this is based on the virtual photon method of relativistic Coulomb excitation [41]. Spin M1 matrix elements can be derived from the M1 cross sections utilizing the unit cross section method and further converted into equivalent electromagnetic B(M1) transition strengths neglecting isoscalar and orbital contributions [35]. The former assumption is motivated by the strong suppression of isoscalar strength due to the anomalous nucleon magnetic moments, while the latter is well justified in the energy region of the spin M1 resonanance [17].

Figure 2 presents the E1 (blue), M1 (green), and E2 (red) GSF in ^{208}Pb [42]. The latter is deduced from the isoscalar E2 strength function given in Ref. [43]. As can be seen, the dominant contribution stems from E1 transitions. The M1 contribution is of the order of a few percent for excitation energies above 8 MeV and reaches at most 10-30% in the peak of the resonance around the neutron threshold. We note that the dominance of the E1 GSF – even in the energy region of the spin M1 resonance – is a common feature in medium-heavy and heavy nuclei.

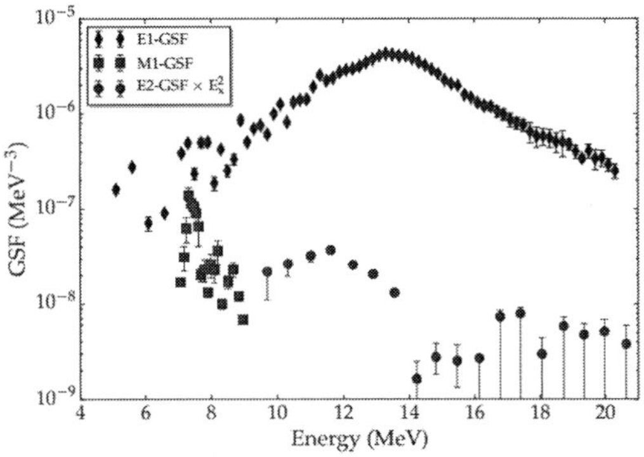

Figure 2. Comparison of E1, M1 and E2 contributions to the GSF of ^{208}Pb. From Ref. [42].

The total GSF summing all contributions is displayed in Fig. 3 (blue diamonds) and compared with data derived with the Oslo method from a ^{208}Pb(^{3}He,^{3}He$'\gamma$) experiment (red circles) [44]. The data set has been reanalyzed recently, for details see Ref. [42]. There are overlapping results from both experiments in the energy region between 5 and 8 MeV (inlet of Fig. 3). The GSF derived from the (p,p′) data is systematically higher in the PDR region although they seem still compatible within error bars in the peak region around the neutron threshold. Between 6 and 7 MeV consistent results are found while below 6 MeV the strong transitions observed in Ref. [30] exceed the average γ strength in the Oslo data by factors 4 to 5. However, one should be aware that single transitions are analyzed for excitation energies E_x < 7 MeV [30] and the level density of 1$^-$ states excited from the ground state is probably too low to discuss an average behavior in the PDR region. Rather the upward GSF is dominated by Porter-Thomas intensity fluctuations preventing a conclusion on the BA hypothesis.

A better suited case to test the BA hypothesis in the energy region of the PDR is ^{96}Mo, where the LD should be sufficiently high because of the considerable ground state

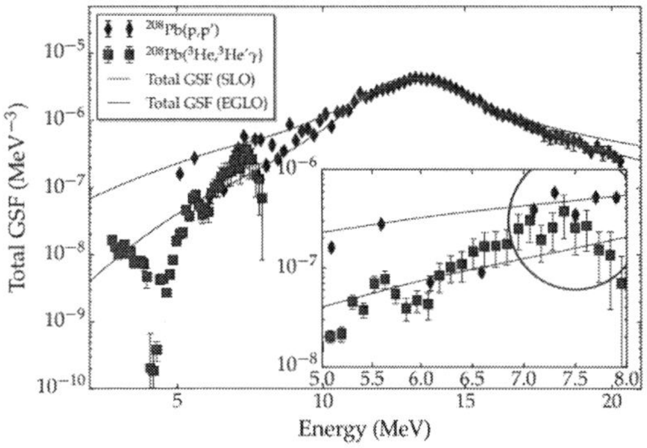

Figure 3. Total GSF of ^{208}Pb from the (p,p′) data [29, 30] in comparison to the reanalyzed [42] results from the Oslo experiment [44]. The inlet shows an expanded view of the low-energy region 5 – 8 MeV. From Ref. [42].

deformation. The case of ^{96}Mo is of particular interest because of a large discrepancy of the GSF deduced from the Oslo method [22, 23] and from (γ,γ') experiments [24]. The GSF function deduced from the $(\vec{p},\vec{p}')$ experiment discussed above [37] is displayed in Fig. 4 together with data from other experiments. Fig. 4(a) gives an overwiev for for the full range $E_\gamma = 6-20$ MeV. The result is compared with (^{3}He,^{3}He$'\gamma$) [22, 23] (open circles), (γ, xn) [45] (grey

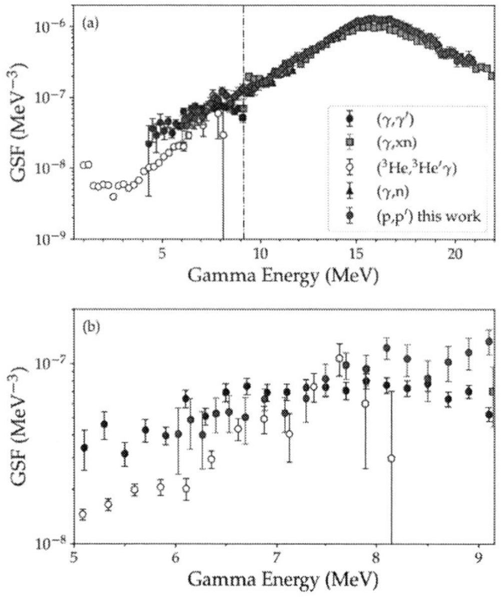

Figure 4. GSF of ^{96}Mo in the energy region from 5 MeV to neutron threshold from the present work (red circles) compared with (^{3}He,^{3}He$'\gamma$) [22, 23] (open circles), and (γ,γ') [24] data including a statistical model correction for unobserved branching ratios (black circles). From Ref. [37].

squares), (γ, n) [46] (blue upward triangles), and (γ, γ') data corrected for unobserved branching ratios [24] (black circles).

Above threshold, there is overall fair agreement with the data from Refs. [45, 46] except that the present experiment finds somewhat larger photoabsorption cross sections around the maximum of the IVGDR. In the energy region between 5 MeV and the neutron threshold [Fig. 4(b)] the results lie in between the two other measurements but agree within error bars with the Oslo experiment except for the lowest energies where statistics are poor.

4 Level densities

Since only the product of GSF and LD is measured by the Oslo method [14], it relies on external data for their decomposition. An independent check of the LD results is thus of special interest. The good energy resolution of the present data permits an extraction of the LD of $J^{\pi} = 1^{-}$ states applying a fluctuation analysis to the fine structure of the IVGDR. Details of the method can be found in Refs. [36, 47, 48].

The LD of $J^{\pi} = 1^{-}$ states in ^{208}Pb between 11 and 16 MeV [36] is displayed in the inlet of Fig. 5 in comparison with different parametrizations [44, 49, 50] of the phenomenological backshifted Fermi gas (BSFG) model. The BSFG parameters deduced from the RIPL-3 data base [49] provide a good description over the full energy range covered by the Oslo experiment and the present data. The parameterization of Ref. [50] works well at higher E_x but undershoot the data at lower E_x, while the situation is reverse with the parameter set of Ref. [44]. An analog analysis of ^{96}Mo data shows similar results [37]. For the two cases studied so far, the LDs deduced from the Oslo and the present method can be consistently described.

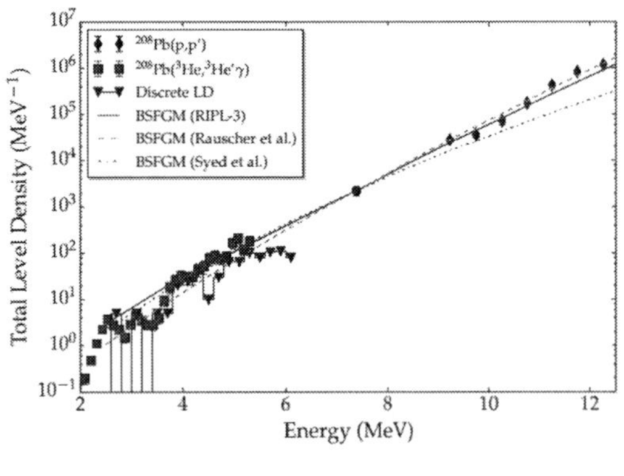

Figure 5. Total LD from the (p,p') data [29, 30] in comparison to the reanalyzed results (cf. Ref. [42]) from the Oslo experiment [44]. The black downward triangles are results from counting the levels identified in Ref. [51] in 200 keV bins. The magenta dashed-dotted, green solid, and cyan dashed lines are BSFG model predictions with the parameters of Ref. [44], Ref. [49], and Ref. [50], respectively. From Ref. [42].

5 Concluding Remarks

A new approach to test the Brink-Axel hypothesis is presented based on a study of the $(\vec{p}, \vec{p}')$ reaction at 295 MeV and extreme forward angles. While no conclusion can be drawn for ^{208}Pb, the extracted gamma strength function ^{96}Mo agrees with results of compound nucleus γ decay experiments [22, 23] indicating that the BA hypothesis holds in the energy region of the PDR, in contrast to results from the (γ, γ') reaction [24] and the claims of Ref. [27]. This is an important finding since the BA hypothesis constitutes a basic assumption of astrophysical reaction network calculations. The high energy resolution and selectivity of the data permits an extraction of the LD at excitation energies above the neutron threshold hardly accessible by other means. A consistent description of the LD with those of the γ decay experiments can be achieved with BSFG models providing independent confirmation of the methods used to separate GSF and LD in Oslo-type experiments.

Acknowledgements

We are grateful to all collaborators of experiments E282 and E376 for their contributions to the present work. This work was supported by the DFG under contract SFB 1245, by the JSPS, KAKENHI Grant No. JP14740154, and by the MEX,T KAKENHI Grant No. JP25105509.

References

[1] M. Arnould, S. Goriely, and K. Takahashi, Phys. Rep. **450**, 97 (2007).
[2] M. B. Chadwick *et al.*, Nucl. Data Sheets **112**, 2887 (2011).
[3] M. Salvatore and G. Palmiotti, Prog. Part. Nucl. Phys. **66**, 144 (2011).
[4] D. Savran, T. Aumann, and A. Zilges, Prog. Part. Nucl. Phys. **70**, 210 (2013).
[5] S. Goriely, E. Khan, and M. Samyn, Nucl. Phys. **A739**, 331 (2004).
[6] E. Litvinova, H. P. Loens, K. Langanke, G. Martínez-Pinedo, T. Rauscher, P. Ring, F.-K. Thielemann, and V. Tselyaev, Nucl. Phys. **A823,** 26 (2009).
[7] I. Daoutidis and S. Goriely, Phys. Rev. C **86**, 034328 (2012).
[8] M. Wiescher, F. Käppeler, and K. Langanke, Annu. Rev. Astron. Astrophys. **50**, 165 (2012).
[9] D. M. Brink, Ph.D. thesis, Oxford University (1955).
[10] P. Axel, Phys. Rev. **126**, 671 (1962).
[11] P. F. Bortignon, A. Bracco, and R. A. Broglia, *Giant Resonances: Nuclear Structure at Finite Temperature* (Harwood Academic, Amsterdam, 1998).
[12] C. W. Johnson, Phys. Lett. B **750**, 72 (2015).
[13] N. Quang Hung, N. Dinh Dang, and L. T. Quynh Huong, Phys. Rev. Lett. **118**, 022502 (2017).
[14] A. Schiller, L. Bergholt, M. Guttormsen, E. Melby, J. Rekstad, and S. Siem, Nucl. Instrum. Methods Phys. Res., Sect. A **447**, 498 (2000).

[15] M. Guttormsen, A. C. Larsen, A. Görgen, T. Renstrom, S. Siem, T. G. Tornyi, and G. M. Tveten, Phys. Rev. Lett. **116**, 012502 (2016).

[16] D. Bohle, A. Richter, W. Steffen, A. E. L.Dieperink, N. Lo Iudice, F. Palumbo, and O. Scholten, Phys. Lett. B **137**, 27 (1984).

[17] K. Heyde, P. von Neumann-Cosel, and A. Richter, Rev. Mod. Phys. **82**, 2365 (2010).

[18] M. Guttormsen *et al.*, Phys. Rev. Lett. **109**, 162503 (2012).

[19] C. T. Angell, R. Hajima, T. Shizuma, B. Ludewigt and B. J. Quiter, Phys. Rev. Lett. **117**, 142501 (2016).

[20] A. Voinov, E. Algin, U. Agvaanluvsan, T. Belgya, R. Chankova, M. Guttormsen, G. E. Mitchell, J. Rekstad, A. Schiller, and S. Siem, Phys. Rev. Lett. **93**, 142504 (2004).

[21] A. C. Larsen *et al.*, EPJ Web of Conferences **66**, 07014 (2014).

[22] M. Guttormsen *et al.*, Phys. Rev. C **71**, 044307 (2005).

[23] A. C. Larsen and S. Goriely, Phys. Rev. C **82**, 014318 (2010).

[24] G. Rusev *et al.*, Phys. Rev. C **79**, 061302 (2009).

[25] C. Romig *et al.*, Phys. Lett. B **744**, 369 (2015).

[26] B. Löher *et al.*, Phys. Lett. B **756**, 72 (2016).

[27] C. T. Angell, S. L. Hammond, H. J. Karwowski, J. H. Kelley, M. Krtička, E. Kwan, A. Makinaga, and G. Rusev, Phys. Rev. C **86**, 051302(R) (2012).

[28] Erratum to Ref. [27], Phys. Rev. C **91**, 039901(E) (2015).

[29] A. Tamii *et al.*, Phys. Rev. Lett. **107**, 062502 (2011).

[30] I. Poltoratska *et al.*, Phys. Rev. C **85**, 041304(R) (2012).

[31] A. M. Krumbholz *et al.*, Phys. Lett. B **744**, 7 (2015).

[32] T. Hashimoto *et al.*, Phys. Rev. C **92**, 031305(R) (2015).

[33] J. Birkhan *et al.*, Phys. Rev. Lett. **118**, 252501 (2017).

[34] *Topical Issue on Nuclear Symmetry Energy*, edited by Bao-An Li, A. Ramos, G. Verde and I. Vidaña, Eur. Phys. J. A **50(2)**, (2014).

[35] J. Birkhan, H. Matsubara, P. von Neumann-Cosel, N. Pietralla, V. Yu. Ponomarev, A. Richter, A. Tamii, and J. Wambach, Phys. Rev. C **93**, 041302(R) (2016).

[36] I. Poltoratska, R. W. Fearick, A. M. Krumbholz, E. Litvinova, H. Matsubara, P. von Neumann-Cosel, V. Yu. Ponomarev, A. Richter, and A. Tamii, Phys. Rev. C **89**, 054322 (2014).

[37] D. Martin *et al.*, Phys. Rev. Lett. **119**, 182503 (2017).

[38] A. Tamii *et al.*, Nucl. Instrum. Meth. A **605**, 3 (2009).

[39] T. Suzuki, Prog. Theor. Phys. **321**, 859 (2000).

[40] F. T. Baker *et al.*, Phys. Rep. **289**, 235 (1997).

[41] C. A. Bertulani and G. Baur, Phys. Rep. **163**, 299 (1988).

[42] S. Bassauer, P. von Neumann-Cosel, and A. Tamii, Phys. Rev. C **94**, 054313 (2016).

[43] D. H. Youngblood, Y.-W. Lui, H. L. Clark, B. John, Y. Tokimoto, and X. Chen, Phys. Rev. C **69**, 034315 (2004).

[44] N. U. H. Syed, M. Guttormsen, F. Ingebretsen, A. C. Larsen, T. Lönnroth, J. Rekstad, A. Schiller, S. Siem, and A. Voinov, Phys. Rev. C **79**, 024316 (2009).

[45] H. Beil, R. Bergère, P. Carlos, A. Leprêtre, A. De Miniac, and A. Veyssiere, Nucl. Phys. **A227**, 427 (1974).

[46] H. Utsunomiya *et al.*, Phys. Rev. C **88**, 015805 (2013).

[47] Y. Kalmykov *et al.*, Phys. Rev. Lett. **96**, 012502 (2006).

[48] I. Usman *et al.*, Phys. Rev. C **84**, 054322 (2011).

[49] R. Capote *et al.*, Nucl. Data Sheets **110**, 3107 (2009).

[50] T. Rauscher, F.-K. Thielemann, and K.-L. Kratz, Phys. Rev. C **56**, 1613 (1997).

[51] A. Heusler, R. V. Jolos, T. Faestermann, R. Hertenberger, H.-F. Wirth, and P. von Brentano, Phys. Rev. C **93**, 054321 (2016).

Photoneutron Reaction Data for Nuclear Physics and Astrophysics

Hiroaki Utsunomiya[1,*], *Therese* Renstrøm[2], *Gry Merete* Tveten[2], *Ioana* Gheorghe[3], *Dan Mihai* Filipescu[3], *Sergey* Belyshev[4], *Konstantin* Stopani[5], *Hongwei* Wang[6], *Gongtao* Fan[6], *Yiu-Wing* Lui[7], *Dmytro* Symochko[8], *Stephane* Goriely[9], *Ann-Cecilie* Larsen[2], *Sunniva* Siem[2], *Vladimir* Varlamov[5], *Boris* Ishkhanov[5], *Tudor* Glodariu[3], *Mateusz* Krzysiek[3,10], *Daiki* Takenaka[1], *Takashi* Ari-izumi[1], *Sho* Amano[11], Shuji Miyamoto[11]

[1] Department of Physics, Konan University, 658-8501 Kobe, Japan
[2] Department of Physics, University of Oslo, N-0316, Oslo, Norway
[3] ELI-NP, "Horia Hulubei" National Institute for Physics and Nuclear Engineering (IFIN-HH) , 077125 Bucharest-Magurele, Romania
[4] Department of Physics, Lomonosov Moscow State University, Moscow 119991, Russia
[5] Skobeltsyn Instutute of Nuclear Physics, Lomonosov Moscow State University, Moscow 119991, Russia
[6] Shanghai Institute of Applied Physics, Shanghai, 201800, China
[7] Cyclotron Institute, Texas A&M University, Texas 77843, USA
[8] Technische Universität Darmstadt, 64289 Darmstadt, Germany
[9] Institutd'Astronomie et d'Astrophysique, Université Libre de Bruxelles, 1050 Brussels, Belgium
[10] Institute of Nuclear Physics Polish Academy of Sciences, PL-31342 Krakow, Poland
[11] Laboratory of Advanced Science and Technology for Industry, University of Hyogo, Hyogo 678-1205, Japan

Abstract. We discuss the role of photoneutron reaction data in nuclear physics and astrophysics in conjunction with the Coordinated Research Project of the International Atomic Energy Agency with the code F41032 (IAEA-CRP F41032).

1 PHOENIX Collaboration for the IAEA-CRP F41032

The Coordinated Research Project of the International Atomic Energy Agency with the code F41032 (IAEA-CRP F41032) was launched in 2016 as a 5-years** project. The goal of the project is to publish two compilations of an updated photonuclear data library and a reference database of photon strength functions in 2020.

We have established an experimental collaboration called the PHOENIX (PHOto-Excitation and Neutron emIssion cross [X] sections) Collaboration for the IAEA-CRP in the γ-ray beam line GACKO (GAmma Collaboration hutch of KOnan university) of the NewSUBARU synchrotron radiation facility in Japan [1]. The PHOENIX Collaboration is hosted by the Konan University; the collaboration involves the University of Oslo, ELI-NP (Extreme Light Infrastructure – Nuclear Physics), Moscow State University, Shanghai Institute of Applied Physics, Texas A&M University, Technische Universität Darmstadt, Université Libre de Bruxelles, and University of Hyogo.

In this collaboration, we are measuring (γ,xn) cross sections with x=1–4 for the first compilation (photonuclear data) and (γ,1n) cross sections near one-neutron threshold for the second compilation (photon strength functions). A list of nuclei to be measured in the PHOENIX Collaboration is given in Table 1.

Table 1. Nuclei covered by the PHOENIX Collaboration.

	Photonuclear data	Photon strength functions
2015	^{209}Bi, ^{9}Be	^{205}Tl, ^{203}Tl, ^{89}Y
2016	^{197}Au, ^{169}Tm, ^{89}Y	^{192}Os, ^{185}Re, ^{138}Ba, ^{137}Ba, ^{64}Ni, ^{61}Ni, ^{60}Ni, ^{58}Ni, ^{13}C
2017	^{181}Ta, ^{165}Ho, ^{59}Co	^{184}W, ^{183}W, ^{182}W, ^{68}Zn, ^{66}Zn, ^{64}Zn
2018	^{159}Tb, ^{139}La, ^{103}Rh	^{160}Gd, ^{158}Gd, ^{157}Gd, ^{156}Gd

We provide (γ,xn) cross section data to resolve the long-standing discrepancy between the Livermore and Saclay data of partial photoneutron cross sections. In many cases, Saclay provided larger cross sections for (γ,1n) reactions than Livermore and vice versa for (γ,2n) reactions [2]. The ratio of the Saclay data to the Livermore data of integrated total cross sections, $R^{\rm int} = \sigma_S^{\rm int} / \sigma_L^{\rm int}$, varies from 0.65 to 1.35 over 19 nuclei from ^{51}V to ^{238}U. As shown in Fig. 1, the ratio for

* Corresponding author: hiro@konan-u.ac.jp
** The data acquisition is limited to 3 years for data evaluations and model calculations.

integrated partial cross sections averaged over the 19 nuclei is very different; it is $<R^{int}(1n)> = 1.07$ and $<R^{int}(2n)> = 0.84$ for the (γ,1n) and (γ,2n) reactions, respectively [3].

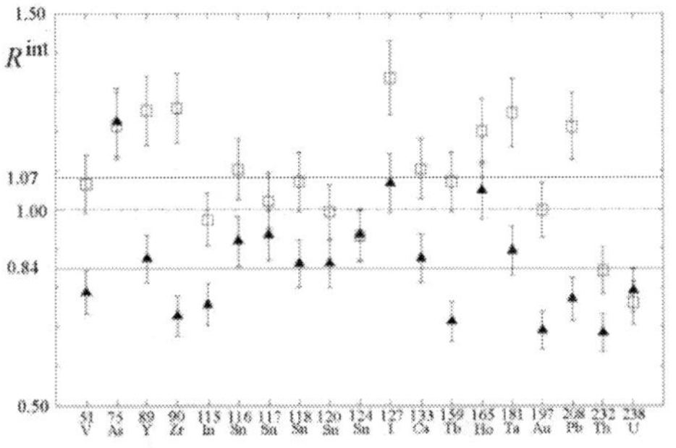

Fig. 1 The ratio of the Saclay data to the Livermore data of integrated cross sections, $R^{int} = \sigma_S^{int} / \sigma_L^{int}$, for (γ,1n) (open squares) and (γ,2n) (solid triangles) reactions.

We also provide (γ,1n) cross section for specific nuclei near the one-neutron threshold for an absolute normalization of the photon strength function below neutron threshold obtained with the Oslo method [4]. Figure 2 shows the resulting photon strength function for ^{117}Sn constructed experimentally, showing the presence of an extra strength over the neuron threshold at 6.94 MeV that is attributed to a pygmy E1 resonance [5].

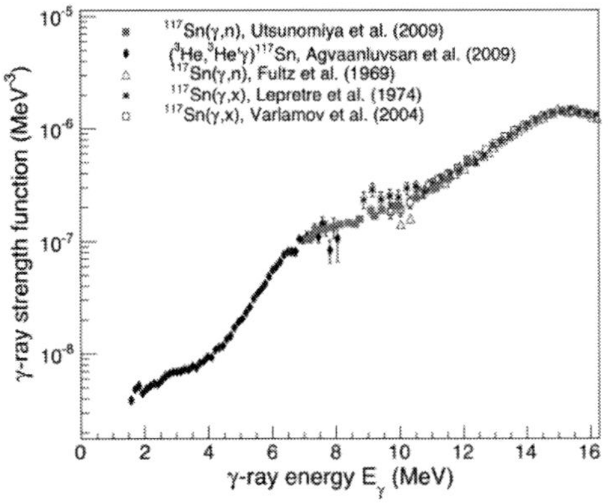

Fig. 2 (color online) Photon strength function for ^{117}Sn constructed from the photoneutron data and the data of the Oslo method.

We also provide (γ,1n) cross sections systematically over an isotopic chain to reach a unified understanding of (γ,n) and (n,γ) cross sections with the γ-ray strength function method [6,7]. This method effectively deduces (n,γ) cross sections for radioactive nuclei involved in the isotopic chains as a result of this unified and comprehensive approach.

The PHOENIX Collaboration will continue after the IAEA-CRP. Nuclei for which (γ,xn) cross sections will be measured after 2018 may include ^{112}Sn, ^{116}Sn, ^{120}Sn, ^{124}Sn, ^{208}Pb, ^{232}Th, and ^{238}U.

2 Direct Neutron-Multiplicity Sorting with a Flat-Response Detector

We have long used a high-efficiency neutron detector of Livermore type to measure (γ,1n) cross sections near the one-neutron threshold. This detector consisting of three rings of total twenty ^{3}He proportional counters embedded in a polyethylene moderator has a strong dependence of the detection efficiency on neutron kinetic energies. We use the so-called ring-ratio (RR) technique developed at Livermore [8] to determine the mean kinetic energies of neutrons. However, because of the RR technique, we limited ourselves to (γ,1n) cross section measurements below the two-neutron threshold, where the number of detected neutrons is equal to the number of (γ,1n) reactions. In the one-neutron channel, the ring ratio can surely determine the mean kinetic energies of neutrons produced in the (γ,1n) reaction. However, when we go above the two-neutron threshold, we immediately encounter an essential problem with the RR technique because there is no way to know in what reaction neutrons are produced and detected.

Let us consider measurements of (γ,xn) cross sections with neutron-multiplicity x=1-3 with a neutron detector of Livermore type by using a pulsed γ-ray beam. The frequency of the pulsed beam is typically 1ms because of the long neutron-moderation time of the polyethylene. During the 1ms time-interval between two successive γ-ray pulses, we detect neutrons. According to the number of detected neutrons (n), let us name neutron events as single (N_s), double (N_d), and triple (N_t) events corresponding to n=1, 2, and 3, respectively. The neutron events are expressed as follows.

$$\begin{aligned} N_s = {} & N_1 \cdot \varepsilon(E_1) + \\ & N_2 \cdot \varepsilon(E_{21}) \cdot (1-\varepsilon(E_{22})) + \\ & N_2 \cdot \varepsilon(E_{22}) \cdot (1-\varepsilon(E_{21})) + \\ & N_3 \cdot \varepsilon(E_{31}) \cdot (1-\varepsilon(E_{32})) \cdot (1-\varepsilon(E_{33})) + \\ & N_3 \cdot \varepsilon(E_{32}) \cdot (1-\varepsilon(E_{31})) \cdot (1-\varepsilon(E_{33})) + \\ & N_3 \cdot \varepsilon(E_{33})(1-\varepsilon(E_{31}))(1-\varepsilon(E_{32})) \end{aligned} \quad (1)$$

$$\begin{aligned} N_d = {} & N_2 \cdot \varepsilon(E_{21})^2 + \\ & N_3 \cdot \varepsilon(E_{31}) \cdot \varepsilon(E_{32}) \cdot (1-\varepsilon(E_{33})) + \\ & N_3 \cdot \varepsilon(E_{31}) \cdot \varepsilon(E_{33}) \cdot (1-\varepsilon(E_{32})) + \\ & N_3 \cdot \varepsilon(E_{32}) \cdot \varepsilon(E_{33}) \cdot (1-\varepsilon(E_{31})) \end{aligned} \quad (2)$$

$$N_t = N_3 \cdot \varepsilon(E_{31}) \cdot \varepsilon(E_{32}) \cdot \varepsilon(E_{33}) \quad (3)$$

Here N_x (x=1,2,3) is the number of (γ,xn) reactions induced by γ-ray pulses and E_{ij} is the kinetic energy of the j-th neutron emitted in the (γ,in) reaction. Note that the neutron detection efficiency (ε) depends on the neutron kinetic energy.

In the triple neutron event, obviously all of the first-, second-, and third-emitting neutrons are detected when the (γ,3n) reaction takes place. The second term of the double neutron event means that the first- and second-emitting neutrons are detected, while the third-emitting neutron is undetected when the (γ,3n) reaction takes place. The single neutron event consists of six terms in which the third term means that the second-emitting neutron is detected, while the first-emitting neutron is undetected when the (γ,2n) reaction takes place. Note that kinetic energy of the *j*-th emitting neutron is, in principle, different from that of the *k*-th ($k \neq j$) emitting neutron. Thus, the RR technique, which is applied to the experimental quantities, N_s, N_d, and N_t, does not work to determine neutron kinetic energies.

One can immediately see that the only solution to determine the number of (γ,xn) reactions is to make the neutron detection efficiency (ε) independent of neutron kinetic energies. In this case, the neutron events are expressed as follows.

$$N_s = N_1 \cdot \varepsilon + N_2 \cdot {}_2C_1 \cdot \varepsilon \cdot (1-\varepsilon) + N_3 \cdot {}_3C_1 \cdot \varepsilon \cdot (1-\varepsilon)^2 \tag{4}$$

$$N_d = N_2 \cdot \varepsilon^2 + N_3 \cdot {}_3C_2 \cdot \varepsilon^2 \cdot (1-\varepsilon) \tag{5}$$

$$N_t = N_3 \cdot \varepsilon^3 \tag{6}$$

Now one can solve a set of Eqs. (4), (5), and (6) to determine N_1, N_2, and N_3, the numbers of (γ,xn) reactions (x=1,2,3) induced by the γ-ray beam from the experimental quantities, N_s, N_d, and N_t. Note that ε is now a constant. It is straightforward to determine (γ,xn) cross sections (σ_{xn}) by $\sigma_{xn} = N_x/(N_\gamma \cdot N_T)$ with the number of γ-rays (N_γ) incident on target nuclei (N_T) per unit area.

We have developed a flat-response neutron detector (FRD) for (γ,xn) cross section measurements in the IAEA-CRP [9]. The detector consists of three concentric rings of 4, 9, and 18 ^{3}He counters embedded in a 46 cm (horizontally) x 46 cm (vertically) x 50 cm (along the beam axis) polyethylene moderator at the distances of 5.5, 13.0 and 16.0 cm from the γ-ray beam axis, respectively. Figure 3 shows a photograph of the flat-response detector.

The total and partial detection efficiencies of the FRD are shown in Fig. 4 along with results of the calibration with a ^{252}Cf source and GEANT4 simulation.

The emission rate of the ^{252}Cf source was calibrated at the National Metrology Institute of Japan. Results are plotted at the average energy (3*T*/2) of the ^{252}Cf neutron spectrum corresponding to the temperature *T* =1.42 MeV [10], where the fission spectrum is given by the Maxwell-Boltzmann distribution, $P(E) \propto E^{1/2} \cdot \exp(-E/T)$. The GEANT4 simulation was performed for monochromatic neutrons (dotted lines) and neutron-evaporation spectra (solid lines) expressed by $P(E) \propto E \cdot \exp(-E/T)$ with the Weisskopf-Ewing model [11]. Results for the evaporation spectra are shown at the average neutron energy (2*T*). The total efficiency for the evaporation spectra average over the energy range of 0.01 – 5.0 MeV is 36.5 % with one standard deviation of 1.6%. The neutron energy range was investigated for ^{209}Bi(γ,xn) reactions at 40 MeV by means of a Monte Carlo statistical model calculation [12]. The calculation shows that the mean neutron energy for the (γ,xn) channel with x=2-5 lies within the energy range.

Fig. 3 (color online) Flat-response neutron detector.

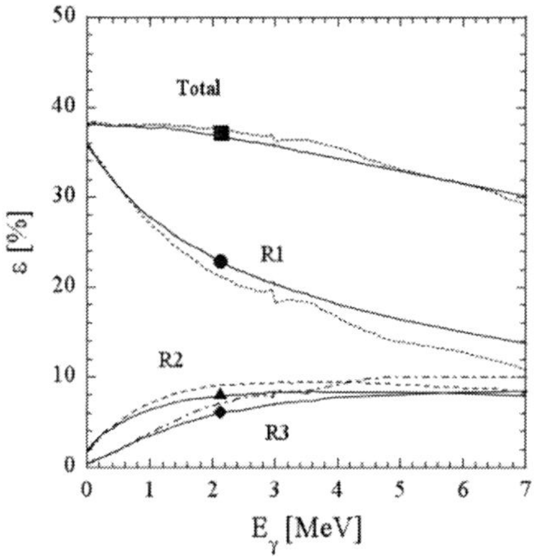

Fig. 4 Total and partial detection efficiencies of the flat-response neutron detector for neutron-evaporation spectra (solid lines) and monochromatic neutrons (dotted lines).

The neutron angular distribution effect on the detection efficiency was investigated for isotropic s- and anisotropic p-wave neutrons with respect to the linear polarization of the laser Compton-scattering γ-ray beam. It was confirmed that the FRD has nearly identical efficiencies for s- and p-wave neutrons, where the strong spatial anisotropy in the p-wave neutron emission is smeared out during the thermalization of neutrons in the moderator. This is a common feature of moderation-based neutron detectors of the Livermore type.

Figure 5 shows distributions of the arrival-time of neutrons at the FRD for single, double, and triple neutron events in the ^{209}Bi(γ,xn) reaction at 34 MeV. Time distributions of the 1st- and 2nd-arriving neutrons and the 1st-, 2nd-, and 3rd-arriving neutrons are also shown for the double and triple neutron events, respectively. It was found that the sum distribution is best fitted with a linear combination of two exponential functions with time constants, ~10μs for the fast component and ~100μs for the slow component, respectively. One can see that the major part of the 1st-arriving neutrons is well characterized by the fast component, while the 2nd and 3rd components by the slow component. The fast component may represent neutron detection with the first ring located at 5.5cm, while the slow component neutron detection with the second and third rings at 13 and 16cm. The arrival-time distribution is a good subject to investigate by a Monte Carlo simulation.

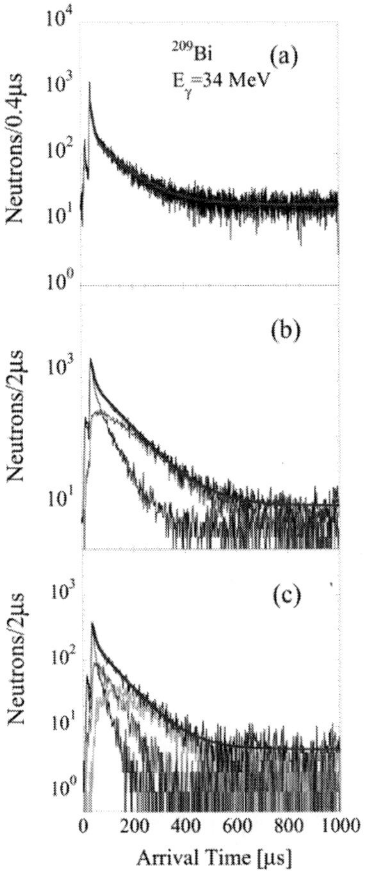

Fig. 5 (color online) Arrival-time distributions of neutrons detected with the flat-response detector for single (a), double (b), and triple (c) neutron events in the ^{209}Bi(γ,xn) reactions at 34 MeV. For the double neutron events, the arrival-time distribution is shown for the 1st-(blue) and 2nd-arriving (green) neutrons as well as the sum neutrons. Similarly, for the triple neutron events, the arrival-time distribution is shown for the 1st- (blue), 2nd- (green), and 3rd-arriving (orange) neutrons as well as the sum neutrons. The best fits to the single (a), double sum (b), and triple sum events (c) are shown by the red lines.

3 New ^{209}Bi(γ,xn) data

The technique of the direct neutron-multiplicity sorting was first used to measure ^{209}Bi(γ,xn) cross sections [13] in the IAEA-CRP. Figure 6 shows partial photoneutron cross sections obtained in the new measurement in comparison with the TALYS [14] calculation of the giant-dipole resonance (GDR) and quasi-deuteron component [15,16]. The GDR cross section was obtained within the Hartree-Fock-Bogolyubov plus quasiparticle random phase approximation approach on the basis of the D1M Gogny interaction both for the E1 [17] and M1 [18] components. See Ref. [13] for details.

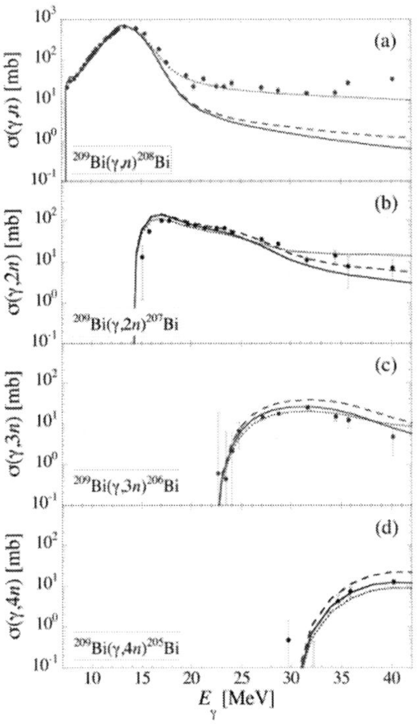

Fig. 6 (color online) Comparison with the present partial photoneutron cross sections with the TALYS calculation with the default preequilibrium parametrization (solid lines), the modified Levinger parameter *L*=20 with (dotted lines) and without (dashed lines) surface effects (E_{surf}= 1 MeV).

3.1 Puzzle of the ^{209}Bi(γ,1n) cross section

While the (γ,xn) cross sections with x=2, 3, and 4 are reasonably well reproduced by the TALYS calculation with the default preequilibrium parametrization (solid lines), the large (γ,1n) cross section above 25 MeV is highly puzzling. The standard statistical model predicts small (γ,1n) cross sections (3–4 mb) in this energy range. Furthermore, it is kinematically considered that kinetic-energies of neutrons emitted in the (γ,1n) channel are very large; they are outside the energy range (0.01 – 5.0

MeV) in which the flat-response is assured.

Although the (γ,1n) cross section can be artificially enhanced by tuning the Levinger parameter involved in the quasi-deuteron model or even by tuning the statistical model parameters within a reasonable range, the large (γ,1n) cross section is most likely experimental; it may be caused by secondary γ–rays produced by electrons and positrons resulting from the pair production and Compton scattering of the γ-ray beam in passing through a thick (10cm) ^{209}Bi target. A simulation shows that contaminant (γ,1n) reactions are induced by secondary γ-rays most effectively around the peak of the GDR, thus significantly reducing the relevant (γ,1n) cross section at high energies. We experimentally investigate this effect by changing the thickness of ^{209}Bi targets. We will write an erratum paper after a thorough investigation in experiment and simulation.

The authors thank P. von Neumann-Cosel for his comment on the neutron kinetic energy expected for the ^{209}Bi(γ,1n) reaction in relation to the large cross section at high energies. I.G., M.K., and D.F. acknowledge the support from the Extreme Light Infrastructure Nuclear Physics (ELI-NP) Phase II, a project co-financed by the Romanian Government and the European Union through the European Regional Development Fund - the Competitiveness Operational Programme (1/07.07.2016, COP, ID 1334). S.G. acknowledges the support from the F.R.S.-FNRS. H.U. acknowledges the support from the Premier Project of the Konan University. K.S. and S.B. acknowledge support from the Russian Foundation for Basic Research (Grant No. 15-02-05939). G.T. acknowledges the support from the Research Council of Norway under contract No. 262952. D.S. acknowledges the support of Deutsche Forschungsgemeinschaft through grant No. SFB 1245. This work was supported by the IAEA under Contracts No. ROM-20476, No. RUS-20501, and No. JPN-20564 for the CRP F41032 on Updating the Photonuclear Data Library.

References

1. http://www.lasti.u-hyogo.ac.jp/NS-en/
2. V.V. Varlamov, B.S. Ishkhanov, Phys. Part. Nucl. **35**, 459 (2004).
3. V.V. Varlamov, B.S. Ishkhanov, V.N. Orlin, K.A. Stopani, Eur. Phys. J. A **50**, 114 (2014).
4. A. Schiller *et al.*, Nucl. Instrum. Phys. Res. A **447**, 498 (2000).
5. H. K. Toft *et al.*, Phys. Rev. C **81**, 064311 (2010).
6. D.M. Filipescu *et al.*, Phys. Rev. C **90**, 064616 (2014).
7. H.-T. Nyhus *et al.*, Phys. Rev. C **91**, 015808 (2015).
8. B.L. Berman and S.C. Fultz, Rev. Mod. Phys. **47**, 713 (1975).
9. https://www-nds.iaea.org/CRP-photonuclear/
10. *Properties of neutron sources*, IAEA-TECDOC-410 IAEA (1987) March.
11. V. Weisskopf, Phys. Rev. **52** (1937) 295.
12. T. Kawano, Private communication.
13. I. Gheorghe *et al.*, Phy. Rev. C **96**, 044604 (2017).
14. A. J. Koning and D. Rochman, Nucl. Data Sheets **113**, 2841 (2012).
15. M. L. Terranova, D. A. De Lima, and J. D. Pinheiro Filho, Europhys. Lett. **9**, 523 (1989).
16. M. B. Chadwick, P. Oblozinsky, P. E. Hodgson, and G. Reffo, Phys. Rev. C **44**, 814 (1991).
17. M. Martini, S. Peru, S. Hilaire, S. Goriely, and F. Lechaftois, Phys. Rev. C **94**, 014304 (2016).
18. V. Tselyaev, N. Lyutorovich, J. Speth, S. Krewald, and P. G. Reinhard, Phys. Rev. C **94**, 034306 (2016).

Evaluation of neutron capture cross section on ^{205}Pb with photonuclear data

Nobuyuki Iwamoto[1,*] and *Toshiyuki* Shizuma[2]

[1]Nuclear Data Center, Nuclear Science and Engineering Center, Japan Atomic Energy Agency, Tokai, Ibaraki 319-1195, Japan
[2]Tokai Quantum Beam Science Center, National Institutes for Quantum and Radiological Science and Technology, Tokai, Ibaraki 319-1106, Japan

Abstract. The neutron capture cross section of long-lived radioactive ^{205}Pb is derived by using the nuclear reaction calculation code CCONE, based on photonuclear data. The present result is smaller than that of TENDL-2015 by a factor of 4. The derived Maxwellian averaged capture cross section (MACS) is the smallest compared to the existing data. The produced amount of ^{205}Pb is explored with a simulated neutron flux in the Pb-Bi eutectic (LBE) target. The continuous use of the system in 25 years creates ^{205}Pb with about 6 kg at maximum in the LBE (including natural Pb of 10^3 kg). The impact of the derived MACS on the stellar nucleosynthesis is investigated. It is found that the abundance of Tl is slightly enhanced due to the increase in the remaining abundance of ^{205}Pb.

1 Introduction

The use of Pb-Bi eutectic (LBE) as a target and coolant is under consideration for the accelerator-driven system (ADS) of Japan Atomic Energy Agency (JAEA). The minor actinides are installed and transmuted into short-lived nuclides in a subcritical reactor [1]. The proton beam is injected into the LBE target. Then, ^{205}Pb is formed from the neutron capture reaction on ^{204}Pb in the target. The radioactive ^{205}Pb has long half-life of 17.3 Myr, and thus it has long-lasting radiotoxicity. The ^{205}Pb amount produced in the operation of reactor should be taken into account, in order to use the target for a long period of time, although the natural abundance of ^{204}Pb is only 1.4 % in Pb. For this purpose, it is necessary to know the reliable neutron capture cross section on ^{205}Pb as well as that of ^{204}Pb.

The neutron cross sections of ^{205}Pb are only included in evaluated nuclear data libraries of JEFF-3.2 [2] and TENDL-2015 [3]. Koning et al. [4] evaluated the nuclear data of Pb for investigating their impact on ADS design. However, the data were not based on experimental data. Its direct measurement has difficulty in a sample production. Instead, a photonuclear measurement on ^{206}Pb was planed for our group to derive the information of photon strength function (PSF) strongly related to photon absorption and emission, since the accurate determination of PSF is important to obtain reliable capture cross section.

On the other hand, the study of an slow neutron capture process (*s*-process) in a low-mass asymptotic giant branch (AGB) star is interested in the Pb-Bi region [5]. The timescale of neutron capture reaction on ^{205}Pb gives an influence on the remaining abundance of ^{205}Tl after the termination of the *s*-process. Hence, for accurate prediction of abundance pattern observed at the surface of extrinsic AGB stars, the determination of neutron capture cross section of ^{205}Pb is also requisite.

2 Photonuclear experiment

Photon scattering experiment was performed at the γELBE facility of Helmholtz-Zentrum Dresden-Rossendorf (HZDR). The bremsstrahlung photons were generated by electrons with kinetic energy of 10.5 MeV, hitting Nb foil. The ^{206}Pb target was used with mass of 3940 mg and enrichment of 99.3 %. The photons scattered from the target were measured by 4 HPGe detectors placed at 90 and 127 degrees. The measurements at the two angles were needed to separate spin states ($J = 1$ or 2) excited from the 0^+ ground state by photons. As a result, 85 and 3 excited levels were identified as the $J = 1$ and $J = 2$ states, respectively, below 8.2 MeV from the measured photon spectra. The integrated cross sections were derived from the intensity of spectrum and were used for the evaluation of photonuclear cross sections. The details of the experiment and analysis are found in Ref. [6].

3 Calculation models

Evaluation of photon and neutron-induced cross sections was made by the nuclear reaction calculation code CCONE developed at JAEA [7]. The neutron optical model was adopted with spherical potential for Pb isotopes [4]. The potential parameters for ^{205}Pb were assumed to be the same as those of ^{206}Pb, since the parameters are considered not to have a large change among the neighbor isotopes. The statistical process of the CCONE code is based on the Hauser-Feshbach model. The excited

*e-mail: iwamoto.nobuyuki@jaea.go.jp

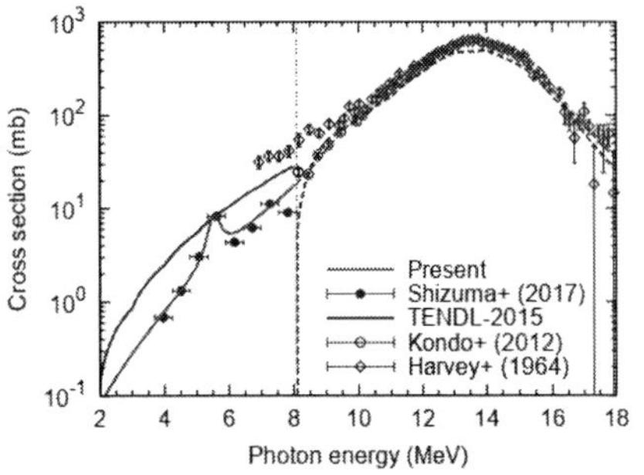

Figure 1. Comparison of the evaluated photonuclear data for ^{206}Pb with available experimental data. The dashed and solid lines represent the (γ,n) reaction and photoabsorption cross sections, respectively. The vertical line corresponds to the threshold energy of (γ,n) reaction.

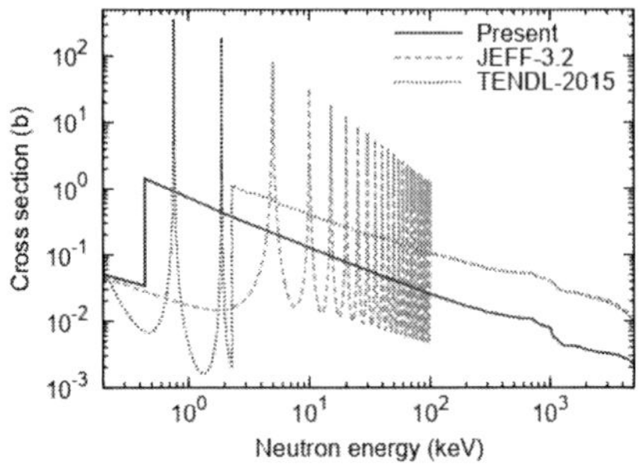

Figure 2. Comparison of the evaluated neutron capture cross section for ^{205}Pb with those of available nuclear data libraries.

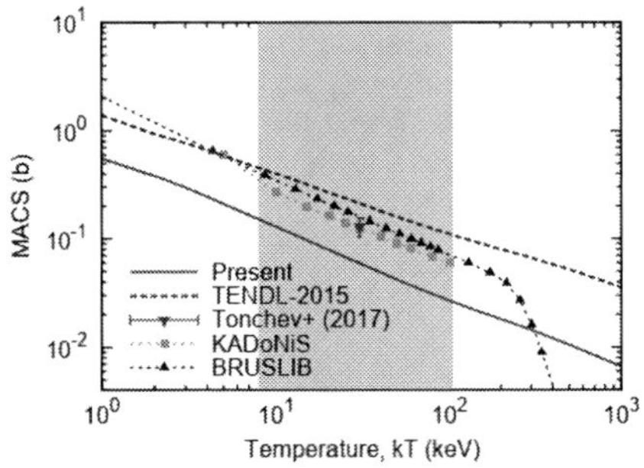

Figure 3. Comparison of the Maxwellian averaged neutron capture cross section for ^{205}Pb with available MACS data.

energy, spin-parity and de-excitation photon energy of discrete levels were retrieved from the RIPL-3 database [8]. Above the adopted excited level, the Gilbert-Cameron formalism [9] with the Fermi-gas model [10] was applied to describing a nuclear level density. The PSF is important to determine the excitation and de-excitation intensities by photons. The PSF form for the E1 radiation was employed from the standard Lorentzian model with energy-dependent damping width. The standard Lorentzian model was adopted for M1 and E2 radiations [11]. The two-component exciton model was also incorporated for the preequilibrium process [12, 13].

4 Results

4.1 Photonuclear cross sections

There are experimental data for two types of photon-induced reaction on ^{206}Pb below 14.8 MeV. They were used for the evaluation of photonuclear data. The ^{206}Pb$(\gamma,1nx)$ reaction cross sections have been measured with annihilation and laser-Compton scattering photons [14, 15]. The data of Ref. [14] were multiplied with a factor of 1.22, recommended by Ref. [16]. The photoabsorption cross section was evaluated below 8.1 MeV which is the threshold energy of (γ,n) reaction. In this region the integrated cross section of (γ,γ') reaction obtained at the γELBE facility was averaged with 550 keV energy bin and was employed to the photonuclear reaction evaluation. The parameters of giant dipole resonance (GDR) were fixed from these measured data. The enhancement of cross section was observed around 5.5 MeV. It was reproduced by giving an additional contribution to the GDR parameters.

Figure 1 shows the comparison of the present results with measured data and TENDL-2015[1] for the $(\gamma,1nx)$ and photoabsorption reaction cross sections. The present evaluation provides the cross sections consistent with both of the measured reaction data. The data of TENDL-2015 has smaller and larger cross sections, compared with measured data of $(\gamma,1nx)$ and photoabsorption reactions, respectively.

[1]The production cross section of ^{205}Pb was used since TENDL-2015 does not include the partial cross sections for $(\gamma,1nx)$ reaction. The (γ,np) and $(\gamma,n\alpha)$ reaction cross sections are negligible in this energy region.

4.2 Neutron capture cross sections

The fixed PSF was applied to calculate the neutron capture cross section of ^{205}Pb. The obtained result is shown in Fig. 2, in which the comparison is done with the cross sections of JEFF-3.2 and TENDL-2015. It should be noted that the resonance cross sections in both libraries are artificially generated. The present cross section is by a factor of 4 smaller than that of TENDL-2015 at 100 keV.

The Maxwellian averaged cross section (MACS) was further derived from the present neutron capture one. This MACS is compared with the data of KADoNiS [17], BRUSLIB [18] and Tonchev et al. [19] as well as that calculated from TENDL-2015 in Fig. 3. It is found that the present MACS is the smallest among the data mentioned above. The difference between the MACSs of Tonchev et al. and ours is unexpectedly large despite having the similar idea for the derivation of the cross section. Tonchev et al. incorporated the contributions of pygmy dipole resonance and core polarization in the gamma-ray transition strength for calculating the neutron capture cross section. Their strengths are considered to be higher than that of the additional contribution introduced in the present work.

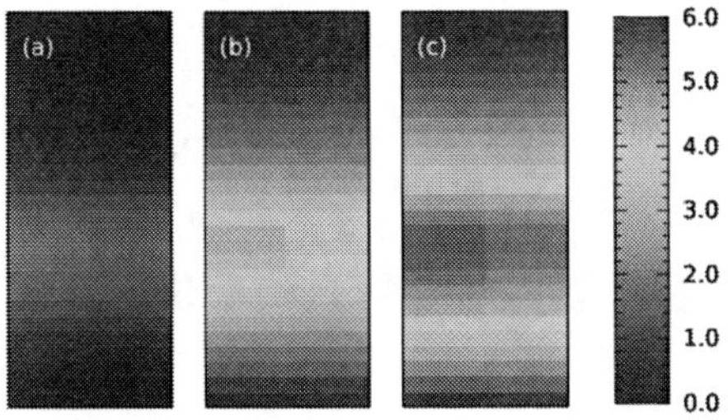

Figure 4. Produced amount (in unit of kg) of ^{205}Pb in the LBE target in the irradiation periods of (a) 5 yrs., (b) 15 yrs. and (c) 25 yrs.

5 Discussion

5.1 ADS

The LBE material is planned to be used as the target in the ADS of JAEA. The neutron capture on ^{204}Pb produces a radioactive ^{205}Pb when the target is bombarded with proton beam and the spallation reaction generates neutrons. A simulated neutron spectrum in the system was used to estimate the produced amount of ^{205}Pb. The condition of the proton beam current is assumed to be 8.7 mA. In this case the neutron flux was 10^{14-16} cm^{-2} sec^{-1}. Figure 4 shows the produced amount of ^{205}Pb in the LBE in the irradiation periods of 5, 15 and 25 years, when the LBE is fixed at the same position in this work. It is found that the 25 years-irradiation makes the ^{205}Pb amount by about 6 kg at maximum in the target with natural Pb of 10^3 kg.

5.2 Stellar nucleosynthesis

The neutron capture reaction on ^{205}Pb is relevant to the *s*-process nucleosynthesis in an AGB star. In this subsection, the impact of the smallest MACS on the s-process was investigated with a canonical s-process model. The MACS library was taken from data based on JENDL-4.0 [20], supplemented with KADoNiS. When the MACS of ^{205}Pb was changed to that of KADoNiS and TENDL-2015, the abundance difference is only found in ^{205}Tl because ^{205}Pb left after the termination of neutron exposure decays to ^{205}Tl. The uncertainty of [Tl/Fe] abundance coming from the difference of the adopted MACS is 0.03 dex (decimal exponent), which is smaller than observational uncertainties (note that $[A/B] \equiv \log_{10}(N_A/N_B) - \log_{10}(N_A/N_B)_{\odot}$, where N_A and N_B are the number of elements A and B, respectively. The symbol $\odot$ stands for the number in the Sun). Hence, it may be still difficult to judge which MACSs of ^{205}Pb are valid from observations of stellar surface composition.

6 Summary

Neutron capture cross section of long-lived radioactive ^{205}Pb (17.3 Myr) was evaluated with photonuclear data by applying the CCONE code. The PSF of ^{205}Pb was fixed, comparing the photonuclear cross sections with the experimental data. The evaluated neutron capture cross section on the basis of the determined PSF is by a factor of 4 smaller than TENDL-2015. In the ADS point of view, the produced amount of ^{205}Pb is about 6 kg at maximum in an installed LBE target (natural Pb of 10^3 kg loaded) even in 25 years-use. In the astrophysical point of view, the impact of the neutron capture cross section can be seen in Tl abundance, but is small. The present evaluation result is planned to be included in the next general purpose library of JENDL.

The authors would like to thank Drs. T. Sugawara and R. Katano of JAEA for providing the neutron flux of an ADS reactor.

References

[1] H. Iwamoto et al., J. Nucl. Sci. Technol., **53**, 1585 (2016)
[2] OECD Nuclear Energy Agency Data Bank. Available from http://www.oecd-nea.org/dbforms/data/eva/ evatapes/jeff_32 (2014)
[3] A.J. Koning et al., Available from https://tendl.web.psi.ch/tendl_2015/tendl2015.html (2015)
[4] A.J. Koning et al., Nucl. Sci. Eng. **156**, 357 (2007)
[5] U. Ratzel et al, Phys. Rev. C **70**, 065803 (2004)
[6] T. Shizuma et al., Phys. Rev. C, submitted (2017)
[7] O. Iwamoto, N. Iwamoto, S. Kunieda, F. Minato, and K. Shibata, Nucl. Data Sheets **131**, 259 (2016)
[8] R. Capote et al., Nucl. Data Sheets **110**, 3107 (2009)
[9] A. Gilbert, A.G.W. Cameron, Canadian J. Phys. **43**, 1446 (1965)
[10] A. Mengoni, Y. Nakajima, J. Nucl. Sci. Technol. **31**, 152 (1994)
[11] J. Kopecky, M. Uhl, Phys. Rev. C **41**, 1941 (1990)
[12] C. Kalbach, Phys Rev. C **33**, 818 (1986)
[13] A.J. Koning, M.C. Duijvestijn, Nucl Phys. A **744**, 15 (2004)
[14] R.R. Harvey, J.T. Caldwell, R.L. Bramblett, S.C. Fultz, Phys. Rev. **136**, B126 (1964)
[15] T. Kondo et al., Phys. Rev. C **86**, 014316 (2012)
[16] B.L. Berman et al., Phys. Rev. C **36**, 1286 (1987)
[17] I. Dillmann, R. Plag, F. Käppeler, T. Rauscher, EFNUDAT Fast Neutrons, EUR 23883 EN -2010, p.55 (2010)
[18] Y. Xu et al., Astron. Astrophys. **549**, A106 (2013)
[19] A.P. Tonchev et al., Phys. Lett. **B773**, 20 (2017)
[20] K. Shibata et al., J. Nucl. Sci. Technol., **48**, 1 (2011)

EPJ Web of Conferences **178**, 06005 (2018) https://doi.org/10.1051/epjconf/201817806005
CGS16

Developing reliable reaction gamma-ray data

Paraskevi Dimitriou[1,*], *T.* Belgya[2], *Y-S.* Cho[3], *D.* Filipescu[4], *R.* Firestone[5], *S.* Goriely[6], *N.* Iwamoto[7], *T.* Kawano[8], *J.* Kopecky[9], *M.* Krticka[10], *V.* Plujko[11], *S.* Siem[12], *R.* Schwengner[13], *H.* Utsunomiya[14], *V.* Vlarlamov[15], *M.* Wiedeking[16], *R.* Xu[17]

[1]International Atomic Energy Agency, Wagramerstrasse 5, A-1400 Vienna, Austria
[2]Centre for Energy Research, Hungarian Academy of Sciences, Budapest, Hungary
[3]Nuclear Data Center, Korea Atomic Energy Research Institute, Daejeon, Rep. of Korea
[4]Horia Hulubei National Institute, Bucuresti-Magurele, Romania
[5]University of California at Berkeley, Berkeley, California 94720, USA
[6]Universite Libre de Bruxelles, CP-226, Campus de la Plaine, 1050 Brussels, Belgium
[7]Japan Atomic Energy Agency (JAEA), Tokai-mura, Ibaraki 319-1195, Japan
[8]Los Alamos National Laboratory, New Mexico, 87454, USA
[9]JUKO Research, Kalmanstraat 4, Alkmaar 1817, The Netherlands
[10]Charles University in Prague, 18000 Prague 8, Czech Republic
[11]Taras Shevchenko National University, Kiev 01601, Ukraine
[12]University of Oslo, PO Box 1048, Oslo 0316, Norway
[13]Helmholtz Zentrum Dresden-Rossendorf (HZDR), Dresden, Germany
[14]Konan University, 8-9-1 Okamoto, Higashinada, Japan
[15]Lomonosov Moscow State University, 119991 Moscow, Russian Federation
[16]iThemba LABS, PO Box 2600, Pretoria Gauteng 0001, South Africa
[17]China Institute of Atomic Energy, PO Box 275(41), Beijing, China

Abstract. We report on efforts to develop reliable photonuclear cross section and photon strength function data by measuring, compiling, assessing, evaluating the available data, and producing tables of Giant Dipole Resonance parameters and global models for use in basic sciences and applications.

1 Motivation

Gamma-ray data from nuclear reactions are important for a large range of applications, as well as for basic sciences. In particular, gamma-ray data to extract photon strength functions and photonuclear cross sections are necessary for energy, safety and medical applications as well as for nuclear physics and astrophysics.

There is a wealth of reaction gamma-ray data related to photonuclear reactions and photon strength functions published in recent years that needs to be compiled and evaluated, and made available to researchers worldwide.

In this work we report on the international effort under an IAEA Coordinated Research Project (CRP) [1] to compile all the reaction gamma-ray data and assess their reliability on the basis of measurement techniques and systematic comparisons with theoretical models.

* Corresponding author: p.dimitriou@iaea.org

2 Photonuclear cross sections

The IAEA Photonuclear data library (http://www-nds.iaea.org/photonuclear) includes photoabsorption cross sections, total and partial photo-neutron reaction cross sections, and neutron emission energy spectra for 164 isotopes. Since its release in 1999, new data have been measured, improved evaluation techniques and theories have been developed and new measurement techniques aiming at resolving the existing discrepancies have emerged.

The current IAEA CRP is updating the photonuclear data library by revising the existing evaluations and adding 37 new evaluations of isotopes that were not included before. Additionally, photo-production of isotopes of interest to medical applications are also being considered, such as isotopes produced by $\gamma+^{39,41}K$, ^{186}Re produced via $^{187}Re(\gamma,n)^{186}Re$ for which data exist.

Existing evaluation approaches and models implemented in the different reaction codes are being assessed and their impact on the evaluations is explored. A systematic comparison of the correction functions F_i -often used as criteria for reliability of (γ,xn) data [2]-produced by six nuclear reaction codes used in the evaluations is shown in Fig. 1. The figure clearly indicates that this approach depends on both the reaction models and input parameters used in the calculations, and that the differences between them vary with energy. The F_i evaluation approach should therefore be applied by considering these model-dependent uncertainties.

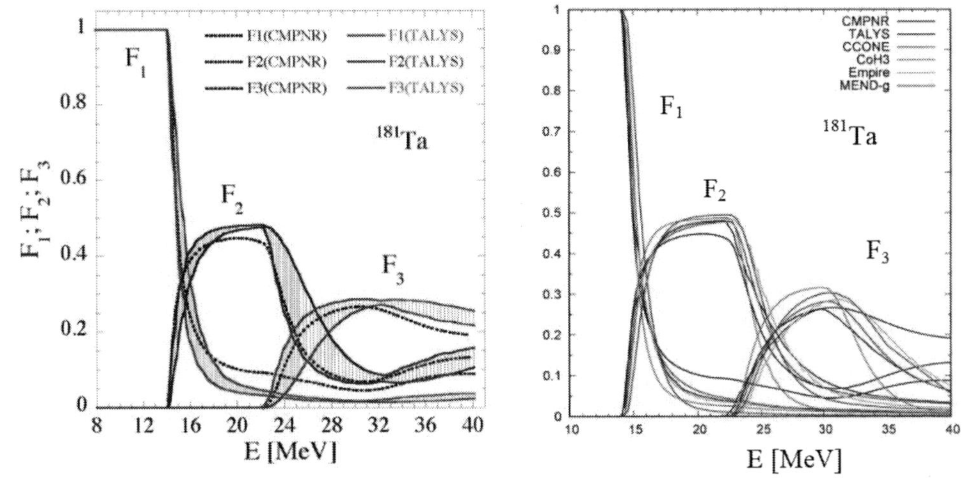

Fig. 1. F_1, F_2, F_3 correction functions for partial photoneutron cross section (γ,in) on ^{181}Ta. Left panel: shaded areas correspond to TALYS calculations using different nuclear level densities and photon strength functions. Right panel: coloured curves correspond to results from CMPNR, TALYS, CCONE, CoH3, Empire 3.2, MEND-g codes. For more details see [1].

In parallel with the evaluations, the new multiplicity counting technique used at the NewSUBARU facility with the aim of resolving the long-standing discrepancies between the existing photoneutron multiplicity data is also producing interesting results [3].

Finally, GDR parameters have been extracted by fitting Lorentzian-type curves to a revised database of experimental photoabsorption and photoneutron cross-section data published up to June 2017. The new compilation is an extension and improvement of earlier compilations and also includes recommendations [4].

3 Photon Strength Functions

Photon strength functions extracted from different measurements have been compiled and assessed. The compilation includes data in the GDR region and in the range near and below the neutron separation energy, including average resonance capture data [5]. The data are being assessed to ensure compatibility near the neutron thresholds, as well as to resolve discrepancies observed between different types of measurements and extraction methods, such as (γ,γ) and data obtained with the Oslo method. Global models (Lorentzian-type SLO/SMLO [4], QRPA [6]) have also been developed in the entire energy range. Figure 2 shows such an example where photon strength data from different sources are compared. All these data will be made available through a user-friendly web interface provided by the IAEA.

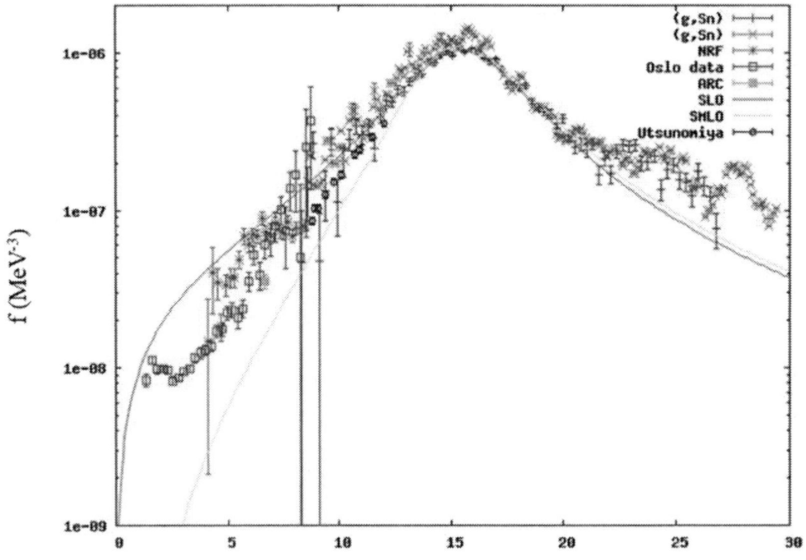

Fig. 2. Photon strength functions (MeV^{-3}) for ^{98}Mo from (γ,abs), average resonance capture data, (γ,n), (γ,γ) and (^{3}He,^{3}He) reactions compared with empirical SLO/SMLO functions (see [1] and references therein).

References

1. S. Goriely, M. Wiedeking and P. Dimitriou, Summary Report of the 2nd RCM of the CRP on Updating the IAEA Photonuclear Data Library and Generating a Reference Database for Photon Strength Functions, IAEA Report INDC(NDS)-0745, Vienna 2017; http://www-nds.iaea.org/publications/indc/indc-nds-0745.pdf
2. V.V. Varlamov, B.S. Ishkhanov, INDC(CCP)-433, Vienna 2002; http://www-nds.iaea.org/publications/indc-ccp-0433.pdf
3. I. Gheorghe et al., Phys. Rev. C 96, 044604 (2017).
4. V.A. Plujko, O.M. Gorbachenko, R. Capote, P. Dimitriou, ADNDT, in press.
5. J. Kopecky, Atlas of Average Resonance Capture Data, INDC(NDS)-0738, Vienna 2017; http://www-nds.iaea.org/pubications/indc/indc-nds-0738.pdf
6. S. Goriely, S. Hilaire, S. Péru, M. Martini, et al., Phys. Rev. C 94, 044306 (2016).

AUTHOR INDEX

AUTHOR INDEX

AUTHOR INDEX

AUTHOR INDEX

AUTHOR INDEX

AUTHOR INDEX